The
Evolving Continents

The
Evolving Continents

Brian F. Windley
Department of Geology
University of Leicester

JOHN WILEY & SONS
NEW YORK · LONDON · SYDNEY · TORONTO

Library of Congress Cataloging in Publication Data:

Windley, Brian F.
  The evolving continents.

  Includes index.
  1. Continental drift. 2. Geodynamics. I. Title.
QE511.5.W56   551.1'3   76-56416

ISBN 0 471 99475 8 (Cloth)

ISBN 0 471 99476 6 (Paper)

Printed in the United States of America

To
Judith

Preface

In Universities today students tend to be taught separate courses in selfcontained packages by different members of staff. In order to avoid a pigeon-holed view of the earth sciences, there is therefore an increasing need for an integrating course in which many themes are woven together. Before we, the staff, let the students graduate at the University of Leicester we give them such a course in which the integrating medium is earth history. This book has grown out of that course.

In the last decade or so two major developments have taken place. Firstly, the well-known revolution in earth sciences during the 1960s was based on new geophysical knowledge of the world's ocean floor and plate mosaic. Since the beginning of the 1970s there has been a growing attempt to work out the structure and evolution of Phanerozoic fold belts in the light of the current concepts of sea-floor spreading and continental drift. Secondly, there has been a great advance, mostly in the 1970s, in the understanding of the Precambrian Shield regions, assisted considerably by improvements in age dating techniques. The combination of these two factors means that whilst the 1960s was the decade of the oceans, the 1970s is the decade of the continents. I have been fortunate in that these advances took place during the preparation of this text.

Important aspects of this book are as follows. Firstly, most books on earth history pay little attention to the Precambrian. It seems to me that sufficient is now known about the Archaean and Proterozoic for them to play a major role in any such text. For this reason the first half of this volume is a review of the Precambrian. Secondly, many books on plate tectonics tend to concentrate on geophysics and the oceanic crust and the principles of plate motion; few deal largely with the 'geology' of plate tectonics from a continental standpoint. With this in mind the second half contains a review of the geology of continental break-up and key Phanerozoic fold belts according to plate tectonic theory. In the last chapter I have brought together the reasons for many of the long-term changes in the rock record—they tell the story of the evolving continents. And, finally, I have summarized in all chapters the types of mineralization characteristic of different tectonic belts and periods of earth history because so often metallogeny is ignored by academics. Thus throughout I have emphasized those subjects that the reader will not easily find in synthesized book form elsewhere.

I have not made an exhaustive review of all shields and fold belts throughout the world; rather I have selected classic regions which are representative of the rock groups or tectonic belts developed at particular times in earth history. In this way it is possible to characterize with well-documented examples the successive stages in development of the continental crust.

On the whole I have tried not to present just a descriptive account of 'what is where', but rather to bring out as many ideas about genetic processes and tectonic models as possible—facts without ideas can be very dull. There is current debate about the interpretation of much data and so I have pin-pointed many such areas of controversy because they enable the student, undergraduate or graduate, to focus in on the debates of today.

It seemed to me repetitious to give a description of each fold belt or continent in turn because so often the geology is broadly similar, and therefore I have synthesized the relevant contemporaneous component parts of fold belts, cratons or time periods as much for the Precambrian (such as greenstone belts, dyke swarms and tillites) as for the Phanerozoic (such as island arcs, palaeoclimatic indicators and Pangaea). But for many chapters I found that there are no recent adequate or suitable syntheses in the literature and so I have brought together in the form of new reviews data and ideas on, for example, Archaean high-grade regions, models for Archaean and Proterozoic crustal evolution, Proterozoic belts and geosynclines, and the break-up of Pangaea. Often different authors have produced their own models for a particular fold belt, or they have synthesized just one rock component such as tillites, Cordilleran volcanics, island arc sediments, evaporites, anorthosites or alkaline complexes. So I have endeavoured to bring together the models and syntheses for the different rock components of the various fold belts and time periods in order to portray in an interdisciplinary way a comprehensive picture of continental development.

For the Phanerozoic I have taken a frank plate tectonic approach. For the Precambrian I have trod more cautiously, documenting opposing ideas and points of view which reflect the state of the science today, but at the same time trying to emphasize those areas of Precambrian geology which are amenable to plate tectonic interpretation.

The book is intended for senior undergraduates who have already received a grounding in the main subjects of geology, post-graduate students who may be able to see their own specialization in a better perspective, and professional earth scientists who are interested in an interdisciplinary over-view of recent advances in understanding of continental evolution. I have not hesitated to give a generous list of references as they contain in their data and ideas the building blocks with which the science has developed; they may save the student, young and old, many hours of library search.

I have benefited from discussions over the years with many friends who have assisted me in understanding the intricacies of a variety of subjects, but in particular I wish to thank the following colleagues who read and improved individual chapters: F. B. Davies, J. D. Hudson, C. H. James, M. A. Khan, M. J. Le Bas, P. C. Sylvester-Bradley and J. H. McD. Whitaker. However, I alone am responsible for any shortcomings or omissions. I wish to acknowledge with thanks the receipt in 1975–76 of a Research Fellowship from the Leverhulme Trust Fund which enabled me to complete the writing free of lecturing duties. Last, but by no means least, for the patient typing of seemingly endless drafts and for her indispensable editorial assistance, I wish to thank my wife, Judith.

Leicester   Brian F. Windley
September, 1976

Contents

Chapter 1

Archaean High-grade Regions

The Archaean (pre-2500 my old) regions of the world (Fig. 1.1) contain two types of terrain: those whose rocks were metamorphosed largely to a high metamorphic grade, the subject of this chapter, and those that are well preserved in a low-grade state, the greenstone belts of Chapter 2. Some regions contain both types and their mutual relationships, as well as the general problems of their possible evolution, are reviewed in Chapter 3.

Most high-grade regions went through a major supracrustal-plutonic event in the

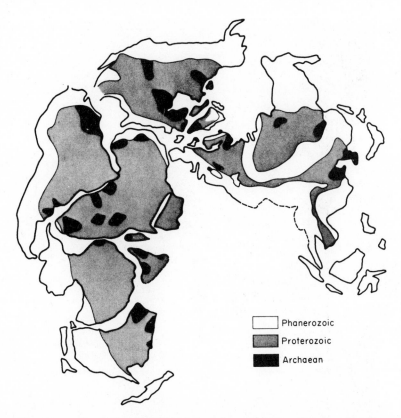

☐	Phanerozoic
▨	Proterozoic
■	Archaean

Fig. 1.1 A Permian pre-drift map of the continents showing the distribution of Archaean regions within Proterozoic cratons surrounded by Phanerozoic mobile belts

1

period 3100–2800 my and a few, i.e. West Greenland and Labrador, have a history that started 3800–3600 my ago.

The commonest rocks are granulite to upper amphibolite facies gneisses; special problems are attached to their mode of origin and to the geochemistry of orthopyroxene-bearing types that might be termed charnockites or felsic granulites. The gneisses contain the remains of some of the earliest sediments and volcanics, about which little is known in detail, and also of layered igneous complexes with calcic anorthosites. The rocks are typically folded into large-scale interference patterns and traversed by major shear belts.

The most important recent advances in the understanding of the evolution of Archaean high-grade regions have been in the North Atlantic craton (Fig. 1.2) that includes NW Scotland, Greenland and the Labrador coast (Bridgwater, Watson and Windley, 1973c). In

this chapter, therefore, data from these regions will be prominent.

Distribution

The Archaean high-grade rocks occur in a wide variety of environments: extensive areas such as the North Atlantic craton and the Aldan Shield, linear belts like the Limpopo mobile belt of southern Africa, small areas between greenstone belts (e.g. Rhodesia), and in small isolated remnants such as in the Lofoten Islands of Norway.

When plotted on a pre-Mesozoic, pre-continental drift map the occurrences are grouped into one or two areas as indicated by Hurley and Rand (1969); this grouping has been used as evidence against extensive continental drift movements between the Archaean and the late Phanerozoic. The regions concerned are surrounded by younger mostly Pro-

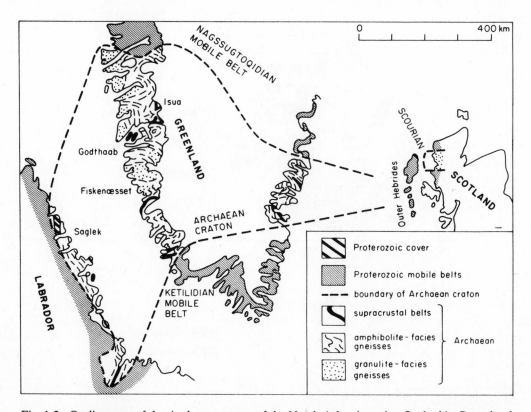

Fig. 1.2. Outline map of the Archaean craton of the North Atlantic region flanked in Greenland by the Nagssugtoqidian and Ketilidian mobile belts of Proterozoic age (after Bridgwater, Watson and Windley, 1973c)

terozoic belts, often formed by remobilization of the Archaean rocks, or partially covered by younger deposits; it is certainly safe to regard the present Archaean areas as just minimal remains of once larger regions. With regard to Africa, for example, D. S. Wood (1970) considers that in Archaean times there existed a block of continuous continental crust of aerial extent at least equal to that of the present continent.

Rock Units

There is a succession of different rock types in these high-grade regions. It is usual to find that major layers are mutually conformable and thus it is possible to map out an intercalation of rock units. In some parts of the world the gneisses are regarded as recrystallized sediments or volcanics and thus the present succession is thought to represent a remnant supracrustal pile. However, the results of V. R. McGregor (1973) from West Greenland demonstrate that the intercalation there is due not to supracrustal deposition but to deformation and conformable intrusion. In future it will be necessary to re-evaluate the reasons for the succession of rocks in other regions in the light of McGregor's relationships which are unequivocal.

Now let us look at the nature of the most important of these interlayered rocks and some of the problems connected with their origin.

Quartzo-feldspathic Gneisses

The most common type is a quartzo-feldspathic gneiss containing biotite and/or hornblende, hypersthene with or without diopside, and plagioclase, potash feldspar and quartz in various proportions; such rocks may constitute as much as 80–90% of these regions (Bridgwater, Watson and Windley, 1973c; Kalsbeek, 1976). Plagioclase-rich types with high Na/K ratios are particularly common (Tarney, Skinner and Sheraton, 1972; Barker and Peterman, 1974).

The gneisses are typically foliated and may have felsic and mafic bands up to about 1 cm wide. Detailed mapping may reveal an intercalation of gneiss types based on variations in composition, alternation of banded or more homogeneous types, and the presence or absence of inclusions of a particular type such as amphibolites or anorthosites; they may thus be described as veined, banded, homogeneous and agmatitic types (Rivalenti and Rossi, 1972).

The gneisses contain a great variety of other rocks. Most prominent are thick layers of amphibolite, mica schist, quartzite and meta-anorthosite; rare, but locally thick, marbles and banded iron formations are a surprising feature in such old rocks; in places there are amphibolite dykes and inclusions of meta-ultramafics. Pegmatites tend to be common in amphibolite facies but rare in granulite facies areas.

Due to their relatively low recrystallization temperature the gneisses have commonly undergone variable degrees of partial melting with the result that any of the rocks mentioned above may be penetrated by remobilized granitic veins and, in advanced stages of break up and veining, migmatites (mixed rocks) are formed.

One of the major problems of Precambrian geology is what is the origin of these gneisses. Whilst it is possible to work out satisfactorily the origin of the amphibolites as volcanic (with pillows), or as dykes (with apophyses and discordances), the anorthosites as igneous (with textures, chromitite-layering, geochemistry) and the quartzites and marbles as sedimentary, many decades of search have provided very few unequivocal ideas on the mode of origin of the gneisses—in particular because they usually contain no field relations of any diagnostic value. There are three main possibilities for the origin of these rocks.

Meta-sediments A few gneisses, containing for example sillimanite and cordierite, are no doubt metamorphosed sediments, but the bulk are rather homogeneous horn-blende/biotite- or hypersthene-bearing types considered to be *in situ* recrystallized sediments such as arkoses and greywackes (Andrews, 1973; Cheney and Stewart, 1975).

4

There is also a prominent Russian school that considers sedimentary processes in the early Precambrian to be similar to those of today and the metamorphosed sediments, including the gneisses, to be amongst the oldest rocks in the crust (Salop, 1964; Sidorenko, 1969; Pavlovsky, 1971). According to this model the layers in the gneisses of undoubted meta-sedimentary rocks, such as marble, quartzite and iron formation, are chemically distinctive 'resisters' to the gneissification process. Smithson, Murphy and Houston (1971) believe that this interpretation 'fits the facts and explains the puzzling occurrences of vast augen-migmatite terrains with small amounts of "resistent" meta-sedimentary rocks concordantly interlayered in Precambrian Shields'. Likewise Kalsbeek (1970) regarded the present bulk composition of granitic gneisses in SW Greenland to be a reflection of the isochemical recrystallization of sedimen-

tary rocks of greywacke-to-arkose type. However many would express the opinion that the fact that the gneisses contain a small percentage of meta-sedimentary and meta-volcanic layers and lenses is insufficient proof that the gneisses themselves (which constitute more than 80% of the present surface area) are of supracrustal origin.

It should be possible to resolve this problem with the use of certain chemical parameters. Tarney (1976) has shown that the TiO_2–SiO_2 abundance levels of Archaean gneisses are substantially different from those in both Archaean and post-Archaean sediments and meta-sediments: 'the distinction is sufficiently good to indicate that the Archaean gneisses were not of sedimentary derivation' (Fig. 1.3).

But one of the best pieces of evidence against a sedimentary origin for the bulk of the gneisses is the fact that where they come in contact with other rock groups, such as the

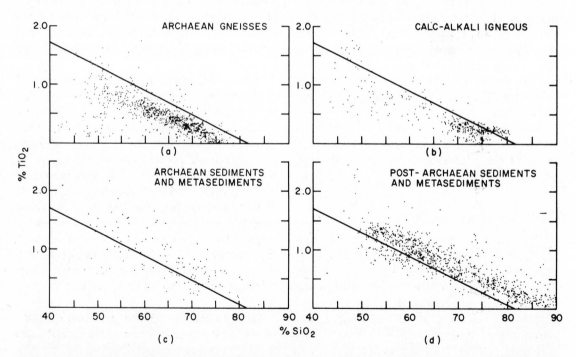

Fig. 1.3. TiO_2–SiO_2 distribution in Archaean gneisses. A: Archaean gneisses from Scotland, E and W Greenland. B: Continental margin calc-alkaline igneous rocks, Late Precambrian to Lower Palaeozoic age, England and Wales. C: Archaean clastic sediments and metasediments, Scotland, Greenland, W America and Australia. D: Late Precambrian to Palaeozoic sediments and metasediments (including migmatitic metasediments), Britain. Diagonal line for reference. Data sources numerous (after Tarney, 1976; reproduced by permission of J. Wiley)

anorthosite–gabbro complexes, they are clearly intrusive. This is not just a remobilization phenomenon due to veins formed by partial melting; sheets of the actual gneiss invade the complexes which can be seen to be in all stages of break up (Myers, 1976; Windley and Smith, 1976).

Meta-volcanics On geochemical grounds Sheraton (1970), Bowes, Barooah and Khoury (1971) and Bowes and Hopgood (1973, 1975) concluded that the Scourian granulite facies gneisses were largely derived from a calc-alkaline volcanic suite of average andesitic–dacitic composition. Bowes (1972) gives the average bulk composition of the gneisses as:

Major elements (standard cell)		Trace elements (ppm)	
K	1·7	Ba	937
Na	7·6	Ca	48
Ca	5·5	Co	45
Mg	4·5	Cr	92
Mn	0·06	Cu	47
Fe	4·0	Ga	17·5
Al	16·6	La	19
Ti	0·4	Li	10
(O 151·1 [OH] 8·5) 160		Nb	5·5
Si	55·2	Ni	60
		Rb	10
		Sc	14
		Sr	560
		V	111
		Y	9
		Zn	79
		Zr	195

They point out that with the exception of K, Rb, U, Ga and H_2O the rocks correspond chemically to the Archaean basalts, andesites, dacites and rhyodacites (or their pyroclastic equivalents) of the greenstone belts.

However, rare earth element similarities are not so convincing. Whereas the rare earth patterns in modern island arc and oceanic tholeiites and Archaean greenstone belt volcanics are characteristically flat (Schilling, 1971; White, Jakes and Christie, 1971; Jakes and White, 1972; Jahn, Shih and Murthy,

1974), they are highly fractionated (depleted in heavy REE) in Archaean gneisses (O'Nions and Pankhurst, 1974; Tarney, 1976). As REE are known to be rather immobile during metamorphism, a volcanic derivation for the gneisses would seem to be highly unlikely.

With respect to the Scourian gneisses Holland and Lambert (1975) point out the following features which militate against a volcanic origin:
1. The Scourie assemblage shows surprisingly little increase of K compared with Si.
2. The trend of variation of K, Na, Zr, Sr, Zn and Cr differs significantly from that of Archaean volcanics.
3. The abundances of Ti, total Fe, Mn, Ba and Ni also are appreciably different from those of Archaean volcanics.

Barker and Peterman (1974) considered the common plagioclase-rich gneisses of many Shield areas to be recrystallized low potash-high silica dacites. They found that there was a lack of gneisses of andesitic composition, which contrasts with the chemical evidence from the Scourian cited above.

Apart from the problem that the meta-volcanic model has no field data to support the chemical analogies, the model suffers from three disadvantages:
1. It is often considered that the granulite-facies metamorphism was responsible for a major depletion in the lower Archaean crust of many elements, in particular K, Rb, Y, Th, U, Cs, Ce, La and Pb (Sheraton, 1970; Tarney, Skinner and Sheraton, 1972). Comparison with less metamorphosed greenstone belt volcanics is thus made difficult by lack of original compositions.
2. If the volcanic rocks were of the pyroclastic type, they could have been intermixed with appreciable sedimentary material, as suggested by Kalsbeek (1970) for hornblende-biotite gneisses from West Greenland.
3. A chemical analysis cannot easily distinguish between a volcanic rock and its plutonic equivalent; many of the so-called meta-dacites could equally be meta-tonalites.

Meta-granitic Rocks (*sensu lato*) A popular concept is that most of the gneisses represent either reworked older basement gneisses or deformed granites (Watson, 1967; McGregor, 1973; Bridgwater, Watson and Windley, 1973c).

V. R. McGregor (1973) established without any reasonable doubt that many Amîtsoq and Nûk gneisses in West Greenland were derived by deformation of homogeneous granites. The leucocratic bands of the gneisses were formed partly by elongation of feldspar megacrysts, and partly by segregation during metamorphic differentiation of quartzo-feldspathic material during shearing. McGregor's results are fundamental and impressive as they demonstrate for the first time exactly how a high-grade granitic gneiss of widespread extent is derived. In a similar way Bridgwater and coworkers (1975) demonstrated that the Uivak gneisses in Labrador are derived by deformation of porphyritic granites. In so far as many Amîtsoq, Nûk and Uivak precursors are homogeneous, feldspar megacryst 'granitic' rocks, they have the appearance of typical post-tectonic granites. Both contain remnants of earlier meta-sediments and meta-volcanics, and thus cannot be regarded as the remains of some primordial granitic crust (cf. Wynne-Edwards, 1972a).

Because most gneisses have a general calc-alkaline chemistry and because there are serious arguments against a simple sedimentary or volcanic derivation, an origin by deformation of tonalitic–granodioritic rocks for many gneisses is most likely (e.g. Bridgwater and Collerson, 1976). The TiO_2–SiO_2 distribution of Archaean gneisses is certainly comparable to that of the British Caledonian continental margin calc-alkaline igneous trend (Tarney, 1976). The Amîtsoq gneisses have a lanthanide distribution pattern remarkably similar to that of the mesozoic Bonsall tonalite from the Peninsular Range batholith in southern California and their feldspar-related mineralogy and chemistry is comparable to that in calc-alkaline igneous suites (Lambert and Holland, 1976). Spooner and Fairbairn (1970) and Holland and Lambert (1975) suggested that calc-alkaline magmas were intruded into deep levels of the continental Archaean crust

and crystallized directly into granulite-facies gneisses..

The preceding discussion on the composition and origin of the quartzo-feldspathic gneisses is very relevant to the problem of whether or not they include more than one generation of gneiss; if so, they contain elements of a much older segment of earth history. There has clearly been much concern about the bulk composition of the gneisses, but this has often been at the expense of considering the time factor in their evolution. If one is not sure of the origin of the gneisses, but opts for a meta-supracrustal source rock, one is confined to a mental straightjacket that cannot see beyond a series of cover rocks. If the options are kept open to include the possibility of a deformed basement model, there is the chance of breaking through to a much older phase of earth history; and in this case some of the gneisses might be of meta-supracrustal, and others of meta-granitoid, origin. The fact that the present stratigraphy in some high-grade terrains is of tectonic origin (shown, for example, by the evidence of V. R. McGregor, 1973, and Collerson, Jesseau and Bridgwater, 1976) negates the hypothesis of simple recrystallization of a supracrustal pile for these terrains and supports the concept of the incorporation of supracrustal rocks into an old basement or earlier sialic crust.

Supracrustal Rocks

Amphibolites Within the gneisses in many areas there are conformable layers of amphibolite from a few centimetres to at least 1 km thick. Here we consider only the thicker layers, which are probably of volcanic origin. Rarely, as in Godthaabsfjord, West Greenland, the amphibolites contain relic pillows showing that they were lavas deposited under water; this Greenland example provides proof that water existed on the earth at least 3000 my ago. Commonly, the thicker amphibolites are unmigmatized by penetrative granitic–gneissic material and thus their present chemical composition may well be close to their original composition (assuming that they have not been chemically depleted or otherwise changed during high-grade metamorphism).

Kalsbeek and Leake (1970) concluded from a detailed geochemical survey that amphibolite layers in West Greenland were probably basic tuffs and lavas; they also made the general point that the common amphibolites in metamorphic regions have never been shown chemically to be derived from sediments but are usually of basaltic origin.

Amphibolites form prominent layers within gneisses in the Ancient Gneiss Complex of Swaziland (Hunter, 1970a), the Minnesota River Valley (Grant and Goldich, 1972), and in the pre-Barberton gneissic basement of Rhodesia (Stowe, 1971). Barker and Peterman (1974) attach special significance to this common bimodal association of plagioclase gneiss and amphibolite in the earliest Precambrian regions, suggesting that they are metamorphosed dacitic and tholeiitic volcanics resulting from consumption of primordial hydrous oceanic crust.

Amphibolites in the Fiskenæsset region of West Greenland are of two types, most comparable in major and trace element chemistry with modern tholeiitic and alkali olivine basalts of oceanic domains. Both display iron enrichments and have low contents of incompatible elements and high Cr, Ni, Mg and Fe (Rivalenti, 1976). Using modern analogies such basalts may have formed in two possible environments: early rifted continental margins or abyssal ocean floor. However it is possible that many or most Archaean tholeiitic rocks were low in potash whatever their tectonic environment (see Brooks and Hart, 1974). This view is strengthened by the fact that there is a secular decrease in the K_2O content of dolerite dykes with increasing age in North America (Fig. 19.7) (Mueller and Rogers, 1973), the most likely explanation for which is that the depth of melting of basic magmas was lower in the Archaean than in later times (Condie, 1976b). As Gill and Bridgwater (1976) point out, the chemistry of Archaean lavas and dykes may be more related to depth of melting in the mantle and relative speed at which the magmas ascended to the surface than to crustal environment.

Mica Schists One of the most prominent meta-sedimentary rocks is mica schist which forms layers up to a few hundred metres thick and contains minerals such as garnet, cordierite and sillimanite; they are most reasonably meta-pelites. However it is the relatively common occurrence of graphite that makes these rocks particularly interesting. Examples occur in:

Malene supracrustals, Godthaabsfjord, West Greenland (V. R. McGregor, 1973)
Lofoten Islands, NW Norway (Heier and Griffin, 1973)
Graphite System, Malagasy (Besairie, 1967)
Nain Province, Labrador (Collerson, Jesseau and Bridgwater, 1976)
Scourian, Scotland (Coward and co-workers, 1969)
Khondalites, Madras (Subramaniam, 1959)

No organic chemistry studies have been made to indicate whether the graphite is of biogenic or inorganic origin. The amount of graphite in these early Precambrian rocks is considerable and in the Graphite System of Malagasy it is enormous. This graphite warrants detailed investigation due to its implications on the origin of Archaean organisms.

Marbles It may seem surprising that marbles are so common in such old terrains. Many would regard them as meta-sediments (Sidorenko and coworkers, 1969), although to some they may be recrystallized calcareous volcanic clays (R. Mason, personal communication).

Usually the marbles range from a few metres to a few tens of metres thick and are often bordered by quartzites and meta-volcanic amphibolites. One extraordinary occurrence is worthy of special mention. In the gneisses of the Sankaridrug area of southern India the marble is up to 250 m thick and extends along strike for 30 km (Naidu, 1963). An interesting feature of this marble is that its MgO content is less than 4%, which makes it suitable for the cement industry (V. S. Krishnan, personal communication). Since it is common knowledge that early to mid-Precambrian carbonate sediments were mostly dolomites, and that limestones did not evolve in abundance until later times, this occurrence of an extremely old dolomite-free marble is unique.

Other high-grade marbles occur in the Lofoten Islands of NW Norway (Heier and Griffin, 1973), the Limpopo belt of southern Africa (Söhnge, Le Roex and Nel, 1948), the Scourian of Scotland (Coward and coworkers, 1969), the Androyan and Graphite Sequences of Malagasy (Besairie, 1967), the Nain Province of Labrador (Bridgwater and co-workers, 1975; Collerson, Jesseau and Bridg-water, 1976), and the Fiskenæsset region of West Greenland (Windley, Herd and Bowden, 1973). There is considerable scope for research into the geochemistry of these marbles.

Quartzites Quartzites are of special importance in many high-grade regions. In the Limpopo belt they reach 3 km thick and are closely interbedded with the marbles (Söhnge, Le Roex and Nel, 1948) and in the Aldan Shield there is an aggregate thickness of 2·8 km (Salop and Scheinmann, 1969). They form thin layers in the Lofoten Islands of NW Norway (Heier and Griffin, 1973), the Androyan and Graphite Sequences of Mala-gasy (Besairie, 1967), the Scourian (Coward and coworkers, 1969) and southern India (Naidu, 1963). The quartzites often contain magnetite or fuchsite, the former grading into iron formations. These rocks are usually massive with little marked bedding; they appear to be recrystallized orthoquartzites.

Iron Formations Some of the quartzites contain so much magnetite (over 50%) that they constitute iron formations and in a few places, like India, they are of economic value. These iron formations invariably have a quartzite association and locally they are banded with alternating quartz and magnetite. In the Limpopo belt they reach 20 m in thick-ness (Söhnge, Le Roex and Nel, 1948), and in southern India 30 m (Saravanan, 1969); other relict iron formations occur in the Lofoten Islands, NW Norway (Heier and Griffin, 1973), the Beartooth Mountains in Montana–Wyoming (Casella, 1969), Swaziland (Hunter, 1970a), and Sierra Leone (Andrew-Jones, 1966). In the Guyana Shield of Venezuela there are economic iron formations (itabarite) in the Imataca gneisses (Kalliokoski, 1965) which are older than 3000 my (Hurley and coworkers, 1968).

One iron formation is particularly impor-tant. In the Isua area of West Greenland a gneiss dome is mantled by a greenschist- to amphibolite-grade sequence with volcanics, calcareous quartzites and an economic magnetite-banded iron formation (Fig. 1.4) (Allaart, 1976). The BIF has yielded a Pb/Pb age of 3760 ± 70 my (Moorbath, O'Nions and Pankhurst, 1973, 1975b). There are no other comparable BIF in West Greenland, nor are similar rocks of this age known at present from any other continent. Discovery of BIF as old as this places new constraints on the model of Cloud (1968a)—that BIF are acceptors of oxygen produced by photosynthesizing organ-isms. Either the model is incorrect or blue-green algae flourished 3800 my ago.

There are relics of BIF in the Godthaab area of West Greenland that are especially interest-ing. The Amîtsoq gneisses, which have a Rb/Sr metamorphic age of 3700–3750 my (Moorbath and coworkers, 1972), were derived by deformation of porphyritic granites which contain inclusions of older rocks (McGregor, 1973). Amongst these are quartz–magnetite–grunerite rocks, probably the remains of extremely old BIF, that are as yet undated.

Finally, what about the environment in which these sediments were laid down? The orthoquartzites, marbles and quartzite-facies BIF are suggestive of an epicontinental platform/shelf sequence, namely shallow water and stable tectonic conditions (Salop and Scheinmann, 1969; Sylvester-Bradley, 1975; Sutton, 1976); if the graphite in the mica schists is biogenic a near-shore environ-ment is probable. It is important to emphasize

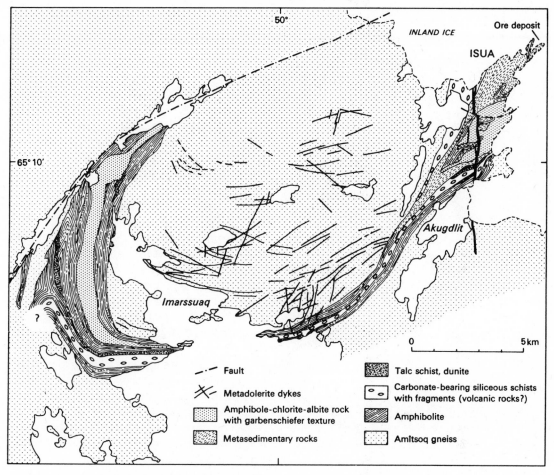

Fig. 1.4. Map of the Isua supracrustal belt, W Greenland; for location see Fig. 1.2 (after Allaart, 1976; reproduced by permission of The Geological Survey of Greenland)

that these sediments are very different from those in the Archaean greenstone belts (clastic – greywacke – flysch – conglomerate – shale association) which accumulated in unstable turbidite-type eugeosynclinal environments. The marked difference in these sediment types makes it impossible for the high-grade Archaean terrains to be simply highly metamorphosed greenstone belt sequences (see further in Chapter 3).

Layered Igneous Complexes

In many high-grade Archaean regions there are tectonic lenses and layers of a variety of deformed and metamorphosed layered igneous rocks; two types can be distinguished in different tectonic zones.

Anorthosite–Leucogabbro Examples of these plagioclase-enriched complexes include the following (Windley, 1973b):

Labrador (Collerson, Jesseau and Bridgwater, 1976)

W Greenland (especially the Fiskenæsset Complex (Fig. 1.5)) (Windley, 1969; Windley, Herd and Bowden, 1973)

Outer Hebrides, Scotland (Dearnley, 1963; Watson, 1969)

Southern India (Subramaniam, 1956; Leelanandam, 1967; Janardhanan and Leake, 1975; Ramadurai and coworkers, 1975; Windley and Selvan, 1975)

Limpopo Belt, S Africa (Hor and coworkers, 1975)

Malagasy (Boulanger, 1959)

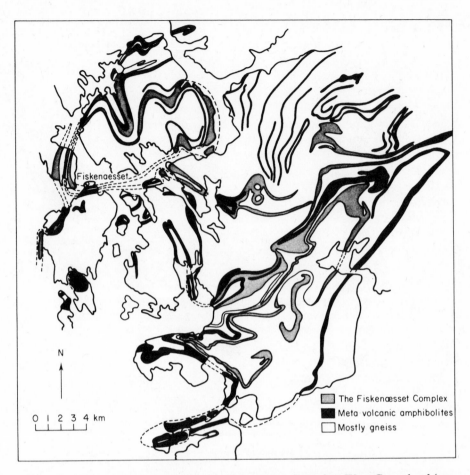

Fig. 1.5. Map showing the anorthositic Fiskenæsset Complex, West Greenland (see Fig. 1.2) bordered by metavolcanic amphibolites within a high-grade gneiss terrain (redrawn from J. Myers, Plate 1, Rapp. Grønlands geol. Unders., **73**, 1976; reproduced by permission of The Geological Survey of Greenland)

Sierra Leone (N. W. Wilson, 1965; Andrew-Jones, 1966; Williams and Williams, 1976)

Belomorides, Baltic Shield (Moshkin and Dagelaiskaja, 1972)

Aldan Shield, etc., USSR (in Bogatikov, 1974)

The complexes range up to about 1 km in thickness but often, due to tectonic disruption and thinning, they are only tens of metres to 100 m or so thick, and where extensively migmatized (invasion by late gneisses) they may be represented by only a few metre-sized pods.

The leucogabbros, which could also be termed gabbro(ic) anorthosites, typically have a prominent cumulate igneous texture marked by subhedral plagioclase megacrysts in a hornblendic matrix. It was once thought by Lacroix, Sorensen and Naidu that the anorthosites were recrystallized sediments but the presence of the cumulate textures in leucogabbros, chemically graded layers and, in key complexes (Fiskenæsset, Limpopo and Sittampundi) chromitite seams demonstrates that these rocks belong to layered igneous complexes. Gabbros form a minor component of some bodies (Fiskenæsset, Sittampundi) but ultramafic rocks are rare; eclogitic types occur at Sittampundi. Sapphirine-bearing rocks are associated with many complexes (Fiskenæsset, Limpopo, Sittampundi and Sakeny, Malagasy).

The crystallization pattern of these bodies is dominated by plagioclase–hornblende; the plagioclase is rather calcic with an An range of 80–100 where best preserved and much lower where retrogressively overprinted, and the hornblende is subsilicic tschermakite–hastingsite–pargasite–edenite. Chromites have a distinctive high FeAl composition ($Cr:Fe = 1:1$), different from that in ophiolites and layered intrusions such as Bushveld. All complexes display calc-alkaline differentiation trends. The original crystallization pattern of the Fiskenæsset Complex has been only weakly modified by metamorphic re-equilibration as igneous cryptic chemical variations are preserved in some mineral species (Windley and Smith, 1974), but in the Limpopo bodies late retrogressive metamorphism has left little intact of the igneous mineralogy (Hor and coworkers, 1975).

The complexes are commonly bordered by supracrustal rocks, either meta-volcanic amphibolites or a shelf-type assemblage of quartzites, marbles and sillimanite mica schists (meta-K pelites), and they are often intruded by the nearby tonalitic gneisses. Although it has been suggested that some might be remnants of ophiolite complexes (obducted oceanic crust) (Garson and Livingstone, 1973), Windley and Smith (1976) point out that they are most similar to layered igneous complexes with calcic plagioclase–subsilicic hornblende (gabbros to anorthositic gabbros or anorthosites) that occur within the Mesozoic tonalite batholiths along active Cordilleran-type continental margins, as in southern California (see further in Chapter 3).

Mafic–Ultramafic In the northern marginal zone of the Limpopo belt of southern Africa (I. D. M. Robertson, 1973, 1974) and on the Scourian mainland of NW Scotland (Bowes, Wright and Park, 1964, 1971; F. B. Davies, 1974) remnants of mafic–ultramafic complexes occur as folded and recrystallized lenses and layers up to a few hundred metres thick in the gneisses (Fig. 1.6).

The ultramafics (pyroxenites and peridotites variably altered to serpentinites or tremolite-pyroxene rocks which, in the Limpopo, are chromite layered with high FeAl) are statigraphically overlain by the mafics (meta-norites in the Limpopo and meta-gabbros in the Scourian) which are locally succeeded by thin anorthosites (not more than 20 m in the Scourian and 100 m in the Limpopo). Original igneous structures such as opaque oxide layering and cumulate textures are preserved in places. The bodies are often spatially associated with meta-supracrustal rocks, but some lie entirely within gneiss.

Until 1976 these mafic–ultramafic bodies were generally grouped together with the anorthosite–leucogabbro complexes (e.g. Windley, 1973b), but Windley (in press) proposes that they belong to different parallel zones in the high-grade mobile belts concerned.

Komatiites

Ultramafic rocks which have the same chemical composition as komatiite, i.e. high MgO and CaO/Al_2O_3 (Viljoen and Viljoen, 1969), have so far been described from two high-grade Archaean regions. It is important to bear in mind the point made by Brooks and Hart (1974) that komatiites need to be defined on more than just geochemistry; textural criteria such as quench or spinifex textures are also necessary to demonstrate that the rocks crystallized as silicate melts. The komatiitic-type rocks in these high-grade regions have, of course, been entirely recrystallized with the result that no primary structures or textures, if they ever existed, have survived.

Viswanathan (1975) and Rivalenti (1976) found meta–ultrabasic/basic rocks with komatiitic chemical affinities in India and West Greenland respectively and, in consequence, they suggested that the evolution of the high-grade regions was comparable with that of greenstone belts. In view of the above remark of Brooks and Hart (1974), and because the geological make-up of the two

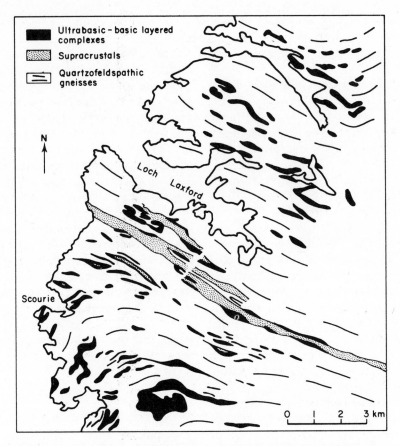

Fig. 1.6. Map showing remnants of ultrabasic–basic layered igneous complexes bordered by supracrustal mica schists in gneisses of the Scourie–Laxford area, NW Scotland (redrawn from F. B. Davies, personal communication)

types of region is so different, the comparison is unreliable.

Eclogites

Eclogites are not common in Archaean high-grade complexes, but they do occur in places.

In amphibolite facies gneisses of Tamil Nadu State, southern India, there is a 70 × 880 m eclogitic band with omphacite pyroxene (Rajasekaran, 1972). Probably the most prominent and extensive Archaean eclogites occur in the Glenelg inlier, NW Scotland (Alderman, 1936). The eclogites are in the form of multitudes of lenses up to several metres thick within garnet–biotite gneisses of probable sedimentary origin. Marble layers and inclusions are common in the nearby gneisses. The provenance or mode of formation of these Archaean eclogites is as yet unknown.

Chronological Relationships

The main problem in erecting a chronology of events in Archaean high-grade regions is that most rock units are mutually conformable. For this reason it is difficult to tell whether the supracrustal rocks represent a metamorphosed cover sequence originally overlying a basement made up now by the adjacent granitic gneisses, or whether the gneisses represent a more highly advanced stage of recrystallization of sediments (arkoses, greywackes) or acid to intermediate volcanic rocks. In other words, what does the conformability mean and what does it hide?

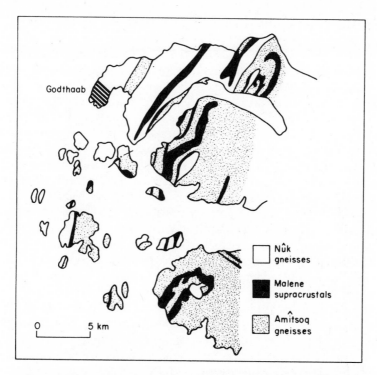

Fig. 1.7. Map of the Godthaab area, West Greenland, showing the distribution of Malene supracrustals and the younger Nûk and older Amîtsoq gneisses (redrawn from McGregor, 1973; reproduced by permission of The Royal Society)

An important breakthrough came with the discovery of V. R. McGregor (1973) that the rock pile in the Godthaabsfjord region of West Greenland did not form at one time (Fig. 1.7). He used the fact that a group of amphibolite dykes (the Ameralik dykes) occur in some older layers (Amîtsoq gneisses) but not in others which are younger (Malene supracrustals, Nûk gneisses) to erect a chronology of events and these have been dated radiometrically, largely by the Oxford Isotope Laboratory. McGregor's method has been successfully applied to the high-grade gneissic terrain of the Labrador coast where a similar tectonic stratigraphy has been found by Bridgwater and coworkers (1975) and Collerson, Jesseau and Bridgwater (1976).

Geochronology and Isotope Data

Radiometric age determinations on rocks from Archaean high-grade regions show an evolution from 3800 to 2500 my ago, about a third of geological time.

Only a few years ago there were no significant ages older than 3000 my and it was generally thought that typical Archaean high-grade regions had been through a single evolutionary stage. It is now realized that several regions have had a multistage history; in particular this refers to West Greenland and Labrador where early gneisses have isotopic ages of 3750 ± 50 and 3622 ± 72 my respectively (Moorbath and coworkers, 1972; Moorbath, O'Nions and Pankhurst, 1975b; Hurst and coworkers, 1975). Late gneisses in Greenland have an age of 3040–3110 my (Pankhurst, Moorbath and McGregor, 1973; Moorbath, 1975a) and in Labrador of 3121 ± 160 my (Hurst and coworkers, 1975). The oldest isotopic ages in most other high-grade regions in the world fall in the period 2700–3100 my (e.g. Guyana, S America, Yilgarn and Pilbara blocks in W Australia, Gwenora in Rhodesia,

Uganda, Wyoming USA, and southern India). Interesting older ages have recently been reported from the Minnesota River Valley, USA (a poorly defined isochron of 3800 my, Goldich and Hedge, 1974), Mashaba, Rhodesia (3580 ± 200 my, Hawkesworth, Moorbath and O'Nions, 1975) and North Norway (3460 ± 70 my, P. N. Taylor, 1975). It is probable, indeed predictable, that older ages will sooner or later be revealed in several of these regions. This applies especially to the ones (i.e. the Scourian, Limpopo and southern India) that have similar tectono-stratigraphic units to those in West Greenland and Labrador (i.e. gneisses and meta-supracrustals associated with layered ultramafic–mafic–anorthositic complexes) and comparable structural and metamorphic histories.

It is widely thought that the radiometric dates obtained from these high-grade gneissic regions are a measure of the age, in general terms, of a regional metamorphism or homogenization or of a late reworking event (cf. Chadwick and coworkers, 1974; Hurst and coworkers, 1975; Chadwick and Coe, 1976). However Moorbath (1975a,b,c, 1976) and Moorbath and Pankhurst (1976) have cogently pointed out that the isotopic data place severe constraints on our ideas about the evolution of these high-grade gneisses. In its simplest form the conclusion is that, rather than indicating any of the above overprint events, Rb/Sr whole-rock isochrons simply tell us the date when the precursors of the orthogneisses were added (or accreted) to the continental crust. Lead isotope studies may also be used to distinguish between the 'reworking' and the 'accretion' models (Moorbath, Welke and Gale, 1969). Moorbath's suggestion has such serious implications for the evolution of early sialic material that it is worth following briefly the basis of his argument.

The $^{87}Sr/^{86}Sr$ ratio of any given rock (or mineral) increases progressively with time because ^{87}Rb is decaying radioactively to ^{87}Sr with a half life of 50,000 my and the growth rate of $^{87}Sr/^{86}Sr$ for a given period of time is proportional to the Rb/Sr ratio in the sample. Now the Amîtsoq gneisses, for example, have

an average Rb/Sr ratio of c. 0·3 and they had an initial $^{87}Sr/^{86}Sr$ ratio 3750 my ago close to 0·701. Because the decay rate of ^{87}Rb to ^{87}Sr is known, these Amîtsoq gneisses must have had an average $^{87}Sr/^{86}Sr$ ratio of 0·715 by 3000 my ago, which is the approximate age of the Nûk gneisses in the same area. If these Nûk gneisses had been derived by remobilization of the Amîtsoq gneisses they should have an initial strontium isotope ratio of 0·715, but in fact they have a ratio of about 0·702. This means that the younger gneisses could not possibly be reworked older 'basement' gneisses and their low initial strontium isotope ratio indicates that their immediate precursors, tonalitic intrusions, were juvenile additions to the continental crust at, or close to, their measured age of 3040–3110 my ago. In other words, the low initial $^{87}Sr/^{86}Sr$ ratio is inherited directly from the upper mantle or low Rb/Sr source region. Because large volumes of granitic gneisses formed in the period 2800–3100 in, for example, Scotland, Greenland and Rhodesia where they have low initial $^{87}Sr/^{86}Sr$ ratios of approximately 0·701–0·702, a major implication of Moorbath's model is that continental growth on a major scale took place during this period.

Although the measured age of a high-grade gneiss could be related to the time of regional metamorphism, Moorbath demonstrates that this takes place not more than 50–100 my after the separation of the gneiss precursor from its source region and its emplacement into the sialic crust. It is interesting to note that this is about the same time span as between the formation of modern oceanic crust at a plate accretion boundary and the formation of a new island arc or continental margin at a plate-consuming boundary.

Metamorphism

Most of the Archaean regions concerned went through a period of high-grade metamorphism, commonly in the granulite facies sometime in the period 2800–3100 my ago. The common isotopic ages from these regions result from this major plutonic phase which, according to the Greenland and Labrador evidence, was late in a long sequence of events. Fig. 8.6 shows the distribution of Precambrian granulite regions, but insufficient have been well dated to show separate maps for Proterozoic and Archaean granulites.

This high-temperature metamorphism of continental rocks suggests that geothermal gradients were steeper than at present; this in turn was possibly a response to an early thin and radioactive crust (Heier, 1973). The earliest semistable granitic crustal fragments would concentrate radioactive species with the result that radiogenic heat production was several times greater 3000 my ago than today (Fig. 19.10). Fyfe (1973) suggested that the earliest radioactive crust may have had a geothermal gradient of about $100°C \, km^{-1}$ and was no thicker than 10 km. This gradient would enable granulites to form at the base of a crust 8–9 km thick. But it is estimated by O'Hara (1975) that by 3000 my ago in the Scourie area of Scotland the geothermal gradient had reached about 24°C/km in a crust that was more than 45 km, perhaps 75 km, thick.

Estimates of pressures and temperatures operational during the late Archaean periods of metamorphism have come mostly from studies of a variety of rocks and chemical reactions in the recrystallized layered igneous complexes:

by R. St. J. Lambert (1976) may be borne in mind: 'unfortunately there are as yet virtually no detailed studies of the mineralogy of truly regionally metamorphosed Archaean rocks, and none which yield closely defined PT conditions to a degree comparable with studies such as those on the Alps and Caledonides'— great scope for students with an electron microprobe!

Deformation Patterns

The intercalated rock units have usually been folded several times with the result that the typical deformation pattern in these high-grade regions is a complex interference structure in which domes and basins are accompanied by refolded isoclines. Such patterns occur on a metre scale in many medium- to high-grade gneisses, but they can also be seen on a regional scale particularly well in the Archaean craton of West Greenland, the Limpopo belt of southern Africa and the Aldan Shield of Siberia.

Fig. 1.5 illustrates a 500 sq km area in West Greenland where the individual folds are up to a few kilometres across. The formation of such a fold pattern is dependent on the varying geometrical relationships between the structures of the early and late fold sets. Most of the folds formed in association with high-grade

Complex	Pressure (kb)	Temp. (°C)	Reference
Fiskenæsset, Greenland	5–6	650–700 (amph. fac.)	Windley, Herd and Bowden (1973)
Fiskenæsset, Greenland	>7	800-900 (gran. fac.)	Windley, Herd and Bowden (1973)
Fiskenæsset, Greenland	9		Platt and Myers (in press)
Sittampundi, India	7–8	850	Chappell and White (1970)
Sittampundi, India	9–10	800	(Leake and coworkers in Yardley and Black, 1976)
Rodil, Scotland	10–13	800–860	Wood (1975)
Scourie, Scotland	8·5–18·5	1025–1075	O'Hara (1975)
S. Harris, Scotland (meta-pelites)	9–11	700–800	Dickinson and Watson (1976)

The above data indicate that pressures during metamorphism were surprisingly high in the late Archaean which implies that the crust was not thin at that time. However caution is advised in discussion of geothermal gradients in the Archaean and the following comment

regional metamorphism late in the history of the regions (e.g. V. R. McGregor, 1973). The first major isoclines in these regions are best displayed by the layered igneous anorthositic complexes (e.g. Windley, Herd and Bowden, 1973; F. B. Davies, 1974; Ramadurai and

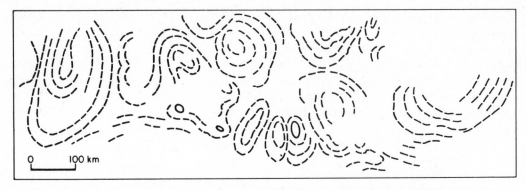

Fig. 1.8. The deformation pattern of the Archaean of the Aldan Shield, USSR (redrawn from Salop and Scheinmann, 1969; reproduced by permission of L. J. Salop)

coworkers, 1975). They may well have formed as nappe-like structures in response to a dominant horizontal tectonic regime (Bridgwater, McGregor and Myers, 1974a; Coward, Lintern ahd Wright, 1976; Myers, 1976).

A somewhat different structural pattern is seen in Fig. 1.8. This half a million square kilometre area of the Aldan Shield is dominated by dome-shaped ovals up to 300 km across in granitic gneisses (Salop and Scheinmann, 1969). There is a close relationship between this style of folding and granitization (partial melting), and the formation of the domes was commonly accompanied by high-grade regional metamorphism and the diapiric ascent of granitic rocks in the dome cores. Salop and Scheinmann (and Salop, 1972b) emphasize that this type of structural style is characteristic of, and exclusive to, the deep levels of the Archaean crust and they thus proposed the term 'permobile' to describe this early stage in earth history.

Some Key Regions

It is convenient at this stage to review briefly what is known about the nature, origin and build-up of some classic high-grade Archaean regions. Evidence of what happened in early Precambrian times is not limited to these regions, but they do provide a useful fund of data as they have generally been studied in some detail.

West Greenland

The stratigraphy, metamorphic grade and tectonic pattern of the rocks in West Greenland are little different from these in other high-grade regions. Because of these similarities the region has been grouped with Labrador, East Greenland and the Scourian of Scotland to form the Archaean craton of the North Atlantic region (Bridgwater, Watson and Windley, 1973c).

Several large hornblende granulite-grade areas occur within a widespread amphibolite-grade terrain. There is commonly evidence that some amphibolite-grade areas were formerly at a higher grade. In the granulite–amphibolite border zones hypersthenes are partially altered to hornblende, in the amphibolite facies gneisses there are relict pods of hypersthene gneisses, and in the bordering amphibolite-grade areas many of the rocks less susceptible to chemical change, such as amphibolites, retain their granulite facies mineral assemblages. On the other hand some extensive amphibolite facies areas, such as Godthaabsfjord and South Fiskenæsset, appear never to have been at a higher grade; they have none of the above features and instead contain prominent layers with a low amphibolite facies mineralogy which appear to have undergone only prograde metamorphism. The current problem is how to distinguish between prograded and retrograded rocks, in particular the gneisses.

Within the quartzo-feldspathic gneisses there are layers of a variety of rocks, in particular amphibolite, aluminous mica schist, anorthosite and associated rocks, and pods and lenses of ultrabasics. All rocks are mutually conformable and their relative age relationships, especially between the older and younger gneisses, are best known in the Godthaabsfjord region (Fig. 1.7) where V. R. McGregor (1973) erected a sequence of events, several of which have since been dated radiometrically. The following chronology for West Greenland is modified after that by V. R. McGregor (1973) and Bridgwater, McGregor and Myers (1974a):

3. Deformation associated with the late high-grade metamorphism decreased the angle of many earlier tectonic or intrusive discordances.

It was the presence of the amphibolite dykes in the Godthaab region that enabled McGregor to break through the conformability barrier; they are present in the layers of older Amîtsoq gneisses but they are absent in the younger rock units. Where the amphibolite dykes are absent, as in most of the Archaean of West Greenland, it is difficult to demonstrate unequivocally the above sequence of events. However the fact that the stratigraphic relationships elsewhere are simi-

my	Events	Method	Reference
2520±90	Emplacement of Qorqut granite, retrogression of granulite facies rocks	Rb/Sr	Moorbath and Pankhurst, 1976
2850±100	Widespread cordierite granulite	Pb/Pb	Black and coworkers, 1973
2900	and amphibolite facies metamorphism. Formation of major fold interference patterns	Rb/Sr	Evenson and Murthy, unpubl.
2850	Formation of Nûk gneisses	Rb/Sr	Moorbath and Pankhurst, 1976
3040±50	Remobilization of earlier rock	Rb/Sr	Pankhurst, Moorbath and McGregor (1973)
3030±20	units, formation of granites and migmatite complexes. Emplacement of Nûk granite (now gneiss) suite. Tectonic interleaving of all earlier rock units. Emplacement of anorthosite–gabbro complexes	Pb/U	Pidgeon, 1973
Not much older than 3000	Deposition of sediments and volcanics (Malene supracrustals)		Hawkesworth, Moorbath and O'Nions (1975)
	Intrusion of Ameralik basic dykes		
	. . . .		
3700–3750	Deformation and metamorphism	Rb/Sr	Moorbath and coworkers, 1972
3650±50	and formation of Amîtsoq gneiss	Pb/U	Baadsgaard, 1973
3780–3800		Rb/Sr, Pb/Pb	Moorbath, O'Nions and Pankhurst (1975b)
3760±70	Isua BIF	Pb/Pb	Moorbath, O'Nions and Pankhurst (1973)
	Emplacement of Amîtsoq granite (now gneiss)		

The present conformability of the major rock layers was caused by a combination of three factors:
1. The interthrusting of the Amîtsoq gneisses (with their Ameralik dykes) and the sediment-volcanic–anorthosite–gabbro suite.
2. The emplacement of the Nûk calc-alkaline granitic suite was tectonically controlled by the earlier thrust planes.

lar to those in the Godthaabsfjord region suggests that the general chronology may be widely applicable.

Important aspects to come out of McGregor's and Oxford's work are:
1. It is necessary to map out an intercalation of rock units in these high-grade areas almost like stratigraphic mapping in Phanerozoic rocks. Subtle variations in gneiss types,

inclusions of rocks such as amphibolite in some layers and not in others, and the intercalation of major rock layers may enable a sequence of events to be established.

2. The present rock intercalation has only tectonic significance. Few original sedimentary–volcanic contacts are preserved; most are interpreted as thrust planes.

3. Within the thrust pile dominated by gneisses, older and younger gneisses can be distinguished on the basis of the absence and presence of amphibolite dykes. The younger Nûk gneisses tend to be slightly discordant.

4. The older Amîtsoq banded gneiss developed by deformation of a homogeneous granite, in places with rimmed rapakivi feldspars. Because of this indubitable relationship geologists in Greenland are not inclined to interpret the quartzo-feldspathic Archaean gneisses as being recrystallized sediments or volcanics.

5. The 2850–2900 my isotopic age reflects a late thermal event that overprinted virtually a thousand million years of earth history. In many high-grade regions throughout the world the oldest radiometric age lies in the range 2800–3000 my; it is probable that many likewise have an older history.

According to Baadsgaard, Lambert and Krupicka (1976) mineral isotopic age relationships in the Amîtsoq gneisses from the Godthaab area suggest that they have been affected by three major events close to 3600, 2500 and 1500 my. In the Isua area the entire gneiss (Amîtsoq)–supracrustal system developed within the approximate time interval 3900–3700 my (Moorbath, O'Nions and Pankhurst, 1975b).

There appears to be more anorthositic material in this Archaean craton than in most others—a few layers reach 6–8 km in thickness but mostly they are up to 2 km thick, and some single layers can be followed along the strike for 100 km. These must be at least 1500 km strike length of anorthositic layers in the Archaean of West Greenland as measured on the present ground surface, of which 500 km make up the Fiskenæsset Complex.

Little has been published about the layers of anorthositic and associated rocks except for

those of the Fiskenæsset Complex, which are the remains of a layered igneous body that is remarkably well preserved in spite of the fact that it has been folded at least three times and metamorphosed to an amphibolite or granulite grade. Fig. 1.5 shows the tectonic pattern of the western half of the complex.

Labrador

In Saglek fjord on the coast of Labrador in the Nain Province there are the remains of a crustal evolutionary sequence similar to that recorded in West Greenland (Bridgwater and coworkers, 1975; Collerson, Jesseau and Bridgwater, 1976). The high-grade complex is made up of the following conformable units:

1. Uivak gneisses (the oldest rocks). A varied group of highly deformed quartzo-feldspathic gneisses derived from homogeneous prophyritic 'granitic' rocks and characterized by the presence of relics of amphibolite dykes (Saglek dykes). They are subdivided into two types: layered granodioritic gneisses (Uivak I) derived from earlier tonalitic igneous parents, and less extensive iron-rich porphyritic granodioritic and ferro-dioritic gneisses (Uivak II) (Bridgwater and Collerson, 1976). They account for at least 50% of the area and both types yield a whole-rock Rb/Sr isochron of 3622 ± 72 my with an initial ratio of 0.7014 ± 0.0008 (Hurst and coworkers, 1975).

2. Saglek dykes. A suite of amphibolite or pyroxene granulite (depending on metamorphic grade) dykes that transect foliation and other structures in, and are confined to, the Uivak gneisses.

3. Upernavik supracrustals are a group of meta-sedimentary schists and gneisses (metapelites, quartzites and marbles) and metavolcanic amphibolites accompanied by layered ultramafic and basic lenses.

4. Young quartzo-feldspathic gneisses which do not contain Saglek dykes. They form two generations of deformed small sheets and veins that intrude the Uivak gneisses and Upernavik supracrustals but which rarely form mappable units. They have a Rb/Sr age

of 3121 ± 160 my (Hurst and coworkers, 1975).

5. Late granites. Some are syntectonic with respect to folds and some were emplaced along shear belts.

The authors concerned conclude that the Uivak gneisses were intruded by a swarm of basic dykes, then tectonically interleaved with a cover sequence, intruded by granitic sheets, metamorphosed and deformed under granulite facies conditions, and finally intruded by syntectonic granites—a chronology essentially similar to that in West Greenland.

Scourian, Scotland

Whilst more has been written on the Lewisian than any other Precambrian high-grade region, there has been little separate treatment of the Scourian and surprisingly few details are known of the early history of this time period.

The 'early Scourian' (or Scourian *sensu stricto*) and 'late Scourian' of Sutton and Watson (1969) are referred to as the Badcallian and Inverian by Park (1970) and Moorbath and Park (1971). Their periods of high-grade metamorphism culminated *c.* 2900 and *c.* 2200 my ago respectively. We are concerned here with the early Scourian.

There are two schools of thought about the chronology of events that gave rise to the early Scourian, this disagreement illustrating well one of the fundamental problems in the interpretation of early crustal gneissic regions the world over (see Table 1.1). According to Bowes, Barooah and Khoury (1971) and Bowes (1976) the Scourian gneisses are largely a metamorphosed volcanic assemblage

of basalts, andesites and rhyodacites (with andesites predominant) that make up the Kylesku Group which is regarded as broadly comparable with the volcanic pile in greenstone belts. The main reasons for these authors' conclusions are the similarity of the bulk chemical compositions of the gneisses and granulites with those of volcanic rocks, the presence of small intercalations of metasediment in the gneisses, and the lead isotope compositions and initial strontium isotope ratios which are interpreted by Moorbath, Welke and Gale (1969) and Moorbath (1975a) as favouring separation of the gneissic precursors from the mantle only a short time before their metamorphism about 2900 my ago.

On the other hand Watson (1973a, 1975), whilst accepting that a minor portion of the gneisses may be derived from supracrustals, considers that the bulk of the gneisses were produced by recrystallization and deformation either of tonalites, granodiorites and granites intruded as the isotopes suggest just before the 2900 my metamorphic event (such intrusive gneisses have since been demonstrated by F. B. Davies, 1975a), or possibly from remnants of a gneissic basement, although the existence of such a basement has yet to be proven.

A new perspective was given to Scourian geology, firstly by Davies (1974) who showed that there are not only a multitude of small ultrabasic and basic fragments in the gneisses (Bowes, Wright and Park, 1964), but also better preserved remains of layered ultrabasic–basic complexes up to 400 m thick and at least 12 km long that are consistently

Table 1.1 Comparative simplified chronologies to explain the formation of the Scourian Complex, NW Scotland

Watson (1975) F. B. Davies (1975a)		Bowes, Barooah and Khoury (1971) Bowes (1976)
Badcallian metamorphism	2900 my BP	Badcallian metamorphism
Major intrusion of tonalite–granodiorite sheets	↑	Intrusion of layered complexes
Intrusion of layered complexes		Deposition of a major, largely volcanic, sequence
Deposition of minor sediments and volcanics	↓	—
Possible? early gneissic basement	Older	No basement

overlain by meta-sedimentary mica schists (Fig. 1.6); and, secondly, by F. B. Davies (1975a) who recognized that there are several generations of gneisses many of which intrude the layered complexes and that therefore this stage in development of the Scourian complex is broadly comparable with the later part of the Archaean of Greenland. Probably the bulk of the Scourian gneisses are similar in type and relationship to the Nûk gneisses in Greenland and the meta-sediments and layered complexes are chronologically comparable to equivalents in Greenland.

Limpopo belt, southern Africa

This high-grade ENE-trending mobile belt separates the Rhodesian and Kaapvaal greenstone belt/granite cratons of southern Africa. The following account of its early history is taken from the review paper by R. G. Mason (1973). As yet it has not been well dated, although it predates the Satellite dykes of the Great Dyke with an age of 2600 ± 120 my (Robertson and van Breemen, 1970), whilst the post-tectonic Bulai and Singelele granites (2690 ± 60, whole-rock isochron, van Breemen, 1970) place a minimum age on its main development.

In general terms the central, main part of the belt consists largely of high-grade gneisses within which there are conformable strips and lenses of metamorphosed sedimentary rocks and igneous anorthosite–basic complexes; these will be considered in turn.

Reworked Basement Gneisses The gneisses are considered to represent an old basement to the sedimentary rocks (Bahneman, 1971). They are typical complexly-deformed tonalitic gneisses with local, more granitic, areas homogenized by partial melting, and they have a partially retrogressed granulite facies mineralogy. One piece of evidence for their basement origin that needs substantiating is that they contain locally discordant amphibolite dykes that are absent in the meta-sedimentary and meta-igneous rocks.

The Messina Formation This contains a predominantly shallow-water meta-sedimentary sequence of meta-orthoquartzites, magnetite quartzites, marbles and dolomites, banded iron formations, and some amphibolites and quartzo-feldspathic paragneisses with garnet, cordierite and sillimanite. Some of the amphibolites might be of volcanic origin.

Within the Messina Formation are layers of metamorphosed anorthosite (mostly bytownite–labradorite) associated with leucogabbro, gabbro, hornblendite, chromite-layered serpentinite and hornblende–pyroxene rock. The presence of the chromitites, and of the plagioclase cumulate texture, grading and layering in the leucogabbros suggests that these rocks belong to layered igneous complexes that were emplaced into the Messina Formation supracrustal rocks (Hor and co-workers, 1975).

India

According to the latest reviews (Pichamuthu, 1971; Sarkar, 1972; Srinivasan and Sreenivas, 1972; Balasunderam and Balasubrahmanyan, 1973; Naqvi, Rao and Narain, 1974; Radhakrishna, 1974) the oldest rocks in India are in a high-grade gneissic state, formed more than 3000 my ago, and they occur in five regions:
1. The charnockite–gneiss province of southern India (Tamil Nadu).
2. The eastern Ghats belt of eastern India (which has had a 1650–1400 my overprint).
3. Rajasthan in the northwest (the Basement or Banded Gneiss Complex).
4. Bihar–Orissa in the northeast (the older Metamorphic Group).
5. The Aravalli–Delhi belt (the Banded Gneiss Complex).

One of the oldest reliable ages is the Rb/Sr isochron date on gneisses from South India of 3065 ± 75 my which has an initial ratio of $0 \cdot 7002$, whilst the granulite-grade metamorphism has a Rb/Sr age of 2580 ± 95 my (Crawford, 1969) or 2950 my (Devaraju and Sadashivaiah, 1969). The amphibolite-grade migmatite gneisses of southern India are dated at 2585 my (Crawford, 1969) and are assigned the term Peninsula Gneiss by Balasundaram and Balasubrahmanyan (1973). In the past there has been too much reliance in India on K/Ar ages, but the rocks offer excellent prospects of very old ages if Rb/Sr and Pb/Pb methods are employed.

All the gneissic rocks in India are currently thought to be older than the lavas and sediments of the main Dharwar Supergroup (isochron age of 2345 my, Crawford, 1969) (see Srinivasan and Sreenivas, 1972, and further in Chapter 2). These low-grade supracrustal rocks belong to the Upper Archaean (2500–3000 my) of Sarkar (1972) and correspond to the Greenstone Belt sequences of other continents.

All the gneissic rocks, whether of granulite or amphibolite grade, contain relict layers of metamorphosed sediments, lavas and plutonic igneous rocks, including magnetite quartzites, khondalites (garnet sillimanite schists and gneisses), calc silicate marbles and calc schists, amphibolites and hornblende schists, pyroxene amphibolites (pyroxene granulites, basic charnockites), meta-gabbros, meta-ultramafics (dunites and peridotites), meta-anorthosites and eclogites. Some regions contain more of one type than another. It is difficult to make a correlation between supracrustal stratigraphies in many regions due to isolation of outcrops.

A Comparative Review

In the foregoing pages of this chapter the rock units and chronological evolution of some classic Archaean regions have been summarized. Several marked similarities will be immediately obvious to the reader. Perhaps one of the most distinctive is the presence in the gneisses of all regions concerned of layers and lenses of meta-basic, meta-ultrabasic and meta-anorthositic rocks that are the remains of layered igneous complexes. Commonly the complexes are bordered by recrystallized shelf-type sediments that include mica schists, quartzites and marbles together with meta-volcanic amphibolites into which they (the complexes) were probably intruded. In West Greenland and Labrador the complexes and supracrustals clearly postdate some gneisses and predate others. Similar age relations with respect to different gneiss generations have been proposed from field evidence, but not yet established radiometrically, in the Scourian and Limpopo mobile belts and they should be looked for in southern India. Several regions have amphibolite dykes in the older but not the younger gneisses and most regions were subjected to high-grade metamorphism and deformation in the late Archaean, about 3000–2800 my, which transformed *inter alia* the intrusive calc-alkaline 'granitic' rocks into gneisses. It is not premature to suggest that these parts of the Archaean crust underwent a broadly similar chemical, tectonic and relative chronological evolution although, naturally, each region has its own distinctive features in terms of, for example, proportions of rock units and chemical differences. Archaean greenstone belts are remarkably similar the world over (see Chapter 2); the high-grade gneissic parts of the Archaean crust likewise share many similarities.

Mineralization

There are not many mineral deposits in Archaean high-grade regions. If the original sedimentary and igneous rocks did contain many mineral concentrations, these might have been destroyed by metamorphic/tectonic processes, or else these early rocks might have been relatively impoverished in ore deposits. In this respect these high-grade Archaean regions contrast markedly with the Archaean greenstone belt–granite terrains which are enriched in ores (Chapter 2). The following are the main types of mineral concentrations. Many are uneconomic or only subeconomic, but they nevertheless provide us with data on the types of mineralization that formed in the early stages of the earth's evolution.

Iron Formations

In several high-grade regions, such as the Beartooth Mountains, USA (Casella, 1969), the Limpopo belt (Söhnge, Le Roex and Nel, 1948) and southern India (Naidu, 1963), there are layers of banded iron formations, usually interbedded with some other meta-sediment such as magnetite quartzite. Most commonly they are just a few metres thick and are of the banded quartz–magnetite type.

The oldest BIF is at Isua, West Greenland (Allaart, 1976). This occurs in a greenschist- to amphibolite-grade sequence of calcareous quartzites and amphibolites (tuffs) mantling a gneiss dome, and the Pb/Pb age of its metamorphism is 3760 ± 70 (Moorbath, O'Nions and Pankhurst, 1973). The BIF is a banded quartz–magnetite type, but haematite–quartz types probably occur under the ice cap.

Cr in Anorthosites and Ultramafics

A chromium metallogenic province exists in these high-grade regions because chromitite seams are common in the anorthosites and ultramafics.

One of the most prominent deposits occurs in the Fiskenæsset Complex, West Greenland (Ghisler and Windley, 1967) where the anorthosites at the top of the layered succession contain chromitite seams, that locally reach 20 m thick, throughout the 500 km strike length of the complex. The chromitites have been metamorphosed, deformed and faulted, and the aluminous chromites have Cr : Fe ratios of about 1 : 1.

There are similar metamorphosed and deformed chromitites in the following bodies:
1. In the Sittampundi Complex, South India (Subramanian, 1956; Ramadurai and co-workers, 1975) where as chromitite seams are up to about 6 m thick and occur towards the top of the stratigraphy.
2. In ultrabasic layers associated with anorthosites in the Kondapalli complex, South India (Leelanandam, 1967).
3. In hornblendite layers in the meta-anorthosites of the Limpopo belt, southern Africa there are chromitite seams up to about 60 cm thick (Söhnge, Le Roex and Nel, 1948; Hor and coworkers, 1975).

Ni, Cu in Amphibolites

Amphibolitic layers in the gneisses commonly contain rusty sulphide-bearing zones but they are rarely economic. One exception is the Ni–Cu sulphide deposit at Selebi–Pikwe in the Limpopo belt, Botswana, which occurs in a 50 m wide amphibolite layer (P. S. L. Gordon, 1973).

Chapter 2

Archaean Greenstone Belts

Archaean greenstone belts are the oldest major group of well-preserved volcano-sedimentary basins and so they give us much direct evidence of early crustal conditions. They occur in many shield areas, vary in age from *c*. 3400 my to 2300 my, and range in size up to 250 km across. Their stratigraphy, general structure, volcanic geochemistry, types of sediments, and ore deposits are remarkably similar and this uniformity allows them to be treated as a single integral group.

The principal occurrences (with recent reviews) are:
1. Swaziland System, Barberton Mountain Land, South Africa (Viljoen and Viljoen, 1969; Hunter, 1974a; Anhaeusser, 1975)
2. Sebakwian–Bulawayan–Shamvaian Systems, Rhodesia (Stowe, 1971; J. F. Wilson, 1973a)
3. Pilbara and Yilgarn blocks, W Australia (Glikson and Lambert, 1973)
4. Abitibi, Yellowknife and many other belts, Superior and Slave Provinces, Canada (Goodwin, 1968; Goodwin and Ridler, 1970; McGlynn and Henderson, 1970; H. D. B. Wilson, Morrice and Ziehlke, 1974)
5. Dharwar System, India (Srinivasan and Sreenivas, 1972; Sreenivas and Srinivasan, 1974; Naqvi, 1976)

The greenstone belts are bordered and intruded by 'granitic' plutons which are not open to dispute. However, the relationship of the belts to nearby high-grade quartzo-feldspathic gneisses is subject to various interpretations, the merits of which are reviewed in the next chapter.

Study of Archaean greenstone belts has reached an interesting stage of development as there is currently a proliferation of contrasting ideas about their mode of development. These vary from the fixist to the mobilist and include models such as the remains of primordial oceanic crust, downsagging basins, pinched downfolds between colliding micro-continents, proto-oceanic ridge systems, terrestrial equivalents of lunar maria, and marginal basins formed in back-arc environments. In this chapter we shall consider the characteristic features of the greenstone belts, and in the next we shall look at how they and the Archaean high-grade regions may have evolved and how they may be interrelated.

General Form and Distribution

The shape of greenstone belts is easier to illustrate than to describe. At their simplest they have a linear plan and a basin-shaped cross-section, at their most complex a cuspate form with a triangular plan. There is a tendency to use the Barberton belt as a model for all greenstone belts—an inclination to be discouraged in view of the following comments.

The traditional idea of the structure of a typical greenstone belt, based on Barberton, envisages a basin-shaped infold or downfold in a sea of granitic material (Anhaeusser and coworkers, 1969). However Stowe (1974) and Coward, Lintern and Wright (1976) demonstrated that some belts in Rhodesia underwent nappe and thrust tectonics and that some

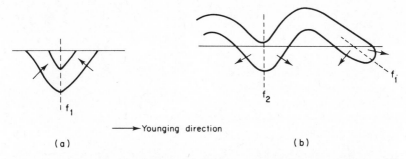

Fig. 2.1. Comparison of the primary structure of greenstone belts. (a) According to Anhaeusser and coworkers (1969). (b) The secondary, post-nappe structure of belts in Rhodesia by Stowe (1974) and Coward and coworkers (1976). (For discussion see text)

synclinal belts actually have downward-facing sedimentary structures: they are inverted sequences in late synforms that refold the overturned limb of earlier nappes (see Fig. 2.1). And yet the stratigraphy of the Barberton belt is directly correlated by Viljoen and Viljoen (1969) with that of the Rhodesian belts. Burke, Dewey and Kidd (1976) also argue that many greenstone belts may have undergone a more complex tectonic evolution than simple downfolding. The two types of structure are, of course, not mutually exclusive.

Greenstone belts vary considerably in size. The Barberton belt is only 40 km across and 120 km long whereas the Abitibi belt is 250 km wide and 800 km long and some belts in the Yilgarn block are up to 1000 km in length. There has been a suggestion that the older (>3000 my) belts (southern Africa) are smaller than younger (<3000 my) ones in Canada, India and Australia.

Looking at greenstone belts on a global scale, Engel and Kelm (1972) pointed out that they have a roughly parallel, linear orientation, although what this proves is debatable.

Geochronology

The most recent radiometric data seem to point towards at least two broad ages of greenstone belt formation.

In the older belongs the Onverwacht Group of the Barberton belt with isochron ages ranging from 3500 ± 200 my (Jahn and Shih, 1974) to 3360 ± 100 my (van Niekerk and Burger,

1969). As yet it is not known whether the Sebakwian rocks in Rhodesia have a similar age, although they are regarded as stratigraphically equivalent by Viljoen and Viljoen (1969). The fact that the upper sediments in the Barberton belt have a Rb/Sr age of 2980 ± 20 my (Allsopp, Ulrych and Nicholaysen, 1968) suggests that the greenstone belt as a whole may have taken at least 500 my to evolve. The Fort Victoria belt in Rhodesia could be older than 3520 ± 130 my (Hickman, 1974).

The younger group comprises the main Rhodesian, the Canadian, Australian and Indian belts that formed after 2900 my ago. Most belts formed in the period 2700–2600 my ago (*viz.* Bulawayan, Rhodesia; Yellowknife and Superior Province, Canada; Yilgarn, W Australia), but the Dharwar rocks may be slightly younger (2345 ± 60 my, Crawford, 1969); according to Burke, Dewey and Kidd (1976) the Birrimian greenstone belts of West Africa are 2100–1800 my old.

The fact that the initial strontium isotope ratios of most of the rocks concerned are near 0·701 indicates that they were derived directly from the mantle rather than the remelting of, or contamination with, older gneissic basement (Hawkesworth, Moorbath and O'Nions, 1975; Moorbath, 1975a,b).

It is important to note that whilst the Bulawayan Group was forming in the Rhodesia greenstone belts about 2720 ± 70 my ago (Hawkesworth, Moorbath and Kidd, 1975), flat-lying cratonic, largely clastic, sequences with volcanics were being laid down in the Dominion Reef–Witwatersrand Systems not far to the north. This demonstrates an overlap between the formation of Archaean greenstone belts and cratonic cover sequences under more stable platform-type conditions typical of the early Proterozoic (Sutton, 1976).

On the basis of zircon dating Krogh and Davis (1972) confirmed a prediction of Goodwin (1971) that greenstone belts in the north of the Superior Province formed 200 my before those in the south. But it has yet to be established whether this reflects a progressive younging southwards or two (or more) ages of greenstone belt formation.

Stratigraphy

General Points

The main subdivision common to many greenstone belts is into a lower, dominantly volcanic, and an upper sedimentary group. The lower group in most belts is further divisible into a lower, primarily ultramafic, and an upper volcanic group in which calc-alkaline, mafic-to-felsic rocks predominate. The typical subunits of this threefold division are shown in Fig. 2.2. The stratigraphy of greenstone belts in general has been studied by Anhaeusser and coworkers (1969), Anhaeusser (1971b) and Glikson (1976a).

The ultramafic group consists principally of ultramafic and mafic volcanics and is noted for the occurrence of chemically distinctive komatiites whose primary diagnostics are a high CaO/Al_2O ratio of >1, high MgO contents of $>9\%$ and a very low potash content of $<0.9\%$ (Viljoen and Viljoen, 1969; Brooks and Hart, 1974). There are peridotitic and basaltic komatiites, some of which are pillow-bearing lavas. This group may also contain layered ultramafic complexes and minor metasediments (aluminous quartz-sericite schists, quartzitic cherts, pelites) which contain little sial-derived detritus.

The calc-alkaline volcanic group is dominated by low-K basalt–andesite–dacite–rhyolite cycles (Goodwin, 1968) and/or bimodal mafic–felsic assemblages (Anhaeusser, 1971b; Glikson, 1976a). Sediments are

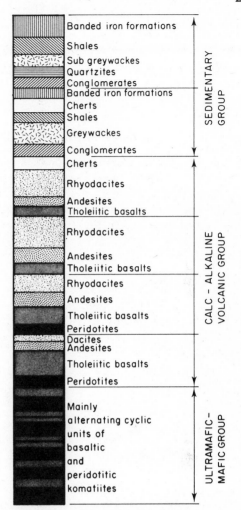

Fig. 2.2 Hypothetical stratigraphic succession for an Archaean greenstone belt based on the Barberton model (modified after Anhaeusser, 1971b; reproduced by permission of The Geological Society of Australia)

mostly chemically precipitated cherts, jaspers and banded iron formations.

The sedimentary, dominantly clastic group consists ideally of a lower argillaceous deeper water assemblage with, in particular, shales, pelitic sandstones and greywackes, and an upper arenaceous, shallow-water assemblage with conglomerates, quartzites and chemically-precipitated limestones and banded iron formations that tend to occupy the tops of cyclic units.

Notwithstanding the overall uniformity amongst the world's greenstone belts, many have their distinctive features. We shall therefore consider now the stratigraphic make-up of the major belts on different continents.

Swaziland System, Barberton, South Africa

The Barberton greenstone belt is well documented by Anhaeusser (1971a), Anhaeusser and coworkers (1968, 1969) and Viljoen and Viljoen (1969) and its main features are summarized in Fig. 2.3. The lower volcanic succession (the Onverwacht Group) is overlain by the sedimentary Fig Tree and Moodies Groups.

The lower part of the Onverwacht Group consists largely of ultramafic and mafic flows and intrusive complexes, divisible into three formations: the Sandspruit, Theespruit and Komati, in upward sequence (Viljoen and Vil-

joen, 1969). The Sandspruit Formation occurs as relict inclusions of serpentinized rocks in bordering intrusive tonalitic gneisses. The Theespruit Formation, comprised principally of metamorphosed ultramafic and mafic lavas, is noted for primitive fossil micro-organisms in carbonaceous cherts (see later this chapter). The Komati Formation contains peridotitic komatiites with pillow structures and quench textures which indicate that a mobile ultramafic magma was subaqueously extruded (Viljoen and Viljoen, 1969). In this and the underlying two formations are pillow-bearing basaltic komatiites, recrystallized to amphibolites and divisible into three chemically distinctive types based on their MgO contents: Barberton ($\pm 10\%$), Badplass ($\pm 15\%$) and Guluk ($\pm 20\%$).

Associated with the above extrusive rocks are three types of layered igneous ultramafic complexes, Kaapmiuden, Nordkaap and Stolzburg, some of which have komatiitic

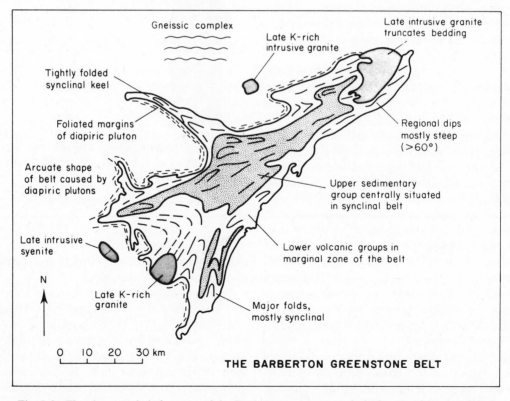

Fig. 2.3. The characteristic features of the Barberton greenstone belt, South Africa (modified after Anhaeusser, 1971a; reproduced by permission of The Geological Society of Australia)

affinities. According to Viljoen and Viljoen (1969) there is a unique magmatic differentiation sequence from peridotitic komatiite, the probable parent magma, to basaltic komatiites of the Badplass, Guluk and finally Barberton types.

The lower ultramafic–mafic part of the Onverwacht Group is separated from the upper calc-alkaline volcanic part by a remarkably persistent 6m thick chert–carbonate horizon, the Middle Marker that has a Rb/Sr age of 3355 ± 70 to 3375 ± 70 my (Hurley and coworkers, 1972).

The upper part of the Onverwacht Group consists predominantly of a cyclic mafic-to-felsic sequence of calcic to calc-alkaline volcanics, typically including rhyolites, rhyodacites, dacites, minor andesites and pillowed tholeiitic basalts together with mafic-to-felsic pyroclastics. Cyclic units are capped with chert and calcareous and ferruginous shales. The upward increase in K_2O and calc-alkaline character of the volcanics in the Onverwacht Group as a whole may reflect the progressive evolution of a thickening sialic crust (Engel, 1966) or an emerging volcanic arc (Condie, 1976b).

The Onverwacht Group is unconformably overlain by the argillaceous Fig Tree Group which passes upwards into the arenaceous Moodies Group. The Fig Tree rocks include rhythmically alternating shale–greywacke sequences, with some chert and banded ironstones and trachytic tuffs.

The Moodies Group contains polymictic conglomerates, quartzites and feldspathic sandstones—a molasse-type sequence. Cross bedding, ripple marks and mud cracks indicate shallow water deposition.

The Swaziland System of the Barberton belt is folded into a synform with the result that the oldest parts of the stratigraphy are prone to invasion by bordering granitic rocks (Fig. 2.3). Deformation tends to be more intense adjacent to tonalite domes (Hunter, 1974a). According to Viljoen and Viljoen (1969) there was little lateral compression in the development of the main regional structures, but they seem to have overlooked the analysis of Ramsay (1963a) who concluded that all deformation episodes involved horizontal shortening as their principal strain component.

The age relationship between the Swaziland System and the oldest nearby granitic/gneissic rocks is much disputed. The ancient gneisses of Swaziland do not come into contact with the Barberton belt, but Hunter (1970a, 1974a) has consistently argued that they predate the development of the greenstone belt. In the gneisses there are discordant, folded amphibolite dykes and Anhaeusser and coworkers (1969) concluded in a since-forgotten statement that 'the inference to be gained from the involvement of the early dykes in the migmatite gneiss palingenesis is evidence in support of this terrain being representative of an ancient, pre-Swaziland System basement'. After this early acceptance of the basement concept we hear no more of these interesting amphibolite dykes which are a key tool in the argument. In contrast, Viljoen and Viljoen (1969) interpret Barberton tonalitic gneisses as diapirs intrusive into the Swaziland System.

Rhodesian Belts

In the Rhodesian Archaean craton there are a great many greenstone, or schist, belts (Fig. 2.4) famous since the classic work of A. M. MacGregor (1951). He emphasized their arcuate, cuspate and synformal shape, their enveloping dome-shaped 'gregarious batholiths' and he recognized and defined the three main stratigraphic divisions, still in use: the Sebakwian, Bulawayan and Shamvaian.

Firstly we shall look at the rocks within these divisions and then consider some of the major problems of Rhodesian geology, such as the interpretation of the age relations between the greenstone belts and high-grade gneisses, and the structural evolution of the belts.

The Sebakwian Group includes serpentinized ultramafics and recrystallized mafics, such as tremolite–chlorite schists, talc schists, actinolite–chlorite schists and hornblende and chlorite schists. There are also arenaceous sediments and banded iron formations. Most of these rocks occur as inclusions within intruded granitic bodies, such as the Rhodesdale

28

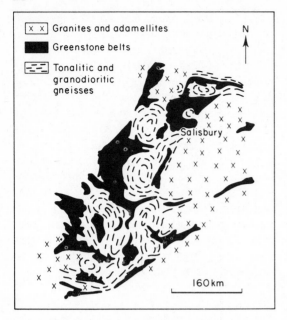

batholith. Viljoen and Viljoen (1969) place emphasis on the lithostratigraphic correlation between the Sebakwian Group and the lower Ultramafic Group of the Barberton belt; in so doing they propose that the definition of the Sebakwian should be revised to incorporate many of the basaltic rocks of probable basaltic komatiite composition in the Bulawayan. With these and other revisions in mind they are able to recognize the same important stratigraphic break that occurs within the Onverwacht Group (*viz*. the Middle Marker horizon), dividing the lower primitive ultramafic rocks from the overlying calc-alkaline volcanics.

The Bulawayan Group locally lies unconformably with a basal conglomerate on Sebakwian rocks and contains metamorphosed basaltic pillow lavas (some are low-K tholeiites), andesites, dacites and calc-alkaline tholeiites; there is a progressive upward increase in the proportion of felsic volcanics and there are conspicuous mafic-to-felsic volcanic cycles, tuffs and agglomerates. The oldest known stromatolites occur in Bulawayan limestones (see later in this chapter).

The Shamvaian Group consists largely of poorly sorted, clastic meta-sediments such as polymict conglomerates (with pebbles of gneiss and granite), greywackes, quartz mica schists and phyllites, together with limestones and banded iron formations.

It was thought by A. M. MacGregor (1951) that all the granite–gneiss 'batholiths' were younger than the greenstone belts. But more recent work shows they are heterogeneous and that areas of early gneiss can be distinguished from granitic plutons, the gneisses being regarded on various grounds to have formed an old basement to the greenstone belts (Bliss, 1969; Stowe, 1971, 1973, 1974; J. F. Wilson, 1973a). However, Viljoen and Viljoen (1969), Anhaeusser (1973) and Glikson (1976a,c) consider all the granites and gneisses to be younger than the early parts of the greenstone belts.

The age problem can be considered in two parts:

1. In the Selukwe area Bliss and Stidolph (1969) and Stowe (1971, 1973, 1974) recognize tonalitic gneisses that underlie the Selukwe schist (greenstone) belt and are pre-Sebakwian in age; associated granites and pegmatites are older than 3300 my (Vail and Dodson, 1969). The gneisses contain inclusions of gneiss, amphibolite, ultramafics, magnesian schists and banded iron formation, and are similar to the ancient tonalite gneisses of Barberton (Viljoen and Viljoen, 1969) and Swaziland (Hunter, 1970a, 1974a). Note that Glikson (1976c) does not accept their pre-Sebakwian age, maintaining that the inclusions can only, or mainly, be distinguished from Sebakwian rocks by their higher metamorphic grade. The superbly exposed Gwenora Dam migmatites, belonging to the tonalitic gneisses, have a Rb/Sr isochron age of 2780±30 my (Hawkesworth, Moorbath and O'Nions, 1975), but the presence of slightly discordant amphibolite dykes at that locality suggests a long period of evolution.

2. Near Shabani there is an unconformity between Bulawayan Group sediments and a tonalitic gneiss basement (Bickle, Martin and Nisbet, 1975), but neither has, as yet, yielded a satisfactory isochron age. Anhaeusser

(1973) does not accept that this unconformity proves that the main greenstone belts are younger than a gneissic basement.

On the Sebakwe River the Bulawayan basal conglomerate lies unconformably on Sebakwian talc schists. Glikson (1976a) emphasizes the fact that the conglomerate contains pebbles of gneiss similar to that of the Rhodesdale pluton, which is 3300 my old and petrologically comparable with the ancient tonalites of the Barberton Mountain Land. Glikson is hinting at the suggestion that all the tonalites, whether gneisses or granites, were emplaced into the lower part of the Onverwacht Group and the Sebakwian, prior to the deposition of the upper part and the Bulawayan.

But the situation may be considerably more complicated as there are probably several ages of greenstone belts in Rhodesia. Hawkesworth, Moorbath and O'Nions (1975) established that the Bulawayan Group volcanics were extruded about 2600–2700 my ago, whilst Hickman (1974) reported a 3520 ± 130 my age for the Mushandike granite which is younger than part of the Fort Victoria greenstone belt (J. F. Wilson, 1973b).

The Rhodesian gneissic–granitic rocks also vary widely in radiometric age. Gneisses at Mashaba have a best-fit regression age of 3580 ± 200 my whilst the Sesombi tonalite and the Gwenora migmatites formed 2690 ± 70 my and 2780 ± 30 my ago, respectively (Hawkesworth, Moorbath and O'Nions, 1975).

Not until more data are forthcoming from southern Africa will the arguments be solved but, in my opinion, the tide is going strongly against the Viljoens/Glikson/Anhaeusser interpretations of the geochronological data of Moorbath, his colleagues, Hickman, and others.

Kalgoorlie System, Western Australia

The Yilgarn block consists of granites and gneisses that enclose a network of northerly-striking greenstone belts. The key to the greenstone belt stratigraphy is in the central Kalgoorlie-Norseman area. There are three volcano-sedimentary sequences, each separated by an unconformity and often consisting of a cycle with ultramafic–mafic lavas overlain by intermediate acid volcanics and clastic sediments. Glikson (1970, 1971a) defined the following three-fold successions (in ascending stratigraphic order): pillowed and massive ultramafics and basalts; basalts, andesites, greywackes, slates, argillites and phyllites; conglomerates and greywackes. Viljoen and Viljoen (1969) and Anhaeusser (1971a) contend that this succession can be correlated with the Barberton stratigraphic model. Some of the lowermost ultramafic extrusive flows contain spinifex quench textures (Nesbitt, 1971) and have komatiitic chemistry (McCall and Leishman, 1971) and there are some differentiated layered ultramafic–mafic intrusions (Williams and Hallberg, 1973).

The ages of granitic rocks in the Kalgoorlie area suggest that the greenstone belts formed in a very short time interval. Granodiorites associated with an early volcanic cycle have an age of about 2700 my (Roddick and co-workers, 1973), whilst granites cutting younger cycles are dated as 2600 my; the regional metamorphism has an age of 2670 my (Turek and Compston, 1971). The metamorphic patterns and development are reviewed by Binns, Gunthorpe and Groves (1976).

Dharwar System, southern India

The Dharwar greenstone belts of southern India occur as well-defined northerly-striking strips up to 450 km long and 250 km across, separated by gneisses and granites. To understand the stratigraphic variations in the greenstone belts it is necessary to unravel their broad structure and age relationships with the widespread gneisses, which themselves contain remnants of older volcano-sedimentary suites large enough to be called greenstone belts (for further discussion of these age problems see the next chapter).

In his recent review Radhakrishna (1974) states that 'a large majority of workers examining Precambrian terrains in India now agree in considering the charnockite–khondalite terrain as the most ancient identifiable part of the primitive crust'

(Aswathanarayana, 1968; Viswanathan, 1969; Balasundaram and Balasubramaniam, 1973; Sreenivas and Srinivasan, 1974; Naqvi, Rao and Narain, 1974). The charnockites have an age of 3100 my (for references see Naqvi, Rao and Narain, 1974), and the amphibolite facies gneisses, which may be a higher level equivalent of the charnockites (Radhakrishna, 1974), and amphibolite xenoliths in the gneisses have a similar age (Crawford, 1969; Venkatasubramaniam, Iyer and Pal, 1971).

Although Pichamuthu (1974) considers all the 'schist belts', to use the Indian terminology, to have formed in the same single basin of deposition, the current popular opinion is that there are two ages of greenstone belt formation in southern India (Srinivasan and Sreenivas, 1971, 1972; Radhakrishna, 1974; Sreenivas and Srinivasan, 1974). The major larger, and well-preserved, Dharwar belts that dominate the geological map of the craton formed in the period 2700–2300 my ago, whilst the few older and smaller belts, termed here pre-Dharwar, occur as remnants in, and are invaded by, the gneisses and are more than 3000 my old. Shackleton (1976a) states that in Mysore the Dharwar rocks clearly rest unconformably on older acid gneisses and that the supracrustal rocks in the high-grade gneissic–granulite areas are pre-Dharwar in age.

There are three pre-Dharwar schist belts, i.e. the Kolar, Bababudan and Nuggihalli (Radhakrishna, 1974) which consist largely of ultramafic and mafic rocks such as serpentinized peridotites, komatiites, high-Mg and low-K tholeiites.

Basaltic komatiites are reported by Viswanathan (1974a) from the Kolar belt. Sreenivas and Srinivasan (1974) refute the komatiitic affinity of these rocks and suggest they are just high-Mg meta-basalts (now amphibolites). Peridotitic and basaltic komatiites occur in the Nuggihalli belt (reported in Viswanathan, 1974b).

The sediments in the pre-Dharwar belts are particularly interesting (Srinivasan and Sreenivas, 1968). The Bababudan is the oldest belt and it contains oligomictic conglomerates, orthoquartzites and shales. The matrix of some of the conglomerates is composed of detrital pyrites with gold and uranium minerals, which makes the rocks comparable with the Early Proterozoic conglomerates of the Huronian, Witwatersrand and Jacobina Systems (see Chapter 5). The main point is that the presence of easily oxidized and weatherable detrital pyrites suggests that the atmosphere was deficient in oxygen at the time these Archaean sediments were deposited; in other words, the sediments have escaped weathering by oxidative processes. These are the earliest conglomerates of this type recorded from any continent.

The major Dharwar greenstone belts include the Chitaldrug (Naqvi, 1973) and Shimoga belts. The meta-volcanics include low-K tholeiites (Naqvi and Hussain, 1973a) and olivine/quartz tholeiites to andesites that have high contents of Fe, Mg, Co, Ni, Cr and Mn (Naqvi and Hussain, 1973b). Meta-sediments are enriched in Mg, Ca, Fe, Ti, Cr, Co, Ni and V, indicating that rocks of basic composition predominated in the source area (Naqvi and Hussain, 1972; Satyanarayana and coworkers, 1973)—possibly in the pre-Dharwar belts. The predominant sediments are meta-greywackes with low K_2O/Na_2O ratios, polymictic conglomerates with pebbles of granite, gneiss, ferruginous quartzite, quartzite and chert, banded iron formations and banded pyritic chert (Naqvi, 1967). The Dodguni Formation of the Chitaldrug belt contains abundant graphite-bearing limestones, the presence of which might be taken as an indication that an oxygenic atmosphere had developed by this time, a suggestion corroborated by the presence of blue–green algal filaments in associated cherts (Gowda, 1970) which suggests some degree of primitive photosynthetic activity.

The youngest Dharwar sediments are semi-pelitic red beds with ferric/ferrous iron ratios of 3–10, indicating appreciable oxidizing conditions during their sedimentation (Sreenivas and Srinivasan, 1974). Thus, during the course of evolution of the pre-Dharwar and Dharwar belts the atmosphere underwent a significant change from an anoxidizing to an oxidizing state.

The formation of the Dharwar greenstone belts was followed by the intrusion of a variety of granitic plutons which culminated in the well-known Closepet granite 2380 my (Crawford, 1969) to 2000 my ago (Venkatasubramaniam, 1974).

Superior and Slave Provinces, Canada

In the Canadian Shield, greenstone belts form a prominent part of the Archaean Superior and Slave Provinces (Fig. 2.5) but they are also present in the Churchill Province where they have been overprinted by Proterozic (Hudsonian) plutonic activity.

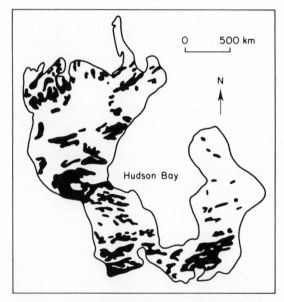

Fig. 2.5. The supracrustal rocks, mostly within Archaean greenstone belts, in part of the Canadian Shield (redrawn after Goodwin, *Geol. Ass. Can. Proc.*, **19**, 1968, 2; by permission of The Geological Association of Canada)

According to Goodwin (1968) the typical volcano-sedimentary pile in a greenstone belt developed sequentially in three main stages:
1. Construction of a thick broad mafic platform by widespread effusion of predominantly tholeiitic basalt.
2. Increasingly felsic pyroclastic eruption leading to erection of high-rising piles upon the mafic platform.
3. Partial denudation of the volcanic piles and construction of volcaniclastic blankets.

In the Superior Province basalts account for 50–60% of total volcanic rocks, 'andesites' about 20–30% and more felsic rocks about 10–15% (H. D. B. Wilson and coworkers, 1965; Baragar and Goodwin, 1969). There are also rare alkaline shoshonitic rocks (Cooke and Moorhouse, 1969; Ridler, 1970; Goodwin, 1972; T. E. Smith and Longstaffe, 1974; Hubregtse, 1976).

Of the multitude of Canadian belts the two most well known (the Abitibi and the Yellowknife) will be discussed here.

The Abitibi belt, reviewed by Goodwin and Ridler (1970), is 800 km long and 200 km wide and thus contrasts markedly with the comparatively small Barberton belt. It is truncated to the east and west by the younger Grenville and Kapuskasing high-grade belts, and so it was originally even longer; it is the largest single continuous Archaean greenstone belt in the world.

The distribution of the volcanic rocks is dominated by the presence of eleven elliptical volcanic complexes (Fig. 2.6), each with mafic-to-felsic extrusive rocks and coeval intrusions and sediments. These complexes lie close to the northern and southern borders (forelands) of the belts, the intervening median part being occupied by uniform tholeiitic bassalts, fine-grained clastics and major granitic batholiths. This pattern probably reflects the original linear distribution of a series of strato-volcanoes.

Basalt flows and gabbroic intrusions predominate in the lower volcanic piles, andesite flows and pyroclastics intercalated with basalts increase in amount upwards, and felsic rhyolites to dacites predominate in the upper levels. The mafic lavas commonly contain pillows, palagonite, variolites, amygdules and hyaloclastites, thereby indicating subaqueous accumulation; primary igneous textures are remarkably well preserved. In the overlying felsic volcanic rocks pyroclastic types are very common.

Komatiitic rocks occur in the Abitibi belt in the Rouyn–Noranda area (Dimroth and co-workers, 1973), as ultramafic flows with

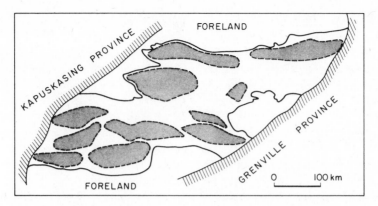

Fig. 2.6. Eleven elliptical volcanic complexes (stippled) in the Abitibi greenstone belt, Canada (redrawn from Goodwin and Ridler, 1970, by permission of The Geological Survey of Canada)

spinifex textures in the Dundonald township (Naldrett and Mason, 1968; Naldrett, 1970) and the Munro Township of Ontario (Pyke, Naldrett and Eckstrand, 1973) and near Wana, Ontario (Brooks and Hart, 1972).

Sediments in the Abitibi belt are particularly of the 'poured-in' turbidite type suggesting rapid accumulation in a tectonically unstable environment. Two principal facies are present:

1. Volcanogenic, comprising greywackes, shales, lithic sandstones, conglomerates and breccias, the constituent clasts being from recognizable volcanic rocks. They are associated with soft sediment deformation structures, chaotic textures, polymictic unsorted materials, graded bedding and abrupt facies changes.

2. Flyschoid, comprising rhythmically bedded greywacke–argillite sequences of uniform composition and construction lacking in lateral facies changes.

The Canadian volcano-sedimentary assemblages contain shallow to deep water transitions which represent remnants of the original basins and trenches (Goodwin, 1973a). A typical transition is marked by:

1. Off-shore thickening of the total assemblage (i.e. all stratigraphic units tend to be thicker nearer the axes of the belts). This fact lends support to the idea that the sedimentation kept pace with progressive opening of the shore lands.

2. A corresponding thick-to-thin and coarse-to-fine clastic transition in the direction of deeper water.

3. In banded iron formations there is a basinward transition from predominantly cherty-oxide to sulphide facies (Fig. 2.11).

Numerous mafic complexes were intruded into the supracrustal rocks, often as differentiated sheets and sills with basal peridotites and pyroxenites passing upwards through mafic gabbros or norites to gabbros, quartz gabbros, and locally to granophyres. The Dore Lake Complex (Allard, 1970) consists largely of alternating layers of anorthosite, pyroxenite and gabbro, and the Bell River Complex of rhythmically banded norite, anorthosite and pyroxenite, both complexes being rich in titaniferous magnetite and ilmenite.

In the Slave structural province the Yellowknife Supergroup occurs in a number of discontinuous greenstone belts (J. F. Henderson and Brown, 1966; McGlynn and Henderson, 1970, 1972; D. C. Green and Baadsgaard, 1971; J. F. Henderson, 1975). The typical stratigraphy is divisible into: (a) a lower volcanic sequence of massive and pillowed meta-basaltic to meta-andesitic flows, intermediate-to-acidic lavas, and tuffaceous rocks including dacites, latites and quartz latites, the acidic rocks occurring in the upper parts of the volcanic piles; and (b) an upper predominantly sedimentary sequence of immature greywackes, shales and mudstone.

Geochemistry of the Volcanic Rocks

The chemical composition of the main types of late Phanerozoic volcanic rocks is related to their geotectonic environment such as oceanic, island arc and continental crust. There is much discussion, indeed disagreement, about the tectonic setting of the greenstone belts in relationship to possible Archaean plate tectonics, and the chemistry of the volcanics therefore provides us with worthwhile additional data which place their own constraints on models of Archaean magma genesis.

In the preceding sections it became clear that whilst greenstone belt successions on different continents resemble each other there is substantial vertical variation within them. In particular, confining one's attention to the predominantly volcanic section, the lower part is often composed of ultramafic and mafic rocks, namely the lower Onverwacht and Kalgoorlie Systems, the Sebakwian and the Nuggihalli, Bababudan and Kolar belts, whereas the upper part is characterized by a calc-alkaline assemblage including mafic-felsic suites and basalt–andesite–rhyolite cycles, *viz.* the upper Onverwacht and Kalgoorlie Systems, the Bulawayan, the main Dharwar belts (e.g. Chitaldrug) and the Superior and Slave Province belts (e.g. Abitibi and Yellowknife). This bipartite division has been emphasized by Glikson (1970, 1971c, 1972, 1976a,c)

These volcanic rock groups of the greenstone belts can be compared with somewhat similar ones of later time.

1. Komatiitic extrusives and intrusives occur in the lower parts of belts in South Africa (Viljoen and Viljoen, 1969), Rhodesia (Bickle, Martin and Nisbet, 1975), Australia (Nesbitt, 1971; McCall and Leisman, 1971), India (Viswanathan, 1974a and b), and Canada (Naldrett, 1970; Pyke, Naldrett and Eckstrand, 1973; Dimroth and coworkers, 1973). Peridotitic komatiites require a very high degree of melting (60–80%, D. H. Green, 1972b) of the mantle and basaltic komatiites extensive (40–60%) melting (Brooks and Hart, 1974). This fact may be explained by shallow depths of melting (Cawthorn and Strong, 1974) which may be consistent with expected high rates of heat flow and steep geothermal gradients in the Archaean.

2. Meta-basalts in the lower ultramafic–mafic parts of the belts have K, Na/K, Sr, Zr and Fe^{3+}/Fe^{2+} similar to, Al and Ti lower than, and Mn, Ni, Cr, Co, Rb and Fe/Fe + Mg (total iron) higher than, modern oceanic tholeiites. Fig. 2.7 shows that the MgO–Ni ratios in meta-basalts etc. are similar to those of modern oceanic ridge basalts etc. implying similar olivine compositions and Mg–Ni ratios in parental mantle material (Gunn, 1976). Chondrite-normalized REE patterns typically show near-flat curves and no La and Ce depletion; such patterns were compared with those of modern island arcs by A. J. R. White, Jakes and Christie (1971) but more recently with those of oceanic tholeiites (Condie, 1975; Glikson, 1976b). Zr–Y–Ti ratios fall in the field of oceanic tholeiites as defined by Pearce and Cann (1971, 1973) and Hallberg and Williams (1972). Archaean K_2O contents are typically in the range 0·14–0·26%, comparable with the range of 0·16–0·22% of oceanic tholeiites (Glikson, 1971c). But such low-K tholeiites are also found in many other modern tectonic environments (Jamieson and Clarke, 1970) such as in island arcs (Kawano, Yagi and Aoki, 1961), continental rifts–flood basalts (Clarke, 1970) and back-arc marginal basins (Hart, Glassley and Karig, 1972). This is an important point in view of the current tendency to conclude that low-K meta-basalts in greenstone belts (e.g. Glikson, 1970, 1971c, 1972), and even low-K meta-volcanic amphibolites in high-grade Archaean regions (e.g. Rivalenti, 1976) are derived from Archaean oceanic crust. Indeed there are reasons for believing that there may have been little or no chemical difference between low-K tholeiites from various Archaean tectonic environments (Brooks and Hart, 1974). An overall shallow depth of melting of mantle material in the Archaean may have been a predominant factor in the generation of these magmas, as noted by Cawthorn and Strong (1974), Condie (1976), and R. C. O. Gill and Bridgwater (1976).

34

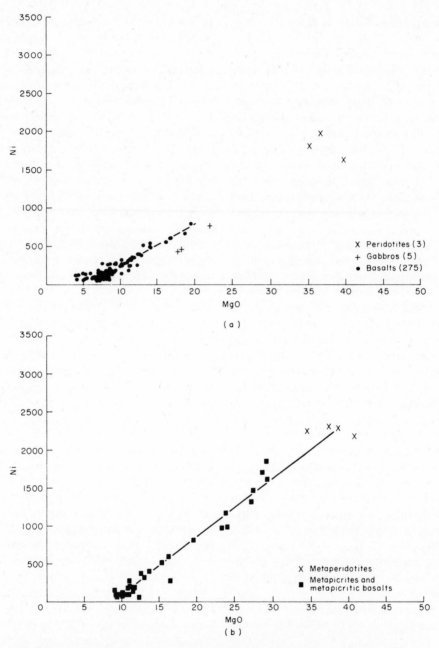

Fig. 2.7. A comparison between the MgO–Ni ratios. (a) 300 modern oceanic ridge basalts and picrites together with associated peridotite and eucrite from Leg 37 drill cores. (b) Archaean metabasalts, metapicrites, peridotites and meta-gabbros. (After Gunn, 1976; reproduced by permission of J. Wiley)

3. Meta-basalts at higher stratigraphic levels belonging to the calc-alkaline series are more differentiated and remarkably similar to island arc tholeiites with respect to their contents of K, Rb, Sr, Ba and rare earth elements (slight light-element enriched) (Condie, 1975; Winchester and Floyd, 1976). Also, the abundance of pyroclastics and rhyolites is compatible with an island arc environment. Although Glikson (1976a,c) concludes that the calc-alkaline

suite evolved within linear troughs in partly cratonized regions, most authors believe they formed in primitive types of island arc systems (H. D. B. Wilson and coworkers, 1965; Goodwin, 1968; D. C. Green and Baadsgaard, 1971; A. J. R. White, Jakes and Christie, 1971; Anhaeusser, 1973; Jahn, Shih and Murthy, 1974). According to Gunn's (1976) analysis the average of large numbers of Archaean andesites is a low-K, low-Sr andesite of *unmistakable* island arc type. In Canadian belts there is a progressive increase in Al, K, Sr and Ba with height in the stratigraphic pile, the rocks showing more and more calc-alkaline features—a trend in time similar to that in many recent island arcs (A. J. R. White, Jakes and Christie, 1971). The Maliyami Formation that overlies the Mafic Formation in the Midlands belt is composed of calc-alkaline tholeiites, andesites and dacites and represents a more advanced stage in the evolution of the arc system (Condie and Harrison, 1976).

However, the Archaean calc-alkaline sequences do have some features different from island arc volcanics (Glikson, 1976a):

a) They contain some ultramafic extrusives unknown in island arcs.

b) Andesites are uncommon in the Kalgoorlie System (Hallberg, 1972), Upper Onverwacht Group (Viljoen and Viljoen, 1969) and Dharwar System (Naqvi, Rao and Narain, 1974), all of which display a pronounced mafic-felsic polarity. The Canadian andesites reported by H. D. B. Wilson and coworkers (1965), also Baragar and Goodwin (1969), are regarded by Gunn (1976) as a misidentification as they are better thought of as high-Na (spilitized) meta-basalts.

c) Most Archaean tholeiites have low-to-intermediate Al_2O_3 contents of 14–15% whereas the bulk of island arc mafic volcanics are calc-alkaline and high-Al basalts.

d) Archaean Ni, Cr and Co abundances are higher by factors of two or three than in island arc basalts (Baragar and Goodwin, 1969; Naqvi and Hussain, 1973a and b), but they are similar to those in modern marginal basins (Tarney, Dalziel and De Wit, 1976).

4. Alkaline-to-shoshonitic volcanics are rare in greenstone belts; they are only known from the Abitibi belt in Canada (T. E. Smith and Longstaffe, 1974; Hubregtse, 1976; plus earlier references given in this chapter). Some Indian meta-basalts have abnormally high P_2O_5 contents (alkaline) (Winchester and Floyd, 1976). The presence of these alkaline rocks is consistent with an island arc tectonic setting but the common absence of such rocks may be indicative of thinner lithospheric plates in Archaean times.

From the above discussion it will be clear that whilst the lower and upper parts of the volcanic sequence of greenstone belts bear most resemblance to modern oceanic and island arc rocks respectively, there are significant departures in both respects. Nevertheless, in the most critical appraisal yet of the petrochemical variations and comparisons, Gunn (1976) concludes that the general similarity between the modern oceanic crust and the lower Archaean volcanic sequences is sufficiently close that the upper mantle must have had essentially the same composition 3000 my ago as today and that the tectonic processes at present generating the magmas at mid-oceanic ridges were probably in operation in the Archaean.

Structure

In this section we shall look at the structural make-up of the greenstone belts, leaving the more theoretical tectonic models for the next chapter.

The classical picture of the structure of all the greenstone belts is one of synforms formed largely by vertical downsinking of volcano-sedimentary rocks (A. M. McGregor, 1951; H. Martin, 1969; Anhaeusser and coworkers, 1969; Glikson, 1970), associated with the diapiric uprise of granite bodies (Stowe, 1971; Anhaeusser, 1973; Glikson, 1971c, 1972). The essential theme running through these ideas, defined or implicit, is that the formation of the basin-shaped structure was largely controlled by vertical movements (Fig. 2.1.).

More recent detailed field work and rethinking have revealed that not all belts have a simple synformal structure and that large-scale lateral displacements have played a significant role in their formation. Ramsay (1963a) was probably the first person to appreciate the importance of the horizontal strain component in the formation of major structures in the belts, a fact corroborated by D. S. Wood (1966) and Coward and James (1974). D. S. Wood (1973) calculated that the dimensional changes associated with the formation of some Rhodesian belts involved a shortening across the subvertical fold axial surfaces of 75% and subvertical extension in the axial surfaces of 300%. The first evidence that horizontal nappe-type movements played a major part in the evolution of greenstone belts came from the Rhodesian craton (Stowe, 1971, 1973, 1974) and such folding has since been tentatively proposed for belts in Botswana (Key, Litherland and Hepworth, 1976). Coward, Lintern and Wright (1976) demonstrate that similar large-scale crustal shortening was responsible for the early development of belts in southern Africa and that some belts are not synclinal as they have downward-facing sedimentary structures; they are late folds formed on the inverted limbs of early nappes (Fig. 2.1). From a theoretical standpoint Burke, Dewey and Kidd (1976) suggest that there may be many more thrust and slides than are so far recognized in greenstone belt successions (implying that the enormous stratigraphic thicknesses of the order of 10–20 km usually listed are an illusion), and that horizontal shortening expressed by vertical cleavage and nappe structures has been an important element in the evolution of many belts. It seems likely that future detailed structural studies will reveal more evidence of early thrusts and nappes in greenstone belts.

Metamorphism

All the sedimentary and volcanic rocks of the world's greenstone belts have suffered some degree of recrystallization, varying from the zeolite to the granulite facies. The most common type of metamorphic imprint has given rise to greenschist facies assemblages—hence the term *greenstone* belt. There are two significant departures from this grade:

1. In many belts there is an increase in grade, mostly to the amphibolite facies, firstly from the centre to the margins and secondly towards intrusive granitic plutons.

2. In a few cases there is an appreciable increase in the grade of regional metamorphism along the strike of the belts. This is most prominent in the main Dharwar belts which show a southward increase from greenschist to high amphibolite or even granulite facies.

Commonly the greenschist metamorphism is of low-pressure type (Glikson, 1971; Hunter, 1974a; Binns, Gunthorpe and Groves, 1976) but the presence of chloritoid in the Yellowknife belt suggests a Barrovian-type facies series (Green and Baadsgaard, 1971).

Ultramafics are usually serpentinized with varieties such as talc schists and tremolite-actinolite schists. Basic volcanics commonly have assemblages like actinolite–albite–chlorite–epidote or, at higher grade, hornblende–andesine–epidote–quartz. In the Dharwar belts meta-sediments in the greenschist facies are schists containing actinolite–chlorite–quartz, sericite–quartz and biotite–chlorite–quartz, whilst in the amphibolite facies they vary from schists to gneisses with pyrope-almandine, garnet–mica–quartz, garnet–pargasite/ferro-richterite–plagioclase quartz, kyanite–biotite–muscovite–quartz and staurolite–biotite–quartz (Sreenivas and Srinivasan, 1974). These assemblages point to a Barrovian facies series, the P/T conditions being in the range:

Greenstone facies	300–350°C/3 kb
Amphibolite facies	350–600°C/4–6 kb

For all the metamorphic variation in greenstone belts, one thing is certain: the high pressure facies types with glaucophane and eclogite are absent.

Mineralization

In this section the mineral deposits that occur in the low-grade Archaean terrains will be summarized, particular regard being paid to the stratigraphic controls on their mode of occurrence (for a general review see Watson, 1976). The types of mineral deposit in different shield regions are remarkably similar and any particular ore is confined to a limited environment. A metallogenic framework thus emerges in which the formation of the mineral deposits can be related to the development of the greenstone belts and their granites, e.g. in the Dharwar craton (Radhakrishna, 1976).

The greenstone belts contain a multitude of economic minerals; they are one of the main depositories in the world of elements such as Au, Ag, Cr, Ni, Cu and Zn. In the past they were termed gold belts and since the gold rush days, which took place in most of the continents concerned around the turn of the last century, they have been the subject of extremely detailed mapping by mining companies and surveys.

The mineral deposits can be related to the major rock groups that make up the greenstone belt/granite terrains as follows:

1. Ultramafic flows and intrusions: chromite, nickel, asbestos, magnesite and talc.
2. Mafic-to-felsic volcanics: gold, silver, copper and zinc.
3. Sediments: iron ore, manganese and barytes.
4. Granites and pegmatites: lithium, tantalum, beryllium, tin, molybdenum and bismuth.

Fig. 2.8 shows the distribution of the main mineral deposits within the volcanic and sedimentary rock groups of the Abitibi belt.

Many elements (e.g. Cr, Ni, Au, Ag, Cu, Zn) can be related to the magmatic differentiation

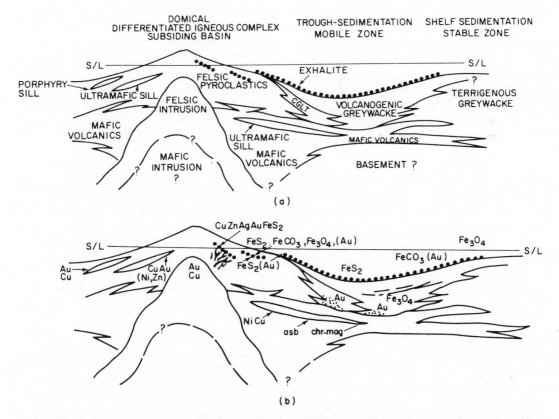

Fig. 2.8. A diagrammatic cross-section of the Abitibi belt. (a) Shows the tectonic–stratigraphic relations with the volcanic–sedimentary complex, and (b) The main mineral deposits. Full width of section 50 miles, maximum vertical thickness 10 miles. (After Hutchinson, Ridler and Suffel, 1971; reproduced by permission of *Trans. Can. Inst. Metall.*)

of the mantle-derived ultramafic–mafic–felsic volcanic rocks. Sedimentary (e.g. Fe, Mn), ultramafic cumulate and volcanic exhalative ores were stratigraphically controlled but the final formation of the majority of the deposits was structurally controlled. The first Sn deposits to form in the evolving continents occur in pegmatites at Bikita, Rhodesia; other minor mineralization in greenstone belts includes barytes, talc, manganese and asbestos.

Chromite

Deposits of chromite are not common in greenstone belts. The most important are in the Sebakwian ultramafics in Rhodesia but there are minor occurrences in the Abitibi belt.

The Selukwe chromite occurs in serpentines and talc–carbonate rocks that are weakly metamorphosed ultramafics belonging to differentiated sill-like lenses conformably intruded into Sebakwian schists. This is one of the largest occurrences of high-grade chrome ore in the world and far exceeds any other Archaean greenstone belt chromite deposit.

Nickel

Surprisingly there are no economic nickel deposits in the Barberton greenstone belt, particularly since the lower ultramafic group is so well developed there.

The most important Archaean Ni deposits occur in SW Australia (Kalgoorlie belt), S Canada (Abitibi belt) and Rhodesia (northwest extension of the Selukwe belt). In these greenstone belts the Ni typically occurs in magmatic segregations at, or close to, the base of mafic to ultramafic sill-like bodies.

There are a number of Ni deposits in the Yilgarn block but these are overshadowed by the Kambalda deposit. Most of the nickel sulphide mineralization at Kambalda occurs at the base of a meta-ultramafic sill intrusion in contact with meta-basalt, but some ore occurs as lenses within the ultramafic rocks (Woodall and Travis, 1969).

It is ironic that in the Canadian greenstone belts, originally thought by Anhaeusser and coworkers (1969) to have no lower ultramafic group, there are so many nickel deposits. Of the 16 economic deposits listed by Naldrett and Gasparrini (1971) the majority occur at the base of differentiated peridotitic bodies, strong evidence that the ore is the result of gravitational settling of sulphide–oxide melt.

Gold

Gold is the most important economic mineral in Archaean greenstone belts throughout the world; for example there are 197 known gold occurrences in the Superior Province of Canada, almost all of which are situated in the volcanic-rich greenstone belts (Goodwin, 1971). The gold distribution pattern has been traditionally related to regional fractures connected with, for example, granitic stocks, lava flows and unconformities, but this structural disturbance has partly obscured the more fundamental stratigraphic control of the gold concentration. Fig. 2.9 shows the gold distribution in Rhodesian greenstone belts, most of which are in the igneous Bulawayan and Sebakwian units.

The gold occurs in the following types of stratotectonic environment:

1. At felsic–mafic volcanic contacts in the upper parts of the central volcanic group, in particular in the Canadian belts and in the Barberton Mountain Land. In the Abitibi belt most gold deposits lie in intermediate to felsic volcanic rocks or their intrusive equivalents.

2. The gold is sometimes located in the carbonate facies iron formations, e.g. in the Abitibi belt, but more typically in the oxide facies as in southern Africa. Anhaeusser (1971b, 1976) points out that the iron formations tend to terminate the sedimentary cycles by overlying the final stage volcanics and pyroclasts; this type of gold mineralization is thus connected with the most highly differentiated exhalative volcanic phases. In reviewing the gold metallogeny of Rhodesia, Fripp (1976) suggests that the gold in Sebakwian banded iron formations closely related to aquagene tuffs was precipitated from sub-

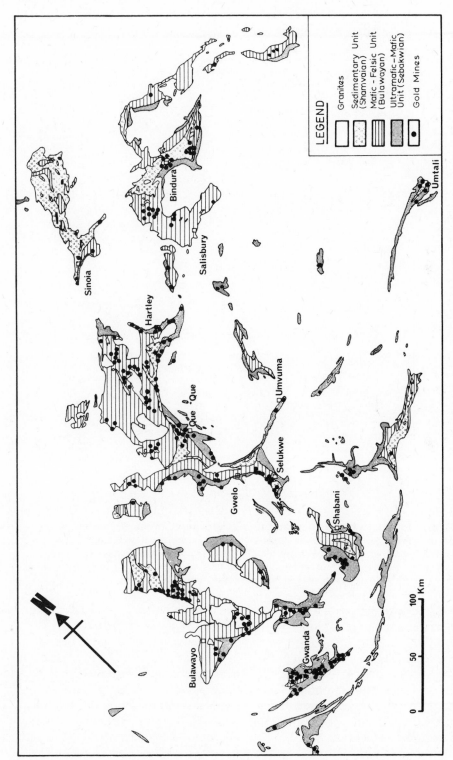

Fig. 2.9. Map showing the distribution of the more important gold mines in relation to the stratigraphy of the greenstone belts in Rhodesia (redrawn after Anhaeusser, 1976; reproduced by permission of *Minerals, Science and Engineering*)

40

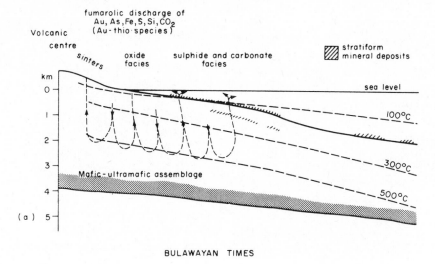

SEBAKWIAN TIMES

fumarolic discharge of
Au, As, Fe, S, Si, CO₂
(Au-thio species)

Volcanic
centre

sinters   oxide   sulphide and carbonate
          facies   facies

stratiform
mineral deposits

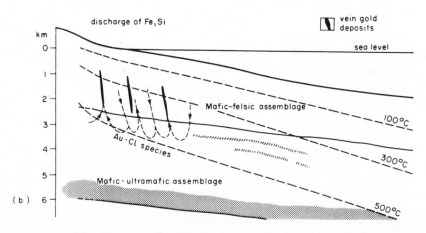

BULAWAYAN TIMES

discharge of Fe, Si

vein gold
deposits

Fig. 2.10. A model for the evolution of gold deposits in the Archaean. (a) Stratiform deposits in Sebakwian iron formations formed by convecting brines leaching the gold from the volcanic pile. (b) Vein gold deposits formed under a lower geothermal gradient in Bulawayan times. (After Fripp, 1976; reproduced by permission of J. Wiley)

aqueous volcanic exhalations of thermal brines in the temperature range of 300–400°C (Fig. 2.10).

3. Basic igneous rocks are commonly the dominant host rocks: in SW Australia the Golden Mile dolerite and the underlying tholeiitic Paringa basalt (Travis, Woodall and Bartram, 1971) and in Rhodesia the Bulawayan greenstone volcanics.

4. In the Abitibi, gold occurs in association with felsic alkaline intrusions in the form of subvolcanic sills, discordant plugs or stocks,

and with their flow and pyroclastic equivalents (Goodwin and Ridler, 1970).

The most important gold mineralization lies in the marginal zone of the greenstone belts near the bordering granitic plutons and this mineralization typically decreases progressively away from the granitic contacts. Thus, although the ultimate source of the gold and associated sulphide mineralization is reckoned by Viljoen, Saager and Viljoen (1970) to be the vast pile of mafic and ultramafic volcanics, the elements concerned appear to have been mobilized by the action of thermal gradients set up by the invading granitic bodies (Anhaeusser and coworkers, 1969; Anhaeusser, 1976). However Anhaeusser and coworkers (1975) produce chemical data showing that komatiitic lavas from Barberton contain surprisingly low contents of Au (1–1·5 ppb) and thus could not be a source for the gold deposits in the greenstone belt.

Silver

Silver commonly occurs with gold in the greenstone belts. In the Abitibi belt it is found typically with Au, Cu and Zn at felsic–mafic volcanic contacts (Goodwin and Ridler, 1970).

Copper–Zinc

Chalcopyrite–sphalerite mineralization typically occurs in the fragmental felsic (rhyolitic to dacitic) volcanics and pyroclastics in the upper parts of the volcanic cycles of greenstone belts, but the world distribution is variable.

In the Superior Province belts there are 185 copper–zinc occurrences (40% of all mineral occurrences) but only seven in each of the Slave and Hudson Provinces (Goodwin, 1971). Lead-bearing sulphide (galena) occurrences are rare in rocks of Archaean age—there are a few in the Abitibi belt (none elsewhere in Canada), the lead usually being associated with Cu–Zn but subordinate to both.

Iron

Banded iron formations are common in Archaean greenstone belts but this ore type reached its peak of development in early Proterozoic basins about 2000 my ago and so will be discussed in more detail in Chapter 5.

Of the four types of iron formation, Algoma, Superior, Clinton and Minette (Gross, 1965), the first, characteristic of the Archaean greenstone belts, is consistently associated with a variety of volcanic rocks, greywackes and black carbonaceous slates.

The iron formations are up to a couple of hundred metres thick and characteristically form short lenses in the volcanic sequence. Some occur between rhyolite–dacite or pyroclastic-andesite flows. In the Michipicoten area in Canada underlying dacite–rhyolite flows are carbonatized to a depth of nearly 200 m. Goodwin (1962) considered this to be caused by the action of fumaroles and hot springs that affected the volcanics during the development of the overlying iron formation. These types of volcanic associations have given rise to the present widely held view that the banded iron formations are chemical

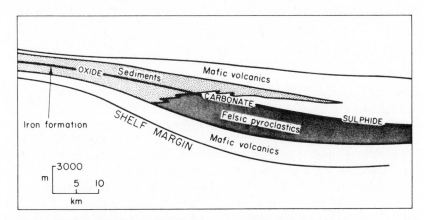

Fig. 2.11. Distribution of iron facies in the Abitibi belt, Canada. The oxide–carbonate–sulphide facies transitions delineate the original shelf-to-basin slope of the 'orogen'. (Redrawn after Goodwin, 1973a; reproduced by permission of *Economic Geology*)

sediments formed by a fumarolic–exhalative process, the iron and silica being derived from a volcanic source.

There are four facies of iron formation, oxide, carbonate, silicate and sulphide, the first and second of which are predominant in the Algoma-type ores. The iron ore bands are usually interbanded with ferruginous chert. H. L. James (1954) first proposed that the oxide, carbonate and sulphide facies formed further from the shoreline in progressively deeper water. Goodwin (1973a) made a considerable advancement on this by demonstrating that the three facies defined the original shelf–basin bathymetry (Fig. 2.11). The shallow water oxide (magnetite) facies grades transitionally through the carbonate (siderite–ankerite–dolomite) facies to the deeper water, basinal sulphide (pyrite–pyrrhotite) facies towards the centre of the orogen.

The Earliest Life Forms

The earliest records of terrestrial life occur in sedimentary rocks in the greenstone belts of southern Africa. These organisms demonstrate that life was in existence at least 3000–3200 my ago and thus represent an important benchmark in the geological history of the continents. There are two types:

Micro-organisms

These have been found in three stratigraphic levels of the Barberton succession.
1. In organic-rich black cherts and shales of the Fig Tree Group (Pflug, 1966; Barghoorn and Schopf, 1966);
2. In chert and argillite of the Upper Onverwacht Group (Engel and coworkers, 1968);
3. In the lowermost sedimentary rocks (cherts) of the Lower Onverwacht Group (Theespruit Formation) only 350–600 m above the contact with the underlying granite gneisses.

Sylvester-Bradley (1975) pointed out that the Lower Onverwacht Group is distinguishable from the younger beds of the System in three ways: in its fossils, carbon isotope ratios and organic chemistry.

The micro-fossils in the Lower Onverwacht are half the size of those higher up the succession—spheroids are 10 mm in diameter and filaments about 7 mm long (Brooks, Muir and Shaw, 1973). Nagy and Nagy (1969) thought that these microstructures 'do not meet the accepted criteria for biological origin' and they classed them as 'organized elements'. In contrast some of the carbonaceous spheroids of the Fig Tree Group resemble algae and others cysts of flagellates; the filamentous forms look like blue–green algae (Schopf, 1974). Barghoorn and Schopf (1966) concluded that the fibrillar and branching form of these micro-organisms is suggestive of a high degree of molecular and polymeric order characteristic of living systems; they appear to be generally similar to degraded plant material.

It is worth mentioning here that the Fig Tree cherts contain much organic material arranged in laminations parallel to the bedding, indicating its original sedimentary deposition. The algal bodies have coatings that mainly consist of compounds of Cu, Fe, Ni and Ca, showing that the organisms were able to precipitate metal salts from water by action of their body processes, whilst the sulphur in the coatings, bound in the metal sulphides, originated from the organic matter of the algae (Pflug, 1966).

The carbon isotope data and the occurrence of micro-fossils from the Barberton succession are shown in Fig. 2.12. The insoluble organic carbons from the Lower Onverwacht have $\delta^{13}C$ values ranging from -14.3 to -18.9%, whilst the kerogens in the younger beds range from -26.1 to -33.0%. The high values in the Lower Onverwacht have received much discussion—they might be due to secondary metamorphic effects (Sylvester-Bradley, 1975). According to Hoering (1967) the ratios in the upper beds are consistent with fractionation by photosynthesis, in which case conditions must have existed soon after 3355 my permitting the growth of photosynthetic plants (Pflug, 1966).

Organic chemical data by Nagy and Nagy (1969) and Scott, Modzeleski and Nagy (1970) indicate that the Onverwacht beds contain a high proportion of aromatics, in contrast to the

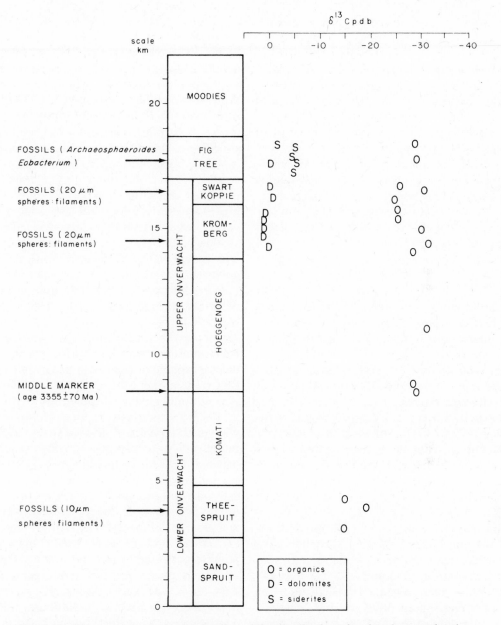

Fig. 2.12. The occurrence of microfossils and the distribution of carbon isotope data in the Swaziland System of the Barberton belt, South Africa (after Sylvester-Bradley, 1975; data from several sources)

Fig Tree which are mostly aliphatics and N-alkanes not found in the Lower Onverwacht. Calvin (1969) suggested that we may be dealing here with a mixture of abiogenic and biogenic compounds and that we may have discovered the time when newly developed life forms were beginning to convert abiogenic materials.

Schopf (1975, 1976) asserts that the bacterium-like bodies and filamentous 'micro-fossils' so far reported are less than convincing evidence of Archaean life. Muir and Grant (1976), however, produce some interesting examples from the Onverwacht Group.

The relationships described in this section may suggest that there was a significant evolutionary burst during the period when the Barberton succession was laid down. Although it has been claimed that this was the time when photosynthesis began and 'proto-life' evolved into 'life', it now seems more likely that these events took place much earlier in time (more than 3700 my ago, Junge and coworkers, 1975), and evidence of them might be better sought in Archaean terrains like West Greenland with a sedimentary history back to more than 3760 my (Sylvester-Bradley, 1975). Some graphitic microstructures have recently been reported but are interpreted as abiotic in origin (Nagy, Zumberg and Nagy, 1975).

Stromatolites

The oldest known unequivocal fossils are laminated stromatolites from limestones of the Bulawayan Group (>2900 my), Rhodesia (Schopf and coworkers, 1971; Bond, Wilson and Winnall, 1973); similar stromatolites occur in limestones and ankeritic dolomites of the younger Archaean (>2500 my) greenstone belt at Steeprock Lake, Ontario (Pannella, 1972). The fact that stromatolites are only known from these two localities is an indication of their restricted development in Archaean times; it was not until after 2000 my ago in Proterozoic times that they became common.

Stromatolites result from accretion of detrital and precipitated minerals on successive sheet-like mats formed by communities of micro-organisms predominated by filamentous blue–green algae. The existence of stromatolites, about 3000 my old, places a minimum age on the time of origin of algal photosynthesis producing gaseous oxygen as a by-product, and of integrated biological communities of primitive micro-organisms presumably including producers (blue–green algae), reducers (aerobic and anaerobic bacteria) and consumers (bacteria, predatory

by absorption) (Schopf and coworkers, 1971).

One of the main obstacles to the evolution of life in Archaean times was, according to Cloud (1968a) the lack of an ozone screen in the upper atmosphere. The present ozone layer prevents the penetration of high-energy ultra-violet radiation in the 2400–2900 (especially the 2600) angstrom wavelength range which would inactivate the life-inducing DNA molecule. The lack of an Archaean ozone screen allowed the damaging radiation to penetrate the surface layers of water, preventing the healthy growth of early life forms which were necessarily restricted to local shielded habitats. It is not surprising, therefore, that primitive life forms are only found in a few isolated localities in Archaean rocks.

There are two fundamentally different types of primitive organism (Cloud, 1968a; Margulis, 1974).

1. Procaryotes, consisting of procaryotic cells which lacked a wall around the nucleus and were incapable of cell division. The nucleus is that part of the cell in which the DNA or genetic coding material is concentrated.

2. Eucaryotes, consisting of eucaryotic cells which had a nucleus enclosed within a membrane and were capable of cell division by which the genetic coding material (DNA) was successively parcelled out among the different cells and descendants of the organism.

The procaryotes were relatively more resistant to ultra-violet radiation than the eucaryotes and so were less likely to be adversely affected by radiation-induced mutations and thus were the first organisms to develop in the early Archaean, when the atmosphere lacked a radiation-protective ozone screen, as typified by the bacteria and algae of the Fig Tree Group. They were the predominant organism up until about 1000 my ago. The eukaryotes are mainly oxygen-using organisms whose appearance was triggered by the development of a relatively oxygenic atmosphere during the mid-late Proterozoic.

Chapter 3

Crustal Evolution in the Archaean

In Chapters 1 and 2 the main features of the Archaean high- and low-grade terrains were reviewed. Here we are concerned with the tectonic evolution of these terrains and so come against the problem of the spatial and temporal relationships between them. There is so much current interest in the early history of the earth that there is no lack of thought about evolutionary models and these will be considered below. All models, of course, reflect, if not the preference and prejudice, the degree of experience of the proposer and this limitation often hinders the viability of the model. A major problem has been that there are insufficient constraints to limit speculation, but there is an excellent prospect that geochemical knowledge will soon be able to place limitations on types of magma generation and derivation. Also, the current tendency to integrate structural, stratigraphic and geochemical data should give rise to more viable hypotheses.

Greenstone Belts

There are as many hypotheses to explain the formation of greenstone belts as once there were for 'granites and granites'. Although much mapping of these belts took place in the early decades of this century, it was not until 1968 that the recent spate of tectonic modelling began. The models are generally of three types: the classical or fixist, the plute tectonic or mobilist, and the 'extraterrestrial'.

Classical Models

1. In 1968 Anhaeusser and coworkers formulated a model which compared the early greenstone belt volcanics with modern ophiolites and the late sediments with the flysch and molasse of geosynclines. This was, however, quickly revised by Anhaeusser and coworkers (1969) who visualized two possibilities:

a) The greenstone belts developed in downwarps at the interface between thin continental crust and oceanic crust (Fig. 3.1a). The belts were localized as linear structures along fundamental fractures in the primitive crust. The model does not involve subduction at the continental–oceanic crust contact.

b) Roughly evenly spaced, strongly oriented, parallel downwarps or fault-bounded troughs in an unstable thin primitive sialic crust were the site of greenstone belt formation (Fig. 3.1b). This model was the one favoured by the authors and by Viljoen and Viljoen (1969). The basins developed as *in situ* depositories by progressive downsagging of the heavy volcanic pile; this is the beginning of the downsagging basin model that was the foundation of the later models of Anhaeusser and Glikson (see below) and D. S. Wood (1973). During the late stages the underlying granitic crust was thickened by some form of underplating (cause unspecified) and the greenstone belts were invaded internally and marginally by diapiric granites that caused compression of the belts (giving rise to their arcuate form) and their increase in metamorphic grade from greenschist at the centre to amphibolite facies along the margins.

45

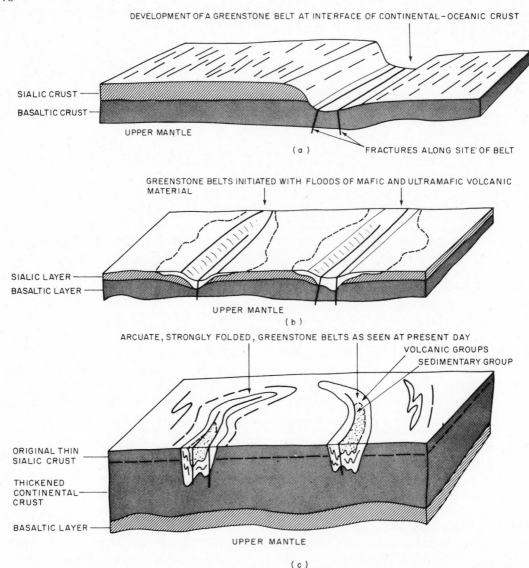

DEVELOPMENT OF A GREENSTONE BELT AT INTERFACE OF CONTINENTAL—OCEANIC CRUST

SIALIC CRUST

BASALTIC CRUST

UPPER MANTLE

(a)

FRACTURES ALONG SITE OF BELT

GREENSTONE BELTS INITIATED WITH FLOODS OF MAFIC AND ULTRAMAFIC VOLCANIC MATERIAL

SIALIC LAYER

BASALTIC LAYER

UPPER MANTLE

(b)

ARCUATE, STRONGLY FOLDED, GREENSTONE BELTS AS SEEN AT PRESENT DAY

VOLCANIC GROUPS

SEDIMENTARY GROUP

ORIGINAL THIN SIALIC CRUST

THICKENED CONTINENTAL CRUST

BASALTIC LAYER

UPPER MANTLE

(c)

Fig. 3.1. Two ways of developing a greenstone belt. (a) In structural downwarps at the interface between continental and oceanic crust. (b) Along oriented downwarps on a thin sialic crust that was later thickened by underplating. (After Anhaeusser and coworkers, 1969; reproduced by permission of The Geological Society of America)

These models were built upon a wealth of field data but little or no geochemical results. The authors were unlucky in that they produced them before plate tectonic revolution had an impact on interpretations of continental development.

2. Anhaeusser (1971a) produced a further development of the downsagging basin model (Fig. 3.2). A limitation of this and the earlier models is that the authors did not entertain the possibility that the nearby high-grade gneisses could represent the basement. Yet this was at a time when pre-greenstone basement gneisses had been or were being suggested in Rhodesia (Bliss, 1969; Bliss and Stidolph, 1969; Stowe, 1971) and South Africa (Hunter, 1970a).

3. Glikson (1970b, 1971c, 1972, 1976a, 1976c) and Glikson and Lambert (1973,

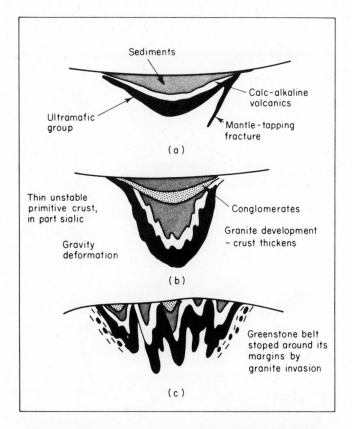

Fig. 3.2. Diagrammatic model showing the evolutionary development of a greenstone belt by progressive down-sagging of a volcano–sedimentary basin into a thin unstable primitive crust (redrawn after Anhaeusser, 1971a; reproduced by permission of The Geological Society of Australia)

1976) produced a set of variations on a basic theme. Important aspects of these models are:

a) They follow previous authors in not accepting evidence for an early gneissic basement. Note, however, that the models are an attempt to explain the development not just of greenstone belts but also of Archaean shields and cratons. The high-grade 'mobile belts' were not considered in the 1970 paper; they appeared as a 'post-script' in 1971 and 1972 (Fig. 3.3) having formed after the greenstone belts, and in later papers they are considered to represent coeval infracrustal roots of the greenstone belts.

b) They emphasize the primitive chemical nature of the lower ultramafic and mafic volcanics (often komatiitic) and so, in the absence of a basement, the formation of a widespread primitive oceanic crust is envisaged as the first stage in proto-continental evolution. In other words, the ultramafic–mafic rocks at the base of greenstone belts are interpreted as relics of 'the primordial crust' (1972) or 'a once extensive simatic crust' (Glikson and Lambert, 1976).

However, seen in historical perspective these suggestions are no longer viable. Viljoen and Viljoen (1969) correctly identified the primitive chemical character of the Barberton komatiites and it is likely that they, and associated lower tholeiites, did form in part of some oceanic crust (although some komatiites lie unconformably on older sialic basement—Bickle, Martin and Nisbet, 1975). But the conclusion that this is a

48

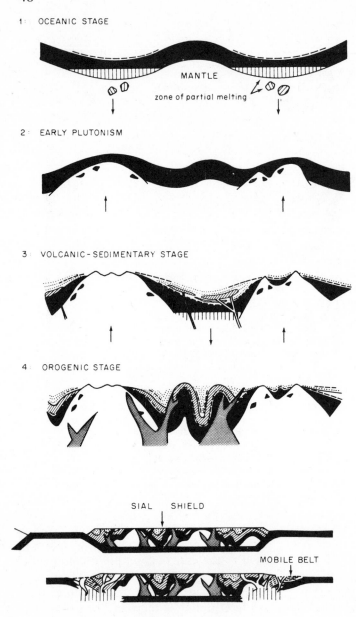

1: OCEANIC STAGE

MANTLE

zone of partial melting

2: EARLY PLUTONISM

3: VOLCANIC-SEDIMENTARY STAGE

4: OROGENIC STAGE

SIAL SHIELD

MOBILE BELT

Fig. 3.3. A model of evolution of Archaean shields. 1: Oceanic stage; mega-rippling of the oceanic crust, minor sedimentation of chert, ferruginous chert, banded iron formations and pelitic sediments derived through erosion of the oceanic crust at structural highs. Partial melting of subsiding eclogite and/or amphibolite segments of the crust. 2: Early plutonism; low degrees of partial melting of the oceanic crust gives rise to sodic acid magmas which rise as diapiric oval-shaped or elongated batholiths. 3: Volcanic–sedimentary stage (greenstone belts); the isostatic rise of the early batholiths associated with the subsidence of intervening tracts of the ocean crust. Further partial melting of the oceanic crust below these troughs gives rise to calc-alkaline volcanism. The erosion of the batholiths as well as of the intrabasinal volcanics results in the accumulation of detrital sediments. 4: Orogenic stage; further subsidence of the volcanic–sedimentary troughs between the granitic nuclei into warmer regions of the crust results in folding and low-grade metamorphism. Low degrees of partial melting at the base of the thickening crust give rise to younger (potassic) granites. Shield formation stage cooling of the volcanic–sedimentary troughs results in the aggregation of the granitic nuclei into shields. The boundaries of these shields with the surrounding oceanic crust constitute the loci of development of high-grade 'mobile belts', possibly above subduction zones. (After Glikson, 1972; reproduced by permission of The Geological Society of America)

remnant of the earth's most primitive or primordial oceanic crust is not acceptable because in more recent years it has been established that just as primitive komatiites occur in younger fold belts, namely the Appalachians in Newfoundland (Gale, 1973), and indeed in a variety of other modern environments (Brooks and Hart, 1974).

c) It is envisaged that parts of the oceanic crust subsided (by downsagging or rifting) and underwent partial melting to give rise to sodic acid magmas which rose upwards to form island nuclei, the greenstone belts forming between them in the intervening troughs. The key to the formation of the sodic granite magmas was the experimental work of Green (T.H.) and Ringwood (1968) which showed that sodic acid liquids can be produced by low degrees of partial melting of amphibolite or eclogite under wet conditions. In the 1970s Green and Ringwood applied their earlier data to subduction zone environments, but Glikson continued with the subsidence model.

d) It is well known that the lowest ultramafic–mafic rocks of some greenstone belts, especially in Rhodesia (the Sebakwian) and at Barberton (the Lower Onverwacht), occur as inclusions within bordering late intrusive tonalites—this is not under dispute. But high-grade gneisses typically contain lots of inclusions and since Glikson and Lambert (1973, 1976) and Glikson (1976a, 1976c) do not believe that the gneisses belong to any form of basement they suggest that they formed late in the greenstone belt cycle and so equate the real greenstone belt inclusions in the tonalites with the inclusions in the gneisses (these are termed 'lower or primary greenstones', Fig. 3.4). In this way they conclude that Archaean gneiss–granulite terrains are the late coeval roots of early greenstone belts that are remnants of some, once extensive, simatic crust. It is this misidentification of the gneiss–granulite inclusions as being equivalent to and having the same origin as the remnants of the lower ultramafic–mafic parts of higher level contempor-aneous greenstone belts now found as inclusions in bordering plutons that makes the Glikson–Lambert model so untenable. Whatever the origin of the greenstone belts, the detailed field, structural and geochronological data of a great number of specialists suggest to them a quite different relationship between greenstone belts and nearby high-grade granulites and gneisses (Bliss, 1969; Bliss and Stidolph, 1969; Hunter, 1970a, 1974a; R. T. Bell, 1971; Stowe, 1971; Bickle, Martin and Nisbet, 1975; Hawkesworth, Moorbath and O'Nions, 1975; Binns, Gunthorpe and Groves, 1976; Coward, Lintern and Wright, 1976; Shackleton, 1976a). For further details on these interrelations see later in this chapter.

e) Glikson and Lambert then consider that the calc-alkaline volcanic part of greenstone belts are 'upper or secondary greenstones' (Fig. 3.4) formed (with the overlying sediments) within linear troughs; these make up the bulk of 'true' greenstone belts. The reason for giving the lower and upper parts of greenstone belts significantly different ages and origins is the fact that locally they are separated by an unconformity and the

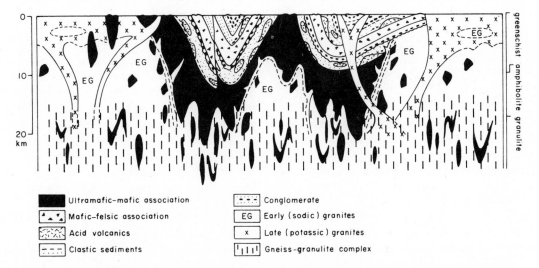

Fig. 3.4. An interpretation of the field relationships between major Archaean rock units in western Australia, South Africa–Rhodesia and India. The figure is meant to portray the general concept of the relations, particularly between lower and upper greenstones and high-grade gneiss–granulite complexes. (Modified after Glikson, 1976a; reproduced by permission of J. Wiley)

intrusion of some Na-rich granites. Thus whilst the primary greenstones 'evolved in environments showing no evidence of pre-existing or proximal sialic crust', the secondary greenstones 'rest on post-granite unconformities and/or include granite-derived arenites and conglomerates' (Glikson, 1976a).

f) As emphasized by Glikson (1976a), the model is applicable if greenstone belts have a synclinal structure, thus enabling the primary greenstones at low stratigraphic levels to be found at deeper crustal levels in the amphibolite/granulite-grade gneisses. However, the recent structural work of Stowe (1974) and Coward, Lintern and Wright (1976) shows that many of the Rhodesian belts have inverted stratigraphic successions and that the present synformal shape is a result of secondary deformation; this structure makes the Glikson model untenable for these belts. It is then interesting to remember that Viljoen and Viljoen (1969) produced reasons to suggest that the stratigraphic make-up of the Rhodesian belts was similar to and correlatable with that of the Barberton belt, and Anhaeusser (1971a) likewise argued that the Barberton belt was stratigraphically similar to the greenstone belts of western Australia. *But* the Glikson model is based on the western Australian belts and said to be applicable to the Archaean of the southern hemisphere (1976a). So how relevant is the model?

Plate Tectonic Models

The following plate tectonic models applied to possible evolution of greenstone belts vary from the implicitly implied to the specific, and analogues with several modern tectonic environments are proposed.

1. Goodwin and Ridler (1970) deduce that the Abitibi belt formed intracratonically upon a thin mafic crust between adjoining forelands of predominantly sialic composition. Fig. 3.5 illustrates this structure which gives rise to three main tectonic units: the primitive sialic cratons, comparatively unstable marginal sedimentary basins and troughs, and orogenically active off-shore volcanogenic belts. The strength of this model lies in the abundant data provided by the volcanic and sedimentary stratigraphy.

The three types of shelf-to-basin transitions in the Abitibi belt, previously mentioned, delineate the broad structure of the 'orogen' (Fig. 2.11). The younger sedimentary assemblages occupy the deeper water areas towards the basinal axis of the belt. Most of the volcanic complexes with their felsic differentiates lie close to the borders of the 'orogen' and, as most mineral deposits are associated with the felsic rocks, the metallogenic patterns are spatially related to the foreland boundaries. The construction of volcanic complexes proceeded in episodic progression, the axis of successive volcanic accumulations shifting away from the forelands so that younger volcanic assemblages lie towards the axial zone. The above data combine to give a sophisticated and convincing evolutionary story. The distribution pattern of sedimentation and volcanism is what would be expected from progressive tectonic spreading or opening of the sialic forelands as a result of incipient ocean floor spreading.

In order to better demonstrate the relationship between greenstone belts and many high-grade (basement) gneisses and in order to take account of the 'oceanic' character of the lower tholeiites (Glikson, 1970, 1971c), Windley (1973a) proposed a model according to which the belts formed in proto-oceanic ridge systems involving only small amounts of (continental) crustal separation. It was envisaged that some belts may have formed in ensialic graben-like rift zones and others, like perhaps the Abitibi belt, in narrow, linear oceanic basins caused by relatively small amounts of opening. A somewhat similar tectonic model based on an ascending mantle plume was applied to the belts of the Kaapvaal craton by Hunter (1974a) and to the Berberton belt by Condie and Hunter (1976) (Fig. 3.6).

The main problem with this model is the vagueness of the rift environment, the rifts forming anywhere in a continental mass. A considerable improvement, therefore, is made by the marginal basin model outlined below, according to which the rifts of essentially simi-

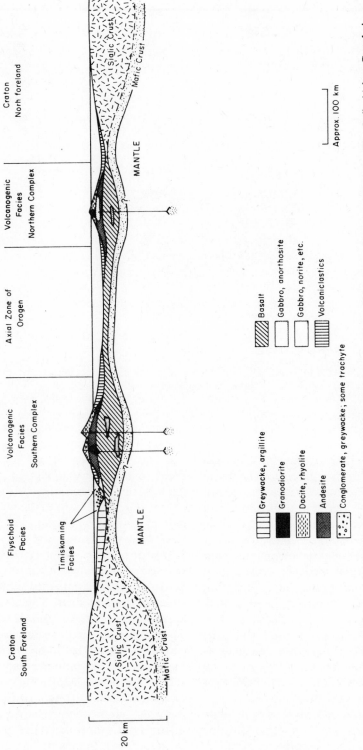

Fig. 3.5. Hypothetical tectonic reconstruction of the Abitibi orogen (for map of the volcanogenic complexes see Fig. 2.6). (After Goodwin and Ridler, 1970; reproduced by permission of The Geological Survey of Canada)

52

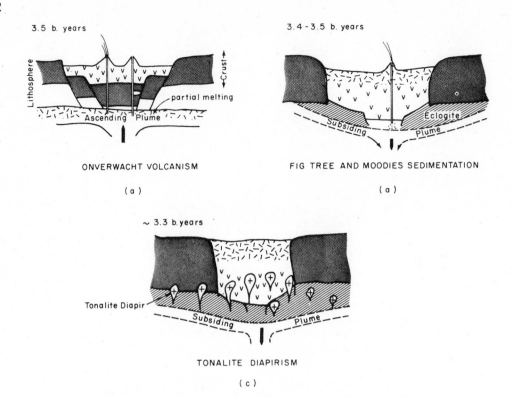

3.5 b. years

Lithosphere

Crust

partial melting

Ascending Plume

ONVERWACHT VOLCANISM

(a)

3.4 - 3.5 b. years

Eclogite

Subsiding Plume

FIG TREE AND MOODIES SEDIMENTATION

(a)

~ 3.3 b. years

Tonalite Diapir

Subsiding Plume

TONALITE DIAPIRISM

(c)

Fig. 3.6. A mantle plume model to explain the evolution of the Barberton belt (modified after Condie and Hunter, 1976; reproduced by permission of Elsevier Scientific Publishing Co.)

lar type are located in a specific back-arc setting.

2. Anhaeusser (1973) re-evaluated the origin of the southern African greenstone belts in terms of an island arc model, several aspects of which, such as the derivation of sodic acid magmas from partial melting of the base of the oceanic crust, were based on Glikson's earlier model. Five stages of development were envisaged: primordial oceanic crust, early volcanism, early plutonism, late volcanism, sedimentation and diapirism, and finally deformation and cratonic nucleation. It is interesting that Anhaeusser used the geochemical data of the Barberton volcanics as evidence for an island arc environment, while Glikson used the same data as an indicator for his downsagging linear trough model.

Although Anhaeusser's model is up-to-date in considering the formation of greenstone belts in terms of plate tectonic theory, it suffers from the same disadvantages as other models

that suppose that there was a primitive oceanic crust which developed into greenstone belts and that the high-grade gneisses and migmatites formed subsequently; it noticeably ignores the evidence from Rhodesia 'of an earlier sialic basement to the main greenstone belt cover' (J. F. Wilson, 1973a) which has been known for some years.

3. According to Talbot (1973) the granitic–gneissic areas between the greenstone belts acted as micro-continents subjected to considerable lateral drift. The greenstone belts are interpreted as vestiges of former oceans. The model is based on the assumption that oceanic crust tended to be not subducted but rather scraped off onto the leading edge of the subducting plate—a form of skimming process along early island arc-type orogenies. Accordingly the greenstone belts occupy the same tectonic position as mélanges on the inner wall of modern arc trenches. The question arises, could the stratigraphy of the oceanic crustal

segment remain unbroken and so well preserved during such an off-scraping process? Modern examples in mélange complexes suggest quite the opposite—that this indeed is the most highly deformed zone in any orogenic belt.

4. Rutland (1973b) formulated a plate tectonic model with reference to the greenstone belts of Australia (Fig. 3.7). The principal points are that the deformation styles of the belts (including steep cleavage due to shortening) are of miotectonic character and that the regular structural trend in the Yilgarn block suggests the operation of a uniform shear system unlikely to be produced by a pattern of small continental nuclei separated by subduction zones. Emphasis is placed on a thin lithosphere (<50 km) and a steep geothermal gradient in the Archaean which gave rise to extensive basic volcanism controlled by parallel mega-fractures. The Archaean greenstone terrain can therefore be regarded as a miotectonic zone developed on older continental crust. Note that Rutland's scheme begins to approach the model outlined below.

5. Several comparisons have been made of the chemical parameters of Archaean volcanic rocks with those of possible modern equivalents. Although few specific tectonic models or schemes of development have been proposed, the comparisons are relevant to the present discussion. As brought out in Chapter 2 the lower volcanic rocks of most belts have chemical features similar to modern rise tholeiites (depleted types with flat REE patterns), whereas those at higher stratigraphic levels are tholeiites exhibiting slight light REE enrichments and generally greater large-ion lithophile element concentrations (Condie,

1975) and these enriched types are similar to modern island arc tholeiites (Jahn, Shih and Murthy, 1974). But in contrast to arc and oceanic ridge basalts only marginal basin basalts have the relatively high Cr and Ni contents and high lithophile element (K, Rb, Ba) values typical of greenstone belt basalts (Tarney, Dalziel and De Wit, 1976).

6. There is a current attempt to explain greenstone belts as fossil marginal back-arc basins. Burke, Dewey and Kidd (1976) expound a stimulating rationale for the concept and Tarney, Dalziel and De Wit (1976) provide a modern comparable example, the 'Rocas Verdes' complex in South Chile (long known by this name which is particularly apposite for a modern greenstone belt!). Condie and Harrison (1976) interpret the chemical relations in the Bulawayan Group of the Midlands greenstone belt in Rhodesia in terms of the evolution of an arc system from oceanic rise tholeiites in a marginal basin to andesites representing a more mature stage in an emerging arc.

Firstly, the theory. There are reasons for believing that the rate of lithosphere production may be proportional to the rate of radiogenic heat production in the mantle, and therefore that the rate of generation of new oceanic lithosphere was substantially higher in the past (Dickinson and Luth, 1971). Because heat generation rates were about three times higher in the Archaean than they are today, thermal gradients were steeper (R. St. J. Lambert, 1976). Today twice as much heat escapes from mid-oceanic ridges as from the remainder of the oceanic crust or the continents and thus it is reasonable to assume that the additional thermal energy in the Archaean was

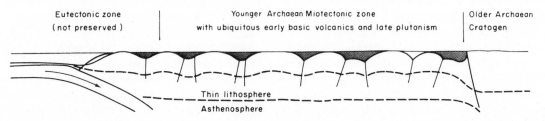

Fig. 3.7. Greenstone belts in Australia formed in a miotectonic zone on an older drifting continental plate (after Rutland, 1973b; reproduced by permission of The Academic Press, London, New York, and San Francisco)

dissipated through greater ridge activity (Burke, Dewey and Kidd, 1976). The total length of ridges was probably longer than today and therefore the total length of subduction zones was equivalently longer so that continental plates were smaller and more numerous than at present. Alternatively, the rate of plate accretion may have been greater in the Archaean. In view of the fact that Archaean basalts are very low in potash this suggestion finds support in the discovery of Sugisaki (1976) that the K_2O content of modern oceanic basalts is lowest in plates with the highest spreading rate. In modern times a correlation can be expected between spreading rate, the intensity of orogeny and the associated magmatic and metamorphic effects such as andesitic volcanism, batholithic intrusion and metamorphism (e.g. Jahn, Chen and Yen, 1976). This is because rapid spreading leads to rapid subduction and consequent rapid new magma generation. Therefore the extensive continental growth in the late Archaean, expressed by widespread extrusion of greenstone belt lavas, intrusion of tonalites etc., especially in high-grade regions, and of almost ubiquitous metamorphism, may well have been due to more vigorous sea-floor spreading than at present.

Reasons for suggesting that the greenstone belts formed in back-arc spreading centres rather than in main oceanic or arc environments are:

1. In a mobile system the main oceanic lithosphere soon 'self-destructs' by subduction and cannot be expected to be preserved to any extent.

2. Even without eventual continent–continent collision (where the uplift process is accentuated) volcanic arcs tend to be uplifted and eroded so that only their plutonic roots are exposed, as in the Sierra Nevada and Peninsular Range batholiths of California. Therefore the greenstone belts with their thick volcanic successions are most likely not to be remnant arcs, but rather fossil marginal back-arc basins which in contrast tend to be well preserved.

Secondly, a modern analogue. In making the above proposal, Burke, Dewey and Kidd (1976) suggest that the Palaeozoic Round Pond area of Newfoundland and a Mesozoic area in the northwestern Sierra Nevada of California bear close resemblance to greenstone belts.

Tarney, Dalziel and De Wit (1976) present a detailed and cogent argument for regarding the Rocas Verdes marginal basin in South Chile as an actualistic counterpart of Archaean greenstone belts (Fig. 3.8). The synclinal form, dimensions, greenschist metamorphic grade, structural style, volcano-sedimentary rock associations and stratigraphy, volcanic geochemistry, relationship with older basement and younger intrusive tonalite–granodiorite plutons of the Rocas Verdes are all remarkably similar to their equivalents in greenstone belts. The main difference is that the younger rock belt, in contrast to the older, contains a mafic dyke complex (100% dykes) typical of ophiolites: but the presence or absence of such a dyke complex is dependent on many factors, such as rate of spreading and supply of magma.

The Rocas Verdes complex has pillow-bearing basalts, pillow breccias and tuffs, shales and greywackes with a high proportion of volcanogenic detritus, and cherts and jasper. Marginal basin basalts have the high Cr and Ni values typical of ridge basalts (but unlike arc tholeiites) and the higher lithophile elements (K, Rb and Ba) of arc tholeiites (unlike ridge basalts) and are slightly light-REE enriched: in all three respects comparable with Archaean greenstone belt basalts. Andesites are not prominent as in most greenstone belts—the so-called andesites of the Canadian belts (H. D. B. Wilson and co-workers, 1965; Baragar and Goodwin, 1969) are regarded by Gunn (1976) as misinterpreted high-Na (spilitized) meta-basalts. The range of bulk compositions and trace element abundances of the plutons and batholiths that engulf the marginal basin rocks (often leaving them as mega-xenoliths in a granitic terrain—cf. Litherland, 1973, for Botswana greenstone belts) are similar to the diapiric plutons that invade Archaean belts.

In my opinion some form of fossil marginal back-arc basin provides the most convincing tectonic environment for greenstone belts that

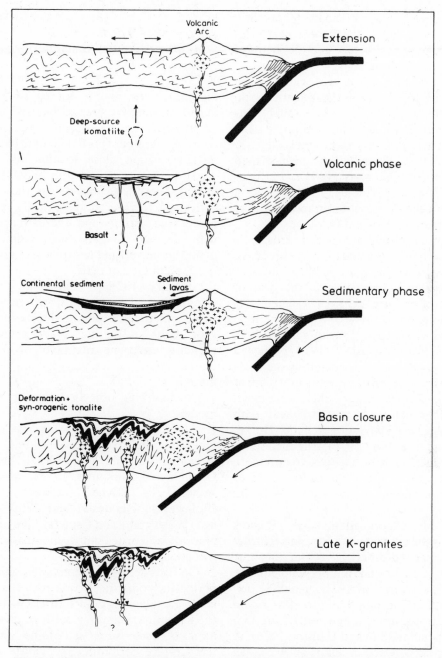

Fig. 3.8. Suggested development of an Archaean greenstone belt in a back-arc marginal basin position on comparison with the 'Rocas Verdes' complex in S Chile. Back-arc extension in an Archaean ductile lithosphere produces a greater component of crustal thinning rather than crustal rifting, as today. Magma production exceeds that required for simple extension. Sediment source is mixed volcanigenic/continental and the sequence may include calc-alkaline andesitic and salic lavas from the adjacent volcanic arc. Later movement of arc towards continent produces deformation and synclinal form of greenstone belt. Andean-type tonalitic-to-granitic plutons (derived from mantle by at least two-stage fractionation process) may be syn- to post-tectonic with compositions dependent on depth of melting of subducted oceanic crust (after Tarney, Dalziel and de Wit, 1976; reproduced by permission of J. Wiley)

has so far been produced. The model satisfies, better than any other, the known field relations, stratigraphy, structure and geochemical features of greenstone belts.

One aspect that needs to be discussed is the common occurrence, as in the Yilgarn block, of several belts that are virtually parallel (Rutland, 1973b). The formation of multiple belts is easy to envisage if they represent vestiges of oceanic crust caught between linear-shaped colliding micro-continents (Talbot, 1973) or parallel rift zones (Windley, 1973a; Hunter, 1974a), but the marginal basin model allows for more than one possibility.

If the arc-trench migrated oceanwards, multiple greenstone belts could form by successive extensions in the back-arc area, a situation directly analogous with that in the present western Pacific (Karig, 1971, 1974) where several marginal basins have developed parallel to each other (Tarney, Dalziel and De Wit, 1976). Uyeda and Miyashiro (1974) proposed that back-arc extension is caused by the thermal effect on the overlying continental plate of a subducting oceanic ridge; if there were much greater ridge activity in the Archaean with smaller plates and longer ridge lengths (Burke, Dewey and De Wit, 1976), subduction of oceanic ridges would be expected to be a more common phenomenon, allowing for a greater ease of formation of successive back-arc basins.

2. From his experimental work Richter (1973) suggested that with steeper thermal gradients a second-order convection can occur giving rise to horizontal rolls perpendicular to the principal convection. The resulting *hot lines* could be responsible for rifting and extension of a continent and could form in a back-arc position (Sun and Hanson, 1975). A series of small convection cells in an Archaean back-arc area could thus give rise to multiple greenstone belts.

Extra-terrestrial Models

D. H. Green (1972b) proposed that greenstone belts may have formed as a result of meteorite impacts and so might be terrestrial equivalents of lunar maria (Fig. 3.9). The basis of the model is that experimental work demonstrates that the temperature of extrusion of peridotitic komatiite magma was $1650 \pm 20°C$ (D. H. Green and coworkers, 1975) and such a liquid requires 60–80% melting of the mantle source composition. Such high temperatures at the earth's surface might imply impact-triggered diapirism and consequent intense partial melting of a mantle of composition close to that of pyrolite. Accordingly the greenstone belts are interpreted as very large impact scars, initially filled with impact-triggered melts of ultramafic-to-mafic composition and thereafter evolving with further magmatism, sedimentation, deformation and metamorphism to give rise to the present type of belts.

Glikson (1976b) suggested that the ultramafic–mafic xenoliths (primary greenstones) in tonalite plutons bordering greenstone belts and in high-grade gneisses and granulites may be remnants of terrestrial maria caused by impact of extra-terrestrial matter.

The following objections pertain to these models:

1. Although there is some evidence of greenstone belt-type sequences (e.g. Isua, W Greenland) as old as 3760 my (Moorbath, O'Nions and Pankhurst, 1973), most greenstone belts formed between 2900 my (Bulawayan, Rhodesia) and 2700–2600 my ago (W Australia and Canada). The major last phase of meteorite bombardment on the moon was about 4000 my ago.

2. On considering the tectonic significance of peridotitic komatiite, Brooks and Hart (1974) made the following points. Rocks with peridotite komatiite chemistry occur in diverse tectonic regimes (such as Alpine and high-temperature peridotites, peridotite nodules, modern sea floor peridotites and ophiolite complexes) throughout most of geological time. The very high temperatures (1600–1650°C) estimated by D. H. Green and co-workers (1975) were for essentially anhydrous melting ($<0.2\%$ H_2O). This temperature may have been lowered by involving significant quantities of water in the melt, and on extrusion this water was released during the

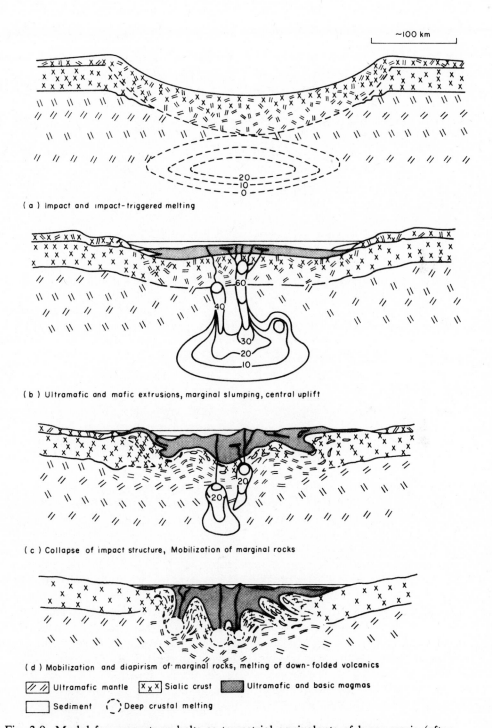

~100 km

(a) Impact and impact-triggered melting

(b) Ultramafic and mafic extrusions, marginal slumping, central uplift

(c) Collapse of impact structure, Mobilization of marginal rocks

(d) Mobilization and diapirism of marginal rocks, melting of down-folded volcanics

Ultramafic mantle   Sialic crust   Ultramafic and basic magmas

Sediment   Deep crustal melting

Fig. 3.9. Model for greenstone belts as terrestrial equivalents of lunar maria (after Green, 1972b; reproduced by permission of Elsevier Scientific Publishing Co.)

rapid crystallization process (dehydration quenching).

3. Too much complicated geology is lumped together under Glikson's 'xenoliths'. Firstly, the xenoliths in the greenstone belt plutons should not be correlated with those high-grade gneisses elsewhere, certainly not until more is known about their type and provenance; secondly, a great many Archaean gneisses formed about 2800–3100 my ago (much later than meteorite bombardment on the moon), and the inclusions within them must be considered separately from those in the much earlier Amîtsoq and Uivak gneisses. Many of the ultramafic–mafic inclusions in the late Archaean gneisses are without doubt derived by fragmentation of layered igneous complexes. There seems no reason to consider these rocks and relationships as results of extra-terrestrial impact events.

The meteorite impact model for the generation of greenstone belts has clearly stimulated much thought and discussion; however, noticeably very few geologists have it taken it up as a viable possibility.

High-grade Terrains

In contrast to the greenstone belts there have been very few attempts to postulate a tectonic model for Archaean high-grade regions. It is clear that as long as the rock units in these regions were thought to have developed as a conformable group little insight would emerge, worthy of realistic consideration, of possible tectonic environments. A major step was made by V. R. McGregor (1973) in breaking through the conformability barrier of part of West Greenland and this led to the development of an isotopically confirmed chronology (e.g. Moorbath and coworkers, 1972; Black and coworkers, 1973; Pankhurst, Moorbath and MacGregor, 1973; Bridgwater, McGregor and Myers, 1974a). A somewhat similar chronology of events has been established in Labrador (Bridgwater and coworkers, 1975; Hurst and coworkers, 1975; Collerson, Jesseau and Bridgwater, 1976), proposed for the Sourian of Scotland (F. B. Davies, 1975a) and for the Limpopo belt of southern Africa (Bahnemann, 1971; R. G. Mason, 1973; Coward, Lintern and Wright, 1976). But it is necessary to go beyond the erection of local chronologies in order to understand the tectonic environment in which these kinds of rocks and relations evolved. Attempts to account for the origin of these high-grade complexes have recently been proposed by Moorbath (1975c), Tarney (1976), Myers (1976) and Windley and Smith (1976). But note that the breakthrough in understanding of the complex chronological development of Labrador, West Greenland and NW Scotland all took place within the one North Atlantic Craton (Bridgwater, Watson and Windley, 1973c). Nevertheless, it seems most likely that the Limpopo belt evolved in a somewhat similar manner and a comparable sequence of events might be established in the high-grade complex of southern India, the lithological and structural make-up of which strongly resembles that of the other regions under discussion.

Now let us look at three ideas related to the tectonic environment of these high-grade terrains.

Recrystallized Greenstone Belts

There is a common impression that Archaean high-grade complexes are simply a pile of recrystallized supracrustal rocks plus some layered intrusions. Such an idea has been applied to the rocks in NW Scotland (Bowes, Barooah and Khoury, 1971; Bowes and Hopgood, 1973, 1975), southern India (Viswanathan, 1974b), the Wyoming Province of the USA (Condie, 1976a), the Ancient Gneiss complex of Swaziland and many other examples (Barker and Peterman, 1974), in Western Australia (Glikson and Lambert, 1976) and the USSR (Pavlovsky, 1971). This principle is dependent on regarding the quartzo-feldspathic gneisses as either metamorphosed sediments (Sidorenko, 1969; Cheney and Stewart, 1975), or volcanics (Sheraton, 1970; Bowes, 1972; Barker and Peterman, 1974). For further details, including opposing arguments, see Chapter 1.

The natural corollary of this approach is to regard the high-grade complexes as just highly metamorphosed greenstone belts (Viswanathan, 1974b; Bowes and Hopgood, 1975); but the lithological differences between the two terrains are so great that this conclusion is hardly likely. Take the sediments, for example: a shelf-type association of quartzites, marbles and pelites in the high-grade and a greywacke–turbidite association in the low-grade regions. A traverse from the Limpopo mobile belt into the Rhodesian or Kaapvaal cratons is sufficient to bring out the very great differences between these two types of Archaean crust (Shackleton, 1976a; Coward, Lintern and Wright, 1976). Tarney (1976) and Lambert, Chamberlain and Holland (1976) conclude from a study of chemical data that the gneissic complexes differ significantly from the greenstone–granitic pluton association.

Oceanic Crust

The supracrustal rocks and layered igneous complexes can be interpreted in terms of a section through the oceanic crust (e.g. Garson and Livingstone, 1973). According to this model the quartzites are correlated with oceanic cherts, the marbles with dolomitic limestones known locally on the mid-Atlantic ridge, pelites with pelagic clays, and amphibolites with low-K oceanic basic volcanics. The anorthosite–leucogabbro complexes (e.g. at Rodil, S Harris, Scotland) are compared with layered igneous complexes, the existence of which under some oceanic tholeiites is inferred (Engel and Fisher, 1975).

However, the following objections make this hypothesis implausible. The quartzites are current bedded where best preserved, as in the Limpopo belt (Shackleton, pers. comm.), and so are better regarded as recrystallized orthoquartzites. The orthoquartzite–carbonate (dolomitic marble)–K pelite (mica-schist) association is a typical shallow-water shelf assemblage found in continental margins (Sylvester-Bradley, 1975; Sutton, 1976). Some of the amphibolites have chemical patterns similar to those of oceanic tholeiites (Rivalenti, 1976). But apart from the problems involved in comparing the chemical composition of Archaean and modern basalts (Gunn, 1976), modern low-K tholeiites occur in a variety of tectonic environments and the distinction between oceanic and island arc tholeiites, so obvious in modern times, may well have been less apparent in the Archaean (Brooks and Hart, 1974). The layered igneous complexes inferred from grab samples to form part of the modern oceanic crust are not at all similar in rock proportions or chemistry to the anorthositic type of Archaean layered complexes; they consist largely of Ti–Fe-rich gabbro and contain only about 5% of Cr-poor labradorite anorthosite (Engel and Fisher, 1975).

Active Continental Margins

When the sedimentary, volcanic and plutonic rock associations and their chemical features are considered together, then the closest modern equivalent to the Archaean high-grade complexes is found in the Mesozoic batholiths of active continental margins along the axis of the Cordilleran fold belts of North and South America (Windley and Smith, 1976).

Taking the Peninsular Ranges of southern California as a model (Gastil, 1975, but see also Cobbing and Pitcher, 1972), the batholith contains three main rock groups:
1. The most abundant are tonalites and granodiorites ranging to quartz monzonites (adamellites) and, rarely, to potassic granites. Some of the earliest 'plutons' are highly deformed and consist of tonalitic gneisses.
2. Within the tonalites there are strips of meta-sediments (the Julian schists) that comprise sillimanite mica schists (meta-K pelites), quartzites and carbonate rocks, an assemblage typical of a shelf–continental margin environment.
3. In the tonalites, and often bordering the above meta-sediments, there are about 100 layered igneous complexes up to 70 sq km in outcrop with distinctive mineralogy. They comprise largely gabbros and anorthositic gabbros (with layering and cumulate textures) which largely formed by precipitation of

hornblende and calcic plagioclase (An_{80-100}) from a wet basaltic–andesitic magma at 900°C and a water pressure of >7 kb (Nishimori, 1974).

Obvious to the reader conversant with Chapter 1 will be the remarkable similarity between the above rock groups and the tonalitic gneisses, the quartzite–carbonate–pelite sedimentary assemblage (locally with metavolcanic amphibolites) and the anorthosite-leucogabbro layered igneous complexes, respectively, in the Archaean high-grade regions. In particular the mineral and rock chemistry of the Peninsular Range layered complexes matches closely that of the Fiskenæsset complex (Windley and coworkers, 1973, 1974) and the bulk and trace element chemistry of the batholithic tonalites, etc., is close to that of the average Archaean tonalitic gneisses (Tarney, 1976); and most important, the rock proportions and relative associations are similar, i.e. in both cases the tonalites intrude and eventually engulf the supracrustals and the layered complexes. Should this comparison be correct, then there seems a strong likelihood that the main period of growth of the continents in the late Archaean took place in a manner not fundamentally unlike accretion at the leading edges of modern continental plates.

Myers (1976) has advocated a Himalayan model for the West Greenland Archaean, largely because of the great crustal depth and width and presence of thrusts in the Himalayas and because the Cordilleran belt has not produced granitoid sheets on the same scale as those in the Greenland Archaean. However, the calc-alkaline volcanism and plutonism of the Cordilleran belt far exceeds that of the Himalayas, the relative scale of development between modern and Archaean rock suites is unimportant since most workers agree that heat flow and thermal gradients were greater in the early Precambrian, and the Cordilleran belt in the Andes has just as great a crustal thickness (more than 70 km, D. E. James, 1971) as the Himalayas and, in the western USA, it has a comparable width. Also, continental collision-type orogens, dependent on depression and partial melting of a continental slab, are expected to give rise to high K_2O/Na_2O ratios in plutonic and volcanic rocks whereas arc- and Cordilleran-type orogens, dependent on subduction of K_2O-poor oceanic lithosphere, typically have low K_2O/Na_2O ratios (Mitchell, 1976a). The constantly low ratio in late Archaean basic volcanic rocks and in the widespread tonalites and granodiorites is inconsistent with a Himalayan-type model. Talbot (1973) may well be right in suggesting that Himalayan-type orogenies characterized by depression and over-riding of rigid continental slabs did not take place in the Archaean because cratons and rigid plates did not evolve until the beginning of the Proterozoic.

Relationship between Low- and High-grade Terrains

Many high-grade terrains occur as discrete entities such as the linear Limpopo belt and the North Atlantic craton. Most low-grade terrains occur as so-called cratons, but between many of the constituent greenstone belts and their associated granite plutons there are commonly areas of high-grade gneiss, e.g. in the Rhodesian craton (Bliss, 1969; Stowe, 1971; Wilson, 1973a). The problem, therefore, that concerns us is not just the broad tectonic relation between a high- and a low-grade 'craton', but the more specific relation between the age, etc., of the gneisses (and indeed of the supracrustal rocks) in the two environments. Are the gneisses in the one earlier or later than those in the other, or are they contemporaneous? Are the two environments representative of a vertical or a horizontal section through the Archaean continental crust?

Because the high-grade complexes had been largely ignored in deliberations on early continental evolution (e.g. Anhaeusser and coworkers, 1969), Windley and Bridgwater (1971) produced a model which saw the high-grade terrains as the deep-seated 'basement' to higher level greenstone belts. This model satisfies some of the radiometric age relations and the fact that parts of some greenstone belts do lie unconformably on high-grade gneisses (Nautiyal, 1965; Radhakrishna, 1967; Stowe, 1971; Bickle, Martin and Nisbet, 1975; Ermanovics and Davison, 1976; Haidutov, 1976).

But the subject is controversial. According to Glikson (1976a, 1976c) the unconformities between the lower and middle parts of greenstone belts such as the Sebakwian and Bulawayan represent fundamental breaks between rock units of very different time and mode of origin. Glikson and Lambert (1973, 1976) and Glikson (1976a, 1976c) suggest that the amphibolite and granulite facies gneisses found either between greenstone belts or elsewhere in the high-grade terrains are the infracrustal coeval roots of the granite–greenstone systems. But these conclusions are disputed by Binns, Gunthorpe and Groves (1976) who are of the opinion that with respect to the rocks in the Yilgarn block of SW Australia, which is the key area for the model of Glikson and Lambert, the 'banded gneisses represent a former sialic basement to the greenstone belts, remobilized during the 2600–2700 my metamorphic episode'.

In his well-balanced appraisal of the relations in southern India and Africa, Shackleton (1976a) concluded that there is structural and metamorphic continuity between the high- and low-grade regions, that the basement gneisses of the Limpopo belt are equivalent to those in the adjacent greenstone belt cratons, that the younger sediments and volcanics of the greenstone belts (with deep-water turbidites and greywackes) correspond laterally to those in the Limpopo belt (the Messina Formation with shallow-water orthoquartzites, marbles and pelites), and that the intrusive Rooiwater layered complex in the lower part of the Murchison Range greenstone belt corresponds to the anorthosite complexes intruded into the Messina Formation. It follows from these observations that the high-grade terrain represents a laterally situated but more deeply eroded level of the crust and it therefore contains far more basement gneisses than the adjacent cratons with their greenstone belts.

The above general conclusions are supported by the results of Coward, Lintern and Wright (1976) who correlated the sediments of the Matsitama greenstone belt of the Rhodesian craton with those of the Limpopo belt (Messina Formation). Detailed structural studies suggested that both groups of sediments were interthrusted or imbricated during some form of nappe-like tectonics. In the Limpopo belt the structurally repeated sediments of the Messina Formation were intruded by the layered anorthosite complexes and tonalitic–granitic rocks. There is a close similarity between this horizontal type of tectonic regime (and chronology of events) and that proposed for the high-grade belt of West Greenland by Bridgwater, McGregor and Myers (1974a).

Future research on the relation between the high- and low-grade parts of the Archaean continents will depend heavily on radiometric age data. For example, recent geochronological results of Hickman (1974) and Hawkesworth, Moorbath and O'Nions (1975) have shown the presence of two ages of greenstone belt in the Rhodesian craton, and on structural–stratigraphic grounds older and younger belts are known in India (e.g. Naqvi, Rao and Narain, 1974; Naqvi, 1976). Consequently, in future when considering any tectonic relationship between a greenstone belt and any granitic or gneissic material, caution must be applied when arriving at conclusions of genetic significance (for example, the late granites of one greenstone belt may form the basement of another) and when making broad generalizations from the particular relationships of one such belt. Quite clearly different age relationships pertain in different places.

A Modern Interpretation for Archaean Crustal Evolution

At this point I must admit to a bias in favour of a proto-plate tectonic model to explain Archaean tectonic evolution. The most viable model, I suggest, basically combines the back-arc marginal basin idea for greenstone belts and the main arc (plutonic batholith) interpretation of the high-grade gneissic complexes. These two models are mutually complementary.

According to the combined model (Fig. 3.10) the main arc would form slightly before the back-arc. Amîtsoq/Uivak-type gneisses formed the 'basement' in small sialic plates. The intrusion of basic dykes (Ameralik/Saglek) was followed closely by the extrusion of tholeiitic lavas (and deposition of shelf/continental margin-type sediments) in the arc area which were probably similar to the low-K tholeiitic series of modern arcs, although it is possible that some andesites were formed. Intensive ridge and subduction activity related to a high heat flow regime in the Archaean gave rise to the intrusion of batholithic proportions of mantle-derived tonalite and granodiorite (the early fractionated series of which were largely hornblende gabbros, leucogabbros and calcic anorthosites); this intrusion was associated with thrusting and nappe formation during a widespread horizontal tectonic regime which produced a tectono-igneous stratigraphic pile that contributed to a major thickening of the crust in late Archaean times. The intrusions probably occurred through a wide depth of crust, from high levels where they engulfed the earlier volcanics to deeper levels where they may have passed directly into foliated gneisses or granulites. These events were synchronous with, or immediately followed by, high temperature metamorphism that transformed all rocks into an amphibolite or granulite grade, whilst deformation converted the 'granitic' rocks into gneisses.

Extension in the back-arc area began before the completion of the above events. Early ultramafic–mafic volcanics may have been extruded onto thinned sialic crust or may have formed in a rift-type oceanic environment; their composition, characterized by a low-K content, was a result of shallow depth of melting of mantle material, to be expected in Archaean times when the geothermal gradient was high (also applicable to basic dykes and volcanics in the arc area). Later volcanics with a calc-alkaline character and much pyroclastic material formed during the emergence of arc-type volcanoes and later sediments derived much clastic material from the erosion of gneissic basement in the adjacent uplifted arc areas. The closure of the back-arc basins, which gave rise to the typical greenstone belt structure, may have taken place contemporaneously with the last deformation and metamorphism in the arc area.

At best, the combined model suggests that the major period of growth of the continents in the late Archaean took place in a manner broadly similar to the accretion taking place at the leading edges of modern continental plates. This implies some form of ridge and subduction activity in the Archaean which, in turn, means that the kinds of rock association, stratigraphy and rock chemistry (bulk and trace element) in the late Archaean and Mesozoic equivalents should be largely similar, which they are. At worst, the combined model provides a basis for comparing future Archaean geochemical results with those from a modern environment whose mode of evolution is better understood. Of course certain parameters must have been different in the past: geothermal gradients were steeper, the continental lithosphere was, perhaps, thinner and probably more ductile, plates were probably smaller and the tectonic regime more horizontal than vertical. Nevertheless, despite these differences an active continental margin-type regime can adequately explain the evolution of the Archaean rocks.

In the Mesozoic continental growth was limited to relatively narrow linear fold belts whereas in the permobile regime of the Archaean it was widespread. This was probably because plates were small in the Archaean and so subduction activity affected the whole plate in contrast to Phanerozoic times when plates were so large that orogenic activity was confined to their leading edges. As Talbot (1973) points out, deformation was not more intensive in the Archaean than at present—it was merely distributed more widely throughout smaller plates.

The proposals made here are consistent with the interpretation of Archaean whole-rock isochron dates of Moorbath (1975a,b) as giving the age of major accretion–differentiation events. On the basis of their low initial strontium isotope ratios the gneisses, greenstone belt volcanics and grani-

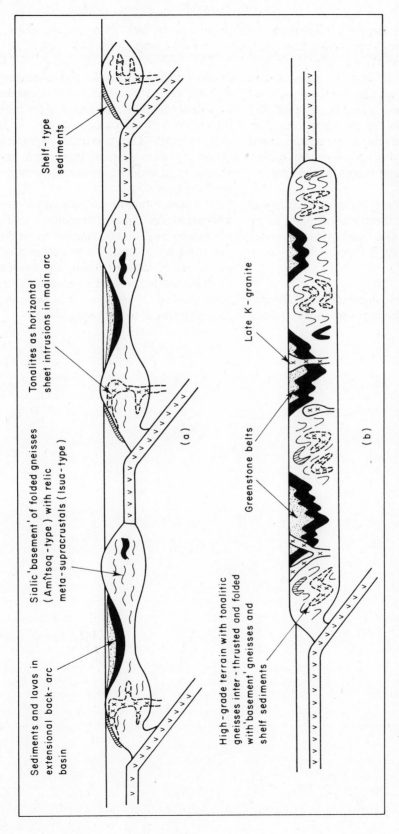

Sediments and lavas in extensional back-arc basin

Sialic 'basement' of folded gneisses (Amîtsoq-type) with relic meta-supracrustals (Isua-type)

Tonalites as horizontal sheet intrusions in main arc

Shelf-type sediments

(a)

High-grade terrain with tonalitic gneisses inter-thrusted and folded with 'basement' gneisses and shelf sediments

Greenstone belts

Late K-granite

(b)

Fig. 3.10. A plate tectonic model to explain the growth of continents in the Archaean. (a) Widespread lateral movement of many early Archaean mini-continental plates with shelf-type quartzites, carbonates and K-pelites and with mantle-derived tonalites in batholithic proportions in proto Cordilleran-type arcs and of volcanics in back-arc environments. (b) Aggregation of mini-continents gives rise to extensive continental plate by the late Archaean consisting of greenstone belts, and high-grade terrains with older and younger gneissic components. Amphibolite- to granulite-grade metamorphism (heat flow) and deformation of tonalites to give rise to tonalitic gneisses takes place in the roots of the main arc

63

64

tic plutons are predominantly juvenile additions from the mantle to the continental crust, at or close to the measured age implying continental growth on a major scale, especially in the mid-late Archaean. This 'early continental accretion episode' took place in a period 'not exceeding 50–100 my in duration, and this time interval is of the same order as that found at present between the most recently formed complementary oceanic ridges, island arcs and continental margins'.

As illustrated in Fig. 3.10, the accretion was related to the aggregation of mini-plates in the late Archaean, to give rise to larger more stable plates by the early Proterozoic.

Postscript

The ideas on Archaean crustal development reviewed in this chapter are more than a documentary list. They are of educational value to the student for two reasons:

1. They show how ideas in this field have evolved in the last decade. It is sometimes easy to criticize early models in the light of the experience of hindsight, but they are the building blocks with which newer ones are invariably created. The historical approach also serves to demonstrate that even the last and/or most widely accepted concept is only a working hypothesis.

2. It is interesting to see how two contrasting uniformitarian (plate tectonic) and non-uniformitarian models can be erected to explain one set of data. Thus, rather than have just a set of data to learn, the student can exercise his judgement in making a critique from a certain standpoint or in making a balanced appraisal of opposing ideas.

Chapter 4

Early to Mid-Proterozoic Basic–Ultrabasic Intrusions

Following the permobile tectonic regime in the Archaean there evolved a new stage in earth history when major cratons stabilized allowing, for example, the deposition of thick supracrustal sequences (as described in Chapter 5), and also the formation of mega-fracture systems which might be regarded as initial attempts to fragment these first-formed extensive continental masses. Magmas were intruded into many of these fractures giving rise to transcontinental dyke swarms and many layered complexes (the subject of this chapter). Subsequently, most regions remained stable for the remainder of geological time and the bodies are therefore still well preserved in Rhodesia, South Africa, East Africa, West Australia, West Greenland, western USA and North Canada. There is, of course, little or no evidence of the bodies in those regions that underwent later tectonic activity, and in others they were covered by later supracrustal deposits. The development of the early Proterozoic bodies, and the basins and geosynclines described in Chapter 5, was clearly diachronous; thus, whilst the Great Dyke was intruded in Rhodesia, the Witwatersrand System was being deposited in South Africa.

The intrusions are divisible into three types: giant dyke-like layered bodies such as the Great Dyke (Rhodesia) and the Widgiemooltha Dyke Suite (West Australia), major swarms of dolerite/diabase dykes, and layered stratiform igneous complexes. The first were intruded into the consolidated greenstone belts and high-level granites of Gondwana-

land, the dyke swarms cut high-grade Archaean gneisses as well as the greenstone belts, and the layered complexes were intruded into older basement or unconformable early Proterozoic cover sediments.

The Stillwater Complex, Montana

The Stillwater Complex is a 2750 my old layered intrusion of basic and ultrabasic rocks that has not been metamorphosed; but it has been tilted into a steeply-dipping position and heavily faulted (Hess, 1960). It was originally intruded as a sub-horizontal sheet into early Precambrian schists and gneisses of the Beartooth Mountains, Montana, but only about 60% of the initial thickness is now visible. The lower part was invaded by a granite 1530–1580 my ago, the upper part was eroded and the remainder tilted in late Precambrian times, and Middle Cambrian sediments were deposited unconformably on the eroded surface of steeply-dipping rocks. The present exposed strike length is about 48 km (faulted at each end) and the maximum stratigraphic thickness about 6000 m. According to Wagner and Brown (1968) the later tectonic events have so modified the original shape of the intrusion that it is not possible to find, in any single traverse across the complex, a fully representative rock sequence.

The Great Dyke, Rhodesia

This is a well-known layered ultrabasic–basic intrusion with a minimum age of 2530 ±

65

30 my and a probable age of 2550 ± 410 my (Allsop, 1965), whilst its Satellite Dykes have a Rb/Sr age of 2600 ± 120 my (I. D. M. Robertson and van Bremmen, 1970).

The dyke has a length of 480 km and an average width of 5·8 km (Fig. 4.1). It is a remarkable and unusual body as it contains four layered lopolithic subcomplexes, Musengezi, Hartley, Selukwe and Wedza, in which the layering dips inwards at a shallower angle than the steeply-inclined dyke contacts (Worst, 1960). According to Bichan (1970) each complex consists of cyclic sequences of ultrabasic rocks which reach a maximum exposed thickness of 2100 m in the Hartley Complex and are overlain by a 900 m thick gabbroic capping. Ideally each ultrabasic cycle has a basal chromite seam, forming a sharp footwall with underlying pyroxenites, followed upwards by peridotites and then pyrox-enites; in the Hartley Complex there are 11 chromite seams. The gabbroic capping has a sharp contact with the underlying ultramafic rocks and consists of a lower zone of anorthositic gabbro followed upwards by gabbros and norites which, at the very top, give way to quartz gabbros. Bichan (1970) concluded that each complex formed by pulsatory injection of magma derived from a parent source. Analyses of a chilled marginal phase and a mean of the Great Dyke Satellites by I. D. M. Robertson and van Bremmen (1970) indicate that the initial magma was of the alkaline olivine basalt or picritic basalt type composition.

Worst (1960) concluded that three major events took place in the evolution of the Great Dyke:
1. Formation of a linear zone of weakness: successive heaves of magma were injected through fissures developed at four positions

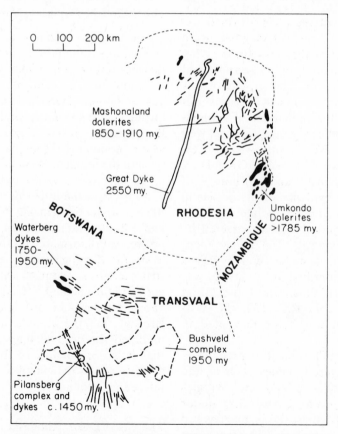

Fig. 4.1. Igneous complexes and dolerite dyke swarms in southeastern Africa (data from Vail and Dodson, 1969; Vail, 1970; Compston and McElhinny, 1975)

along this line until they met a horizontal plane of weakness in the earth's crust where they spread out laterally and differentiated as individual units.

2. Subsidence of the floor and formation of a graben with layers sagging into their present synclinal structure and concomitant shearing of the dyke's contacts.

3. Erosion down to the present level with removal of any lateral extension of the layers beyond the present dyke margins.

Bichan (1970) considered that the dyke was located over the position of a thermal updraft in a mantle convection cell and the waning of the heat flow pattern resulted in slumping of the dyke into its graben.

The Widgiemooltha Dyke Suite, West Australia

This suite includes the Coronation, Jimberlana and Binneringie dykes; the first two have a Rb/Sr age of 2420 ± 30 my. These are giant northeast-trending ultrabasic–basic dykes that traverse the greenstone belt–granite terrain of West Australia.

The Binneringie Dyke has no lopolithic shape, but vertical layering throughout; it is 320 km long and 3·2 km wide (McCall and Peers, 1971). Marginal bronzite gabbros pass inwards to more ferroan augite–pigeonite gabbros and there are minor intermediate and acid phases with granophyric and devitrified acid glass in the form of segregation patches, dykes and dykelets representing infillings by late magmatic phases of shrinkage cracks in the gabbro. In spite of the fact that the layering is vertical there is much rhythmic and cryptic layering as well as graded and cross bedding. In order to explain these unusual features McCall and Peers considered that the crystallization took place within vertically moving convection currents and that all the rhythmic and cryptic effects were caused by extensive marginal heat loss whilst the bedding effects were caused by pressure and supersaturation rhythms. The average composition of the chilled bronzite gabbro is dissimilar to that of the Great Dyke and its satellites and more akin to that of other major layered intrusions.

The Jimberlana Dyke is at least 180 km long and 2·5 km wide and is a small analogue of the Great Dyke of Rhodesia (Campbell, McCall and Tyrwhitt, 1970; McClay and Campbell, 1976). It has a very steep, V-shaped cross section and many internal canoe-shaped subcomplexes along its length, just like the Great Dyke. The intrusion largely consists of cumulate bronzitites and norite gabbros with phase, rhythmic and cryptic layering. It formed by the emplacement of several pulses of magma that gave rise to two magmatic unconformities.

Salt Lick Creek Complex, West Australia

The Salt Lick Creek Complex is one of several layered basic–ultrabasic intrusions in the Lower Proterozoic, East Kimberley region of western Australia (Wilkinson and coworkers, 1975). It has an approximate age of 1800–1900 my BP, its circular shape is 3·3 km in diameter and its maximum stratigraphic thickness 1 km. It is noteworthy for its olivine–plagioclase cumulates overlain by plagioclase–orthopyroxene and then anorthositic cumulates.

The Bushveld Complex, South Africa

The intrusion was emplaced into the upper part of the Transvaal System 1950 ± 150 my ago (Nicholaysen and coworkers, 1958). It is the world's largest igneous body covering approximately 66,000 sq km, it has a vertical thickness of up to 8 km, and it is the largest repository of magmatic ore deposits in the world (see later in this chapter) (Willemse, 1969).

The Complex occupies an elliptical area (Fig. 4.1) with a central part underlain by granite, microgranite, felsite and granophyre, and two marginal lobate belts, the western and eastern, made up of rocks ranging from dunite to norite, anorthosite and ferrodiorite.

Following on the classical work of Hall, Daly, Wagner and many other geologists in the early part of this century, recent research has demonstrated two important facts:

1. The Complex is not a lopolith: the mafic and ultramafic rocks do not extend under the central granites.

2. The Bushveld granites are younger than the mafic and ultramafic rocks, and did not form *in situ* from the magma that yielded these rocks.

1500 my ago. The table below shows prominent examples (see Figs. 4.1 and 4.2).

Particularly impressive is the scale of dyke intrusion at this time in earth history. The

Dyke Swarm	Age (my BP)	Reference
Scourie, NW Scotland	c. 2200	Moorbath and Park (1971) Tarney (1973)
Kangamiut, W Greenland	?	Bridgwater and coworkers (1973a,b) Escher and coworkers (1975, 1976)
Many swarms, W Greenland	?	Berthelsen and Bridgwater (1960) Chadwick (1969) Rivalenti (1975)
Matachewan–NS, Superior Province	2690±93	Gates and Hurley (1973)
Slave–Superior Province–NW	2150–2165	Fahrig and Wanless (1963)
Abitibi–NE, Superior Province	2147±68	Gates and Hurley (1973)
Nipissing Diabases, Canada	2155±80– 2162±27	Fairbairn and coworkers (1969)
Mackenzie (Set III)–NW, Slave Province	1370–1730	Gates and Hurley (1973)
Sudbury–NW swarm	1460±130(?)	Gates and Hurley (1973)
Mackenzie (Set I)–ENE, Slave	2692±80	Gates and Hurley (1973)
Mackenzie (Set II)–NNE, Slave	2093±86	Gates and Hurley (1973)
Mackenzie (Set IV)–NW, Slave	2174±180	Gates and Hurley (1973)
Granite Mountains, Wyoming	1600	Reed and Sartman (1973)
Bighorn Mountains, Wyoming	{ 2826±58 { 2200±35	Stueber, Heimlich and Ikramuddin (1976)
Ivory Coast, W Africa	1740±170 (K/Ar)	J. D. A. Piper (1973b)
Waterberg, southern Africa	1750–1950	Vail (1970)
Mashonaland, Rhodesia	{ 1850±20 { 1910±280	Compston and McElhinny (1975)
Umkondo, Rhodesia	>1785	Vail and Dodson (1960)
Pilansberg dykes, S Africa	c. 1450	Vail (1970)
Soriname, S America	1600–1750	J. D. A. Piper (1973b)
Roraima, Guyana	1500	J. D. A. Piper (1973b)
Hart, Australia	1800±25	J. D. A. Piper (1973b)
Chopan Dykes, India	2370±460	Balasundaram and Balasubrahmanyan (1973)
Other dykes, India	2000	Naqvi, Rao and Narain (1974)

In spite of the fact that much of West Africa was undergoing orogenic activity 1850±250 my ago (Clifford, 1972), the intrusion of the Bushveld complex is indicative of widespread stable conditions in southern Africa at about this time.

Basic Dyke Swarms

Distribution and Extent

Vast swarms of dolerite (diabase) dykes were intruded into the Archaean terrains (both high- and low-grade) and early Proterozoic cover rocks after 2700 my and particularly in the period between 2500 and

Scourie dyke swarm is not less than 250 km across its strike direction (Bridgwater, Watson and Windley, 1973c), the Kangamiut swarm is about 12·5 km across and 240 km long and is one of the world's densest dyke swarms (Fig. 4.3) (Escher, Escher and Watterson, 1975), the whole of the Archaean craton of West Greenland is criss-crossed by numerous intersecting dyke swarms (Berthelsen and Bridgwater, 1960; Chadwick, 1969), the Labrador–Slave swarm in Canada can be followed intermittently along strike for at least 2500 km and probably continues in West Greenland, the Sudbury–Mackenzie swarm is 3000 km long and more than 500 km wide, the

Fig. 4.2. Dolerite dyke swarms in Canada, Greenland and Scotland
with names and ages of individual swarms (data from Fahrig and
Wanless, 1963; Fahrig and Jones, 1969; Gates and Hurley, 1973;
Escher, Jack and Watterson, 1976)

Waterberg swarm at least 200 km long, and
the Umkondo and Mashonaland dykes occur
throughout an area about 400 km by 300 km
in NE Rhodesia (Fig. 4.1) (Vail, 1970). The
intrusion of these enormous swarms represents a new type of igneous and tectonic
activity in the earth, totally different from that
of the Archaean; they demonstrate a major
change in crustal conditions and serve to separate the permobile late Archaean from the
early–mid-Proterozoic when extensive rigid or
semi-rigid plates had formed.

Two features of the dykes are worth considering in some detail: their structural relationships and their chemical composition.

Structural Relationships

Many of the dykes under discussion are
particularly interesting because they tell us of
the conditions operating within, or on the

margins of, the continental plates in early–
mid-Proterozoic times.

Escher, Jack and Watterson (1976) suggest
that there is a genetic link between the major
dyke swarms, with ages somewhere between
2500 and 2000 my, in Scotland, East and West
Greenland, and Labrador. There are two sets,
trending roughly NE–SW and NW–SE (Fig.
4.2). Their close association with ductile shear
zones in the gneissic wall rocks shows that the
dykes represent a conjugate swarm along
shear fractures rather than along tensional
openings. Escher, Jack and Watterson (1976)
considered the possibility that the dyke
swarms are a reflection of plate margin igneous activity, but concluded that they are, in
fact, intracontinental/intraplate phenomena.

The basic dykes in the Lewisian can be used
to separate the Scourian and Laxfordian tectonic events (Sutton and Watson, 1951).
These Scourie dykes show a great variation in

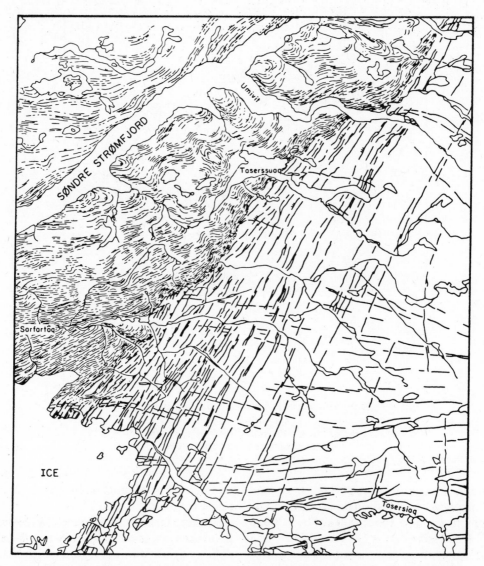

Fig. 4.3. Part of the Kangamiut dyke swarm in West Greenland. Note the two generations of dykes and the more highly deformed zone to the north (after Escher, Escher and Watterson, 1975; reproduced by permission of *Canadian Journal of Earth Sciences*)

structure and petrology throughout the Lewisian—many of these variations are considered to date from the period of intrusion and reflect diversity in their conditions of emplacement at slightly different times and in different parts of the complex (Park and Cresswell, 1972; Tarney, 1973). The dykes were generally intruded at depth into crust that was still hot; Tarney (1963) suggested a temperature of 500°C and a pressure of 5–6 Kb corresponding to a depth of 15–20 km, but O'Hara (1975) calculates more than 8·5 Kb (>30 km depth). In places early members of the suite show evidence of synkinematic intrusion, elsewhere they are regarded as just syn-metamorphic, whilst later members were intruded under cooler, more brittle conditions. Features which suggest that the dykes were intruded at depth into highly ductile crust include: oblique and sigmoidal foliation indicating simple shear of the dyke walls during intrusion, a close relationship with shear zones in the wall rocks, a coarse grain size, and presence of primary hornblende and garnet (Park and Cresswell, 1972; Tarney, 1973). Thus it can be concluded that in these regions the continental plates had not become entirely rigid by the early–mid-Proterozoic.

To what extent early Proterozoic basic dykes elsewhere in the world were intruded under such synplutonic conditions at depth is not known because not many of the dykes have been described in detail. But the Kangamiut and Scourie dykes serve to show that depth of erosion and variable conditions of the host rocks must be taken into account when considering dyke evolution at this time in earth history.

Whether or not the dykes are intracontinental or synplutonic, it seems most likely that the early to mid-Proterozoic basic dykes were intruded in association with stress systems that formed during early abortive attempts to break-up the continental plates.

Chemical Composition

The dykes vary in places from basic to ultrabasic (e.g. Berthelsen and Bridgwater, 1960). The Scourie suite ranges from mafic or noritic to bronzite picrites and olivine gabbros (Tarney, 1973).

Surprisingly few chemical studies have been made of these dykes. One of the most detailed is that of Condie, Barsky and Mueller (1969) who demonstrated that the Wyoming diabases are continental tholeiites depleted in Sr relative to Rb, K, Ca, and usually Ba, and enriched in Rb relative to K and sometimes Ba. Similar depletions and enrichments are apparent in Jurassic continental tholeiites from Antarctica (the Ferrar dolerites) and Tasmania (Compston, McDougall and Heier, 1968). Some dykes contain up to 75% of large plagioclase phenocrysts, and this is consistent with the high K/Sr and Rb/Sr ratios as plagioclase is expected to concentrate Sr relative to the other elements concerned. The rocks define a fractionation trend in which Sr is rapidly depleted relative to Rb and K and enriched relative to Ca (the normal Sr trend). To explain these relationships the authors favoured a model of extensive plagioclase crystallization from a Ca–Al rich primary tholeiite of anorthositic gabbro composition and suggested that the distribution of tholeiite exhibiting this Sr-depletion trend may offer an indirect method of mapping the distribution of Ca–Al rich ultramafic source areas in the upper mantle. Rivalenti and Sighinolfi (1971) and Rivalenti (1975) found that basic dykes from two areas in West Greenland have a composition similar to the Wyoming diabases, especially with regard to a Sr-depletion trend relative to K and Rb.

The Scourie and Kangamiut dykes are high in iron and phosphorus and rather low in potash; the general chemistry of the Scourie dykes is similar to that of the Karroo tholeiites correlated by Cox (1970) with the fragmentation of Gondwanaland (Garson and Livingstone, 1973). The K_2O values for the average Scourie dykes from the Outer Hebrides and the mainland (0·33% and 0·56%), for the Fiskenæsset dykes (0·50%) and for the Kangamiut dolerites (0·34%) are close to the average value of 0·36% for six tholeiites from a volcanic island in the Red Sea median trough (Gass, 1970), whilst the Scourie and Kangamiut dyke chemistry as a whole is close to the range of oceanic tholeiites given by Hubbard (1969), Jakes and Gill (1970) and Cann (1970a), with the exception that the dykes are relatively enriched in iron.

Mineralization

The early Proterozoic dykes and layered complexes have a similar type of mineralization belonging to the Cr–Ni–Pt–Cu association, which occurs in basic-to-ultrabasic host rocks. These rocks and their metals represent a significant influx of material from the mantle into the continental crust at this time.

Intrusion	Host rock	Mineralization	Age (my)
Bushveld, S Africa (Willemse, 1969)	(various)	Cr Fe Ti V Pt Ni	1950
Lynn Lake, Manitoba (Milligan, 1960)	gabbro	Ni Cu	1700
Sudbury, Ontario (Kirwan, 1966)	norite–micropegmatite	Cu Ni Co Au Pt	1720
Sarqâ, South Greenland (Berrangé, 1970)	hornblende peridotite	Pl Au Ag Cr Cu Ni	c. 1800?
Thompson-Moak Lake, Manitoba (Wilson and Brisbin, 1961)	meta-sediments associated with serpentinized peridotite	Ni Cu Co Au Pt	c. 1800
Great Dyke, Rhodesia (Bichan, 1969)	ultramafics	Cr	2550
Stillwater, USA (Jackson, 1969)	ultramafics	Cr	2750
Usushwana, S Africa (Hunter, 1970b)	gabbro	Cu Ni	2870

Considerable evidence, such as shatter cones and quartz fabrics, has come to light supporting a modification of Dietz's original 1964 suggestion that the magmatic events which produced the Sudbury igneous rocks and their ores were triggered off by the fall of a stony meteorite.

Chromite occurs in the Great Dyke as the dominant phase in chromite seams and as disseminated crystals in all olivine-bearing rocks. The ultrabasic cycles ideally have a basal chromite seam and there are eleven such seams in the Hartley complex. Many large layered intrusions have chromite accumulations, and those in the Great Dyke are an indication that mantle fractionation processes responsible for appreciable chromium concentrations were operating by early Proterozoic time.

Some of the most spectacular magmatic ore deposits occur in the Bushveld Complex. According to Willemse (1969) these include:

1. Tin and fluorspar in the granite.
2. Vanadiferous and titaniferous magnetite in the upper layered sequence (anorthosite).
3. Platinum and nickel in the Merensky Reef and in bronzitite pipes.
4. Chromite in the lower layered sequence (pyroxenites, norites and anorthosites).

The Bushveld Granite contains economic tin deposits, largely occurring as disseminated cassiterite and in pipe-like bodies and fissure fillings in the late granites (Hunter, 1973). These are important because, as Watson (1973b) pointed out, the fractionation processes responsible for tin accumulations were rather ineffective and slow to operate in earth history. The Bushveld deposits indicate the processes were operating 2000 my ago and it is probably no coincidence that they were associated with the intensive magma fractionation that gave rise to the Bushveld Complex.

Chapter 5

Early to Mid-Proterozoic Basins and Geosynclines

On the eroded remnants of the Archaean greenstone belts and their granitic rocks and of the high-grade gneisses, there were deposited vast sequences of volcanics and sediments. These rocks herald a new major stage in Earth history as they, and their geotectonic environment, are different from comparable Archaean examples. For the first time asymmetrical fold belts were formed; in fact the first Cordilleran-type geosynclines and aulacogens are recognizable from this period. These are particularly exciting rocks to study as the time boundary between the Archaean and Proterozoic represents the most dramatic and fundamental period of change in the evolution of the continents—far more important than that separating the Precambrian and the Phanerozoic.

Basins in southern Africa

Archaean crustal development in the Kaapvaal craton of southern Africa was completed by 3000 my ago, thus allowing the accumulation of vast thicknesses of unconformable flat-lying sediments and volcanics in broad basins at a time when greenstone belts were still forming in Rhodesia, India, Australia and Canada. Approximately 43 km of sediments and volcanics belonging to the Pongola, Dominion Reef, Witwatersrand, Ventersdorp, Transvaal and Waterberg–Matsap Systems (shown in Fig. 5.1) were laid down between about 3000 my and 1800 my ago; the rates of vertical movement decreased progressively with time from 0·27 (in the Archaean) to 0·023 mm/yr (Table 5.1).

Table 5.1. Estimated rates of vertical crustal movement during deposition of sedimentary–volcanic systems in southern Africa (after Hunter, 1974b; reproduced by permission of Elsevier Scientific Publ.)

Supergroup	Maximum thickness (km)	Time-span (my)	Rate (km/my or mm/yr)
Swaziland	21·3	80	0·27
Pongola	10·6	70	0·15
Witwatersrand/ Ventersdorp	16·7	300	0·05
Transvaal	9	250	0·036
Waterberg	6·5	160	0·023

The Pongola System was deposited in what is now the Swaziland region before 3000 my ago but its age relationship with the Barberton Greenstone Belt is unknown; 2870 ± 30 my ago (Rb/Sr, R. D. Davies and coworkers, 1969) it was intruded by the Usushwana Complex which comprises a layered suite of pyroxenites, gabbros and granophyres in the form of two dyke-like bodies (Hunter, 1970b).

The Dominion Reef, Witwatersrand and Ventersdorp Systems constitute the 'Witwatersrand Triad' (Whiteside, 1970). The major part of the Pongola, Dominion Reef and Ventersdorp Systems consists of extrusive igneous rocks whereas those of the Witwatersrand and overlying *Transvaal and Waterberg Systems* are predominantly sedimentary. In decreasing order of abundance, the volcanic

74

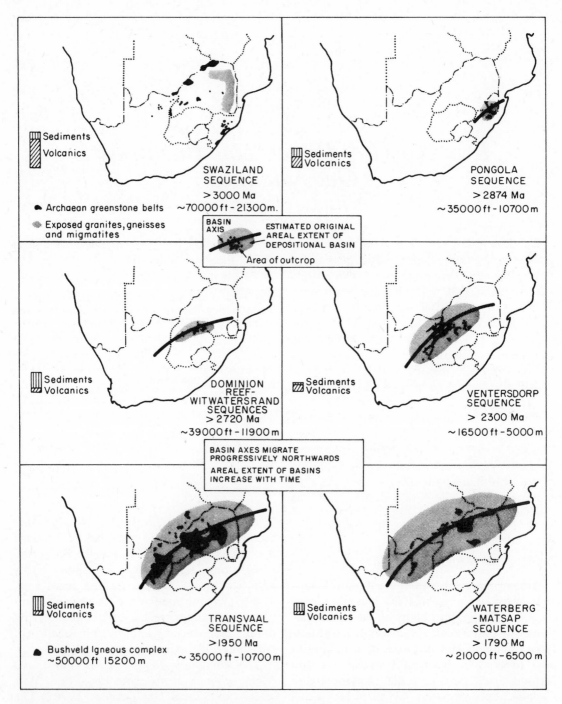

Fig. 5.1. Generalized geological maps depicting the progressive development of cratonic basins from the Archaean to the early Proterozoic in South Africa. Crustal thickening and stabilization results in increased areal extent of each successive stratigraphic sequence. The progressive increase with time in the sedimentary/volcanic ratio is indicated (after Anhaeusser, 1973; reproduced by permission of The Royal Society)

rocks include andesites, rhyolites, trachytes and tholeiitic basalts, and prominent pyroclastic deposits of agglomerate, tuff and tuff breccia. The sedimentary rocks, on the other hand, include conglomerates, orthoquartzites, arkoses, sandstones, shales, dolomites and limestones, cherts and banded iron formations.

Anhaeusser (1973) has shown how the progressive stabilization of the continental crust can be followed indirectly by examining the thickness and areal extent of these basins; their areas of outcrop, estimated depositional boundaries and basin axes are depicted in Fig. 5.1. From the earliest supracrustal rocks in South Africa (the Swaziland System), through the Pongola to the Waterberg–Matsup Systems, Anhaeusser (1973) demonstrated a progressive increase in size of the depositional basins. There is also a noticeable change in the proportion of volcanic to sedimentary rocks, the older Systems having relatively greater volcanic components than the later predominantly sedimentary accumulations. The northwesterly migrating basin axes (see also Hunter, 1974b) might be a reflection of some form of continental accretion about an early Archaean nucleus.

Anhaeusser (1973) demonstrated that strong contrasts emerged between crustal growth in the Archaean and early Proterozoic (Table 19.6). The structural characteristics of the deeply infolded synclinal greenstone belts were not repeated but gave way to broad basins containing flat-lying or gently dipping strata, the main deformation features being mild warping, doming, epeirogenic subsidence, faulting, and dyke invasion; metamorphism was unimportant. The associated sediments are marked by a high proportion of stable platform types, in particular conglomerates, orthoquartzites and sandstones. The volcanic rocks include thick extensive nonsequential varieties embracing continental flood basalts of tholeiitic type, and a calc-alkaline sequence with predominant potash-enriched andesites and rhyolites and important pyroclastics. If a comparison were made with late Phanerozoic volcanics on the basis of general chemistry and relative abundance and association of lava types, this sequence is most akin to the calc-alkaline lavas of the continental margins, such as in the Andes, Cascades and New Zealand, which are also underlain by sialic crust (Baker, 1972). The question arises, what are these types of lavas doing in so-called stable ensialic cratonic basins? Alternatively, could they not be the result of some plate movement and subduction activity not yet recognized?

The Huronian of southern Canada

In the 'southern province' of the Canadian Shield Huronian and Animikie clastic sedimentary and volcanic rocks rest unconformably on Archaean greenstone belt–granite remnants (Fig. 5.2; Card and co-workers, 1972). The deposition of these rocks took place about 2300 my ago; they were intruded by the Nipissing diabases (2150 my), the Mongowin ultramafic pluton (1770 my), the Cutler granitic pluton (1750 my), the Sudbury Nickel Irruptive (1730 my), and many other later igneous bodies. Deformation and metamorphism referred to as the Penokean orogeny took place before the intrusion of the Nipissing diabases; later metamorphism of these diabases has a Rb/Sr isochron age of 1950 ± 100 my, the approximate age of the Hudsonian orogeny.

The sequence is a relatively uniform, southeastward thickening clastic wedge, largely of sandstone, quartzite, pelite and paraconglomerate (tillite) with local basal tholeiitic flood basalts. Current directions indicate that the basins received nearly all of their fill from the adjacent sialic craton to the north, and they therefore do not display the depositional polarity (i.e. platform-derived followed by platform-directed sediments) of Phanerozoic Cordilleran-type geosynclines nor even of the Coronation Geosyncline developed in the period 2100–1750 my ago in NW Canada. They are also marked by a general absence of rocks characteristic of the ophiolite assemblage, and of orogenic and post-orogenic sediments from a tectonic provenance—the greywackes and lithic arenites which characterize flysch and molasse. These unusual features make it difficult to erect a conventional geotectonic environment for the belt. Probably the most peculiar feature is the lack of a bordering eugeosynclinal facies. Frarey and Roscoe (1970) postulate that the Huronian depositional basin was an exogeosyncline in the form of an unstable, tectonically-controlled platform developed along the perimeter of the Archaean protocontinent. The Huronian is thus viewed as representing the first development of platform-type sedimentation due to increased crustal thickening and stability following the Archaean 'mobile phase'—in other words comparable with the Witwatersrand Triad. Dietz and Holden (1966) made the point that the Huronian represents a miogeosyncline developed at a continental crust–ocean crust interface, analogous to the clastic wedge on the continental shelf off eastern America today.

The southern part of the fold belt was affected by the Penokean orogeny prior to the intrusion of the Nipissing diabases. The grade of regional metamorphism and the intensity of deformation generally increase southwards and with increasing age of rock units. Geothermal gradients were steep, metamorphism reaching the almandine amphibolite facies of the low pressure–intermediate temperature type, as indicated by the coexistence of staurolite and andalusite.

The Circum-Ungava Geosyncline, Canada

The geosyncline is made up of the Labrador Trough, the Cape Smith Belt and the Belcher Fold Belt surrounding the Archaean Ungava Craton (Fig. 5.2) (Dimroth and coworkers, 1970; Davidson, 1972); it was interpreted as a suture between colliding continents by Gibb and Walcott (1971). The structure is internally asymmetrical, the sedimentary western zone towards the craton comprising much orthoquartzite, dolomite and iron formations, and the predominantly volcanic eastern zone away from the craton composed of mafic volcanics and intrusives and shales and greywackes. A medial geanticlinal ridge separates these two zones. Sediment input was mainly from the

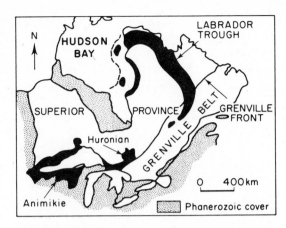

Fig. 5.2. Map showing the distribution of the Huronian and Animikie Systems and the Circum-Ungava geosyncline in SE Canada (compiled from Dimroth and coworkers, 1970; Card and coworkers, 1972; reproduced by permission of The Geological Survey of Canada)

cratonic side during the early stages and from the orogenic side during the later history.

The volcanics are compositionally very primitive, being more basic than the mafic fraction of Archaean volcanic belts. The fact that they contain hardly any andesites, dacites or rhyolites contrasts markedly with the Archaean volcanic rocks which typically evolved towards acidic end fractionates. Frequency distribution curves for differentiation index and for K_2O, Na_2O and TiO_2 contents show that the Labrador Trough lavas have a compositional spread that is closer to modern oceanic tholeiites than to Archaean tholeiitic basalts.

Three cycles of sedimentation and volcanism are recognized in the Labrador Trough. In the western zone these cycles begin with an orthoquartzite–carbonate sequence and terminate with shales and flysch-like sediments. Extensive mafic effusive and hypabyssal rocks interfinger with the shales of the upper parts of the cycles and constitute the bulk of the eastern zone.

Both zones overlie Archaean basement gneisses. On the western side high-grade gneisses have not been affected by the Hudsonian orogeny and are overlain unconformably by the sediments. On the eastern side the granitoid basement gneisses have been reworked during the Hudsonian and are difficult to distinguish from highly meta-morphosed, early Proterozoic sediments.

During the Hudsonian orogeny collision of the two continental plates thrust the geosyncline westwards onto the Ungava foreland with the result that the intensity of deformation and metamorphism increases eastwards across the geosyncline. Seguin (1969) concluded that the width of the Labrador Trough has been shortened by slightly more than 50%. In pelitic rocks with an amphibolite grade, biotite, garnet, staurolite, kyanite and sillimanite are indicative of a high-pressure Barrovian-type facies series, which is consistent with the interpretation of the geosyncline as a continent–continent collision boundary. The main metamorphic period occurred between 1600 and 1750 my ago. The palaeomagnetic data of Irving and Lapointe (1975) suggest that the bordering cratons did not move substantially during the ocean closure and therefore the best modern analogue is a Red Sea-type narrow rifted basin. Kearey (1976) uses gravity data to suggest that a small ocean was closed by subduction beneath the Churchill belt to the east, thus accounting for the eastward increase in metamorphism and deformation.

The Coronation Geosyncline, Canada

In the NW Canadian Shield there are the well-preserved remnants of a northerly-trending orthogeosyncline, developed in the period 2100–1750 my ago, termed the Coronation Geosyncline (Hoffman, Fraser and McGlynn, 1970; Fraser, Hoffman and Irvine, 1972; Hoffman, 1973; Hoffman, Dewey and Burke, 1974). This is a unique structure, its primary components being remarkably similar to those of late Phanerozoic Cordilleran-type geosynclines, and it is the main basis for the suggestion that continent–ocean plate movements were in operation 2000 my ago; it is, therefore, worth looking at in some detail.

The geosyncline is situated between the Slave Province (the craton) to the east, and the Bear Province (orogenic belt) to the west. Stratigraphic thicknesses of 10,700 m are

reached in the geosyncline, with as little as 2000 m over the craton. Fig. 5.3 shows how it is divisible into continental platform, shelf and rise and batholithic segments.

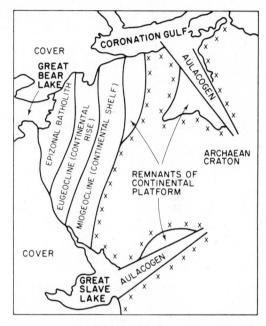

Fig. 5.3. Tectonic subdivisions of early Proterozoic rocks of the Coronation Geosyncline and associated aulacogens in the NW Canadian Shield (compiled from Fraser, Hoffman and Irving, 1972; Hoffman, 1973; reproduced by permission of The Geological Association of Canada)

Hoffman distinguished seven main phases of deposition (Fig. 5.4):
1. Pre-quartzite phase—a basal terrigeneous sequence of mudstones, siltstones and arkoses, with fragmental and pillow basalts, basic ash-fall tuffs and stromatolitic dolomites.
2. Quartzite phase with cross-bedded orthoquartzites, mudstones and quartz pebblestones in a westerly-thickening wedge of mature terrigeneous sediments derived from the platform to the east.
3. Dolomite phase of cyclic cherty stromatolitic dolomite and calcareous mudstone constituting a westward-facing shallow-water dolomite shelf.
4. Pre-flysch phase of black pyritic mudstones—a euxinic starved basin sequence

78

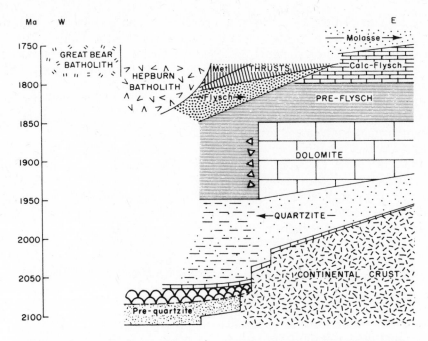

Fig. 5.4. Restored time–distance cross-section of the Coronation Geosyncline showing the 7 main phases of deposition. Arrows indicate the directions of sediment transport. For individual rock types see text (after Hoffman, 1973; reproduced by permission of The Royal Society)

derived from the craton and accumulated during foundering of the continental shelf.

5. Flysch phase—a westward-thickening clastic wedge of coarse greywacke turbidites derived by erosion of the batholithic rocks to the west.

6. Calc-flysch phase of rhythmically interlaminated limestone and mudstone with fine greywacke turbidites derived from the orogenic belt to the west and deposited in a shoaling marine trough.

7. Molasse phase of immature terrigenous sediments derived from the west and deposited in an alluvial environment—crossbedded red sandstone and mudstone.

The early pre-quartzites, quartzites and dolomites were derived from a stable Archaean craton to the east and were deposited in a westward-facing continental shelf. The main turning point in the depositional history of the geosyncline was the foundering of the continental shelf: as it subsided into deeper water it was mantled by starved black mudstones of the pre-flysch phase. The emergence of a tectonic land mass in the orogenic belt to the west caused a reversal of the sediment source direction: the flysch, calc-flysch and molasse phases constitute a clastic wedge foredeep (exogeocline) derived from the west.

During the orogenic stage the supracrustal rocks were tectonically transported toward the platform (Fig. 5.5). In the west, the continental rise and clastic wedge sequences are deformed with a penetrative cleavage and recrystallized by regional low-pressure metamorphism with formation of cordierite, andalusite, and/or sillimanite schists and garnet amphibolites. To the east, unmetamorphosed continental shelf and clastic wedge sequences have been flexurally folded and overthrust eastwards over a basal décollement. East of this thrust zone thin platform deposits are nearly horizontal.

The important point is the similarity between this depositional–structural history and that of Cordilleran-type geosynclines, such as in western North America, which are characterized by early platform-derived sedimentation, by clastic wedges shed from

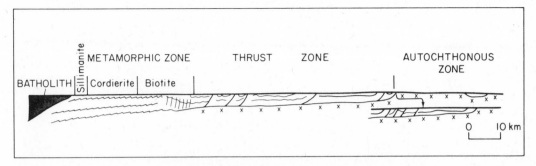

Fig. 5.5. Structural cross-section across the Coronation Geosyncline (after Hoffman, 1973; reproduced by permission of The Royal Society)

magmatic arcs towards the platform on one side and the oceanic trench on the other, and by platform-directed overthrusting. The only elements that are missing in the Coronation Geosyncline are the fossil trench and the trench-directed clastic wedge, presumably buried beneath Palaeozoic cover to the west. A crucial element in the recognition of this type of fold belt is the presence of the synorogenic clastic flysch wedge as this is the sedimentary record of orogenesis in the geosyncline; the molasse was deposited synchronously with the major period of uplift.

The Mount Isa Geosyncline, Australia

This geosyncline lies along the eastern margin of the early Proterozoic craton at the southern end of the McArthur Basin; it contains Carpentarian (c. 1800–1500 my) supracrustal rocks made up of a lower Tawallah Group of orthoquartzites and basic volcanics (Eastern Creek Volcanics) and an upper McArthur Group of fine-grained carbonate sediments. Dunnet (1976) suggests that the district lies in the vicinity of a triple junction. The Paradise Rift is an aulacogen or failed rift arm that received much deltaic sand deposition and basic volcanism, and there is a suture between the cratonic boundary domain to the west and the mobile belt to the east. Dunnet considers that the mobile belt involved only a minor rift-type separation because oceanic crustal material is absent, there is continuity from the Carpentarian platform-type sediments into the mobile belt and because absence of deep marine sediments and a tran-

sition from volcanic-arenite to carbonate reef and evaporite sedimentation through time implies a relatively minor separation period prior to closing of the two plates.

The First Aulacogens

Aulacogens are large, long-lived, graben-like trenches first described by Shatsky (1955) in the Russian platform and by Salop and Scheinmann (1969) in the Siberian platform (Fig. 5.6). They extend from the early Proterozoic geosyncline into the Archaean craton, often with a radial disposition, and received thick sequences contemporaneous with the miogeosyncline flanking the craton.

The formation of aulacogens requires the existence of stable continental platforms and so they could not form in the unstable Archaean terrains; they first appear in early Proterozoic times and constitute an excellent tectonic marker of the progressive cratonization of the continents. They are common in Proterozoic and Phanerozoic platforms and some have formed recently, opening not into a geosyncline but into an ocean, e.g. the Benue Trough (Burke and Dewey, 1973a) (see further in Chapter 14).

Detailed work by Hoffman (1973) and Hoffman, Dewey and Burke (1974) in an aulacogen bordering the Coronation Geosyncline (Fig. 5.3) has revealed the following characteristics and evolution. The aulacogen is a deeply subsiding trough in which sedimentary rocks, much thicker than on the adjacent platform, accumulated during every phase of the geosyncline. The sediment

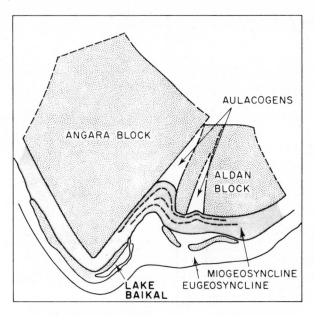

Fig. 5.6. Two early Proterozoic aulacogens in rifted Archaean blocks bordered by early Proterozoic mio- and eugeosynclines, Siberia (redrawn after Salop and Scheinmann, 1969; reproduced by permission of L. J. Salop)

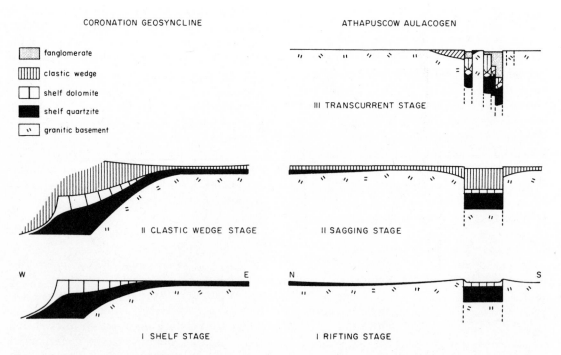

Fig. 5.7. Three-stage evolution of an aulacogen and its relation to the evolution of the Coronation Geosyncline (after Hoffman, 1973; reproduced by permission of The Royal Society)

transport was longitudinal along the length of the trough, the sediments increasing in thickness longitudinally towards the platform margin where basic volcanics occur at five horizons and where tonalite-to-granodiorite laccoliths were intruded, attesting to the great depth of the bounding faults.

During the shelf stage in the geosyncline, the aulacogen was in an incipient rifting stage (Fig. 5.7) during which the lips of the fault–trough stood high and shed thin quartzitic sediments into the trough, and subalkalic basalts were extruded, as might be expected in an active rift with thinned crust. As the continental shelf foundered and was buried by the clastic wedge, the aulacogen passed into a sagging stage during which time the lips were depressed and covered with thick mudstones, greywacke turbidites and redbeds in a broad downwarp during the flysch to molasse phases of the geosyncline. Alkalic basalts were extruded, as might be expected in an area of thickened crust. Finally, the aulacogen was the site of regional transcurrent faulting giving rise to scattered down-dropped basins in which thick alluvial sediments were deposited.

Tillites

The Gowganda Formation in Ontario of Huronian age has been recognized as a glacial deposit since the beginning of this century, but it is only recently that further examples have been demonstrated, indicating a more extensive early Proterozoic glaciation; they are, however, so far only known in the following sequences in North America and South Africa:

Gowganda Formation Ontario	Schenk, 1965; Frarey and Roscoe, 1970
Chibougamau Series, Quebec	Young, 1970
Fern Creek and Enchantment Lake Formations, Michigan	Church and Young, 1970
Reany Creek Formation, Michigan	Puffett, 1969
Padlei Formation, Hurwitz Group, NW Territories	R. T. Bell, 1970
Headquarters Schists, Wyoming	Young, 1970
Pretoria Series, Transvaal, S Africa	Cullen, 1956
Griquatown Series, Cape Province, S Africa	Cullen, 1956
Witwatersrand System, S Africa	Wiebols, 1955

Most of the North American occurrences lie in a belt nearly 3000 km long from Chibougamau to Wyoming (Young, 1970), but the most detailed account of the glacial rocks and structures is from the Gowganda Formation.

The Gowganda paraconglomerates are 2288 ± 87 my old (Fairbairn and coworkers, 1969), 15–200 m thick, and occupy an area of about 20,000 sq km. The primary evidence in support of a glacial origin for these beds is:

1. Massive polymict conglomerates with exotic clasts up to boulder size, mainly of pink granite, variably distributed and concentrated.
2. Finely laminated argillites (varves) commonly with dropped clasts of plutonic rocks.
3. Presence of a striated basement.

The Animikie tillites of Michigan lie directly on the Archaean basement, in contrast to the Huronian Gowganda which overlies 4000 m of older Huronian sedimentary formations. The Reany Creek and Fern Creek Formations are correlated with the Gowganda. In the Medicine Bow Mountain region of SE Wyoming there is an 8 km thick sedimentary succession that includes polymictic conglomerates and laminated dropstone argillites (the Headquarters Schist), correlated with those of the Gowganda Formation. The glacial deposits in both regions are overlain by four very similar sedimentary formations.

With the exception of the Hurwitz Group all the Huronian age glacial deposits lie near the southern border of the Superior Province and palaeocurrent measurements indicate that both the sediments and glacial deposits were derived from the north. Palaeomagnetic results by Symons (1967) suggest that these regions lay in high latitudes in early Proterozoic times. The tillite matrix materials and associated argillites have high soda/potash ratios which are thought to be indicative of lack of chemical weathering due to the existence of glacial conditions (Young, 1970).

Little detail is known about the South African glacial sediments. Those in the Griquatown Series are less than 30 m thick but extend through an area of 20,000 sq km, and there are two or more tillites, 180 m apart stratigraphically, in the Government Reef Series of the Witwatersrand System.

Banded Iron Formations (BIF)

Derry (1961) was one of the first to notice that Precambrian banded iron formations are of about the same age on many continents. He realized in particular that they reached their peak of development in early Proterozoic basins or geosynclines situated near the boundaries of the Archaean cratons—the greatest development occurred between 2600 and 1800 my ago (Goldich, 1973). According to Ronov (1964), the BIF account for 15% of the total thickness of early Proterozoic sedimentary rocks.

Important examples occur in:

Labrador Trough, Canada	Dimroth, 1968
	Gross, 1968
	Fryer, 1972
Animikie Basin, USA	Gundersen and
	Schwartz, 1962
	Trendall, 1968
Hamersley Group, W Australia	McLeod, 1966
	Trendall, 1968
	Trendall and
	Blockley, 1970
	Ayres, 1972
	Trendall, 1973a
Transvaal System, S Africa	Trendall, 1968
Griquatown Formation, Cape Province, S Africa	Trendall, 1968
Mauritania, W Africa	Dorr, 1965
Minas Gerais District, Brazil	Barbosa and
	Grossi Sad,
	1973
	C. H. Maxwell, 1972
Krivoyrog, Ukraine	Tugarinov,
	Bergman and
	Gavrilova, 1973
India	Krishnan, 1973
Baltic Shield	Chernov, 1973
	Frietsch, 1973

For a general review of the character and possible modes of evolution of these BIF see Lepp and Goldich (1964), Gross (1965), Govett (1966) and Stanton (1972).

The Proterozoic BIF belong to the Superior type of Gross (1965) who summarized their main features. They are thinly laminated rocks, mostly belonging to the oxide, silicate and carbonate facies. They rarely contain clastic material and are stratigraphically associated with chert, dolomite, quartzite,

black carbon-bearing shale, slate and volcanic rocks. According to Gross (1965) the sequence dolomite, quartzite, red and black ferruginous shale, iron formation, black shale and argillite, in order from bottom to top, is common with local variations on all continents; Trendall (1968), however, pointed out that in the Hamersley, Animikie and Transvaal basins cherts locally underlie the iron formations, but otherwise the BIF 'do not consistently follow or precede any other sediment type'. They are not associated with as many volcanic rocks as are the Archaean iron formations, but there are normally volcanics somewhere in the succession, although their relative positions in the stratigraphy are highly variable. For example, the Lower Griquatown BIF is overlain by the 1200 m thick Ongeluk volcanics, the Hamersley BIF is underlain by the volcanic Fortescue Group and is interbedded with the 500 m thick Woongarra acid volcanics, and the Ironwood BIF, East Gogebic Range, Minnesota is interbedded with volcanic rocks (Trendall, 1968).

The sedimentary rock associations indicate that the BIF were deposited on continental shelves or in miogeosynclines. They were probably laid down in less than 200 ft of water and maybe even above wave base (Dimroth, 1968). There is general agreement that they are chemically precipitated sediments.

Trendall (1968) quoted the following data on the thickness of some BIF:

	ft
Animikie	
Gunflint Range	450–550
Cuyuna Range	0–450
Gogebic Range	650
Menominee Range	650
Marquette Range (Negaunee BIF)	>2000
Hamersley	
Brockman BIF	>2000
Total aggregate thickness	3000
Lower Griquatown	c.1500
Upper Griquatown	500
Transvaal Dolomite Series	700–800

The BIF typically extend for many hundreds of kilometres (cf. the Archaean BIF) outlining the former extent of the sedimentary basins. Gross (1965) suggested that they were once present around the entire shoreline of the

Ungava craton for a distance of more than 3200 km.

The distinctive character of the internal subdivisions of the BIF enables them to be recognized and correlated over considerable distances. Individual parts of the main Dales Gorge Member of the Hamersley Brockman BIF can be correlated at the 1 inch scale over about 20,000 sq miles (Trendall and Blockley, 1968) and correlations of varves within chert bands can be made on a microscopic scale over 185 miles (Trendall, 1968). According to Gundersen and Schwartz (1962), a 5 ft thick algae horizon in the Biwabik BIF of the Mesabi Range can be followed over 50 miles, and Trendall (1968) stated that in the Griquatown BIF stilpnomelane bands, a few inches thick, can be identified in boreholes 40 miles apart. These examples serve to demonstrate the fact that early Proterozoic sediments are incredibly well preserved and can be intercorrelated, not only on a large, but even on a small scale over enormous distances.

Blue–green algae and fungi have been identified in the non-ferruginous cherts of the Gunflint iron formation of Ontario (Tyler and Barghoorn, 1954) and some of these fossilized structures resemble modern-day iron-precipitating bacteria such as Sphaerotilus, Gallionella and Metallogenium (Cloud, 1965). It is known that these bacteria are able to grow and precipitate ferric hydroxide; they live in ferruginous water, particularly where bog-iron ore is forming. Kuznetsov, Ivanov and Lyalikova (1963) considered that they first assisted in creating a reducing environment in the subsoil by causing reduction of fixed ferric to mobile ferrous iron, and then they assisted in the reoxidation of the ferrous iron when it reached the oxidizing zone. However Trudinger (1971) warned that the evidence for such bacterial participation in the precipitation of sedimentary iron ores is still fragmentary, although the bacteria might be expected to accelerate the process of the iron deposition. We are therefore left with the tempting, but unproven, concept that biological activity may have assisted the deposition of the early Proterozoic, and presumably the Archaean, iron formations. It was during the early Archaean more than 3700 my ago (Junge and coworkers, 1975) that algal photosynthesis began and oxygen became available for the oxidation of ferrous iron in the oceans. The concentration of atmospheric oxygen was extremely low—Berkner and Marshall (1965) calculated that it remained at between 0·001 and 0·0001% of the present atmospheric level for 2000 my after algal photosynthesis began—but the modern iron bacteria are able to oxidize ferrous iron at such low levels of concentration. Primitive iron bacteria therefore may well have accelerated the precipitation of iron deposits at a time when the prevailing low oxygen levels probably restricted the rate of purely chemical oxidation (Trudinger, 1971). But the fact that red arkoses and conglomerates containing andesite pebbles with oxidized weathering crusts, and red pelites underlie the Sokoman BIF in the Labrador Trough means that oxidizing conditions must have been reasonably high during deposition and diagenesis (Dimroth, 1975).

One of the most long-standing controversies regarding the evolution of Precambrian BIF centres around whether the iron and silica were derived by erosion from adjacent land-masses, or by submarine volcanic exhalations.

There is one major obstacle to the continental erosion theory. If enormous quantities of iron and silica were transported from the continents, large amounts of other detrital chemical constituents, in particular aluminous material, must have been either left behind as residual laterite deposits, or transported and dispersed in the sea with, or not far from, the iron deposits. However no such laterites have been recognized, the alumina content of BIF is extremely low, and there are very few nearby aluminous shales. Although it is true, as Stanton (1972) indicated, that the sulphide facies of iron formations is in fact a pyritic shale with 6–7% alumina, this is of relatively minor importance volumetrically compared with the other three facies types.

If the weathering–erosion hypothesis proves untenable, the alternative lies in the familiar volcanicity associated with the BIF.

The most likely modern analogue of the Lower Proterozoic iron formation basin is the Red Sea with its FeMn-rich sediments, volcanics, reef limestones, hot brines and saline evaporites (Degens and Ross, 1969). H. L. James (1969) pointed out that the Red Sea sediments are chemically similar to the haematite-type iron formations and that if they were chemically differentiated through dewatering and diagenetic processes they would yield an end product comparable to the iron formations. Both the Red Sea sediments and iron formations are commonly enriched in manganese. Reef limestones occur beneath and are laterally continuous with the recent sediments and they overlie evaporite beds. Trendall (1973a) suggested that the rhythmic microbands of the iron formations of the Hamersley Group are so close to evaporitic varves that a common origin can hardly be doubted; indeed evaporites are the only comparable type of more recent sediment. He argued that the banding of the iron formations originated by the annual accumulation of iron-rich precipitates whose deposition was triggered by evaporation from a partially enclosed basin with an average water depth of about 200 m. Thus a comparison with a Red Sea environment seems reasonable. This conclusion, based on sedimentological evidence, is corroborated by the palaeomagnetic data of Irving and Lapointe (1975) which indicate the geosynclines of Canada that bear the iron formation (such as the Labrador Trough) formed intracratonically, the most likely modern analogue being a Red Sea-type narrow basin rift system.

But there are other ideas to explain the problematical rhythmic alternation or episodicity of the iron-rich and iron-poor layers of the BIF, and what significance do they have for the crustal evolution at this time? The following explanations have been proposed:

1. They formed as varves in tropical lakes due to annual variations in temperature (Eugster, 1969). Millimetre-scale microbands could be the result of annual variations in precipitation with mesobands reflecting a 25-year climatic periodicity, both variations being controlled by the history of the earth's revolution around the sun (Trendall, 1972).

2. Microbanding may be due to daily photosynthetically-controlled variations in pH and Eh in hot springs (Walter, 1972b).

3. They may be annual varves deposited in a glacial environment (Govett, 1966). This idea was taken up by Cloud (1973) who pointed out that in North America at least there may have been a causal relationship between the BIF and the preceding or contemporaneous tillites (see preceding section of this chapter). Assuming such a possibility the ferrous iron, accumulating in the lower levels of a formerly stagnant hydrosphere, would be turned over and circulated into the photic zone by the onset of a temperature gradient caused by melting of glacial ice. The upwelling iron-rich waters would flood continental margins where episodic growths of their phyto-planktonic microbiota gave rise to the BIF.

The glacial connection may explain a further factor: that BIF largely lack clastics. Continental ice-sheets are responsible for very deep erosion (White, 1973); they preceded the BIF and gave rise to low-level terrains which contributed very little land-derived clastic debris to the site of deposition of the BIF.

Goodwin (1973b) plotted about 100 early Proterozoic BIF on a pre-Cretaceous reconstruction of the continents and found that they are aligned along a distinct curvilinear trend. Because the Permo-Trias distribution of the continents is hardly applicable to the early Proterozoic, I have plotted Goodwin's trend on Piper's (1976b) Proterozoic supercontinent (Fig. 8.6). The trend might possibly indicate the position of an early Proterozoic plate boundary, but if so, we are not sure of the width of the ocean concerned. If it were narrow, the BIF global trend might outline the distribution of a middle Precambrian Red Sea-type rift system. Following on the Archaean–Proterozoic transition to a more stable continental environment there might well have been an incipient development of oceans along primitive rift systems. Alternatively, if the BIF were formed at an accreting plate margin, and if there were subsequently extensive ocean floor spreading, many of them should have been subducted and others, perhaps, obducted at continental boundaries. But the BIF of the Labrador Trough, for example, which apparently formed at a continent–ocean interface, were neither subducted nor obducted. So perhaps the oceans concerned were originally narrow rifts (Irving and Lapointe, 1975), closed by only a small amount of deformation (Kearey, 1976), allowing the BIF to be remarkably well preserved at their original craton boundaries.

Red Beds

During Proterozoic time a new type of sediment evolved: the red bed which is largely absent in Archaean sequences. This is a sandstone with a haematite pigmenting agent commonly forming a coating around the sand grains.

Representations of the earliest red beds, approximately 2000–1700 my old, occur on several continents:

further discussion of their environment, relationship to older and younger igneous rocks, and their role in the evolution of the Proterozoic mobile belt of the northern hemisphere, see Chapter 6.

The last great episode of banded iron formations, about 2000 my ago, marks the excess accumulation of O_2 in the oceans. After this time O_2 began to escape from the hydrosphere and to invade the atmosphere. According to Cloud (1968b), in the absence of an ozone screen at this time ultraviolet light in the range of 2000–2900 Å impinged on the Earth's surface and converted some of the escaping O_2 to O and O_3 (ozone). Since the reaction rates with respect to surface materials of both these products are many orders of magnitude greater than those of O_2, surface oxidation rates would have been high even in such early tenuous oxygenic atmospheres (Berkner and Marshall, 1967). Thus red beds, enriched in ferric oxides, appear from this time.

There is general agreement that red beds

System and Area	Age	Reference for Age
Waterberg System, S Africa	>1790 (U/Pb)	Oosthuyzen and Burger (1964)
Martin Formation, N Canada	1635–1835 (K/Ar)	Fraser and coworkers (1970)
Dubawnt Group, N Canada	1716 (mean K/Ar) 1732 ± 9 (Rb/Sr isochron)	Fraser and coworkers (1970)
Echo Bay–Cameron Bay Groups, N Canada	c. 1700–1800 my (K/Ar)	Fraser and coworkers (1970)

Thick red bed sequences continued to form in mid to late Proterozoic time:

Roraima Formation, Suriname, S America	1599 (Rb/Sr)	Priem and coworkers, 1973
Jotnian, Baltic Shield	1300–1500	
Torridonian, NW Scotland	935–751 (Rb/Sr)	Moorbath, 1969
Bathurst region, N Canada	c. 1200 (K/Ar)	Fraser and coworkers, 1970

The red beds are molasse-type immature sediments deposited in shallow water. They are particularly well developed in northern Canada where they have a wide time range from late Aphebian (c. 1800–1640) to Neohelikian (at least 1200 my). Fraser and coworkers (1970) demonstrated that they were deposited in fault-controlled intracratonic basins, which implies the existence of a broad stable platform throughout most of Helikian time (1640–880 my). In other parts of the world they are likewise thought to have formed under stable cratonic conditions. For

require oxygenous conditions for their formation, but there is disagreement about when the ferric oxides formed. Traditionally it is believed that the soil in most tropical regions is enriched in ferric oxides, which is removed by erosion, transported and introduced to drier lowlands in the amorphous state where it is transformed to crystalline haematite. However, from a detailed study of late Cenozoic red beds in Baja California, Mexico, T. R. Walker (1967) concluded that the formation of the ferric oxide pigment was caused by oxygenated pore solutions during diagenesis.

He demonstrated that the rocks were progressively reddened by the passage of time and therefore that the climate in the erosional source area was irrelevant to the formation of the haematite-cemented sandstones. In comparison with these Phanerozoic examples, very little is known of the oxidation conditions and local environments of Proterozoic red beds.

Mineralization

In this section we shall consider some of the principal early–mid Proterozoic mineral accumulations. These are important, not only because of their economic value, but also because they tell us a great deal about the depositional environments and the conditions operating in the atmosphere and oceans in this critical period in the evolution of the crust—critical because they represent a dramatic change from those operating in Archaean times.

Gold and Uranium in Conglomerates

The Huronian Supergroup in Canada contains major uranium-bearing conglomerates (Roscoe, 1968) and the Jacobina Series in Brazil has conglomerates enriched in gold. All the systems of the Witwatersrand Triad (Feather and Koen, 1975; Pretorius, 1975) and the Transvaal in South Africa have conglomerates enriched in both elements, and another, little quoted, example in this context is the detrital gold in the conglomerates of the Homestake Mine in the Black Hills of South Dakota (Ridge, 1972). Uranium-bearing sedimentary deposits also occur at Rum Jungle in northern Australia and in the Tarkwaian Group of West Africa. These deposits represent an important, even unique, metallogenic event in the early Proterozoic as this type of metal concentration is rare in the Archaean (e.g. the Bababudan conglomerates of India).

The conglomerates contain a great variety of detrital minerals within the rock matrix, in particular pyrite, gold and uraninite. A widely recognized view is that the detrital minerals

were derived by erosion of the earlier greenstone terrains and concentrated by fluviatile and deltaic processes in a shallow-water high-energy environment; minor later modifications gave rise to small ore-bearing veinlets. 'Deposition took place along the interface between a fluvial system that brought the sediments and heavy minerals from an elevated source area and a lacustrine littoral system that reworked the material and redistributed the finer sediments along the shoreline. The goldfields were formed as fluvial fans around the periphery of an intermontane, intra-cratonic lake or shallow-water inland sea'

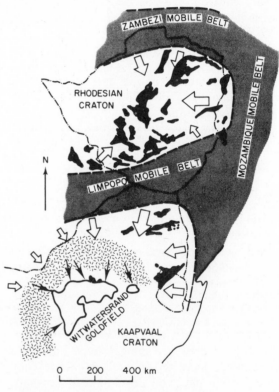

Fig. 5.8. Schematic map of southern Africa showing the provenance area of the Witwatersrand gold field in relation to the gold-bearing Archaean greenstone belts (after Anhaeusser, 1976; reproduced by permission of *Minerals, Science and Engineering*)

(Pretorius, 1975). Fig. 5.8 shows how the Witwatersrand placer gold province is located within the Kaapvaal craton, the entry points of fan conglomerate deposits being related to a provenance area containing uplifted gold-bearing Archaean greenstone belts.

Finally, the pyrite and uraninite give us further information on the early Proterozoic oxygen balance as they have escaped weathering by oxidation. According to Stanton (1972) the most likely explanation for this seems to be a very rapid accumulation and burial of the whole clastic assemblage not far from its erosional source at this time in earth history, when the oxygen content of the atmosphere and the prevailing surface temperature were low.

Manganese in Sediments

There were appreciable accumulations of manganese particularly in carbonate sediments (limestones and dolomites), in the period 2000–2300 my. In the following account it is interesting to see what different geological processes they later suffered.

In South Africa there are large manganese deposits associated with the roughly 2000 my old Transvaal and Loskop Systems (de Villiers, 1960). The manganese concentrations have two main modes of occurrence and host rock. First there are the basal ferruginous shaly beds of the Gamagara Formation of the Loskop System at the contact with the Dolomite Series of the Transvaal System; the second host is a siliceous (cherty) or banded ironstone cataclastic thrust breccia, up to 30 m thick, in close proximity to the Dolomite but formed as a result of post-Waterburg low-angle thrusting. In both cases the origin of the manganese is thought to lie in the Dolomite Series where it was deposited under supergene conditions by replacement of the banded ironstone, shale and breccias. However the manganese ore, although occurring in tectonic breccias, is itself undeformed; thus it is concluded that the dolomite was decomposed and the ore-bearing solutions deposited in the breccias during the Carboniferous ice-age and in post-Karroo times.

In South America manganese deposits occur in:

1. The Serra do Navio Schists (Scarpelli, 1973) of the Amapá Series in the Guiana Shield (Brazil), which underwent a high-grade metamorphic event at least 2000 my ago (Cordani, Melcher and de Almeida, 1968). The main ores are lenses of carbonate or garnetiferous mangano-schist (with spessartine–rhodonite and Mn-rich olivine) within graphitic quartz–biotite–garnet schists, according to information supplied by Industria e Comercio de Mineiros SA (1966). Manganiferous clayey limestones were apparently metamorphosed in the sillimanite–almandine subfacies of the almandine amphibolite facies.

2. In the Minas Gerais district of Brazil where the ores occur in lenses and bands intercalated with 2400 my old schists, gneisses and amphibolites.

Syngenetic manganese deposits (Roy, 1966) occur in the Aravalli Group in India which is dated at 2000–2500 my (Sarkar, 1972). The ores occur as folded beds with phyllites, quartzites and dolomitic limestones, regionally metamorphosed in the chlorite and biotite zones and also, within impure limestones, thermally metamorphosed by a biotite granite. The manganiferous beds are little more than a metre thick but, like the South African and South American ores, they indicate an appreciable accumulation of manganese in a sedimentary environment not long before 2000 my ago.

Lead–Zinc in Carbonates

By 1700 my ago geosynclines similar to those of today were well developed. Cordilleran types had shelf and rise sequences with algal carbonate reefs and platforms. The carbon dioxide content in the oceans had increased to such an extent that thick dolomite sequences were deposited and biological activity was well advanced, caused by the oxygen increase produced by the proliferation of blue–green algae which probably entrapped carbonates. Thus we find the relatively sudden appearance of algal reefs and stromatolitic dolomites, especially in the shallow-water platforms. These features combined to give rise to the first favourable environment for major lead–zinc sulphide accumulations.

The minor concentrations of sphalerite–galena in the 2300 my old Dolomite Series of the Transvaal Supergroup, South Africa, are some of the oldest ores of this type in the world. They occur in gold-bearing veins and as replacement deposits below impervious layers of chert and shale (Haughton, 1969) and Pb and Zn are enriched in the Malmani dolomite (Button, 1975). The lead–zinc–dolomite association did not reach maturity, however, until about 1700–1600 my ago in the Mount Isa Geosyncline, the McArthur Basin in Australia, and the Black Angel deposit in West Greenland.

The McArthur Basin contains at least eight important stratiform base metal deposits including McArthur River (Pb–Zn–Ag) and Mount Isa (Pb–Zn–Ag and Cu). The McArthur River ore member (Cotton, 1965) is a black carbonaceous dolomitic pyritic shale which overlies vitric tuffs. According to Cotton the local sequence contains numerous algal stromatolite biostromes and algal reef dolomites, diagnostic of a shallow marine or intertidal environment, whilst salt crystal pseudomorphs and mud-crack impressions indicate complete marine withdrawal.

The Mount Isa ores occur in the Mount Isa geosyncline (Bennett, 1965). There are two types of ore in a shale–dolomite–siltstone sequence with cross-bedding, convolute bedding and slump breccias: copper in a chalcopyrite-rich silica-dolomite which is an algal reef and reef breccia, and lead–zinc in the famous galena–sphalerite-rich black carbonaceous Urquhart shale which interdigitates with the reef deposits and is regarded as an organic product of off-reef sedimentation. Williams (1969) considered that the ores were deposited in a broadly volcanic environment, the lower part containing the Eastern Creek Volcanics whilst acid tuffs are dispersed throughout the ore-bearing strata. He suggested that the metals are of exhalative-sedimentary origin and were introduced into a basin of deposition by submarine fumaroles.

The ores show a close spatial relationship with a major triple junction and the Paradise aulacogen (Dunnet, 1976).

The Black Angel lead–zinc deposit occurs within the 1·3 km thick metamorphosed tremolite-bearing dolomitic marble of the Marmoralik Formation (which has a minimum age of 1700 my) in the Umanaq area of West Greenland (Henderson and Pulvertaft, 1967). In places the marble has a dark colour due to the presence of finely disseminated graphite. There are no volcanic rocks associated with the ore body or indeed within the Formation, but there are meta-volcanic amphibolite formations below and above it.

In the Lower Purcell Sediments of the Canadian Cordillera (reported in Chapter 7) there is a major Pb–Zn ore body within an argillite at Sullivan in British Columbia. The Sullivan lead is estimated to have an age of 1250 my or 1340 my; this agrees with the fact that the Purcell Sediments began to accumulate about 1450 my ago (Harrison, 1972). There are insufficient volcanic beds in the Purcell Sequence for the ore environment to be classified as volcanogenic, but Sangster (1972) postulated that a roughly funnel-shaped zone of tourmalinized wall rock represented a feeder pipe (analogous to the chlorite alteration pipe beneath massive sulphide ores in volcanic rocks) through which the ore-bearing solutions penetrated to reach the sea floor where they were deposited contemporaneously with the sediments; the ore is thus interpreted as an exhalative type in a non-volcanogenic environment. On the basis of deep seismic reflections, Kanasevich (1968) and Kanasevich, Clowes and McCloughan (1968) proposed that a major mid-Proterozoic rift valley existed beneath southern Alberta and British Columbia. He then proposed a very interesting model: the Sullivan Pb–Zn ore body lies just within the edge of the rift and therefore the ore-bearing solutions may have been emplaced in the rift structure, i.e. in a tectonic environment similar to that of the modern hot metalliferous brines in the Red Sea (Degens and Ross, 1969).

Micro-fossils

The term Proterozoic is derived from the Greek for 'early life'. Although a few micro-fossils and stromatolites developed locally in Archaean sediments, they began to proliferate in early Proterozoic times. The more complex multicellular Metazoa, which require an oxygenous atmosphere for their growth, should also occur in Proterozoic rocks since the rapid increase in the oxygen content in the atmosphere is indicated by the transition from banded iron formations and red beds. However, unequivocal examples only appear in abundance in late Precambrian rocks; the problem of the precise time of their first appearance remains unsolved.

The best known indubitable early Proterozoic micro-fossils are the abundant algal flora from the black stromatolitic cherts of the Gunflint Iron Formation of the Huronian in Ontario (Barghoorn and Tyler, 1965; Cloud, 1965) which formed soon after 2300 my ago. These organisms include a variety of filamentous, spore-like and anomalous forms only a few microns in size; some are morphologically comparable to living blue–green algae, others resemble living bacteria although they are smaller (Cloud, 1968a).

In the Witwatersrand goldreefs there are remarkable fossilized remains of microorganisms that include bacteria, algae, fungi and lichen-like plants. Hallbauer (1975) envisages a carpet-like colony of columnar carbonaceous individuals, each about 0·3–0·55 mm in diameter and up to 7 mm in length. Gold and uranium were extracted from the environment by the organisms in a way similar to modern fungi and lichens.

There are beautifully preserved organic remnants in the well-exposed early Proterozoic rocks in the Graenseland area of southwest Greenland (Bondesen, Petersen and Jørgensen, 1967). The Ketilidian sediments here, about 2000 my old, have suffered extremely little deformation and metamorphism with the result that primary stratigraphy and sedimentary structures are intact. Besides stromatolite-like macro-structures, there are abundant remnants of bacteria-like structures, filaments with an irregular cellular structure, irregular lumps of organic material, parts of threads which occasionally branch and, in particular, spore-like spheres of 0·5–1·5 mm diameter which are described as *Vallenia erlingi* (Bondesen, 1970). But it is not just the existence, but the widespread extent, of the organic remains that is impressive here. Some remnants occur in almost all sedimentary formations and rock types—dolomites, dolomitic shales and cherts are the most favourable hosts, but they even occur in greywackes and quartzites, and finally there is a 1–3 m thick coal–graphite bed. Moreover, the remains have an extensive horizontal distribution, the 2 m thick Vallenia-bearing dolomite having been found at four localities 25 km along strike and it is thought possible to trace it for more than 50 km.

Pedersen and Lam (1968) found that the Vallenia-rich dolomite still contained alkanes, aromatic hydrocarbons and methyl esters, whilst in their 1970 paper they reported that the coal layer contained alkanes, alkyl benzynes, naphthalene, single alkyl naphthalenes, single monoterpenoids and esters of fatty acids—all original organic compounds, indicating extensive biological activity 2000 my ago. Isotopic studies by Bondesen, Pedersen and Jørgensen (1967) suggested that Vallenia may have been photosynthetic and that the host dolomite was precipitated in oxygen-poor water. Also the composition of the coal indicates that it represents marine organic material accumulated under extremely reducing, euxinic conditions. The widespread distribution of the organic remnants suggests that organic material was present under nearly all sedimentary conditions, and that they may have had an influence on processes such as the formation of chert, dolomite and carbonaceous shale. If the organisms were photosynthetic, they would have contributed considerably to the development of an oxygenic atmosphere (Bondesen, 1970).

These early Proterozoic micro-organisms were largely of the procaryotic type, incapable of cell division and relatively resistant to DNA-damaging ultraviolet radiation. It was not until after about 1800 my ago that an effective ozone layer formed in the atmosphere which prevented radiation penetration and thus allowed the eucaryotes to develop (Fig. 19.4).

The evolution of stromatolites during the whole Proterozoic period will be reviewed in Chapter 7.

Chapter 6

The Mid-Proterozoic Mobile Belt in the North Atlantic Region

In the mid-Proterozoic period 1800–1000 my ago a series of extensive medium- to high-grade mobile belts were formed across Laurasia and Gondwanaland. At first sight many have a bewilderingly wide spectrum of rocks. For example, when Dearnley produced his review of orogenic belts in 1966 he found that 'the presently available information is not adequate to define the overall pattern of the Hudsonian regime fold belts. Thus it is only recently that a framework has begun to emerge that might explain their occurrence and evolution.

The most outstanding of these belts extends across the North Atlantic Shields from California to the Ukraine (its broad evolution is described by Bridgwater and Windley, 1973). Prominent within the mobile belt are the regions of reactivated earlier crust, such as the Churchill, Grenville, Ketilidian, Laxfordian and Svecofennian, but the belt is also characterized by its post-tectonic igneous rock suites: andesine–labradorite anorthosites, rapakivi granites, plateau basalts and ignimbrites, alkaline complexes and carbonatites, and basic dykes. Many of these appeared for the first time in earth history as prominent rock groups and therefore the mobile belt responsible for them is an important landmark in the evolution of the continents. Any viable model for this evolution must satisfactorily explain and relate the formation of the reactivated basement terrains and the post-tectonic igneous rock groups.

The Main Components of the Belt

Fig. 6.1 shows the North Atlantic Shield Mobile Belt extending from west to east USA, South Canada and Greenland, northwest Scotland, south Norway, Sweden, Finland, Karelia and the Russian platform in the Ukraine. A similar, but lesser known, belt with prominent anorthosites extends across what is now the southern hemisphere from eastern South America–central Africa–Malagasy–India edge of Antarctica–SW Australia (Fig. 8.6).

The main geological units of the North Atlantic belt include:

Churchill and Grenville	Canada
Ketilidian–Gardar and	
Nagssugtoqidian	Greenland
Laxfordian	Scotland
Dalslandian, Gothian, Jotnian,	Baltic Shield
Svecofennides, Karelides	
East European Platform	Latvia, Lithuania,
	Estonia, E Poland
Ovruch–Volnian	Ukraine

The post-orogenic intrusive–extrusive suite is concentrated in the southern half of the belt; in North America it is narrowly confined to the zone south of the Archaean craton of the Superior Province, whilst it is broader further to the east, in particular in the Baltic Shield.

Other mid-Proterozoic mobile belts 1800–1000 my old include:

Eastern Ghats	India
Kibalian, Bugando–Toro, Ubendian,	Africa
Lukoshian, Birrimian, Kavirondian–	
Nyanzian (Eburnian and Huabian	
orogenies)	

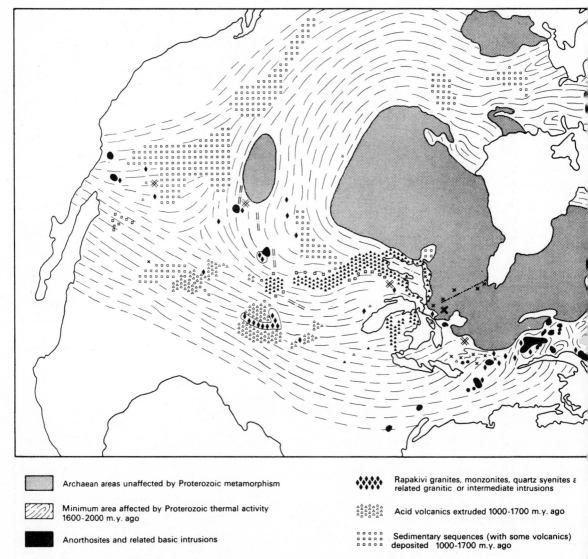

Fig. 6.1. The distribution of the principal rock groups in the North Atlantic Shield belt
formed during the mid-Proterozoic (modified after Bridgwater and Windley, 1973)

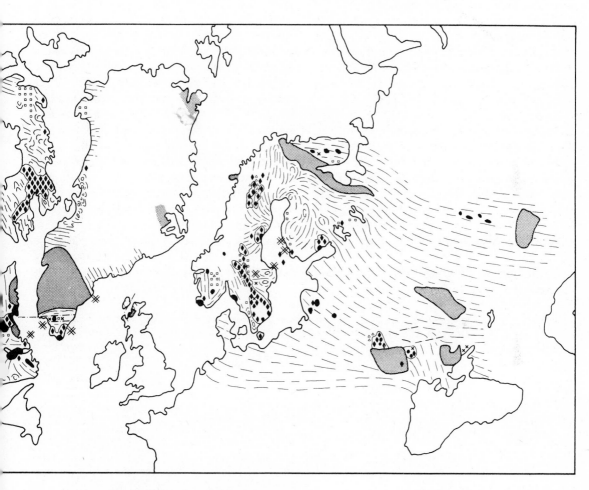

Basic extrusives and associated sediments deposited 1000-1200 m.y. ago

x x x Alkali intrusives emplaced 1000-1700 m.y. ago

⊛ Dykes

＝ Cataclasis

Faults

Chronological Summary

The correlation of the main rock groups throughout the belt is made difficult by the fact that there were several periods of volcanicity and granitic intrusion in many regions. At present it appears that there was a diachronous development of the main rock groups along the belt; for example, the major anorthosites at the western end appear to be some 200–300 my later than those in the east (Bridgwater and Windley, 1973).

Before viewing more detailed relationships let us look at the chronological evolution of the principal events in the four main regions: mid-continental USA and southern Canada, south Greenland and eastern Canada, the Baltic Shield, and the Ukraine. A simplified chronology of events in these regions is given in Table 6.1; for a more detailed chronological analysis see Bridgwater and Windley (1973, Appendix 2).

In the period 2000–1800 my widespread deformation and high temperature meta-morphism affected Archaean basement rocks and unconformable early Proterozoic supracrustals. From 1800 my onwards there was a continuous series of extrusions, intrusions and sedimentation. The anorthosites are the most easily correlatable rock group; since they formed in all regions at about the same time relative to other events, they and their associated rocks form a diagnostic suite. Although there were many types of granite intruded at various times, one stands out as unique: 'the rapakivi granites'. Acid volcanics are a common feature of this mega-belt and they are often associated with the rapakivi granites, being regarded as an extrusive equivalent of the granitic magma. A series of deep-seated faults then controlled the deposition of red continental sandstones and the extrusion of plateau basalts, to be followed by the intrusion of major swarms of basic dykes of transcontinental scale. At about this time several large layered igneous intrusions were emplaced. The last important event was the intrusion of alkaline complexes associated

Table 6.1. Chronologies of events that occurred between 2000 and 1000 my in four main regions of the North Atlantic Shield (after Bridgwater and Windley, 1973)

my BP	Ukraine	Baltic Shield	S Greenland and E Canada	S Canda and Mid-continental USA
1000		Dalslandian 'orogeny'	'Grenville' orogeny	'Grenville' orogeny Rapakivi granites (Nevada) Acid volcanics (Colorado)
1100	Post-tectonic granites	Peralkaline intrusions (S Sweden) E–W dykes with *anorthosite* fragments	Basaltic–alkali magmatism Basic dykes	Keweenawan basic magmatism in graben. Clastic sediments. *Anorthosite* fragments in gabbros
1200			Basic dykes with *anorthosite* fragments. Alkali intrusions, dykes in graben	Basic dykes Acid–basic volcanics
1300		Dykes controlled by graben faults		Granites
1400	Basaltic–alkali magmatism	Jotnian basalts and sandstones (1300–1500 my)	Sandstones in graben	Acid–basic volcanics (Wisconsin) Rapakivi granites Acid volcanics (Missouri) Sioux quartzites
1500		Major dyke swarms Egersund *anorthosite* and granites		Granites (Nebraska, Dakota) Rhyolites
1600		Acid volcanics, porphyries, *anorthosites* and rapakivi granites (1550–1750 my) Basic dykes		Basalts, rhyolites, granites (Arizona)
1700	Koresten rapakivi granite–*anorthosite*–alkaline complex		Adamellite–*anorthosite* complexes Rapakivi granites Basic dykes	Granites (Dakota, Colorado) Rhyolites (Sudbury)
1800	Acid–basic lavas			

with carbonatites, structurally controlled by the same system of faults that influenced the earlier red sandstones.

Cover–Basement Unconformities

One of the largest scale effects of the deformation of a mobile belt that can easily be seen is the modification of an unconformity between bedded supracrustals and relatively homogeneous gneisses. Perhaps the best exposed example in the world is in the Ivigtut region of South Greenland, between a cover of early Proterozoic (c. 2000 my old) supracrustals and Archaean gneisses at the northern margin of the Ketilidian mobile belt (Henriksen, 1969). Fig. 6.2 shows how the unconformity is undeformed in the north and is increasingly deformed southwards as it passes into the Ketilidian mobile belt where it is involved in thrusting, shearing and folding. This deformation went hand-in-hand with a progressive southward increase in the recrystallization of the cover rocks from almost unmetamorphosed to high amphibolite facies, a similar increase in the deformation of metadolerite dykes cutting the basement gneisses, and in the remobilization of the gneisses giving rise to autochthonous and allochthonous granites. In the most highly deformed area the unconformity has been completely obliterated with the result that there are now conformable sheets of, for example, sillimanite schist in granitic gneiss.

Amphibolite Dykes

There are a great number of amphibolite dykes, often confined to linear swarms, in many mid-Proterozoic high-grade complexes. Such dykes may be used for several purposes; at the simplest they provide a means of separating older from younger events and so of erecting a chronology in the basement complex. The older age of the wall rocks may be shown by such features as discordancies at the dyke contacts and inclusions of wall rock within the dykes, whilst the superimposed younger events may include metamorphism of the dyke material, deformation (folding/ boudinage) and rotation of the dykes until they become conformable layers in the usually deformed host rocks. Igneous features useful in deciding whether nearly conformable amphibolite layers or amphibolite fragments are remnants of earlier dykes include chilled margins, apophyses, inclusions of wall rocks, blasto-ophitic textures, and plagioclase aggregates. In the absence of fossils such dyke criteria assist in erecting quite complicated chronologies in high-grade gneiss areas. Proterozoic basic dykes have been particularly useful in working out the evolution of the Lewisian and Ketilidian mobile belts.

The migmatization and agmatitic break-up of the amphibolite dykes provides evidence of the degree and type of reactivation of host granitic rocks (see especially Watterson, 1965). Dykes which were intruded into a basement complex after an orogenic period cut discordantly through deformed granitic gneisses. During a later plutonic period the rocks may be reheated, partially melted, and variably mobilized with the result that granitic veins may introduce the dykes. As the intensity of granitic veining increases, a point is reached when blocks of dyke amphibolite become physically separated, this process taking place without any disorientation of the blocks. The result is that the dyke blocks appear to be 'swimming' in reactivated homogeneous granite; trails of such amphibolite inclusions can be used to define the presence of whole swarms of migmatized basic dykes. Excellent examples from the Julianehaab Granite are shown in Fig. 6.3. Similar dyke fragments are abundant in Laxfordian migmatite complexes in the Lewisian of Scotland (Myers, 1971a) and in Svecofennide gneisses of southern Finland (Sederholm, 1923).

Some amphibolite dykes formed from the quench of hydrous basaltic magma intruded under essentially syntectonic conditions into hot country rocks. Watterson (1968) has shown how syntectonic dykes in the Ketilidian can be recognized by such features as internal sigmoidal foliation and intimate association with shear zones. In the Ketilidian there was a single period of injection of syntectonic basic

96

Fig. 6.2. Map of the Ivigtut area, SW Greenland, showing in the north Ketilidian supracrustals resting unconformably on Archaean gneisses. A progressive southward increase in deformation and metamorphism is also seen in the section (modified after Henriksen, 1969; reproduced by permission of The Geological Association of Canada)

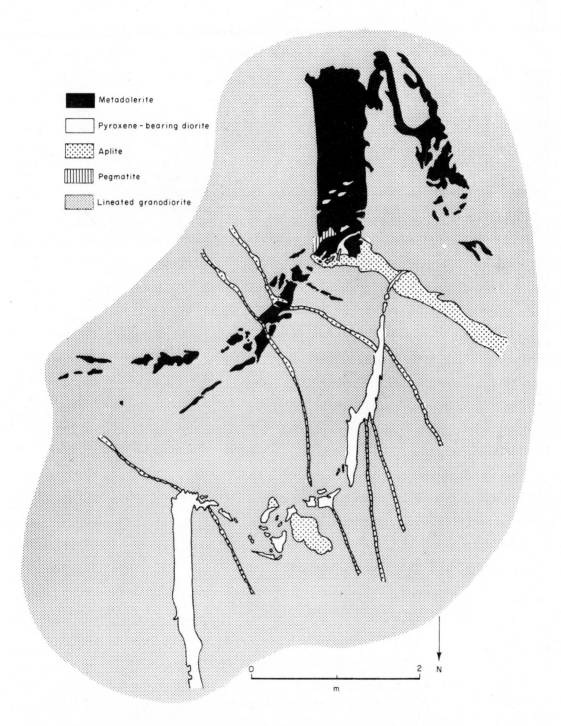

Metadolerite

Pyroxene – bearing diorite

Aplite

Pegmatite

Lineated granodiorite

0   2   N

m

Fig. 6.3. Folded and broken metadolerite and pyroxene-bearing diorite dykes within a shear belt in granodiorite belonging to the Julianehaab granite (after Allaart, 1967)

dykes (Allaart, Bridgwater and Henriksen, 1969) which acts as an important time marker in the long sequence of events that makes up this mobile belt.

Bearing in mind the suggestion of Bridgwater, Escher and Watterson (1973a) that the Ketilidian fold belt may be comparable with Andean-type fold belts, it is interesting to observe that syntectonic hornblende intermediate-to-basic dykes are common in the granitic batholiths of British Columbia (Phemister, 1945; Roddick and Armstrong, 1959), Washington and Oregon (Goodspeed, 1955) and Peru (Cobbing and Pitcher, 1972). These dykes are associated with shear zones in the wall rocks, are noted for the absence of chilled margins and the presence of primary igneous amphiboles, just like the amphibolite dykes in the Julianehaab granite which may also be a mesozonal batholith.

Linear Shear Belts

Linear shear belts were particularly well developed in the North Atlantic region in the period 2500–1600 my ago (Sutton and Watson, 1974; Davies and Windley, 1976). Some are well exposed as ductile shear zones (e.g. Bak and coworkers, 1975), others are inferred from the arrangement of aeromagnetic anomaly patterns (Watson, 1973a).

The shear belts mostly dip steeply, range up to about 50 km in width and several hundred kilometres in length. The longest belt inferred by Watson (1973a) extends for some 2500 km along the front of Hudsonian reworking against the Superior Province in the Nelson River area of Canada.

Earlier structures such as gneiss foliation and dykes are deflected into the linear belts largely with transcurrent displacements. Earlier mineral assemblages and rock fabrics are modified and reconstituted into new metamorphic equivalents showing that there is a close relation between the deformation and recrystallization in the shear zones.

Such high temperature belts of simple shear are well known in NW Scotland (F. B. Davies, 1976; Beach, 1976) and in West Greenland where there are at least four (Bak and co-

workers, 1975), the most prominent of which lies between the Nagssugtoqidian reworked Archaean block and the stable Archaean block to the south (Escher and coworkers, 1975, 1976).

There is evidence that some of the Proterozoic belts were superimposed or evolved on lineaments that originated in the Archaean (F. B. Davies, 1975b); in other words, there is a close relationship between these shear belts and earlier zones of weakness in the continental plate. From a detailed analysis of the strain heterogeneities in the Lewisian complex of the Outer Hebrides, Coward (1973) demonstrated that the fundamental factors controlling the amount of deformation were the composition of the rocks concerned and the metamorphic conditions. Watson (in discussion of Park, 1970) made the interesting point that in reworked basement complexes rocks which suffered low intensity deformation in early stages of reworking underwent similar low intensity deformation during later stages; Coward (1973) showed that such a relationship occurred when the succeeding deformation phases were accompanied by the same metamorphic conditions but that a change in these conditions, for example from granulite to amphibolite facies, brought about a change in mineralogy and thus in the physical properties of the rocks and hence, in turn, a variation in the ductility contrasts between adjacent rocks.

The belts seem to be broadly contemporaneous with the hairpin bends of the polar wander paths for North America and Africa at 1700 ± 1000 my BP (Donaldson and coworkers, 1973; Piper, 1974) and thus may be related to major changes in position of the Proterozoic supercontinents. Sutton and Watson (1974) suggested that the transcurrent displacements may have been caused by a clockwise rotation of the continental mass. The shear belts in West Greenland clearly record internal distortion suffered by a continuous continental lithospheric plate (Bak and coworkers, 1975). Relevant here is the fact that the most recent palaeomagnetic data indicate that at the time the belts were forming Laurentia and Africa were drifting as intact bodies without any loss of coherence (Piper, Briden and Lomax, 1973; Irving and Lapointe, 1975; McGlynn and coworkers, 1975; Irving and McGlynn, 1976). The evidence therefore suggests that the continental plates were not entirely rigid but underwent ductile high temperature internal strain along several discrete zones. M. R. Smith and Jensen (1974) suggested that continents may fracture along latitudes of rotation giving rise to a number of rotating blocks. Compression and shear of one block against another might result in features characteristic of the internal Proterozoic belts with transcurrent ductile displacements. If there are modern equivalents to these Proterozoic shear belts they might be expressed at the present high level of erosion either as intracontinental transform faults or as aulacogens (Badham, 1976a).

Granite–Migmatite Complexes

One of the best documented examples of an *in situ* reactivated granite is the Julianehaab granite which occupies an area of 6000 sq km in the Ketilidian fold belt of South Greenland (Allaart, 1967). It is a relatively homogeneous, medium- to coarse-grained, granodioritic to adamellitic body with evidence of several phases of deformation and remobilization. Much of the 'granite' consists of autochthonous Archaean gneiss but three younger mainly allochthonous granites were generated in connection with basic and intermediate igneous activity. Within the granite there are many discontinuous zones consisting of relict gneisses and meta-supracrustal rocks. One of the most interesting aspects of this body is the relationship between periods of homogenization and mobilization of the granitic rocks and the intrusion of an array of basic to ultrabasic bodies of three main types:
1. Syntectonic amphibolite dykes.
2. Composite net-veined diorite sheets and dykes, and
3. An appinitic suite with pyroxene- and hornblende-bearing diorites, gabbros and hornblendites (Walton, 1965).

The relationships between the emplacement of these bodies and periods of deforma-

tion and migmatization enabled Allaart (1967) to construct a detailed sequence of events as follows (modified after Allaart, Bridgwater and Henriksen, 1969):

Youngest
    Thin tholeiitic dykes in persistent swarms
    Major pegmatite swarms: aplites
    Composite net-veined diorites
    Aplitic granites
    Syntectonic basic dykes
    Porphyritic ('big-feldspar') parautochthonous granites
    Hornblendite, gabbro and diorite intrusions
    Allochthonous granites
    Deformation: local hot-shear belts

The Julianehaab granite is important because it demonstrates what happens to a major piece of high-grade crust when it is reheated and remobilized during a mid-Proterozoic period of plutonism under mesozonal conditions. But there is a contradiction between the field and isotopic evidence as shown by the divergent views of the authors in van Breemen, Aftalion and Allaart (1974). According to Allaart the granite contains much partially reactivated basement material, but the low initial $^{87}Sr/^{86}Sr$ ratio of 0.7022 ± 0.0010 indicates to van Breemen and Aftalion that no basement material is involved.

Granite–migmatite complexes with many pegmatites and aplites form an important part of the late Laxfordian history in NW Scotland. A U–Pb zircon age of 1715 ± 15 my and a Rb–Sr whole-rock age of 1750 ± 34 was obtained by van Bremmen, Aftalion and Pidgeon (1971) from Laxfordian intrusive granites in South Harris in the Outer Hebrides. Myers (1971b) described an excellent, although relatively small (217 sq km), complex in Harris which developed by the irregular intrusion of potash-rich granite accompanied by local homogenization and granitization of Archaean (Scourian) banded gneisses followed by the intrusion of many potash-rich pegmatites.

There was a widespread development of granites and migmatites in the Svecofennian belt in southern Finland (Fig. 6.4) (Härme, 1965; Hietanen, 1975) and southern Norway

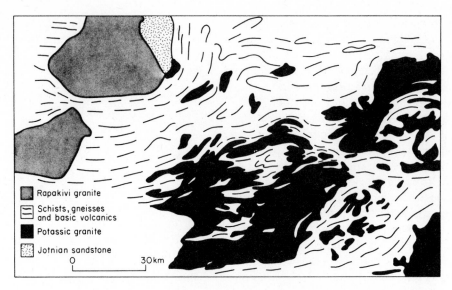

Fig. 6.4. Rapakivi granites and potassium granites in migmatites in SW Finland (redrawn after Härme, 1965; reproduced by permission of The Geological Survey of Finland)

(Smithson, 1965). In Norway the granitic magmas, derived by remobilization of underlying granitic gneisses, rose to higher levels in the crust to form granitic domes, diapirs and plutons in overlying supracrustal rocks. Usually these high-level plutonics were emplaced late in the evolution of the mobile belt; late orogenic potassic granites in northern Sweden have a Rb/Sr age of 1535–1565 my (Padget, 1973), whilst many granites associated with the main plutonic period formed 1700–1900 my ago (Welin, Christiansson and Nillson, 1971). Hietanen (1975) suggests that early trondhjemites and late potassic granites in southern Finland (Fig. 6.5) were generated at a Cordilleran-type active continental margin in a manner analogous to the formation of the Mesozoic 'granitic' batholith in the northern Sierra Nevada.

Anorthosites

Anorthositic plutons are confined to two linear belts in the northern and southern hemispheres when plotted on a pre-Permian continental drift reconstruction (Herz, 1969). They are related spatially and temporally to post-tectonic granites and acid volcanics, and genetically to the formation of the North Atlantic Shield mobile belt (Fig. 6.1). For general reviews see Isachsen (1969), Middlemost (1970), Michot (1972) and de Waard, Duchesne and Michot (1974).

In the northern hemisphere some anorthosite bodies lie within the 100 my old part of the fold belt (the Grenville of Canada and the Dalslandian and Gothian of southern Norway and Sweden); others, which lie outside this zone, are in areas affected by the 2000–1800 my early Proterozoic orogenic activity: Poland, Latvia, Lithuania and Estonia (all known from borehole data), Finland, Ukraine, central Sweden and western USA. But many recent geochronological studies have established that there was an early Proterozoic thermal event in the areas affected by the 1000 my overprint. The anorthosites are thus concentrated within a broad belt of crust affected by early Proterozoic metamorphism and some were recrystallized 1000 my ago.

Fig. 8.6 shows that an anorthosite belt extends from Capivarito in Brazil (Sighinolfi and Gorgoni, 1975), through southern Angola (Simpson and Otto, 1960), Tanzania, Malagasy (Boulanger, 1959), Queen Maud Land in Antarctica, eastern Ghats (De, 1969), and Bengal (Chatterjee, 1936) in India to the Musgrave Range in Australia.

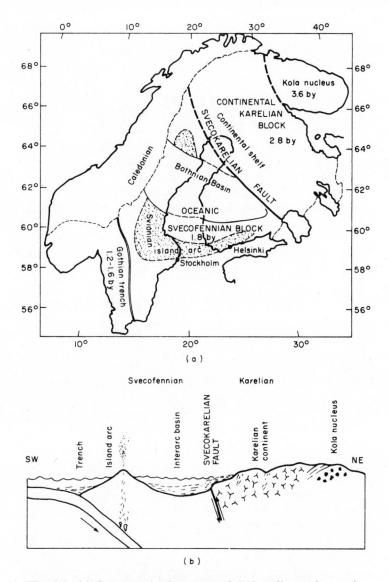

Fig. 6.5. (a) Geotectonic elements and their radiometric ages in the Fennoscandian Shield. Svecokarelian fault zone separates the older (2800 my) continental Karelian block from the younger, Svecofennian block which was transformed from oceanic to continental during the Svecofennian orogeny (1900–1700 my ago). (b) Cross-section through A from SW to NE showing the possible island arc system in the shield 2500 to 2000 my ago. Deposition of quartz sand and pelite on the Karelian continental shelf, deposition of greywacke-type sediments in the Bothnian interarc basin and volcanism in the Svionian island arc were coeval. (After Hietanen, 1975)

In the USSR there are many Proterozoic anorthosites (Moshkin and Dagelaiskaja, 1972; note also Bogatikov, 1974, for several reviews in Russian). Besides those marked on Fig. 6.1, they occur in the Aldan–Stanovoy Shield, the Baikal region, the Anabar Shield and the Mongolo-Okhotsk region (Fig. 8.6).

Individual bodies are commonly conformable with adjacent rocks and they vary considerably in size—most fall within the range 100–10,000 sq km—accounting for some 20% of the surface area of the Grenville Province in Canada.

Little geochronological work has been applied to this rock suite. K/Ar dates on Labrador anorthosites give minimum ages around 1400 my, the Duluth gabbroic anorthosite has an age of 1125 my (U/Pb on zircons) and the South Rogaland anorthosite in southern Norway one of 1000 my (U/Pb on zircons—Pasteels and Michot, 1975), while bodies from South Finland and the Ukraine were probably emplaced about 1700–1750 my ago. Although an average age of emplacement of the anorthosites is about 1600 ± 200 my, those in South Greenland (Bridgwater and Harry, 1968) and Minnesota (Taylor, 1964) continued to form at depth until at least 1200 my ago; in general those at the western end of the mobile belt formed some 200–300 my after those in the east (Bridgwater and Windley, 1973). Recent data suggest that in SE Canada (plus the Adirondacks) there were two periods of emplacement: 1500–1450 and 1200–1100 my ago; only one body has an intermediate age (Baer, 1976a). In the eastern Ghats belt associated charnockites have whole-rock Rb/Sr ages of 1300–1500 my (Aswathanarayana, 1968), a minimum age of the southern Angola anorthosite is 1269 ± 90 my (Simpson and Otto, 1960), and associated granulites in the Musgrave Ranges have a Rb/Sr isochron age of 1390 ± 130 my (Compston and Arriens, 1968).

The anorthosites consist of more than 90% plagioclase, and associated gabbroic, noritic and troctolitic anorthosites have 78–90% plagioclase. Most rocks contain, in addition, hypersthene with or without olivine and augite. The composition of plagioclase most commonly lies in the range An_{45-55} (Romey, 1968, puts the average at $50 \cdot 9$ An).

According to Anderson and Morin (1969) there are two types of massif anorthosite, differentiated by the composition of their plagioclase and iron–titanium oxides:
1. Labradorite type with plagioclase in the range An_{45-68} and titaniferous magnetite or the oxidized equivalent magnetite plus ilmenite. The main rock type is gabbroic, noritic to troctolitic anorthosite rather than true anorthosite.
2. Andesine type with An_{25-48} plagioclase (commonly antiperthitic) and haemoilmenite. Here anorthosite predominates and gabbroic (or noritic) anorthosite forms a subordinate border facies. The labradorite type occurs as xenoliths in, and is cut by dykes of, the second type showing that the former is the older.

Genetic relationships associated with the anorthosite suite are highly controversial. It is impossible here to go into the details of the opposing hypotheses (see Isachsen, 1969); however it is worthwhile, firstly, to consider how the formation of these rocks played a role in the evolution of the continents during this period, and secondly, to outline the two main points of contention.

Tectonic environment and mode of emplacement

The plutonic character of the anorthosites is complicated by the fact that those lying within the Grenville belt have been affected by a period of plutonism—and these are largely the ones that have been studied the most. It is also clear that they formed throughout an enormous extent of crust and over a considerable time period and so it would not be surprising if it were found that they did not all form in the same way.

There are two proposed environments. Some of those that are undeformed have been shown to be post-tectonic intrusions: Nain (E. P. Wheeler, 1960), Kiglapait (Morse, 1969), and Michikamau (Emslie, 1970). However, where the anorthositic bodies have been deformed and metamorphosed, their original relationships are less clear; some are thought to be syntectonic intrusions—the South Rogaland complex in South Norway, including the well-known Egersund–Bjerkrem–Sogndal bodies (Michot, 1969), and some of the anorthosites in the eastern Ghats of India (De, 1969).

Gravity surveys indicate that the andesine-type Adirondack Marcy anorthositic body is not underlain by a great depth of complementary basic and ultrabasic rocks and thus that it forms a slab-like sheet (Simmons, 1964); it was probably emplaced along the contact between the basement and a cover series (de Waard and Walton, 1967). The labradorite-type Michikamau and Mineral Lake bodies in Canada are also thought to have a sheet-like form (Emslie, 1970). The Morin anorthosite, Quebec, rose through the gneissic crust as a diapiric dome part of which spread laterally in a nappe-like horizontal lobe on encountering supracrustal rocks (Martignole and Schrijver, 1970a).

Origin of associated mangerites and charnockites

Charnockites and mangerites are plutonic igneous rocks which have the bulk composition of granite and monzonite respectively, the charnockite containing orthopyroxene or fayalite and the mangerite orthopyroxene or fayalite plus quartz (de Waard, 1969). The andesine-type anorthosites are closely related to the mangerite–charnockite suite (monzonite, syenite, quartz syenite, adamellite, rapakivi granite) characterized by perthitic feldspar and orthopyroxene.

The mangeritic–charnockitic rocks or granulite facies gneisses commonly border the anorthosites in an aureole (e.g. the Borgia body, Baer, 1976b), are intrusive into them, contain xenoliths and xenocrysts of anorthosite, and their frequency decreases away from the anorthosite contact (Isachsen, 1969). In places they are intimately interlayered with the anorthosites, locally there is apparently a complete lithological gradation from anorthosite through noritic anorthosite and norite to

mangerite–charnockite and rapakivi granite (de Waard and Romey, 1969), and there is a chemical similarity between the silica-rich members of the anorthosite and the rapakivi suites (Emslie, 1973). Much controversy centres on the genetic relationship between the anorthosites and the mangerite–charnockite suite (de Waard, 1969)—is it a comagmatic differentiate of the anorthosites or was it derived independently? (For a list of data consistent with both hypotheses, see Isachsen, 1969.)

Rapakivi Granites and Acid Volcanics

The most outstanding rocks associated in space and time with the anorthosites are the rapakivi granites with their diagnostic megacrysts of potash feldspar mantled with a rim of oligoclase (de Waard, Duchesne and Michot, 1974). They occur widely throughout the North Atlantic mobile belt from the Ukraine to the Baltic Sea, Finland (Fig. 6.4), Sweden, South Greenland, Labrador, Baffin Island, Wisconsin, Idaho, Nevada to California. Examples, together with ages, are:

1050 my	Nevada	Volborth (1962)
1350 and 1650	Sweden	Lundqvist (1968)
1400	Wisconsin	Elders (1968)
1450	Berdyaush, Urals	Salop (1968)
1700	S Finland (e.g. Wiborg)	Vorma (1975)
1720	Koreston, Ukraine	Gorokhov (1964), Moshkin and Dagelaiskaja (1972)
1740 and 1755	S Greenland	Gulson and Krogh (1975)
1786	S Greenland	van Breemen Aftalion and Allaart (1974)

Some occur independently of the anorthosites (South Greenland, Wisconsin), others with them (Sweden, Ukraine, South Finland, Labrador). Fig. 6.6 illustrates a model relating the formation of the anorthosites to the rapakivi granites and associated volcanics and graben structures described below.

Bridgwater and Windley (1973) point out that the thermal effect of the rapakivi granites (and also the anorthosites) on the surrounding country rocks varies considerably from place

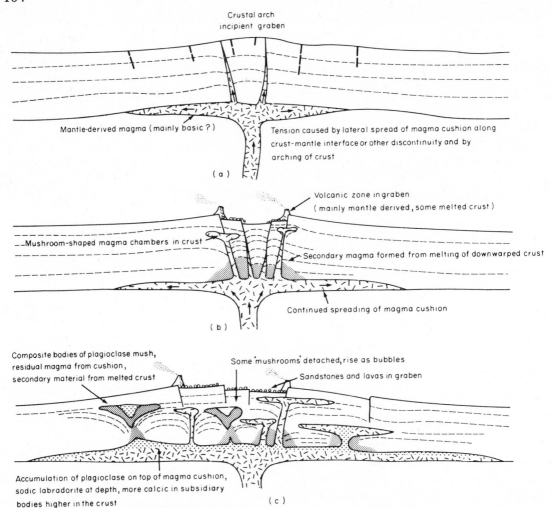

Fig. 6.6. A model that relates anorthosites, rapakivi granites, acid and basic volcanics and graben structures formed in continental crust (after Bridgwater, Sutton and Watterson, 1974b). (a) Formation of crustal arch above magma cushion. (b) Continued spread of magma cushion, formation of graben and intrusion of mantle material into the crust along graben fractures. Some volcanicity. Downwarping of crustal material and melting near the base of the crust. (c) Continued spreading of magma cushion and further formation of graben structures. Accumulation of plagioclase on top of magma cushion at depth and in smaller chambers in the crust. Rise of plagioclase cumulates through the crust in a manner similar to salt domes, partly lubricated by residual material from magma cushion and partly by remobilized crustal material. Basic and acid volcanicity on surface, deposition of intermontane sediments

to place, reflecting differences in present erosion, in the original depth at which the bodies were intruded, and, perhaps most important, in the temperature of the country rocks into which these bodies were emplaced. In Finland, where the upper parts of rapakivi granites are exposed, the adjacent country rocks show little effect of thermal metamorphism (Eskola, 1963). On the other hand in South Greenland,

where there is considerable relief, the upper parts of the granites show little effect on their wall rocks, while their root zones, some 2–3 km lower down in the crust, are surrounded by 1–2 km wide areas of reheated and remobilized country rocks commonly with granulite facies mineralogy (Bridgwater and Watterson, 1967). These zones are locally so large that the crust must already have been close to melting when the rapakivi granites were emplaced.

Chemically the rapakivi granites are rich in potassium, iron and fluorine, expressed by the presence of potash feldspar, fluorite and iron-rich mafic minerals such as fayalite (G. P. Wheeler, 1965; Simonen and Vorma, 1969; Frisch and Bridgwater, 1972). Many are close to quartz monzonite in composition and are thus similar to the mangerites; in fact in the Wiborg area of Finland rapakivi granite grades into mangerite or quartz monzonite (Kranck, 1969).

Fig. 6.1 shows that there are the remains of extrusive acid volcanic rocks in the North Atlantic Shield. Although they were extruded at different times and in different places in the period 1700–1000 my ago, there is a close similarity in age and petrochemistry between several volcanic sequences and the rapakivi granites.

According to Kranck (1969), in the Wiborg area of Finland there are labradorite porphyry lavas, chemically similar to the rapakivi granites, in association with rhyolite flows. But the classic area for seeing these interrelationships is in the Loos–Hamra region of Finland described by von Eckermann (1936) and Lundqvist (1968). Here there is a spectacular sequence of acidic volcanic rocks grading from syenites to rhyolites which occur together with a rapakivi granite of identical composition. Apparently the granitic rocks were intruded almost contemporaneously with the volcanics and thus there are gradations between them. And, on Suursaari island, the rapakivi magma extruded as a lava and formed quartz porphyry and agglomerates (Eskola, 1963). In the Kiruna region of northern Sweden there are quartz porphyries and syenite porphyries which have Rb/Sr ages of 1605 ± 65 my and 1635 ± 90 my (Padget, 1973). These lavas are similar in age to the Dala porphyries of central Sweden (1670 my). From the above account it is clear that this prominent phase of acidic volcanism commonly gave rise to porphyritic lavas.

Clearly the rapakivi granites concerned crystallized from magma at high crustal levels, as suggested by Savolahti (1962); Kranck (1969) considered this problem and concluded that the rapakivi granites, labradorite porphyry lavas and rhyolites were derived from a similar magma and that this was related to the magma that gave rise to the anorthosite-mangerite suite (see also de Waard, Duchesne and Michot, 1974).

A further interesting fact is that some of the earliest ignimbrites formed at this time in Earth history. In central Sweden in the sub-Jotnian succession there are extensive porphyritic rhyolites (cut by rapakivi granites) with typical eutaxitic textures and, in spite of the fact that they have been completely de-vitrified, flattened shards and other structures characteristic of welded tuffs can still be recognized (Hjelmqvist, 1956). Similar tuffs with well-preserved glass shards occur in the flinty acid volcanics (hälleflintas) of the Leptite Formation of the Svecofennides of Sweden (Geijer, 1963).

Continental Sandstones and Basalts

In the North Atlantic Shield towards the end of the mid-Proterozoic there was extensive faulting which controlled the location of sedimentation and volcanism as well as of later alkaline intrusions in graben-like structures. This was a period of high-level tectonics associated with surface deposition and post-tectonic igneous events which generally occurred after the emplacement of the anorthosites and rapakivi granites (Bridgwater and Windley, 1973). The formation of these structures implies the existence of wide platforms on stabilized late Precambrian cratons within which epicratonic troughs were sited (Stewart, 1976). Examples are:

1000–1450 my	Belt–Purcell Supergroup, western USA and Canada
1100	Keweenawan Trough, mid-USA
1100	Seal Lake Formation, Labrador
1100–1400	Apache and Grand Canyon Series, Arizona
1300–1500	Jotnian, Sweden and Finland
1400	Gardar Province, South Greenland
1600–1800	Lake Onega beds, Baltic Shield
1600–1800	Palaeo-Helikian basins, NW Canadian Shield

The sediments are mostly coarse clastic red continental sandstones of molasse type, comparable with modern, shallow-water, piedmont facies flood plain deposits, and the volcanics mostly vary from tholeiitic to alkalic basalts.

Many of these deposits were laid down in elongate basins or troughs controlled by major faults which were active for a considerable time. For example, Asklund (1931), Polkanov (1956), and Laurén (1970) considered that the rise of the rapakivi granites in southern Sweden and the subsequent deposition of the Jotnian sandstones were controlled by the same fault lineaments. In Finland the rapakivi granites are generally pre-Jotnian, but some in Sweden are post-Jotnian. Eskola (1963) suggested that once the rapakivi magmas formed they remained in the liquid state for sufficiently long, encapsuled in their deep-seated reservoirs, for them to be intruded diachronously, some just before some immediately after the deposition of the Jotnian sediments and volcanics. In the Gardar Province of South Greenland the east–west fault system, that controlled the deposition of the first sandstones and lava extrusion about 1400 my, were active long beforehand (Henriksen, 1969) and continued to be intermittently so, localizing the intrusion of the major alkaline complexes until 1000 my ago. Although the main movements on these east–west faults were horizontal, there were vertical displacement components of as much as 2 km.

Soon after 1800 my great thicknesses of red arkosic sandstones and conglomerates were deposited in northwest Canada in fault-controlled basins (e.g. Martin Formation, Echo Bay and Cameron Bay Groups, etc.). The extreme compositional and textural immaturity of these sediments reflects a continuously rising rugged source terrain. The basins are partly filled with acidic-to-basic lavas, the extrusion of which was probably triggered by faulting. This fault–basin cycle was followed by deposition of thick ortho-quartzites of the Athabasca, Thelon and Tinney Cove Formations and of the lower Hornby Bay Group (Fraser and coworkers, 1970).

There are close similarities in lithology and environment between some successions in different parts of the North Atlantic Shield. The Jotnian basalts in Scandinavia overlie sandstones comparable to the Hornby Bay Group sandstones of northwest Canada, and these in turn overlie unconformably pre-Jotnian intermediate-to-acid porphyries similar to those of the Echo Bay and Cameron Bay Groups; in both places the last groups are intruded by porphyritic (rapakivi-type) granites (Donaldson and coworkers, 1973). Moreover, the sub-Jotnian and Jotnian, taken as one unit, is comparable with the Keweenawan succession of the USA.

However, the continental deposits differ considerably in age. For example, they range from the Lake Onega sediments of the Baltic Shield and the Red Beds of the northwest Canadian Shield, deposited in the period 1600–1800 my ago (Kratz, Gerling and Lobach-Zhuchenko, 1968; Fraser and coworkers, 1970), to the sediments and lavas of the Keweenawan Trough in the mid-USA with an age of c. 1100 my (W. S. White, 1966). The widespread deposition of such fault-controlled, epicratonic coarse clastic sandstones in many parts of the North Atlantic Shield during the later part of mid-Proterozoic time suggests that the high-level conditions which controlled their formation were broadly similar over a major portion of the earth's crust (J. H. Stewart, 1976).

A remarkable structure that is worth considering in a little more detail is the Keweenawan Trough. In the Lake Superior region continental red clastics and basic amygdaloidal lavas of the Keweenawan Group (900–1135 my) rest unconformably on early Proterozoic and Archaean rocks. The most prominent gravity high in North America suggests that the 20 km thick succession continues within a major graben or rift zone, nearly 160 km wide, beneath Phanerozoic cover rocks for 1500 km from Lake Superior to Kansas (Chase and Gilmer, 1973) (Fig. 8.3). The Trough has the character of an intracontinental aulacogen infilled with basic volcanics, marginal fanconglomerates, mainly alluvial sediments and, because of blocked drainage, some lake deposits (Pettijohn, 1970).

Intrusions into Cratons

Basic Dyke Swarms

During the period from about 1200–1000 my vast numbers of basic dykes were introduced into the continents, mostly in intense swarms. Some may have been feeders to higher level basalts, such as those described above.

1000–1200 my	Baffin Island, Canada	Fahrig and Wanless (1963)
1150	Arizona, USA	Livingstone and Damon (1968)
1150–1170	Logan sills, Lake Superior	Hanson (1975)
	Asby, Sweden (post-Jotnian)	Geijer (1963)
	Finland (post-Jotnian)	Eskola (1963)
1200	Death Valley, Calif., USA	Wright and co-workers (1974)
	Gardar Province, S Greenland	van Breemen and Upton (1972)

Like the basic dyke swarms intruded in the period 2000–2500 my ago (Chapter 5) many of these late Precambrian swarms are of considerable size; indeed, some are transcontinental. In the Gardar Province of South Greenland (Emelus and Upton, 1976) there are hundreds of basic dykes, most of which belong to a northeast-trending swarm (Fig. 6.7). Individual dykes are of considerable width, 50 m being not uncommon although the main range is probably from a few hundred to one hundred metres. In Finland there are many dykes in all the Jotnian areas where they traverse the Jotnian sedimentary rocks, rapakivi granites and related mafic rocks, and are regarded by Eskola (1963) as the feeders of the plateau basalt extrusions.

Compositionally the dykes are mostly quartz dolerites and olivine dolerites (or diabases), but where they occur in the late alkaline provinces such as the Gardar they are closely associated with varieties such as trachydolerite, gabbro and gabbro-syenite. There is thus an overlap here with the dykes described in the later section in this chapter on alkaline complexes.

These late Precambrian dykes which are usually vertical are commonly faulted—a criterion which helps to distinguish them from unfaulted late Phanerozoic dykes in the same regions. They may be sheared and mylonitized especially along their margins, but the most obvious feature is that they may be displaced by faults for distances of up to a few kilometres.

It is possible that the dykes are comparable to the Mesozoic dykes bordering the North Atlantic ocean which are an early expression of the continental rifting and ocean floor spreading. As far as can be seen at present major swarms situated well within continental masses cannot be so related to continental drift but may be more an expression of subcontinental convection cells operating within continents that were undergoing rifting elsewhere (P. M. Clifford, 1968).

Layered Complexes

At about the same time as the basic dykes were intruded and the plateau basalts extruded several large igneous complexes were emplaced; the Muskox and Duluth bodies in Canada will be considered here.

The *Muskox Intrusion*, situated in the Canadian Northwest Territories, is a layered dyke-like body with a funnel-shaped cross-section, an exposed length of 118 km and a maximum outcrop width of 13 km. It has been dated at 1100–1200 my by Rb/Sr methods on micas and whole rocks; this is the same age as the nearby Coppermine basalts. The intrusion was emplaced just below the unconformity between basement gneisses of the Epworth Group and relatively undeformed quartzites of the Hornby Bay Group. The shape of the intrusion was probably controlled by the unconformity which dammed the magma causing it to spread out into a funnel shape (C. H. Smith, 1962). The body was emplaced at least 3·3 km below the surface of the Earth, as this is the maximum thickness of the overlying Hornby Bay sediments. Aeromagnetic and Bouguer gravity anomalies show that the intrusion continues for at least 120 km beneath its roof rocks.

The intrusion consists of three units. A feeder dyke of bronzite gabbro with internal zones of picrite, two marginal zones that grade

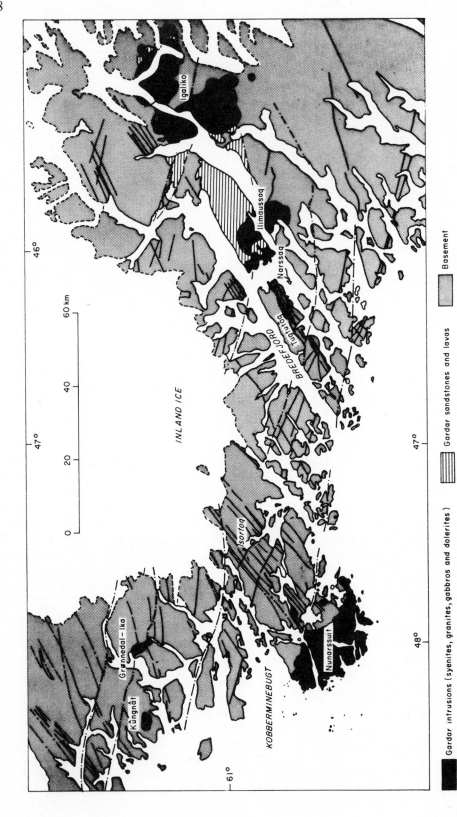

Fig. 6.7. Map of the Gardar Province of SW Greenland showing the main alkaline intrusions, dyke swarms and lavas and sandstones (after Watt, 1966). Reproduced by permission of The Geological Survey of Greenland

inwards (upwards) from bronzite gabbro to peridotite, and a layered cumulate series, 2·8 km thick, ranging generally from dunite at the base, through peridotite and various pyroxenites and gabbros, to granophyric gabbro, and a local capping of granophyre at the top. The chemical evolution of the body is described by Irvine and Smith (1969) and Irvine (1970).

The *Duluth Complex*, Minnesota, is dated as 1150 my and was intruded largely into Keweenawan lavas and sediments (R. B. Taylor, 1964). It mainly consists of a 5 km thick sequence of gabbroic anorthosites, and not gabbro as it is usually termed (Phinney, 1970). There are several separate intrusions, the main rocks of which range from troctolite, gabbro and ferrodiorite to ferro-hedenbergite granophyre—a sequence displaying an iron enrichment trend (Wager and Brown, 1968).

Although the Duluth Complex is listed here as an intrusion emplaced towards the end of Atlantic Shield (Fig. 6.1), mostly near the end of the mid-Proterozoic. The complexes commonly occur in the same fault-controlled blocks and rift valleys (aulacogens) that contain the continental deposits.

The type of magmatism varied considerably from alkaline and peralkaline to tholeiitic. This is expressed by a remarkable variation of rock types from alkaline gabbros and granites, quartz syenites, syenites, nepheline syenites, and peralkaline rocks such as foyaites, naujaites, lujavrites and kakortokites, to carbonatites, lamprophyres, camptonites, gabbros and dolerites. 'As a general rule the alkaline magmatism appears to be restricted to distinct fault-controlled belts and to have persisted for a longer period of time than the tholeiitic magmatism, suggesting perhaps deeper more localised control of magmatic activity later in the period' (Bridgwater and Windley, 1973).

Some of the main examples are:

1000 my	Haliburton–Bancroft, Ontario	Emslie, in Stockwell and coworkers (1970)
>1000	Several complexes in Grenville Province	McGlynn, in Stockwell and coworkers (1970)
1000–1100	Carbonatite complexes, Ontario and Quebec	Gittins, MacIntyre and York (1967)
1000–1275	Gardar Province, S Greenland	Sørensen (1966) and van Breemen and Upton (1972)
1050	New Mexico	Kelley (1968)
1100	Norra Karr, Sweden	Eckermann (1968)
1100	Seal Lake, Labrador	Brummer and Mann (1961)
1285	Blue Mountain, Ontario and St Hilaire, Quebec	Emslie, in Stockwell and coworkers (1970)
1500–1750?	Oktyabryski and Tersyanski, USSR	Semenenko and coworkers (1968)
1650–1750	Carbonatite complexes, Ontario and Quebec	Gittins, MacIntyre and York (1967)

mid-Proterozoic time, it has obvious affinities with the somewhat earlier anorthosite bodies referred to earlier in this chapter. Because the complex contains inclusions of anorthosite up to 8 m across, and many other dykes, sills and flows in the area include anorthosite accumulations (Phinney, 1970), it seems that a sizeable anorthosite mass exists at depth.

Alkaline Complexes

Following closely on the deposition of the continental deposits was the intrusion of alkaline complexes in many areas of the North This table shows that most of these bodies formed in the period 1300–1000 my ago. However this so far gives too simplified a picture of the magmatic history of the period. In some regions there was a long and complex sequence of events extending over several hundred million years; nowhere is this better illustrated than in the Gardar igneous province of South Greenland which is extremely well exposed and documented.

A summary of the magmatic evolution of the Gardar Province is given by Sørensen (1966), Upton (1974) and Emelus and Upton (1976). Fig. 6.7 shows the major intrusions

Table 6.2. Chronology of main events during the Gardar period in SW Greenland (after Allaart, Bridgwater and Henriksen, 1969; van Breemen and Upton, 1972)

Gardar subdivision	Main events	Isotopic age (my)
Late	Camptonitic dykes	
	Agpaitic intrusions	1020
	Saturated and undersaturated syenite intrusions, gabbros and granites	1180 ± 37
Middle	Major NE-trending basic dyke swarms (several generations) trachytes and syeno-gabbros	
	Some major intrusive centres	
Early	ESE- and local ENE-trending troctolitic dykes and early syeno-gabbro dykes	1187 ± 9
	Lamprophyric dykes (generally NE trending)	
	Nepheline syenite and carbonatite intrusions	1245 ± 16
	Basic and trachytic lavas and sandstones	
	Faulting (ESE-, ENE- and NS-trending sets), continued intermittently throughout Gardar time	

and the main dykes and faults, together with the sandstones and volcanics referred to in the last section, and Table 6.2 gives the Gardar chronology. The establishment of the main fault systems was followed by volcanism and sedimentation, and then by eight intrusive phases.

The intrusive rocks can generally be divided into dykes and plutonic complexes. There are a great number of dykes of several generations, the general emplacement order being lamprophyres and trachytes, dolerites and olivine dolerites (the most common type), granophyres, alkali microgranites and microsyenites, and finally trachytes and tinguaites. Two types of giant dykes are outstanding: composite dykes up to 500 m wide with marginal gabbro and central syenite that formed by a non-dilatational stoping mechanism (Bridgwater and Coe, 1970), and secondly troctolitic gabbro dykes up to 800 m wide.

Most of the plutonic complexes post-date the dykes. The predominant rocks are augite syenite, alkali granite and nepheline syenite, together with subordinate calc-alkaline granite and gabbro (Fig. 6.8). Probably the most well known is the Ilimaussaq Intrusion (Ferguson, 1964). Generally the bodies in the west of the province are saturated and akaline, whereas those in the east are undersaturated or peralkaline. Many exhibit spectacular rhythmic igneous layering (Ferguson and Pulvertaft, 1963), all are composite having been formed by two or more pulses of magma, and were emplaced by cauldron subsidence or stoping (Sørensen, 1966). Several major intrusions are located at the intersection of fault and dyke zones.

Many of the alkaline complexes are associated with carbonatites and these likewise lie in or near major faults. Individual bodies commonly have a central core of carbonatite bounded by nepheline syenite and zones of pyroxene and carbonate-rich rock. There is a 800–1000 my old carbonatite at Mountain Pass, California (Olson and Pray, 1954) together with seven shonkinite–syenite–granite stocks with fenites (Ridge, 1972), in the Gardar Province there is the Grønnedal–Ika nepheline syenite–carbonatite complex (Emeleus, 1964), and of twelve alkaline–carbonatite complexes in Quebec and Ontario eight have K/Ar ages of between 1005 and 1112 my, one of 1560 my, and three of between 1655 and 1740 my (Gittins, MacIntyre and York, 1967).

The evolution of the carbonatites was closely tied to that of the associated alkaline rocks. According to Parsons (1961) the sequence of events was first the formation of a

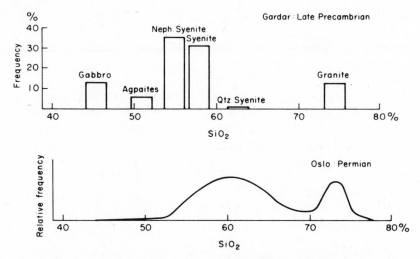

Fig. 6.8. Histogram of the relative abundance of the different intrusive rock types in the Gardar Province in comparison with the Permian sub-volcanic rocks of the Oslo Province (after Watt, 1966). Reproduced by permission of The Geological Survey of Greenland

central diatreme by explosive gases and the ejection of alkaline volcanic rocks, accompanied by alkaline metasomatism (fenitization) of the country rocks, followed by the carbonatization of the volcanic neck rocks and emplacement of carbonatite and syenite—an evolution reminiscent of that of the carbonatite volcanoes of the East African Rift Valley. In the Canadian bodies the early volcanics have since been eroded, but at Qagssiarssuk, South Greenland, minor carbonatite intrusions are still associated with early alkaline volcanics, amygdaloidal flows of carbonatized melilite rock, pyroclastic cones and tuffisite diatremes (J. W. Stewart, 1970).

With the exception of the isolated Palabora carbonatite in South Africa tentatively dated at 2060 my (Holmes and Cahen, 1957), the 1650–1750 my old Canadian bodies are the oldest known carbonatites in the world. Two interesting facts emerge: firstly, it took until 1750 my ago for the earth's crust to evolve sufficiently for the first major group of carbonatites to form, and secondly, these oldest carbonatites are little different from the recent ones in East Africa. An interesting feature is the grouping of the bodies at 350–400 my, 1000–900 my, and 1650–1800 my, corresponding to the age of the Caledonian-

Appalachian, Grenvillian, Hudsonian and Svecokarelian orogenies (Vartiainen and Wooley, 1974).

In the previous sections it has been demonstrated that some anorthosites are spatially associated with rapakivi granites, which in turn are related to fault-controlled acid volcanics, and that the intrusion of alkaline complexes locally took place in the same fault blocks. The close association of these rock suites in both time and space is nowhere better demonstrated than in the Korosten Complex in the Ukraine (Semenenko and coworkers, 1960). The complex includes several multiphase plutons, the intrusion of which was associated with faulting, subsidence and uplift of separate blocks of the Shield. The plutons consist of three rock suites: the first, forming the border zones, includes labradorite anorthosites, gabbro–anorthosites, gabbro–norites and gabbro–monzonites; the second, intruded at a later stage in the upper part of the plutons, includes rapakivi granites, rapakivi-like biotite–amphibole granites and biotite granites; and the third and youngest suite, forming the centres of the plutons, consists of alkaline rocks such as quartz–aegerine syenites. Whereas these rock suites formed at somewhat different times and often in different places in many parts of the North Atlantic Shield belt, they were telescoped together in the Ukraine to form this remarkable Korosten Complex.

We thus arrive at a point in time, 1000 my ago, when the final stages took place of some 700–800 my of igneous activity in the North Atlantic Shield mobile belt. In a way it was a culmination as it is significant that the last magma was chemically the most highly fractionated giving rise to, for example, the Ilimaussaq peralkaline intrusion with its 130+ minerals.

Mineralization

In the North Atlantic Shield mobile belt mineral deposits formed largely in association with the volcanic and intrusive rocks, and were thus dependent upon the prior formation of a relatively stable continental crust. However, in other parts of the world geosynclinal belts were forming at this time bordering stable cratons—for example, the Mount Isa Geosyncline, Australia, with the famous lead–zinc deposits which are 1600 my old.

One of the problems inherent in providing a synthesis of mineral deposits is that few have been dated isotopically with any reliability. Also, because of the diachronous development of many or most rock groups, it is impossible to make distant age correlations based on similarities with other bodies. This problem is well brought out in that excellent treatise of annotated bibliographies of mineral deposits of the western hemisphere by Ridge (1972).

U, Li, Be, Sn, etc. in Granites and Pegmatites

The partial melting processes that gave rise to the reactivated granites about 1800–1600 my ago in the Svecofennian, Karelian, Laxfordian and Ketilidian belts were largely unable to mobilize or concentrate elements sufficiently to produce mineral deposits of any economic value, although a few late granite plutons and pegmatites are economically endowed. In the Beaverlodge district of Saskatchewan there are more than 3000 uranium-bearing veins and pegmatites about 1800 my old in the Tazin Group of metasedi-

ments and associated granites which are regarded as largely granitized portions of the Tazin rocks (Tremblay, 1967). There are also the famous Black Hills pegmatites, associated with the 1620 my old Harney Peak granite stock in South Dakota (Page and coworkers, 1953), which contain economic concentrations of lithium minerals, beryl, mica, feldspar and tin.

Ti, Fe in Anorthosites

The major anorthosite bodies commonly contain ilmenite–titaniferous magnetite-haematite deposits (Gross, 1967). These oxide minerals are disseminated through the plagioclase, often forming layered rocks or intrusive masses. Titaniferous magnetite is more prevalent in gabbroic anorthosite or noritic gabbro, and ilmenite–haematite in cores of anorthosite bodies. The gabbroic phases frequently contain more iron and titanium oxides than do the anorthosite phases. Iron–titanium deposits occur, for example, in anorthosites at:

Bergen, Egersund, Inner Sognefjord, South Norway	Geis, 1971
Lofoten, North Norway	Geis, 1971
St Urbain and Allard Lake, Quebec	Rosé, 1969
Iron Mountain, Laramie Range, Wyoming	Hagner, 1968
Duluth, Minnesota	Lister, 1966
Sanford Lake, Adirondacks, NY	Gross, 1968

Porphyry Copper

According to Badham, Robinson and Morton (1972) porphyry intrusions at Great Bear Lake in northern Canada are associated with Cu (chalcopyrite) mineralization within the Echo Bay Group andesitic lavas (1800 my). It is interesting to speculate whether these rocks formed in a way similar to the porphyry copper deposits associated with modern active continental margins. The analogy is strengthened when one considers that the Echo Bay Group was deposited on the subsiding northern edge of the Slave Craton in a volcanic island belt (Badham, Robinson and Morton, 1972) that lay on the hinge line between the miogeosynclinal sediments of the carbonate–clastic platform to the south (the edge of the craton) and the Cordilleran eugeosyncline to the northwest (Fraser and coworkers, 1970).

Cu and U–Ag in Post-tectonic Volcanics

There are several distinctive mineral deposits in basic volcanics and tuffs extruded onto, or at the edge of, stable cratons in the period 1500–1000 my. We see examples here of the transfer of appreciable quantities of metals, presumably from the mantle, to the upper parts of the continental crust in this period (Watson, 1973b).

The most prominent is copper which is concentrated in basalt flows in the Coppermine River Group (1100 my), NW Canada, and in the Keweenawan basalts (1100 my) in Michigan (also in the lavas of the slightly younger Bukoban System in East Africa, c. 800–1000 my). The ores consist of either native copper or chalcocite in amygdules within the tops of flows or in quartz–carbonate veins and breccias.

U, Th, Be, etc. in Alkaline Intrusions

Several 1300–1000 my old alkaline complexes have a variety of element/mineral concentrations:

Ilimaussaq, South Greenland	U, Th, Nb, Be, Zr, Li, Rare Earths
Ivigtut, South Greenland	Cryolite, Pb, Ag
Blue Mountain, Ontario	Nepheline
Seal Lake, Labrador	Be, Nb, Th
Mountain Pass, California	Barytes, Rare Earths
Bancroft, Ontario	Nepheline, Corundum, Uranium, Radium
St. Hillaire, Quebec	

One of the most interesting complexes is Ilimaussaq in which there is a progressive enrichment of the U, Th, Nb, Be, Zr, Li and rare earths, especially lanthanum and cerium. The radioactive and rare elements were concentrated in the residual fraction of the highly differentiated magma entering hydrothermal solutions in the last stage of solidification.

The Ivigtut granite (Rb/Sr age of 1250 my) pluton contains an impressive cryolite (Na_3AlF_6)–siderite (75–20%) deposit with marmatitic sphalerite and silver-rich galena.

An unusual body is the 800–1000 my old carbonatite at Mountain Pass, California, which has remarkably large concentrations of barium and rare earths (Olson and Pray, 1954).

Chapter 7

Mid–Late Proterozoic Basins, Dykes, Glaciations and Life Forms

The late Proterozoic period is taken here to extend from 1000 my ago to the start of the Cambrian, about 600 my or 570 my ago. This period coincides with the Hadrynian Era in the Canadian time scale (Stockwell and coworkers, 1970) and the Epiproterozoic–Eocambrian of Salop (1972a) but is not coincident with the late Precambrian subdivision used in Australia (the Adelaidean, 1400–570 my) (Brown, Campbell and Crook, 1968; Walter and Preiss, 1972), or in the USA (Precambrian 2: 800 my to base of the Cambrian) (H. L. James, 1972). In the USSR the term Riphean is used for the period extending from 1600 ± 50 my to 680 ± 20 my, and Vendian for the 680 to 570 ± 10 my period (Sokolov, 1972).

There is some overlap of geological processes operating in the mid–late Proterozoic with those of the early Palaeozoic because the early stages of development of, for example, Appalachian–Caledonian fold belt (continental rifting, miogeoclinal deposition, early seafloor spreading) took place in the period 900–600 my. The deposition of the Torridonian, Moinian and early Dalradian sequences belonging to that fold belt will, consequently, be treated in Chapter 11.

This period is of special significance in the evolution of the continents for several reasons. A few rock groups, such as sedimentary copper deposits, appear for the first time, but more important is the fact that many rock groups, although they appeared in minor amounts in earlier periods, developed more abundantly in the late Proterozoic (e.g. red beds, tin and manganese deposits). The period is notable for the appearance of the eucaryotes and sexual cell structures and the rapid diversification of the first metaphytes and metazoans. During this time global glaciation occurred with the formation of widespread tillites. Cordilleran-type geosynclines reached a new peak of development and from this period our present pattern of cratons or shields and mobile belts began to emerge (Sutton, 1963).

Depositional Environment

In the mid–late Proterozoic there was a change from epicratonic basinal deposition in the preceding period (see Chapter 6) to a miogeoclinal type of sedimentation along rifted continental margins, notably in the Cordilleran–Appalachian–Caledonian geosynclines (A. D. Stewart, 1976). Glacial tillites commonly occur in these sequences. Schermerhorn (1975) reviews the regular depositional environments at this (and succeeding Cambrian) time. A typical succession may be: shallow-water quartzites or sandstones, shallow platform carbonates, pre-flysch deep-water laminated fine-grained mudstones, flysch, molasse sandstones. Interpreted in plate tectonic terms this sequence is similar to that in typical Cordilleran-type geosynclines in Phanerozoic fold belts. The deposition of early terrigenous quartz sands is followed by the building of a carbonate bank in the miogeosyncline; with the onset of deformation the carbonate bank collapses and is covered by deep-water mudstones starved of clastic material. Next a flysch sequence of greywacke turbidites is succeeded by lithic sandstones (molasse). The Amadeus Basin has approximately this succession which is fundamentally similar to that in the early Proterozoic Coronation Geosyncline in Canada. The implication to be drawn from this particular type of sedimentary development is that the late Proterozoic sequences started to form at the trailing edge of stable continental margins which were converted by the onset of subduction, mostly by early Palaeozoic times, into tectonically active margins.

As Schermerhorn points out, individual basins have their peculiar characteristics; thus the Adelaide Geosyncline lacks greywacke turbidites, flysch is absent in the Caledonian succession but present in the Appalachians, whilst the Brioverian of NW France has no carbonate stage. The platform deposits are usually thin compared with those in the geosynclines and do not extend for more than a few tens of kilometres from the margins of the late Precambrian geosynclines. This distribution suggests that the shields stood above sea level at this time (Sutton, 1967).

Let us now look at some mid–late Proterozoic sequences.

The Belt–Purcell Supergroup, western North America

The Belt–Purcell Supergroup (the term Belt is used in the USA and Purcell in Canada) extends from Idaho to British Columbia and is a 23 km thick sequence of remarkably homogeneous fine-grained clastics, dominantly siltstone and argillite with considerable interstitial carbonate; conglomerates and angular unconformities are rare. Radiometric data by Obradovich and Peterman (1968) on glauconite indicate that sedimentation took place intermittently from 1300 to 760 my (1450 to 850 my according to Harrison, 1972). Sedimentary rocks of Belt–Purcell age extend locally within the Cordilleran Structural Province to California and eastern Alaska

(Wheeler and Gabrielse, 1972; Wheeler and coworkers, 1974).

Ross (1970) concluded that most Belt Palaeoslopes were gentle, currents were slow, and the source terrain low with the result that thin marginal deposits covered large areas. Much of the fine-grained sediment was deposited on mud-flats with algal heads projecting above the water in a very shallow-water environment with an ill-defined shoreline, indicated by the lack of distinctive near-shore sediments and the presence of stromatolites and halite pseudomorphs. It is envisaged by Ross that the sequence was laid down in shallow salt water in a broad basin that was once essentially a large uninterrupted expanse of water and mud-flats. This environment required an extensive flat-lying stable platform of highly eroded basement rocks. From a consideration of its geometry and sedimentation Harrison (1972) concluded that the Belt Basin was an epicratonic re-entrant of the sea to the west dominated tectonically by slow gentle downward warping.

In Canada, the Lower Purcell sequence consists of shallow-water siliceous clastics in the east and deeper water turbidites in the west, and the Upper Purcell of basaltic lavas and shallow water sandstones, shales and minor carbonates with stromatolites (Wheeler and Gabrielse, 1972; Wheeler and coworkers, 1974).

Some of the Purcell rocks were affected by the East Kootenay Orogeny (maximum sillimanite grade) with a K/Ar thermal age of 750–850 my (corresponding to the low-grade Racklan orogeny in the northern Cordillera), possibly caused by the initiation of a short-lived easterly-dipping subduction zone (Fig. 7.1b).

The Windermere System of Canada was laid down unconformably on the deformed Belt–Purcell rocks. In Windermere time (800–600 my) there was an abrupt change to coarse, poorly bedded sediments suggesting that source areas of significant relief may have been caused by uplift by the preceding orogenies at the edge of the continent. The Windermere assemblage comprises in the lower part poorly sorted, rapidly deposited

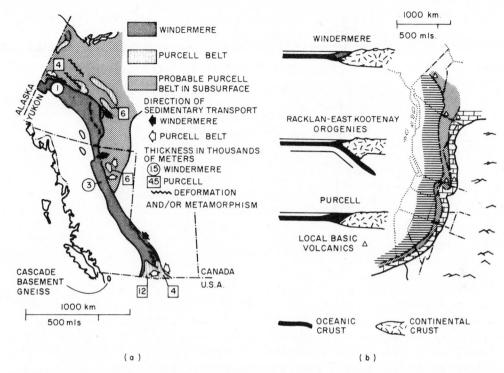

Fig. 7.1. A: Map of the Belt–Purcell and Windermere groups in western North America. B: A plate tectonic model to explain their occurrence according to Wheeler and Gabrielse (1972); reproduced by permission of The Geological Association of Canada

quartzo-feldspathic grits, sandstones and shales, with thick tholeiitic volcanics (conglomerates may be glaciomarine tillites according to Aalto, 1971) and, in the upper part, pelites and distinctive carbonate units. Like those of the Purcell System, the sediments thicken appreciably and rapidly to the west (Wheeler and coworkers, 1974).

In plate tectonic terms Wheeler and Gabrielse (1972) proposed that the Purcell–Windermere 'Basin', with the similar succeeding Lower Palaeozoic strata, evolved between oceanic crust to the west and a North American craton to the east (Fig. 7.1b). Sedimentation took place during a miogeoclinal stage when westward prograding continental terrace wedges formed a continental shelf–slope–rise in which the sediments were derived consistently from the craton to the east. Features consistent with the continental terrace wedge are:

1. The constant polarity of sedimentary facies revealing a one-sided cratonal source.

2. A relatively uniform total thickness of stratigraphic column, presumably controlled by initial depth of water and maximum amount of downbuckling of oceanic crust caused by sedimentary loading.

3. The common occurrence of basic volcanic sills and flows in strata near the outer edge of the terrace—possibly the result of tension in an oceanward-tilted wedge.

4. Continued oceanward tilting of the sedimentary wedge as shown by unconformities that bevel towards the craton.

It is interesting to note that there are deposits of phosphorite in the Belt Series. Phosphate deposition typically occurs between latitudes 40°N and 40°S along the western sides of continents and is caused by the upwelling of cold polar currents as they move equatorwards along western continental margins (Sheldon, 1964). The occurrence of phosphorites in the Belt Series is corroborative evidence that these rocks were deposited on a westward-facing miogeoclinal shelf.

Table 7.1. Stratigraphy of the Adelaide System showing the relative position of the two major late Precambrian tillites (after D. A. Brown, Campbell and Crook, 1968)

Cambrian	
Marinoan	The Pound Quartzite containing a medusoid–octocoral–annelid fauna
	Shales, siltstones, quartzites, dolomitic limestone with stromatolites
	Tillite boulder bed with sandstones. Upper glacials
Interglacials:	Siltsone, dolomite, muddy sandstone and local iron formation
	Silty shale with minor arkose
Sturtian	Tillite boulder bed with quartzites. Lower glacials
Torrensian	Argillite, dolomite, limestone, magnesite, quartzite, arkose
	Basal sandstone–conglomerate (1·95 km)
Willouran	Trachyte and minor andesite and rhyolite (600 m)
	Siltstone, slate, dolomite, quartz sandstone (3·6 km)

A refinement of this model was proposed by J. H. Stewart (1972) who concluded that the time of continental separation at the beginning of the Cordilleran geosyncline was marked by deposition of the unconformable Windermere Group on a miogeoclinal shelf along the trailing edge of the continent less than 850 my ago. The thick tholeiitic basalts near the bottom of the sedimentary sequence were a result of volcanic activity related to the thinning and rifting of the crust during early stages of separation. However, in an alternative model J. H. Stewart (1976) suggested that most of the Belt–Purcell sediments were deposited in epicratonic troughs and, consequently, that an ocean did not lie to their west.

The Adelaidean System, Australia

This system includes those beds deposited in the period 1400–570 my. The most well-known sequences occur in the Adelaide Geosyncline, the Amadeus Basin and the Kimberley Basin (Brown, Campbell and Crook, 1968).

The Adelaide Geosyncline The sequence in this 700 km long geosyncline exceeds 15 km thickness, is divided into four time–rock series (Willouran, Torrensian, Sturtian, and Marrinoan) and is notable for its lack of turbidites, its shallow-water marine clastics and carbon-

ates, its few volcanics, its two tillite horizons which reach a maximum thickness of 2·1 km, and its famous Ediacara fauna. The main stratigraphy is shown in Table 7.1.

The Amadeus Basin This 3·6 km thick, dominantly shallow-water sequence is situated in central Australia between high-grade granite gneiss complexes to the north and south (Arunta and Musgrave). The lower sedimentary quartzitic unit has been metamorphosed, part of the sequence has been recumbently folded, and there are prominent tillites associated with glacial erratics and striated blocks and correlated with the Sturtian glacials. The sequence is probably best known for the fact that a well-preserved, and biologically diverse, late Precambrian microflora occurs in the Bitter Springs Formation (Schopf, 1972). The major unconformity at the base of the overlying Cambrian, together with the deformation and metamorphism mentioned above, are taken to indicate a late Adelaidean 'orogeny'.

The Kimberley Basin This basin contains the most complete Adelaidean sequence and comprises three groups: the lowest (550 m thick) is a shallow-water sequence of sandstone, subarkose with minor black shale and siltstone; the intermediate (1200 m) consists of siltstones, sandstones and subarkose with a prominent boulder bed (the Landrigan Tillite) containing polished and striated cobbles, pebbles and boulders; and the uppermost group (4000 m) commences with an arkose–siltstone–sandy limestone sequence, followed by a second tillite, which rests on a polished and straited pavement and succeeded by quartz-rich arenites, siltstone and shale.

Basic Dykes

The intrusion of swarms of dolerite (diabase) dykes and sills on several continents in the period 1000–600 my was a reflection of the continued widespread, relatively stable, tectonic conditions that prevailed at this time. Many of the dyke swarms may have formed in connection with early rifting along continental margins, like the early Mesozoic dykes bordering the present Atlantic ocean (May, 1971). So far these basic intrusions are best documented in northeast, east and southeast Africa and north Canada, but they are also noted in the Baltic Shield and England.

Surprisingly few of these dyke swarms have been well dated by the Rb/Sr isochron technique. We have to rely on K/Ar ages which will remain suspect until corroborated by more sophisticated methods, and the age of many dykes is guessed only by reference to dated older and/or younger formations.

The most abundant dykes occur in northeastern Africa. According to Vail (1970) the late Precambrian terrain bordering the Red Sea in eastern Egypt, Sinai peninsula, western Saudi Arabia and northeast Sudan, is one of the most intensely dyke-intruded areas of the world; Fig. 7.2 shows that dykes occur throughout an area nearly 2000 km long and 500 km wide. Only one K/Ar age is available (740 ± 80 my) from a single dyke, but many of the dykes are associated with granitic bodies, some of which have Rb/Sr dates (quoted by Vail, 1970): Aswan Granite, Egypt (600 ± 20 my); Gattarian Granites, eastern desert, Egypt (500 my); Arabia (500–600 my). Some dykes are younger and some older than the granites, some have different strike directions, and there are dykes of acid and basic composition. It is not clear whether some of the dykes

are of Palaeozoic age, but they do predate the Mesozoic cover. It is unfortunate that so little is known about this remarkable Red Sea igneous province.

In Rhodesia some Umkondo dolerites have K/Ar ages of 650–1150 my and Waterberg dolerites likewise of 600–1130 my (Jones and McElhinny, 1966). This largely late Precambrian age is provisionally confirmed by the most recent palaeomagnetic results (J. D. A. Piper, 1973b).

The weakly metamorphosed and folded Bukoban System in Tanzania formed between 800 and 1000 my ago (upper basaltic lavas at 800 my and lower sandstones at 900–1000 my: J. D. A. Piper, 1973b). There are considerable numbers of dolerite dykes in this part of East Africa, some of which are associated with the Bukoban rocks whilst others are at a distance from them. Some swarms are 300 km long and 200 km wide. Harpum (1955) considered that, with their associated basalts, many of the dykes constituted a Bukoban igneous subprovince. Although preliminary K/Ar dates on some dykes have yielded ages of c. 2500, 1900 and 900 my (Snelling and Hepworth, quoted in Vail, 1970), their significance is unknown or uncertain and so the dykes remain provisionally grouped with the Bukoban rocks.

About 675 my ago (Fahrig, Irving and Jackson, 1971) a spectacular swarm of diabase dykes was intruded in northern Canada (Fahrig, Irving and Jackson, 1973). These 'Franklin diabases' crop out in a huge zone extending from Great Bear Lake and Coronation on the west, eastward to northern Ungava Bay and Baffin Island where they are particularly concentrated. Palaeomagnetic data suggest that the dykes were emplaced at low latitudes—a conclusion corroborated by the fact that late Proterozoic sediments of similar age contain gypsum, anhydrite and stromatolites which are indicative of a warm depositional environment. Also in the Bear Province diabase sheets and sills intruded into the Shaler Group on Victoria Island and the overlying Natkusiak lavas have a 'best estimated' K/Ar age of 640 my, and two groups of dykes near the Great Bear Lake have dates of

120

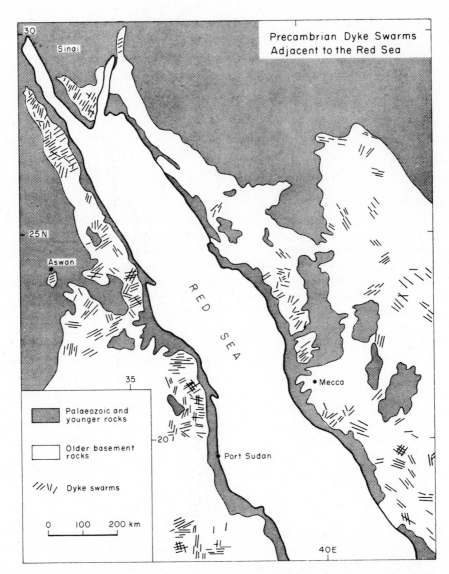

Fig. 7.2. Late Precambrian basic dyke swarms in the Red Sea region (after Vail, 1970; reproduced by permission of J. R. Vail)

614–619 my and 630 my (Fraser, Hoffman and Irvine, 1972). In the Canadian Slave Province shallow-dipping gabbro sheets that cut Goulburn and older rocks have radiometric ages of 650–700 my (McGlynn and Henderson, 1972), and in the Grand Canyon of Arizona the Cardenas lavas have a Rb/Sr age of 1090 ± 70 my (McKee and Noble, 1974).

Although their precise age is not known, there are many basic dykes of presumed late Precambrian age cutting mid-Precambrian rocks of southern Norway. According to Barth

and Reitan (1963) they postdate 1000 my old granitic rocks and the somewhat earlier anorthosite-norite suite of the Egersund Province.

The Precambrian inliers in central England, Wales and southeast Ireland contain undated small dyke swarms that Baker (1971) regarded as equivalent in age. They occur in the Rosslare Complex in County Wexford, the Uriconian Group of Shropshire, the Stanner–Hanter Complex of Radnorshire, the Johnston Series in Pembrokeshire, and the

Malvern Complex and the Warren House Series of Worcestershire.

Tillites and Global Glaciation

The most extensive period of glaciation(s) in earth history took place in the period between 1000 my and 600 my and the effects were more widespread than those of the early Proterozoic, Permo-Carboniferous and Quaternary glaciations.

The main rock resulting from glacial action is *tillite* which is a consolidated till, whilst the term *tilloid* is used for a tillite-like rock of doubtful origin (Harland, Herod and Krinsley, 1966). The general term *diamictite* includes boulder beds, clays and sands, pebbly sandstones and mudstones, tilloids and tillites (Flint, Sanders and Rodgers, 1960), and corresponds to the *diamict* of Harland, Herod and Krinsley (1966) and the *mixtite* of Schermerhorn (1966).

Distribution

Most continents contain some evidence of Late Precambrian glaciation, the main exception being Antarctica (for a detailed listing of many of these localities and their references see Harland, 1964a; Harland and Herod, 1975). Fig 7.3 illustrates the distribution of late Proterozoic tillites, the main type of glacial deposit.

The principal occurrences are as follows:

Europe: Scotland, Ireland, Norway and Spitsbergen, Sweden, Normandy (France), Czechoslovakia.

North America: many localities in the Western Cordillera from California to the Yukon and in the central Appalachians of the USA extending to Newfoundland in Canada; East Greenland.

South America: Brazil.

Australia: many localities extending in a broad arc from the southeast across

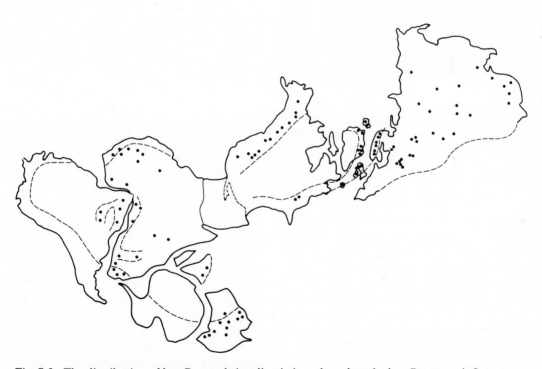

Fig. 7.3. The distribution of late Precambrian diamictites plotted on the late Proterozoic Supercontinent of Piper (1976b); reproduced by permission of The Royal Society. Also marked are the boundaries of late Precambrian 'geosynclines' (except in Asia). (Data from many sources, in particular Harland, 1964; Dunn, Thomas and Rankama, 1971; Schermerhorn, 1975; Stewart, 1976)

central to northwest Australia. Two glaciations—Marinoan or Egan and Sturtian.

Africa: Equatorial, southern and southwest Africa are the main areas (Congo–Katanga–Angola, Zambia, Zaire, Southwest Africa, South Africa).

USSR: Belorussia, the Urals, Khazakhstan, the Tien-Shan, central and northeast Siberia, Lake Baikal, Bashkirian Highlands.

China: little detailed recent knowledge is available, but late Precambrian glacial deposits are quite common, especially in central and southern China (Sinian tillites), Sinkiang, Turkestan, west Shansi, Hunan and Anhui.

Age

It is not easy to date accurately a period of Precambrian glaciation; however, several tillites have been radiometrically dated and some have been given an approximate age within a small time range on the basis of stratigraphic relations with beds above and below which have been isotopically dated. Steiner and Grillmair (1973) suggested that there were three late Precambrian glacial episodes with mean ages of 616 ± 30 my (Eocambrian), 777 ± 40 my (Infracambrian I), c. 950 ± 50 my (Infracambrian II). The following occurrences are assigned to these periods:

1. Eocambrian

my	
650	Egan–Marinoan glaciation, Australia
668 ± 23 (660–680)	Varanger Ice Age, northern Norway
600	Mortensnes ($\equiv$ Varangian) tillite, Finnmark, Norway
600–640?	Sveanor tillite, Spitsbergen, Norway
600	Upper tillite. Eleonore Bay Formation, East Greenland
560–630	Granville tillite, Upper Brioverian, Normandy
570 ± 10– 675 ± 25	Vendian period (tillites of Europe and Asiatic USSR)
570–600	Conception Group with tillite, Newfoundland, Canada
620–650	West Saharan tillite, Algeria
653 ± 70 (max.)	Nama–Damara tillite, SW Africa
600+	Lavras tillite, Brazil
570	Cambrian–Eocambrian glaciation, China

600 (or 660–720)	Nantou glaciation, China
650	Wushsingsham glaciation, China

2. Infracambrian I

740–750 ± 40	Moonlight Valley–Sturtian glaciation, Australia
715	Rybachy tillite of Kola Peninsula, European USSR
810	Volhynian tillite, Russian platform
747–810	Chingasan tillite, Siberia, USSR
750	Hsiho glaciation, China
750–850	Diamictites, West American Cordillera (California to the Yukon)
800 ± 50	Glaciogen Toby conglomerate, Windermere System, British Columbia
< 820	Southern Appalachian tillites, USA
750 ± 50	Petit Conglomérat, Zaire
719 ± 28	Bushmannslippe glacial sediments, SW Africa

3. Infracambrian II

950 ± 50	Grand Conglomérat, Zaire, lower Congo and Katanga
950 ± 50	Lower tillite of Tien-Shan near base of Late Riphean, USSR
950	Huishan glaciation, China

Tillites of probable late Precambrian age are found in Ghana, Togo, Dahomey, and the Anti-Atlas of Morocco (J. D. A. Piper, 1973a) and Sierra Leone and Guinea (Harland, 1964a).

Stratigraphy

In Norway, Australia, west central and southwest Africa, East Greenland and the USSR there are two prominent tillite horizons separated by interglacial sediments. In Australia there is one tillite formation in the Amadeus Basin but two in the Adelaide Geosyncline (lower Sturtian, Upper Marinoan—see Table 7.1), the Kimberley Basin (lower Landrigan and upper Walsh or Egan), and the Sturt Platform, equivalent to those of the Kimberley Basin (Brown, Campbell and Crook 1968). In west central Africa the lower and upper tilloids of the West Congo System are equated with the Grand and Petit Conglomérats (tillites) of Zaire (Schermerhorn and Stainton, 1963). In southwest Africa the fluvo-glacial sediments in the Orange River area in the Upper Hilda Formation of the Gariep Group, correlative with the Bushmannslippe glacial sediments, predate the tillites in the Nama and Damara Groups

(J. D. A. Piper, 1973a). In central east Greenland there are two tillites in the Cape Oswald Formation at the top of the Eleonore Bay Formation (Harland, 1964a).

The principal sediments occurring with the tillites are sandstones (in places red with ripple drift and cross-stratification), quartzites, shales and dolomites (locally stromatolitic), whilst the interglacial sediments include siltstones, mudstones, dolomites, and proximal and distal turbidites in Finnmark. Evaporites, red beds and iron formations also characterize many glaciogenic sequences (G. E. Williams, 1975). The tillites vary considerably in thickness from, for example, 7 m (lower tillite of Finnmark) to 2·1 km (Sturtian tillite in Adelaide Geosyncline). According to Spencer (1975) the North Atlantic Varangian tillites range in thickness from 2 m to 1 km and they mostly lie from 50 m to 800 m beneath fossiliferous Lower Cambrian beds; the ice age probably lasted about 10–30 my (660–680 my).

Dunn, Thomson and Rankama (1971) suggested that the late Precambrian tillites might eventually be adopted as stratigraphic time markers, but Crawford and Daily (1971) disputed this as almost certainly incorrect due to the implication of synchroneity of glaciation; they suggested that when accurate data become available, areas of glaciation will be found to lie along the late Precambrian polar wander curve. However, Piper (1973a) pointed out that if it can be demonstrated from palaeomagnetic data that the glaciation reached low latitudes, it will be legitimate to regard them as stratigraphic marker horizons. Also the correlation of glaciations with variable galactic parameters enhances, within limits, the probability of valid intercontinental Precambrian time–stratigraphic correlations (Steiner and Grillmair, 1973).

Characteristics

The essential character of a tillite is a wide range in grain size, from the fine-grained matrix to the conspicuous clasts or stones of various sizes from boulders to pebbles. The establishment of criteria diagnostic of glacial deposition has proved difficult, as tillites can easily be confused with 'mixtites' formed by non-glacial processes such as submarine sliding and mudflows. The equivocal interpretation of some features is made even more difficult by the fact that some tillites have themselves undergone subaqueous mass movement. These difficulties in interpretation have caused some genuine tillites to be misidentified as mudflows, and some non-glacial conglomeratic beds as tillites. However, these problems are gradually resolving themselves, particularly with the help of definitive criteria such as those laid down by Harland (1964a,b) and Harland, Herod and Krinsley (1966). The pendulum has swung to the present position where not only is the glacial origin of many mixtites accepted, but also detailed models of different types of glacial mechanisms are proposed, such as ice rafting, sedimentation from floating or grounded ice shelves, and deposition from continental ice sheets (Reading and Walker, 1966; Spencer, 1971). Harland (1964b) pointed out that few criteria are in themselves decisive of a glacial origin and usually a combination of features has to be sought for a definitive answer. He listed the diagnostic sedimentary features which are grouped according to their value as glacial indicators.

From a critical evaluation of the characteristics of mixtites Schermerhorn (1974, 1975) concludes that the evidence favours an origin by mudflow sedimentation in an active tectonic environment in contrast to a glacial origin.

The Glacial Environment

Two of the most sophisticated models explaining the complicated development of late Proterozoic glaciations are by Reading and Walker (1966) and Spencer (1971) for the Finnmark (Norway) and Port Askaig (Scotland) tillites respectively.

Spencer showed that the formation of the Port Askaig tillite involved seventeen successive ice advances and retreats (meltings). 'Many are recorded by the cycle: base (marine?) sediments deposited during a rise of

124

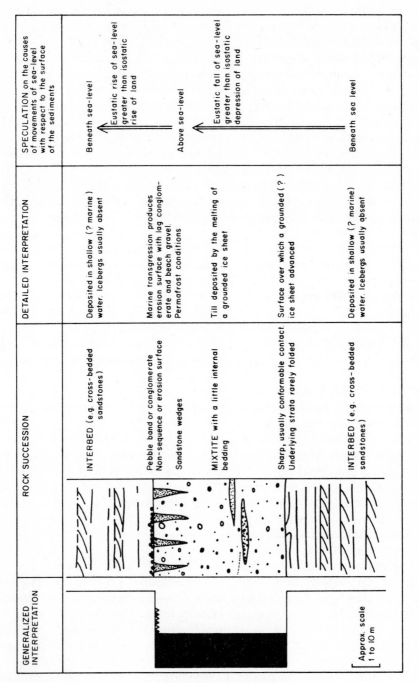

GENERALIZED INTERPRETATION		ROCK SUCCESSION	DETAILED INTERPRETATION	SPECULATION on the causes of movements of sea-level with respect to the surface of the sediments
		INTERBED (e.g. cross-bedded sandstones)	Deposited in shallow (? marine) water. Icebergs usually absent	Beneath sea-level
		Pebble band or conglomerate Non-sequence or erosion surface Sandstone wedges	Marine transgression produces erosion surface with lag conglomerate and beach gravel. Permafrost conditions	← Eustatic rise of sea-level greater than isostatic rise of land
		MIXTITE with a little internal bedding	Till deposited by the melting of a grounded ice sheet	Above sea-level
		Sharp, usually conformable contact. Underlying strata rarely folded	Surface over which a grounded (?) ice sheet advanced	← Eustatic fall of sea-level greater than isostatic depression of land
Approx. scale 1 to 10 m		INTERBED (e.g. cross-bedded sandstones)	Deposited in shallow (? marine) water. Icebergs usually absent	Beneath sea level

Fig. 7.4. The hypothetical glacial advance–retreat cycle characteristic of the Port Askaig tillite (after Spencer, 1971; reproduced by permission of The Geological Society of London)

sea-level, glacial mixtite, sub-aerial permafrost conditions (sandstone wedges), beach conditions (recording a transgression), marine sediments etc. (top).' (Fig. 7.4 summarizes the ideal glacial advance–retreat cycle.) The most useful time indicators in the sequence are the sandstone wedges which have a polygonal cross-section, interpreted as replacements of ice-wedge polygons formed under ice-free permafrost conditions, and which register as many as 27 periglacial periods. The tillite sequence, up to 750 m thick, contains granite and sedimentary fragments (the largest of which has the enormous size of $320 \times 64 \times 45$ m) occurring within 47 mixtite beds ranging in thickness from 50–65 cm and separated by sedimentary interbeds from a few centimetres to 200 m in thickness. As Spencer pointed out, the tillite is considerably thicker than analogous Pleistocene sequences and 'represents a glacial period of comparable, or even greater magnitude than that of the Pleistocene Ice Age'. If this is true, think of the significance of the Sturtian tillite sequence in Australia which reaches $2 \cdot 1$ km in thickness.

Palaeomagnetism and Palaeolatitudes

Because the geomagnetic and rotational poles are closely associated, geomagnetic latitudes should be equivalent to palaeogeographical latitudes and therefore palaeomagnetic data may be used to indicate the position of continental masses with respect to former geographical poles and equator (Runcorn, 1961). The interesting result of much recent palaeomagnetic work on rocks associated with Eocambrian tillites is that they were formed near the palaeo-equator of the time; this contrasts with the Permo-Carboniferous and Pleistocene glaciations which were near-polar.

Harland (1964a) reviewed his data with Bidgood and concluded that tillite horizons in southern Norway and eastern Greenland were deposited in near-equatorial latitudes. Likewise J. D. A. Piper (1972, 1973a) produced much data indicating that Africa lay in low latitudes from before 1000 my until postlower Cambrian times; in other words, Africa lay close to its present position during this period. Finally, Irving and Park (1972) produced an apparent polar wander curve which indicated that the whole of North America was situated less than 20° from the palaeo-equator in the period 600–800 my ago, suggesting that the late Proterozoic North American tillites were formed in low latitudes.

As Harland and Rudwick (1964) pointed out, our usual concept of an ice age is so moulded by what we know of the Pleistocene glaciation that we initially find it difficult to envisage ice in the tropics; but since the late Precambrian ice was clearly far more extensive than the Pleistocene ice, we have to imagine a drastically different situation. Land ice was probably situated at high latitudes at the start and end of the glacial epoch, and extended to middle latitudes during the coldest periods; it is however unnecessary to postulate land ice at low latitudes as 'floating ice and drifting icebergs could have carried material far beyond the boundaries of the continental ice sheets, depositing glacial sediments of similar composition over wide areas of the sea floor. Such deposition would account for the existence of tillites in the Infra-Cambrian tropics' (Harland and Rudwick, 1964).

However the above conclusion · of an extensive synchronous worldwide glaciation essentially at low latitudes was disputed by Crawford and Daily (1971) and McElhinny, Giddings and Embleton (1974). Their palaeomagnetic results indicated that very large polar shifts took place across almost all continents during the late Precambrian, from which they concluded that the poles migrated rapidly over different parts of the globe leaving records of glaciated regions as they passed over the various continents and that these glaciations took place at high latitudes.

H. Williams (1975) pointed out that the glaciogenic sequences contain substantial evidence for a marked seasonal inequability of climate. To account for this he postulated a considerably increased obliquity of the ecliptic in the late Precambrian, the effect of which would be to weaken climatic zonation and so allow ice sheets and permafrost to form in low and middle latitudes. So at the time of writing there is a controversy regarding the extent and place of the late Precambrian glaciations—it will be interesting to follow new palaeomagnetic data on this subject.

Mineralization

The following types of mineralization are found in sediments deposited during the late Proterozoic.

Copper in sediments

There are anomalously high concentrations of copper in late Proterozoic sediments in many parts of the world. They are important from our point of view because, as Watson (1973b) suggested, they represent the first major sedimentary accumulations in earth history. Their formation was dependent on a supply of copper eroded from earlier rocks, most probably basic volcanics. The host sedimentary sequences were laid down in aulacogens and along rifted continental margins.

The Katanga System in Zambia contains well-known economic sedimentary copper deposits. The main ore formation is a copper sulphide-enriched shale up to 70 m thick within the lower Roan Group which rests unconformably on a granite–schist basement (Mendelsohn, 1961). Sedimentary structures and lithologies indicate that near-shore shallow-water sedimentation gave rise to sandy pebble beds and carbonate deposits, largely of algal reef type.

The sediments of the Belt Basin in the northwestern USA contain anomalously high amounts of copper (Harrison, 1972). At least 100 ppm values occur in almost all formations throughout most of the thousands of square kilometres of the Belt terrain but there are no major economic deposits. The copper minerals occur along bedding planes, particularly in sandy or silty laminae. The occurrences also tend to contain anomalous amounts of silver and mercury.

According to Watson (1973b) the copper-bearing sediments of the Katanga System fringe the southern border of the large late Proterozoic craton on which the copper-bearing Bukoban volcanics were extruded about 800 my ago, and the cupriferous Belt sediments lie on the western edge of the Superior craton on the south side of which the copper-bearing Keweenawan lavas were extruded about 1000–1200 my ago. As the sediments were most likely derived by erosion of the cratons themselves, it is a reasonable assumption that the copper was supplied by the recently erupted lavas.

Further examples of mid–late Proterozoic sedimentary stratiform copper deposits are listed by Sawkins (1976):

Keweenawan Trough, Lake Superior
Coppermine River, NWT, Canada
Seal Lake, Labrador
Southern Appalachians
Adelaide Geosyncline
Damara Belt, SW Africa

Manganese in Sediments

The first major manganese deposits formed in early Proterozoic geosynclines where they are particularly associated with carbonate rocks. Conditions favourable for Mn deposition returned in late Proterozoic times when thick sedimentary sequences were laid down on, or at the border of, the broad cratons that had stabilized by this time.

Important concentrations occur in India and South West Africa The Sausar Group was deformed and metamorphosed in the Sausar orogenic cycle that ended 869–996 my ago (Sarkar, 1972). The manganese ore belt lies in the States of Madhya Pradesh and Maharashtra in central India (Roy, 1966). The term 'gondite' is locally used for the spessartite–quartz ore derived from non-calcareous argillaceous and arenaceous manganiferous sediments originally in an oxidizing environment. The main ore horizons occur in the Mansar Formation, consisting largely of muscovite schists, whilst the remainder of the Sausar Group contains thick dolomites and marbles together with quartzites and quartz schists.

At Otjosundu in South West Africa the Damaran System contains manganese ore deposits in three major horizons, at the contacts of quartzite, iron formation marble and biotite schists (Roper, 1956).

Tin in Granite Plutons

Watson (1973b) pointed out that, although tin is concentrated in the Archaean Bikita pegmatites in Rhodesia, forms detrital grains in the Dominion Reef Formation in South Africa and isolated but economic deposits in the Bushveld granophyre, tin mineralization did not appear in major quantities in the crust until the late Proterozoic period when it was associated with high-level alkaline and peralkaline anorogenic granites and pegmatites.

The post-orogenic stages of the Karagwe–Ankolean, Irumide, Kibaran–Burundian and Namaqualand–Natal mobile belts of Africa were characterized by the emplacement of widespread granites about 800–1000 my ago. Tin–tungsten–tantalum deposits are associated with many of these granites at various localities from Uganda to South West Africa and Natal (Fig. 19.13). One of the principal regions in this belt is in Zambia where there are a hundred main deposits in intrusive cassiterite pegmatites related to late granites (Legg, 1972). There is a belt of stanniferous pegmatites 40 km long and 1·5–3 km wide in the circumcratonic gneisses around the Rhodesian craton, which Hunter (1973) suggested may be about 1000 my ago. Tin deposits are associated with a group of 900 my old high-level alkaline granite plutons in the Rondonia district of western Brazil (Priem and coworkers, 1971).

The Development of Stromatolites

The occurrence of the first stromatolites in Archaean greenstone belts was reported in Chapter 2, but these were only locally developed. The organisms responsible for these structures reached an advanced stage of development in Proterozoic times, being particularly abundant in the period 1700–600 my BP. Stromatolites are the commonest fossil in the Precambrian and are used as an intercontinental Proterozoic zonal fossil in the USSR and Australia.

Stromatolites are layered stratiform, conical or columnar, biogenic sedimentary structures formed by sediment-binding algae. Studies on modern stromatolites show that they grow in the intertidal and shallow subtidal zones where they are associated with cross-bedded sediments deposited in extensive tidal flats. Characteristically they occur in cherts, dolomites and limestones which are often associated with banded iron formations in early–mid Precambrian environments.

Cloud (1968a) suggested that the binding of the micro-organisms in the stromatolites took place within originally gelatinous silica and mucilaginously entrapped calcium carbonate which provided a means by which the organisms could remain anchored below an ultra-violet shielding layer of sediment. The stromatolite-building blue-green algae are more resistant to DNA-damaging ultra-violet light than bacteria and other organized cells. This resistance, coupled with the shielding effect of the trapped sediments, was probably the reason why the blue–green algae adapted so well to the intertidal zones in the Precambrian; here they lived for the remainder of geological time, the effect of the ultra-violet radiation having been removed by the build-up of an ozone screen in late Proterozoic time.

The record that emerges indicates that stromatolite environment has scarcely changed during the last 1000–2000 my. They are commonly associated with interstitial breccias and ooliths, suggesting turbulent water, with contraction-cracked desiccated sediments and truncated flat-topped ripple marks, implying exposure to the atmosphere (Cloud, 1968b), and with flat pebble conglomerates suggestive of a palaeo-beach zone (Vidal, 1972). According to Cloud (1968b) the maximum height (or amplitude) reached by many stromatolites at their maturity is a reflection of the tidal range. Compared with the modern examples at Shark Bay, Western Australia, which reach a maximum amplitude of 0·7 m, he pointed out that Precambrian stromatolites are often considerably larger ranging from about 2·5 m to 6 m in amplitude; some of the largest occur in the late Proterozoic Belt Series in Montana, and the Otavi

128

Series, South West Africa, and Cloud (1968b) inferred from this that the Precambrian tidal range was much greater in the past and hence that the lunar orbit was closer to earth; however, Walter (1970) challenged this as an incorrect extrapolation. He pointed out that early–middle Cambrian stromatolites near Lake Baikal are up to 15 m high and that Devonian stromatolites in western Australia grew in water as deep as 45 m, and he therefore suggested that many fossil stromatolites may have formed subtidally. This was supported by Awramik (1971) who concluded, from the lack of evidence indicating a periodically-exposed intertidal-to-supratidal environment as well as the enormous size of some forms referred to above, that most columnar stromatolites grew in the subtidal zone.

An assessment of the number, widths and groupings of laminations in modern stromatolites compared with those in fossil forms provides an interesting means of using stromatolites as palaeontological clocks. Pannella (1972) demonstrated that the growth patterns may be of daily, tidal, monthly, seasonal and annual periodicity; he concluded that, since some 2200 my old stromatolites in the Great Slave Supergroup have tidal bands, the earth–moon system must have been in existence since the early Proterozoic, although there is a tendency for the number of days to have decreased per tidal band.

Biostratigraphy of the Proterozoic

It was first realized by Russian biostratigraphers that stromatolites underwent sufficient evolutionary changes for them to be used as zone fossils, enabling the Riphean period (1600 ± 50–680 ± 20 my) to be subdivided into three major units (Raaben, 1969); furthermore, the three main members of the late Riphean (950–680 my) in the USSR are characterized by specific types of stromatolites that can also be recognized in Spitzbergen. The discovery that distinctive stromatolite assemblages are recognizable in their correct stratigraphic position in widely separated parts of the USSR was a remarkable breakthrough, providing a means of correlating late

Proterozoic formations on an intercontinental basis.

Cloud and Semikhatov (1969) showed that many columnar stromatolites have intercontinental distributions (Karelia, Aninikil Group, Labrador Geosyncline, Witwatersrand and Transvaal Systems) and occur over roughly the same stratigraphic ranges as defined in the USSR. These authors also demonstrated that the pre-Riphean part of the Proterozoic (c. 1600–2500 my) and the terminal or post-Riphean Vendian (c. 680–570 my) are characterized by certain diagnostic stromatolites, which means that much of the Proterozoic period may be subdivided on the basis of stromatolite types.

The stratigraphic distribution of stromatolites in Proterozoic sequences in Australia is the most comparable to that in the USSR (Walter, 1972a; Walter and Preiss, 1972). They are distributed throughout eight different basins or geosynclines in sediments ranging from 2200–570 my old. In particular, three groups of stromatolites have a distinctive distribution, occurring in early Proterozoic, late

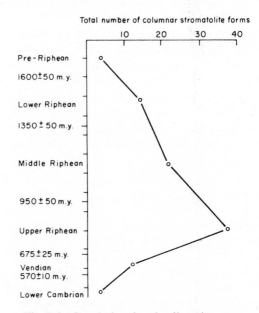

Fig. 7.5. Graph showing the diversity curve for Proterozoic columnar stromatolites (after S. M. Awramik (1971), *Science*, **174**, 825–827; copyright 1971 by the American Association for the Advancement of Science)

Riphean and Vendian sediments (Walter, 1972a).

The stromatolites were most prolific and reached their peak of diversity in the late Riphean, after which they declined appreciably in the Vendian and Palaeozoic. Awramik (1971) and others ascribed this rapid decline to the appearance of the Metozoa, similar to the Ediacara and Nama faunas in the Vendian; it is likely that these early animals not only ate the algae, but also destroyed the stromatolites by burrowing into them and inhibiting their growth by feeding on bottom deposits. Fig. 7.5 illustrates the sharp drop in the diversity of stromatolites in the Upper Riphean period just before the start of the Cambrian.

The First Metazoa

Metazoa are multicellular animals that require an oxygenous environment for their growth. They reached their peak of development in Phanerozoic time when many evolved protective shells and hard skeletons. During the last decade impressions of many soft-bodied animals have been found in late Precambrian rocks on several continents, providing evidence of early Metazoan life that existed before the appearance of hard parts; the origins of multicellular life lie somewhere in earlier Proterozoic sequences.

Quartzite, or Sandstone, the uppermost formation of the Marinoan Series of the Adelaide Geosyncline, 500 m below the unconformable basal Cambrian. The Pound Quartzite contains similar fauna at several localities up to 110 km from Ediacara. The age range of the Ediacara fauna is probably 590–700 my (Glaessner, 1971).

The fauna includes, in particular, jelly fish (Medusoids), worms (Annelids), sponges, soft corals related to the living sea pens, and several types of creatures unlike any known organisms. Glaessner (1971) lists the following assemblage:

Coelenterata (67% of specimens):
  Medusoids (3 species)
  Hydrozoa (3 species in 3 genera)
  Conulata (1 species)
  Scyphozoa (2 species in 2 genera)
  Anthozoa (Pennatulacea—4 species in 3 genera)
Annelida (25% of specimens):
  Polychaeta (5 species in 2 genera)
Anthropoda (5% of specimens):
  Trilobitomorpha (1 species)
  Crustacea (1 species)
Plus Tribrachidium of unknown affinities.

Fossils similar to the Ediacara fauna occur in late Precambrian rocks in several parts of the world. Awramik (1971) cites the following occurrences and ages:

Location	Rock unit	Age (my)
Ediacara, S Australia	Pound Quartzite	
Flinders Range, S Australia	Pound Quartzite	
Punkerri Hills, S Australia	Punkerri Sandstone, Officer Basin	
Deep Well, N Australia	Arumbera Sandstone	
Charnwood Forest, England	Woodhouse Beds	>574–684
South West Africa	Kuibis Quartzite, Nama Series	>510
Torneträsk, N Sweden		c. 600
Podolia, Ukraine, USSR	Bernashov Sandstone	590
Yarensk, NE of Moscow, USSR	Gdov laminarites	c. 590
Olënek, N Siberia, USSR	Khatyspyt Formation	550–675
Rybatschii Peninsula, USSR		670–900
Central and E Russian Platform		650–675
SE Newfoundland, Canada	Conception Bay Group	?574 ± 11

The type of locality of these late Precambrain metazoans is at Ediacara in the Flinders Range, 320 km north of Adelaide in south Australia. The fossiliferous beds lie in the current-bedded and ripple-marked Pound

The Appearance of Sex

One of the 'most consequential innovations to have occurred during the course of biological evolution' was the ability of micro-organisms to change their mode of reproduction from asexual mitosis (simple cell division) to sexual meiosis (splitting of a body cell into two germ cells followed by their union, or fertilization). Schopf (1972) records evidence of this adaptation in fossil microflora in 900 my old black carbonaceous cherts of the Bitter Springs Formation in the Amadeus Basin, central Australia. The population consists of 50 species including bacteria, eucaryotic algae and probably fungi, but it is dominated by blue–green algae very similar to extant forms. Schopf made the imaginative analogy that through much of geological time the blue–green algae have suffered from the 'Volkswagen syndrome—little or no evolution of external form concealing marked changes of internal machinery'. The critical evidence for sexual reproduction by this time lies in the fact that some eucaryotic algae have unicells in varying stages of mitotic cell division, whilst others have a tetrahedral arrangement of four spore-like cells, similar to the spore tetrads of living plants formed by sexual meiotic division. Thus by 900 my ago primitive micro-organisms had apparently evolved advanced techniques of sexual cell reproduction; in this respect the Bitter Springs microflora may provide an important benchmark in early biological evolution.

It is said that 'the rise of the eucaryotic cell from its procaryotic ancestors was the single greatest quantum step in evolutionary history'. This step involved the development of a sheath around the nucleus enabling cell division to take place and so for the DNA code to be passed on to the daughter organisms. It is usually assumed that the earliest evidence of possible eukaryotic remains is found in the 1300 my old Beck Springs dolomite in southern California and that the oldest assured eukaryotic organisms are in the Bitter Springs Formation. These occurrences give a minimum age of the evolution of the eucaryotic form of life. However this age has been challenged by Knole and Barghoorn (1975) who found from experiments that partial degradation of certain modern algae simulated exactly the entire range of morphological variations including tetrahedral arrangements

said to occur in the Bitter Springs Formation. They interpret the latter as just blebs of degraded protoplasm within undecomposed sheaths. If this conclusion is correct, it follows that the eucaryotic cells may not have developed until the very end of the Precambrian, near the time of appearance of the Ediacara fauna and, secondly, that the tetrahedral structures referred to above are no more than pseudotetrahedral arrangements, implying that sexual reproduction evolved at a later date.

The Precambrian–Cambrian Boundary

Many estimates have been made of the age of the end of the Precambrian—570 my is probably the most popular current figure (Geol. Soc. London Phanerozoic Time-scale 1964); we are concerned with the evolutionary changes that took place across this important time boundary. Harland (1974) gave the following critique of the possible factors that might have been responsible for the sudden development of life forms at this time.

Sedimentary structures suggest widespread shallow water conditions and the presence of glauconite indicates a marine environment. It has been suggested that the Precambrian sea water was too acid to allow growth of calcareous animal hard parts, too alkaline to allow sufficient concentration of Ca for precipitation of carbonate, and to have had too high a Mg/Ca ratio to allow precipitation of calcium carbonate. But the fact that calcareous algae flourished in the late Precambrian negates these hypotheses. Also a major change in CO_2 or O_2 pressures in the oceans seems unlikely as calcareous algae and carbonate sediments continue unchanged across the time boundary.

Harland and Rudwick (1964) and Harland (1974) suggested that warming of the seas and melting of the ice after the last major period of Eocambrian glaciation, about 600 my ago, gave rise to flooding of peneplaned land and variable salinities with mixing melt waters, and this new environment provided ideal conditions for rapid biological evolution. The sudden increase in marine tides probably favoured protection from wave action or intertidal desiccation.

One of the main factors responsible for the evolutionary explosion was probably the increase in oxygen content in the atmosphere to 1% of the present level (Berkner and Marshall, 1967).

Although there are some unconformities at the base of the Cambrian and although there was locally tectonic activity at this time (e.g. the Katangan orogeny at 620 my and the Cadoman phase of the Assyntian (Baikalian) orogeny at 570 my) the large number of successions that continue across the boundary without marked diastrophism suggests that the continental crust was not particularly mobile at this time (Harland, 1974). Thus there seems to have been little tectonic control, in the broadest sense, of biological evolution.

About 570 my ago the Metazoa evolved hard skeletons and shells in a process that concerned the formation of collagen, the main structural protein in Metazoan tissues. Its synthesis required molecular oxygen and, according to Towe (1970), the organisms that were first capable of using oxygen for its synthesis were the most primitive: those that had the fewest organs, the least demanding muscles and epidermal respiration. This concept is consistent with the fact that the earliest Ediacara Metazoans lacked hard parts, but presumably did contain collagenous tissue making their soft bodies more easily preservable. Thus the beginning of the Phanerozoic defines not the appearance of life but the rapid development of fossils with hard parts.

Chapter 8

Continental Evolution in the Proterozoic

The application of plate tectonic theory to Proterozoic high-grade fold belts has reached a very interesting controversial stage. On the one hand there is the plate tectonic model, according to which belts formed by collision and welding of two continental plates, and on the other there is growing evidence that the bordering 'plates' have undergone little or no independent horizontal motion.

There has never been a universally accepted hypothesis to explain the mode of formation of high-grade mobile belts that contain large areas of reactivated basement. One of the problems involved in falling back on the classical orogeny model based on an early geosynclinal stage of development is that many of these mobile belts have little or no evidence of an initial phase of sedimentation or volcanism. This means that the main chronological events that have come out of very detailed studies include partial melting and mobilization of earlier gneisses and granites with formation of younger granites followed by intrusion of a number of igneous rocks of different types. For example, according to Kroner (1976) many mobile belts in southern Africa developed largely by remobilization of earlier Archaean rocks, and Clifford (1972) concluded that 'the main effects of Eburnian–Huabian orogenesis are recorded as remobilization of pre-existing crystalline basement. The widespread absence of geosynclinal sediments and volcanics of that orogenic cycle represents one of the most impressive hiatuses in the African record'—and from this he further concluded that we are now looking at such a deep section of the crust that the sedi-

ments and volcanics have been eroded leaving the underlying basement rocks exposed. This reasoning leads to a noteworthy point: the Proterozoic mobile belts in Africa and the Grenville belt in North America are floored by older crystalline continental crust and contain no examples of geosynclinal sequences deposited on extensive areas of oceanic crust (Clifford, 1972; Wynne-Edwards, 1972b). According to this viewpoint ocean floor spreading did not operate during the last half of Precambrian time and hence this part of earth history largely involved intracontinental, not marginal, orogenesis.

A new perspective was added to the understanding of these mobile belts by the application of the continent–continent collision plate model by Dewey and Burke (1973). This model viewed the main components of the mobile belt as having formed in a thickened overriding continental plate, and the present erosion surface to be a deep-level section of the plate exposing plutonic rocks such as granulites, anorthosites and mesozonal granites. They suggested that the Grenville belt is comparable with the Hercynian and Himalayan fold belts which share, amongst other things, the following characteristics: a steep geothermal gradient and low pressure–high temperature metamorphism; large amounts of syn- and post-tectonic granites and migmatites; considerable basement reactivation; and a particularly wide orogenic belt. A crucial point about the model is that it does not require early deposition of a supracrustal series to start off the 'orogenesis'. We should not forget that Phanerozoic Cordilleran-type fold belts are developed upon an older sialic basement.

Having laid out the problem and the two main evolutionary models, let us now look at them in more detail, with particular regard to the palaeomagnetic and structural evidence in North America and Africa. In the final part of this chapter we shall consider the possible development of a supercontinent in late Proterozoic times and its rifting and break up which were initial stages in the formation of the Cordilleran–Caledonian–Appalachian fold belts.

North America

The apparent polar wander curve for the Precambrian Shield of North America, Greenland and NW Scotland (Laurentia) through Proterozoic time is given in Fig. 8.1. The sharp turning points in the loops are termed *hairpins* and are considered to record a major change in direction of horizontal movement relative to the pole. The polar paths between the hairpins are referred to as *tracks*,

which correspond to periods of time called *super-intervals* (Irving and Park, 1972) during which the continent underwent a major change in latitudinal position.

The main palaeomagnetic results of Donaldson and coworkers (1973), Irving and Lapointe (1975), McGlynn and Irving (1975), Pullaiah and Irving (1975) and Irving and McGlynn (1976) regarding the Proterozoic development of North America are as follows:

1. Palaeomagnetic poles from several places within the Laurentian Shield from 2200 to 1000 my ago (and especially 2200–1800 my) all lie on a single polar path (tracks 2, 3, 4 and 5 of Fig. 8.1); this suggests on face value that the bulk of the Shield was intact and behaved as a single entity during that time interval. The main implications of these results are, firstly, that during the Hudsonian orogeny (when a large percentage of the Shield was isotopically readjusted) there was no significant sea-floor spreading, subduction and plate convergence within the area now occupied by the Hudsonian Province; and, secondly, that the early

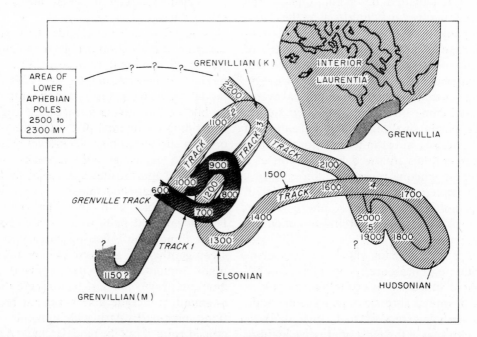

Fig. 8.1. Polar wander paths from 2200 to 600 my BP relative to Laurentia with the five main tracks indicated. Hairpins are named by the orogeny to which they are thought to correspond (after Irving and Lapointe, 1975; reproduced by permission of *Geoscience Canada*)

Proterozoic Circum-Ungava Geosyncline (Labrador Trough, etc.) cannot be the remnant of a *wide* ocean.

2. Concordant results from a variety of rocks dated at 1400 my (e.g. Michikamau anorthosite, Ontario; Cober Island Complex, Missouri) indicate that the intervening part of North America has remained intact within a single crustal plate since that time.

From palaeomagnetic evidence and age determinations Fahrig and Jones (1969) suggested that the NW-trending diabase dykes of the Slave Province (the Mackenzie dykes), the Muskox Intrusion, the Coppermine lavas and the Sudbury dykes (the southern extension of the Mackenzie swarm) were products of closely related igneous activity termed the Mackenzie igneous episode. Rb/Sr data by Gates and Hurley (1973) suggest an age of between 1370 and 1730 my for the Mackenzie dykes and 1460 ± 130 my for the Sudbury ones (Chapter 4). Concordant palaeomagnetic data on the dykes, lavas and the Muskox Intrusion spread from the Slave Province to Sudbury, indicate that the intervening terrain has remained in one piece since the igneous rocks were emplaced (Donaldson and coworkers, 1973).

3. The fact that the poles for Grenville rocks shown in Fig. 8.1 do not fit the main polar wandering curve suggests that in the interval 1150–1000 my the Grenville Province underwent a major horizontal displacement with respect to the remainder of the Precambrian Shield (Irving, Emslie and Ueno, 1974b).

Palaeolatitude evidence should correlate with the types of sedimentary rocks of the relevant age indicating the environment of deposition of the sediments. The following evidence is pertinent for the North American Proterozoic:

1. From 2500–1950 my ago thick dolomitic carbonate sequences with prominent algal reefs and stromatolites were deposited in low palaeolatitudes, e.g. Coronation Geosyncline. Likewise red beds (1800–1400 my), stromatolitic limestones, red beds and evaporites (1000–570 my), and banded iron formations (2100–1800 my) were deposited at a time when the palaeolatitudes were generally less than 30° (Donaldson and coworkers, 1973; Irving and Lapointe, 1975).

2. When the Huronian tillites were formed (2100–2200 my) the relevant southern part of the Shield was in high latitudes, about 70°N (Irving and Lapointe, 1975).

The Grenville Belt

This mobile belt extends in a northeasterly direction from approximately southeast USA to the southern coast of Labrador and probably continues in southern Scandinavia as the Dalslandian–Gothian Belt. Five major models are currently being considered for its tectonic mode of development.

1. The classical model of Wynne-Edwards (1972). Firstly there was deposition of up to 10 km of carbonates, quartzites and shales of platform type belonging to the Grenville Supergroup on an Archaean and early Proterozoic cratonic basement. This unconformable sequence was then deeply buried and intrusions of anorthosite tended to spread laterally along the cover–basement junction. The main peak of metamorphism was subsequently reached 1200 my ago (Rb/Sr and Pb–U determinations) and the rocks cooled through the isotherm at which they retained argon about 950 my ago (the widespread K/Ar Grenville age). Erosion has removed most of the supracrustal cover exposing the catazonal level of the orogenic belt with the result that it now consists largely of reactivated older basement with recognizable structures of Kenoran and Hudsonian age. This model is thus similar to that of Clifford (1972) with regard to the African orogenic belts but its major drawback is that it lacks a *reason* for the orogenesis.

2. The Dewey and Burke (1973) continental collision model (illustrated in Fig. 8.2).

Between two colliding continental plates intervening oceanic lithosphere was largely destroyed, to be preserved only at a high level as ophiolites and blueschists. Extensive convergence was accommodated by thickening of the leading overriding plate causing geoisotherms to rise and partial melting to start in the lower part of the thickening plate.

136

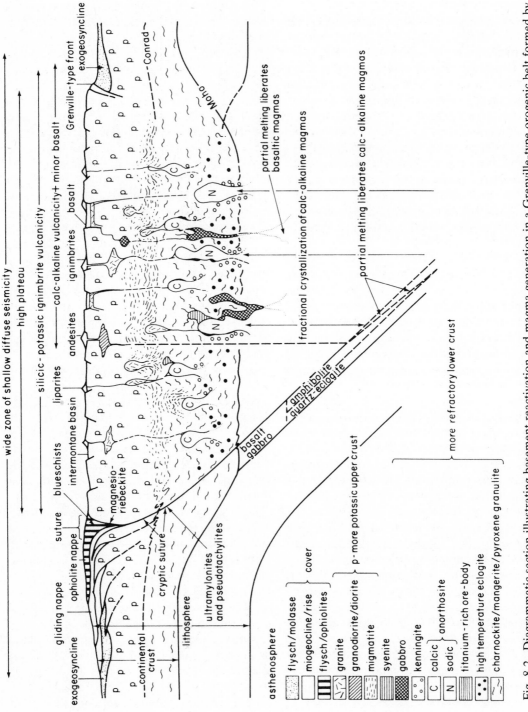

Fig. 8.2. Diagrammatic section illustrating basement reactivation and magma generation in a Grenville-type orogenic belt formed by collision of two continental plates (after Dewey and Burke, 1973)

This process was assisted by the fact that the plate already had a higher than normal heat flow as the events took place immediately after the high-grade plutonism of the early Proterozoic.

The anorthosites remained in the lower catazonal levels of the crust, the more sodic andesine types rising to a slightly higher level than the labradorite type. Some granitic rocks (mangerites–charnockites and rapakivi granites) remained close to the anorthosites, but others rose to higher levels and so were separated from them. Some granitic liquids reached the earth's surface to give rise to potassic acid volcanics such as labradorite porphyry lavas and porphyritic rhyolites. This volcanism was a late event in the history of the belt, at a time when continental convergence and thickening were nearing completion and isostatic recovery of the orogenic belt was marked by the development of high-level gravity faults which controlled the formation of intermontane epicratonic troughs associated with the acid volcanism and deposition of molasse-like coarse clastics. These closing stages in the evolution of the mobile belt were accompanied by the formation at deep levels of alkaline magmas which rose upwards through the crust, controlled by the late major faults, to be intruded at high levels in graben-like structures.

The Grenville Front cannot be the site of the main Benioff Zone or intercontinental suture because the distinctive iron formation of the Labrador Trough continues for at least 240 km south of the front into the Grenville Province. Dewey and Burke (1973) consequently suggested the main suture must lie somewhere southeast of the present Grenville Belt.

3. Palaeomagnetic evidence by Irving, Emslie and Yeno (1974b), Irving and Lapointe (1975) and Irving and McGlynn (1976) supports a plate tectonic model for the Grenville orogeny. Poles from the Grenville Belt are displaced from those from the rest of Laurentia suggesting that the southern part of the belt was displaced about 5000 km 1150 my ago. The main part of the original belt moved northwestwards (according to present-day coordinates) between 1125 and 1000 my to collide eventually with the Laurentian Shield. This model requires that the suture zone lies just to the southeast of the Grenville Front within zone S shown on Figure 8.3. Thomas and Tanner (1975) suggest that a 1200 km long linear negative Bouguer anomaly may indicate the suture position, but no surface geology has yet been found to confirm it.

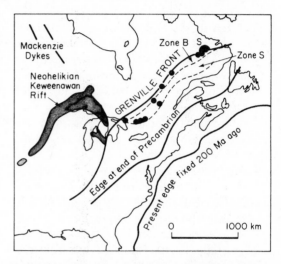

Fig. 8.3. Successive southeastern margins of North America. The Grenville suture should lie between zones B and S. The Seal Lake rift system (S), the Mackenzie dykes and related Coppermine–Muskox volcanism and the Keweenawan rift and mid-gravity high are parts of an early Grenville rift system. Dots are alkaline complexes (after Irving and McGlynn, 1976; reproduced by permission of The Royal Society)

However J. D. A. Piper (1975) emphasizes that Grenville-type palaeomagnetic directions are now known from within the Laurentian Shield (500 km northwest of the Grenville Front—Halls, 1975) and this fact would make the plate tectonic hypothesis virtually untenable for the Grenville Belt.

4. Baer (1976a) interpreted the geology of the Grenville Belt in terms of a plate tectonic model. Three evolutionary stages are envisaged:

a) Emplacement of anorthositic rocks into pre-existing sialic basement in the period 1500–1400 my ago (but what caused such anorthosite intrusion?).

b) About 1300 my ago a 'proto-Atlantic ocean' developed along NE-trending fractures to the southeast of the present belt. Associated rifting on the northwest continental plate gave rise to the Seal Lake graben, the Mackenzie dyke swarm and related Coppermine–Muskox volcanism (c. 1250 my), the Keweenawan rift filled with basic volcanic rocks (1200–100 my old—see Baragar in Baer and coworkers, 1974), an aulacogen near Bancroft–Renfrey in which the Grenville Group of marbles, clastics and volcanics was deposited (1310 ± 15–1250 ± 25 my) and, finally, a chain of nepheline-bearing alkaline complexes on one side of the aulacogen (1280 my).

c) The ocean closed and the resulting continental collision gave rise to reactivation and deformation of the Grenville Province about 1100 my ago. Some anorthosites and mangeritic plutons were remobilized, some re-intruded and others deformed marginally. According to this model the suture lies hidden somewhere in the Appalachians.

5. The millipede model of Wynne-Edwards (1976), according to which ductile spreading followed by contraction of the crust was the orogenic mechanism responsible for crustal reworking from about 2500 my to perhaps 600 my ago, and accounting in particular for the formation of the Grenville, Namaqualand and Damaran mobile belts. In the case of the Grenville, continental crust crept northwards like a millipede at a rate of about 10^{-1} cm/yr over an easterly trending spreading centre. The crust underwent thinning and extension as it became ductile over the zone of thermal upwelling, allowing the formation of an epicontinental sea and corresponding cover sediments. The ductile crust and cover then moved off the spreading system and were subjected to deformation as they were compressed against the unheated craton. Aligned trains of plutonic complexes like the anorthosites provide relative movement vectors which track the motion of the plate as it passed across the spreading system.

At the present stage of research none of the five models completely satisfies everyone; evidence can be found to support each but at the same time objections can be made to all. However, since it is one of the world's most intensively investigated mobile belts it may not be long before new constraints are found to limit the choice of working models.

Africa

In his general assessment of African mobile belts Clifford (1972) concluded that the Proterozoic belts formed during the Eburnian and Huabian (1850 ± 250 my ago), Kibaran (1100 ± 200 my ago) and Pan-African (550 ± 100 my ago) orogeneses developed on an older floor of sialic crust. His main evidence for this conclusion was based on the fact that the geosynclinal (supracrustal) parts of the belts can be seen today resting unconformably on older crystalline basement:

1. The Kibara Group (1300–2000 my old) rests on Ubendian, Lukoshian and Buganda–Toro rocks.
2. The Katanga, Outjo and western Congo Systems (late Precambrian) rest on pre-2700 my old basement in central Africa, SW Africa and the Lower Congo respectively.

Clifford emphasized that the belts commonly retain the recognizable remains of older mobile belts—for example, the Mozambique belt contains rocks from the Nyanzian in Kenya, the Bulawayan in Rhodesia and the Dodoman in Tanzania.

Shackleton (1973b) pointed out that if it is possible to follow, without offset or displacement, earlier oblique tectonic lineaments through a younger mobile belt, the latter must have formed ensialically. For example, he showed that the fold trends and northeast border of the Ubendian–Ruzizi belts (>1800 my old) could be traced continuously and without offset through the 1000 my old Kibaran and Irumide (Fig. 8.4) belts in central Africa.

In a traverse from the Voltaian basin in Ghana to the Buem-Togo fold belt, Grant (1969) demonstrated that the sediments in the border zone of the belt can be correlated with those on the foreland platform of the craton.

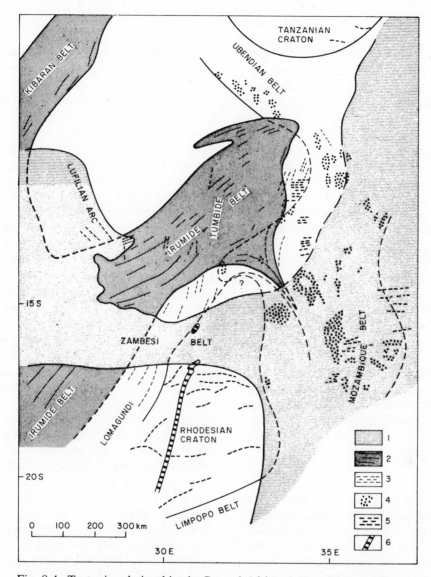

Fig. 8.4. Tectonic relationships in Central Africa. 1. Pan-African (Mozam-
biquian, Zambesi and Lufilian) fold belts (±500 my). 2. Irumide and Kibaran
fold belts (±1000 my) with fold trends. 3. Lomagundi and Tumbide trends
(±1700 my). 4. Granulite facies rocks in Ubendian and Mozambique belts
and supposed connection with Limpopo belt. 5. Archaean trends (>2500 my)
in Rhodesian and Tanzanian cratons and possible Archaean trends within
younger fold belts. 6. Great Dyke of Rhodesia and its possible continuation on
north side of Zambesi belt. (After Shackleton, 1973b; reproduced by permission of the Academic Press, London, New York and San Francisco)

This means that the fold belt–craton boundary
in this region cannot be a major suture
between two colliding continental plates.

Kroner (1976) asserted that all the Pre-Pan-
African Proterozoic mobile belts of southern
Africa are of ensialic origin; they developed
on pre-existing older granitoid crust and
involved no large-scale plate movements. In
particular he showed that in the Namaqua belt
of South West Africa three periods of tectogenesis, 3000–2500, 2300–1650 and 1400–
950 my ago, were all superimposed on earlier

Archaean sialic crust. He concluded that the Proterozoic evolution of southern Africa is characterized by plate 'destruction' rather than by 'accretion'.

In reviewing the application of palaeomagnetism to Proterozoic tectonics Briden (1976) pointed out firstly that the palaeomagnetic method has reached a sufficient degree of sophistication to be able to detect relative motions between cratons or plates of more than 15° in translation or rotation. Secondly, that present palaeomagnetic data indicate that the vector sum of relative movements between the West African, Congo and Kalahari cratons was less than 15° within the periods 2300–1950, 1950–1100, 1100–700 and 700–500 my, and also that the sum of all these relative motions from 2300 my to the present was less than about 15°. The clear implication of this evidence is that either the intervening deformed belts formed by some kind of intra-plate ensialic mechanism, or by lateral plate movements with some subduction but in such a way that the two plates returned to almost their original position—possible perhaps if only narrow rift-like oceans were involved.

Pan-African Belts

About 550 ± 100 my ago an important 'plutonic' event affected a large part of Africa and adjacent parts of Gondwanaland (Fig. 8.6); it has been referred to as the Pan-African Orogeny (Kennedy, 1964). However, Grant (1973) prefers the neutral term *thermotectonic event* and Hepworth (1976) points out that the greater part of the Pan-African area has been determined only on the single parameter of radiometric age determinations. The Pan-African belts contain metamorphosed and deformed supracrustal rocks (Katanga, Outjo and Western Congo Systems) and prominent areas of partially reactivated basement, especially in the Mozambique and Zambesi belts, Malagasy, Cameroons, Nigeria, etc.

There is much current controversy regarding the mode of formation of the Pan-African belts and, indeed, it is proving to be one of the most interesting tectonic problems being debated today on the evolution of the earth's

crust (Hurley, 1972, 1973). There are two extreme possibilities: they were formed by ocean floor spreading, subduction and drift of continental masses, or by *in situ* upwelling of hot material from the mantle with little or no horizontal movement of the bordering cratonic areas.

1. According to the plate tectonic model the belts formed largely in thickened colliding continental plates and they are comparable to a 30 km deep section through the Himalayas and to the present erosion level through the Grenville belt (Dewey and Burke, 1973). Evidence in support of this tectonic origin includes the presence of Barrovian-type intermediate pressure facies series and granulite-grade metamorphism implying much erosion and therefore considerable crustal thickening, and positive isostatic gravity anomalies, basic volcanic rocks and serpentinites along the margins of belts indicating that they are major crustal boundaries (Shackleton, 1976b). The deep-level crustal section required by the model adequately accounts for the intense reactivation of basement material, the common absence of ophiolites and the common lack of supracrustal sequences. The collisional plate model has recently received renewed support from the discovery of ophiolite sequences in the Anti-Atlas (Leblanc, 1976) and in the Damara belt of South West Africa (Watters, 1976).

2. The following palaeomagnetic and structural evidence may be cited in favour of an intra-plate origin for the Pan-African belts.

The basis of the palaeomagnetic argument is: 'If the cratonic nuclei had moved as separate plates, contemporaneous palaeomagnetic poles from different cratons would differ widely and it would not be possible to define any single polar-wander curve for Africa. If, however, the orogenic belts are intracratonic it is likely that the amounts of contraction across them would not be palaeomagnetically detectable' (J. D. A. Piper, 1973b).

Piper (1973b) demonstrated that palaeomagnetic poles from most African cratons separated by younger mobile belts define a single polar-wander curve and he therefore concluded that the bulk of Africa behaved as an integral unit during the period 2300–400 my and thus that the Proterozoic mobile belts formed intracratonically or ensialically. His main lines of reasoning were:

a) The poles from rocks of West Africa (2200–2100 my minimum) lie close to the pole of lavas from South Africa (*c*. 2300 my), but these areas are separated by 3000 km occupied by several later orogenic belts, in particular the eastern part of the Orange river belt (1100–1000 my) and the Damaran belt belonging to the Pan-African System (650–400 my).

b) The poles from late Precambrian lavas in SW Africa lie close to the pole of the Bukoban System in Tanzania (all are between 800 and 1000 my old); but these regions are on either side of the Damaran–Katangan mobile belt (Pan-African, 650–400 my).

c) The poles for Malawi rhyolites and the Kaimur Series (*c*. 750 my) in India are close to those from Bukoban igneous rocks (*c*. 804 my) in Tanzania. These come from opposite sides of the Mozambique mobile belt in the pre-drift (Mesozoic) configuration of Africa and India.

Piper (1973b) also collected palaeomagnetic results from radiometrically dated Proterozoic rocks from Australia, India and South America and found that the majority of poles fell close to the African polar-wander path for their quoted ages, when the continents were restored to their pre-Mesozoic drift configuration. It seems, therefore, that the Gondwanaland continents did not undergo independent relative horizontal movements during the Proterozoic, probably remaining as a coherent group until the Mesozoic fragmentation. More compelling palaeomagnetic evidence was

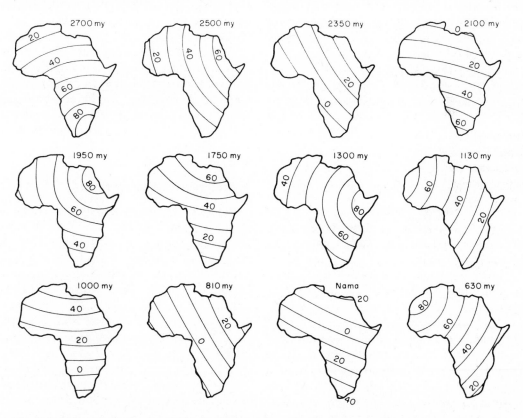

Fig. 8.5. Palaeolatitudes for Africa in Precambrian times based on palaeomagnetic data (after Piper, 1973b; reproduced by permission of the Academic Press, London, New York and San Francisco)

cited by Piper, Briden and Lomax (1973) and J. D. A. Piper (1974, 1976a,b) for the existence of a supercontinent since the early Proterozoic in which the Pan-African belts formed ensialically (see later).

Structural evidence favouring the ensialic model is listed by Shackleton (1973b, 1976b). The Zambesi and Mozambique belts truncate earlier oblique structures without any apparent displacements (Fig. 8.4), a trail of granulite facies rocks through the Mozambique belt suggests tectonic continuity with bordering fold belts, and there is a lack of ophiolites, mélanges and blueschists (except for small occurrences in southern Africa—Kroner, 1976); Hurley and Rand (1969), Engel and Kelm (1972) and Goodwin (1973b) pointed out that the coherent grouping of Archaean age provinces and granulite terrains, the continuity of Archaean tectonic patterns and the linear distribution of early Proterozoic banded iron formations across the Gondwana continents were evidence of the ensialic nature of most Proterozoic orogenic belts, including the Pan-African.

Accepting that Africa remained intact as one plate at the time, Piper (1973b) used palaeomagnetic and palaeoclimatic data to chart the drift of the continent across the palaeolatitudes throughout the Proterozoic (Fig. 8.5).

Geosynclines

Traversing the continents there are a number of low-grade, well-preserved early Proterozoic geosynclines (see Chapter 5). They are of two types:
1. The Coronation Geosyncline of Canada has a stratigraphy and structure that make it comparable with a Cordilleran-type succession at a modern leading plate margin (Hoffman, 1973) and a similar example may be the Mount Isa Geosyncline of Australia (Dunnet, 1976). It is likely that such belts formed at the marginal edge of the drifting Proterozoic 'plates'; they correspond in time to the large changes in direction of motion of the plates expressed by the hairpin bends of the polar wander curves.

2. A large number of internal geosynclines are located within the early Proterozoic continents and these are characterized by the presence of banded iron formations, for example, the Labrador Trough, Animikie basin, Hamersley Group, Griquatown formation, etc. (for details see Chapter 4).

Two points are relevant here with regard to a possible tectonic environment for these geosynclines:
1. Detailed study of the structure and stratigraphy of the Hamersley Group iron formations has led to the suggestion that they were laid down as seasonally varved evaporitic chemical precipitates in a narrow, partially enclosed basin (Trendall and Blockley, 1970; Trendall, 1973a). The occurrence of diagenetic riebeckite and stilpnomelane in the BIF suggests that Na and K were present. Saline evaporites are typically restricted to the Phanerozoic and, in particular, to Red Sea-type rifts characteristic of the early stages of fragmentation of continents (Burke, 1975; Kinsman, 1975a).
2. Palaeomagnetic and geophysical evidence from Laurentia makes it likely that the internal geosynclines of the Labrador Trough, Cape Smith belt etc. formed intracratonically from narrow rifts like the Red Sea which never opened into wide oceans (Irving and Lapointe, 1975; Kearey, 1976).

A Proterozoic Supercontinent

Evidence has been accruing over the last few years to indicate that several of the present continents were grouped together for long periods in the Proterozoic and that a supercontinent may have existed during this period of time (J. D. A. Piper, 1974, 1976a,b). If correct, this conclusion has wide implications for earth history as a whole and so we shall consider here the main evidence on which it is based.
1. Palaeomagnetic data are consistent with the proposal that several Proterozoic fold belts formed by ensialic processes without involving appreciable continental drift. For example, data from the West African and Rhodesia-Kaapvaal cratons for the time range 2300-

1900 my and from the Sudan, Tanzania, Zambia, Malawi, South and SW Africa for the range 900–500 my lie on single polar wander paths indicating that the intervening orogenic belts (e.g. the 1150 ± 200 my Kibaran and Irumide, and the widespread 750 ± 450 my Pan-African) could not have formed by large-scale lateral motions and convergence (Piper, Briden and Lomax, 1973). The results of Irving and Lapointe (1975), McGlynn and coworkers (1975) and Irving and McGlynn (1976) from the Laurentian Shield show a coherence of poles implying that the four major blocks (the Superior, Slave, Beartooth and Nain Provinces) did not move by very large amounts relative to one another from 2200–1800 my, and therefore that the Shield remained essentially intact during the formation of the Hudsonian high-grade belts and the lower grade Circum-Ungava geosyncline. The data also suggest that the boundary between the Churchill and Superior provinces was not

an early Proterozoic suture as postulated by Gibb and Walcott (1971) and Gibb (1975).

The significant point about the above results is that the North American and African continents can each be considered to have behaved as coherent plates during Proterozoic time.

2. Palaeomagnetic data suggest that during the Proterozoic (a) Africa and South America were part of a single continent and that North America belonged to this continent until about 1000 my ago (Piper, Briden and Lomax, 1973), and (b) the Canadian Shield was joined to Greenland and NW Scotland (A. D. Stewart and Irving, 1974; Irving and Lapointe, 1975). J. D. A. Piper (1974, 1976b) found, firstly, that the polar-wander paths of Africa and North America can be closely superimposed over their whole lengths from 1000 to 2200 my and possibly 2700 my, and when this is done the Afro-Arabian and North American regions form a continuous continental body; and, secondly, that the Gondwanaland

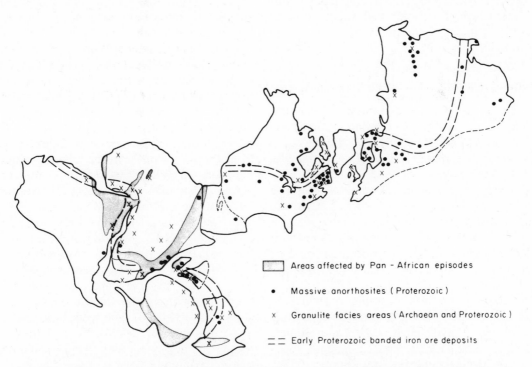

☐ Areas affected by Pan – African episodes

• Massive anorthosites (Proterozoic)

× Granulite facies areas (Archaean and Proterozoic)

= = Early Proterozoic banded iron ore deposits

Fig. 8.6. The late Proterozoic Supercontinent of Piper (1976b) showing the distribution of massive anorthosites (from Mem. 18, NY State Mus. Sci. Serv.), Pan-African belts (from Piper, 1976b; reproduced by permission of the Royal Society), Precambrian granulite facies areas (mostly from Oliver, 1969) and early Proterozoic banded iron ore deposits (after Goodwin, 1973b; reproduced by permission of the Academic Press, London, New York and San Francisco)

continents occupied their Pangaea positions during the Proterozoic (at least as early as 750 my ago according to McElhinny and Embleton, 1976). The combination of these results is the configuration shown in Figs. 7.3, 8.6, a supercontinent, stable from at least 2200 my to shortly after 1000 my BP. This is an interesting assembly, particularly with regard to the separation of North and South America, the position of Arabia and the distribution of Precambrian granulites, banded iron formations and anorthosites. Sutton and Watson (1974) and Davies and Windley (1976) point out that Proterozoic megalineaments are aligned across substantial parts of the supercontinent, thus suggesting a high degree of coherence of the continental blocks (see further in Chapter 6).

Note, however, that McGlynn and coworkers (1975) state that the palaeomagnetic evidence does not support the joining of Laurentia and West Gondwanaland in the interval 2200–1800 my. Also important in this construction of a Proterozoic supercontinent is the problem of the mode of origin of the Grenville fold belt.

Rifting and Break-Up of the Supercontinent

The postulated supercontinent underwent rifting and break-up in the late Proterozoic from about 1200–1000 my until 850 my ago.

The earliest manifestation of the rifting was the emplacement of the igneous rocks reviewed in Chapter 6—basalts, dolerite dykes, alkaline complexes, carbonatites and layered stratiform complexes. Many of these were mutually related and also associated with the formation of prominent faults in rifts or aulacogens. Burke and Dewey (1973a) suggest the existence of eleven 1100 my old aulacogens, the best example of which is no doubt the Keweenawan Trough/Mid-Continent Gravity High (Chase and Gilmer, 1973). These late Proterozoic igneous rocks and their associated structures are very similar

to those developed during the break-up of Pangaea during the Mesozoic–Cenozoic as documented in Chapter 14. Sawkins (1976a) summarized much evidence for widespread continental rifting in the period 1200–1000 my ago and suggested that when a supercontinent forms, heat dissipation from the mantle via spreading-ridge systems can only occur in oceanic environments geographically remote from the central parts of the supercontinent. The lack of access to heat dissipation systems, caused by the thermal blanketing by the extensive continental lithosphere, necessitates the creation of new centres and mechanisms for dissipation of mantle heat in the form of rifts. One implication of this idea is that the life expectancy of this Proterozoic supercontinent (at a time when radioactively generated heat flow was greater than in the mid-Phanerozoic) was similar to or less than Pangaea which survived for only about 200 my. Note, however, that J. D. A. Piper (1975, 1976b) implies that the Proterozoic supercontinent was in existence for a much longer period of time.

By 850 my ago separation of some continental blocks had given rise to trailing plate edges whose sedimentary development was markedly different from that in the epicratonic troughs of the period 1700–850 my (Stewart, 1976) as documented in Chapter 7. The detrital sequences in these new environments grade from thin sandstones on the cratons to thick miogeoclinal wedges towards the continental margins in North America (J. H. Stewart, 1972; Rankin, Espenshade and Shaw, 1973) and NW Scotland (Dewey, 1969). The late Proterozoic miogeoclinal sequences represent the initial deposits in the Cordilleran and Appalachian–Caledonian geosynclines and consequently pass upwards into Cambrian sequences with similar depositional patterns (J. H. Stewart, 1976). The scene is now set for us to continue the development of the rock record in the Caledonian–Appalachian fold belt in Chapter 11.

Chapter 9

Palaeomagnetism and Continental Drift

Palaeomagnetists are able to study the variation of the earth's magnetic field through geological time because permanent magnetization takes place when many igneous rocks and sediments form. It is possible to determine the palaeomagnetic pole responsible for the alignment of magnetic particles, and then a succession of pole positions for rocks of different age from one area. When the resulting polar wander curve of one continent differs from that of another there must have been independent drift of the continents. In this chapter we shall see how palaeomagnetic data have been used to chart the course of continental movements throughout the Phanerozoic. For recent reviews of the application of palaeomagnetism to tectonic problems see Hospers and van Andel (1969), Creer (1970a), Tarling (1971b), McElhinny (1973) and Thompson (1974).

Palaeomagnetic Poles

It is not intended to give here either a review of the historical developments in the field of palaeomagnetism or of the basic principles of the method: these aspects are well treated elsewhere. But it will be worthwhile to mention briefly some of the main points about palaeomagnetic poles which are relevant to continental drift.

1. Secular variations of the geomagnetic field can be followed from records in observations for the last 400 years, in pottery furnaces for 1000 years, in deep-sea sedimentary cores for 10,000 years, in lake sediments for 15,000 years and in lavas for up to 20 my. These

observations show that the average pole positions are centred on the earth's present rotation axis around the geographic rather than the geomagnetic pole. This pattern suggests that the field has been one of a geocentric axial dipole; in other words, like a dipolar magnet centred at the middle of the earth and aligned parallel to the earth's spin axis. This is important because the method of determining palaeomagnetic pole positions is based on the theory that the earth's magnetic field is aligned with the rotation axis. As a consequence, palaeoequators can be calculated to lie normal to palaeorotation poles and palaeogeographic maps with their palaeolatitudes can be constructed.

Two measurements are made. Firstly, one of palaeomagnetic declination giving the direction of the magnetic pole which, as argued above, defines the geographic pole; secondly, one of the magnetic inclination from which the palaeolatitude can be calculated because the tangent of the inclination equals twice the tangent of the latitude. A consequence of the method is that ancient latitudes relative to the present geographic pole can easily be determined from the inclination, but ancient longitudes cannot be determined because ancient declinations may be explained either by a rotation of the continent about a point within it or by translation of the continent along a small circle, or by some combination of the two. Although this is unfortunate, it is not too serious since it is only the palaeolatitudinal position that controls the formation of climatically sensitive sediments and glacial deposits (except for alpine types) and for the most part faunal and floral distributions. Also, the geometrical fit of continental shapes and the matching of geological structures from one continent to another can be used to assist the reassembly of past continents, and oceanic magnetic anomaly patterns can help to reposition the continents as far back as the early Jurassic.

2. By measuring the angles of inclination and declination in a magnetized igneous or sedimentary rock, a palaeopole can be determined and if a sequence of poles are calculated from earlier and later rocks these can be plotted on á projection to define a polar-wander curve for the area concerned. At this point it may not be clear whether the locus of points making the curve was caused by actual wandering of the pole or by lateral movement of the sampling site. However, when the poles are plotted for two continents for rocks of known age from, shall we say, the late Mesozoic to the present, it is clear that two independent curves are defined that converge and meet at the present pole. Because the geomagnetic field is only dipolar this means that drift of the continents must have taken place, and this is the crux of the palaeomagnetic argument validating the continental drift theory.

Polar-wander Paths

It has not taken long since Creer, Irving and Runcorn (1954) constructed the first polar-wander curve (for British rocks) for palaeomagnetism to become a cogent force substantiating the continental drift hypothesis. In order to follow the movements of continents throughout the Phanerozoic it is thus necessary first to consider the palaeomagnetic evidence. In a review of this type it would, of course, be possible to treat the individual polar wander paths for each continent in turn; but we shall take a broader approach and consider, via their polar wander curves, the relative movement of continents belonging to the two landmasses of Laurentia and Gondwanaland.

Laurentia

Runcorn (1956, and later in 1962 and 1965) realized that the polar paths for North America and Europe were appreciably offset, that they eventually converged on the present pole and that this was significant palaeomagnetic evidence for drift. Fig. 9.1(a) gives an up-to-date portrayal of Runcorn's two curves, and Fig. 9.1(b) shows that with the two continents rotated 40° they became essentially congruous. This amount of rotation has the effect of closing the North Atlantic ocean and compares remarkably well, as McElhinny (1973) notes, with the 38° required by Bullard, Everett and Smith (1965) to obtain the best geometrical fit between North America and Europe using the 500 fathom submarine contour.

The current interpretation of the polar paths in Fig. 9.1(b) is that their convergence from the Cambrian to the Silurian reflects the drift together of the two continents during the closure of the proto-Atlantic ocean. They remained in a co-polar situation from the Silurian to the Trias when the continents formed part of Pangaea, and then diverged in the Trias when the continents broke away from each other and drifted independently until the present. This sequence of events is consistent with information derived from magnetic anomaly patterns, palaeoclimatic indicators and geological correlations from one continent to the other.

For further debate on the most problematical Palaeozoic part of the Laurentian polar paths related to the history of the proto-Atlantic ocean, see Chapter 11.

Gondwanaland

Recent information on palaeomagnetic apparent polar paths of the continents of Gondwanaland is given by the following: South America (Creer, 1970b; Valencio, 1974); Africa (Briden, 1970a); India (Wensink, 1973); Australia (McElhinny and Embleton, 1976); Antarctica (McElhinny, 1973). For general reviews of Gondwanaland

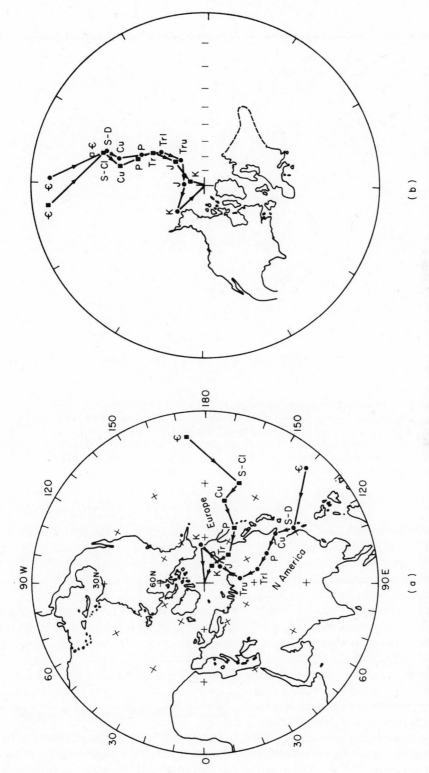

Fig. 9.1. (a) Comparison of the apparent polar wander paths for North America (circles) and Europe (squares) from the Cambrian to the present. (b) The two polar wander paths after rotation of 40° to the Bullard, Everett and Smith (1965) fit of the North Atlantic (after McElhinny, 1973; reproduced by permission of Cambridge University Press)

148

Fig. 9.2. Generalized diagram showing the path of the South Pole during the Phanerozoic. The two super-continents of Gondwanaland and Euramerica combine into Pangaea approximately during the Silurian and there then exists a common polar wander path until various times during the Mesozoic (M). The paths for the constituent parts of Pangaea then diverge as shown (after McElhinney, 1973; reproduced by permission of Cambridge University Press)

palaeomagnetism see McElhinny and Luck (1970), Tarling (1971a) and Hailwood (1974).

Palaeomagnetic pole positions show that the southern continents were loosely grouped together in the early Palaeozoic and continued to remain so until the Mesozoic. Briden (1967) followed the movement of the continents concerned during the Palaeozoic and realized that the distribution of their pole positions is irregularly spaced as they are grouped together for certain periods of time. This implies that the Gondwanaland continents were stationary or 'quasi-static' in the intervals Cambrian–Ordovician, Devonian–early Carboniferous and late Carboniferous–Permian; between these periods there was rapid apparent polar wandering.

When Pangaea fragmented, the continents of Gondwanaland went their separate ways in every direction and their apparent polar wander paths are therefore very different through the Mesozoic and Cenozoic (Fig. 9.2); moreover, their paths are markedly dissimilar from those of North America and Europe at this time, providing additional evidence of continental drift.

Continental Drift during the Phanerozoic

There have been many attempts to reposition the continents in the past taking into account their drift based on a variety of evidence:

Wegener, 1929 (visual geometrical fit to form Pangaea)

du Toit, 1937 (geological correlations across Gondwanaland)

Bullard, Everett and Smith, 1965 (computer fit across the Atlantic using 500 fathom submarine contour)

Dietz and Holden, 1970 (Pangaea fit using oceanic magnetic and palaeomagnetic data, and its fragmentation until the present)

Smith and Hallam, 1970 (Gondwanaland computer assembly using 500 fathom submarine contour)

Tarling, 1971a (visual fit of Gondwanaland using palaeomagnetic data and the 100 fathom contour)

Seyfert and Sirkin, 1973; Smith, Briden and Drewry, 1973; Briden and coworkers, 1974 (Phanerozoic maps of all the continents based on palaeomagnetic and geological data)

In the final part of this chapter let us follow, with the aid of the palaeogeographic maps in Figs. 9.3(a)–(h), the history of continental drift throughout the Phanerozoic. We start in the Cambrian when there was a somewhat odd distribution of the continents—odd in the sense that we do not see this kind of distribution again in the ensuing 500 my. The southern continents were loosely clustered together to form a proto-Gondwanaland, but North America, Europe and Asia were separate. There was a small proto-Atlantic or Iapetus ocean between Europe and North America. The south pole lay somewhere in northwest Africa. The lack of longitudinal control is illustrated by the marked difference in size of the ocean between Asia and Australia–India in the maps of Seyfert and Sirkin (1973) and Briden and coworkers (1974). Note also the wide discrepancy in the relative orientation of the Gondwana continents; according to Seyfert and Sirkin (1973) they have roughly the same orientation as in the Ordovician and later periods, but on the maps of Smith, Briden and Drewry (1973) and Briden and coworkers (1974) they are almost 'upside down'. How these continents arrived at these relative posi-

tions in the Cambrian is not certain. Did they drift there after the fragmentation of the late Proterozoic Supercontinent of J. D. A. Piper (1974)? According to McElhinny and Embleton (1976) Gondwanaland was a unit as far back as 750 my ago.

During the Ordovician, tillites were deposited in northwest Africa in the south polar region. According to Briden and coworkers (1974) there was a dramatic redistribution of the continents between the early Ordovician 510 ± 40 my and the early Devonian 380 ± 35 my ago, although in the opinion of Seyfert and Sirkin (1973) there was very little. An important development was contraction of the Iapetus ocean in the Silurian and its closure by the mid-Devonian, giving rise to the Caledonian–Appalachian fold belt. Continuing convergence caused the northwest corner of Gondwanaland to impinge on Euramerica, the final collision being responsible for the formation of the Hercynian–southern Appalachian belt in the late Carboniferous to Permian, and at the same time Asia collided with Europe removing the intervening ocean and producing the Ural Mountains. Note that Seyfert and Sirkin (1973) indicate that the Ural Mountain suture formed between the Silurian and the Devonian; this is inconsistent with geological evidence (Hamilton, 1970) and is a major departure from other models for the origin of this fold belt (e.g. G. A. L. Johnson, 1973; Briden and coworkers, 1974). Between the early Devonian and the Permian Gondwanaland rotated in a clockwise sense and moved northwards about 30°; Euramerica moved in a northeasterly direction but Asia remained almost stationary. By 250 my ago in the mid-Permian the Pangaea landmass was created together with the re-entrant Tethyan ocean.

Widespread deposition of evaporites and red beds took place in Laurasia during the Devonian, tillites in Gondwanaland in the Permian, and coal measures in Laurasia during the Carboniferous and in both continental masses in the Permian. These climatic indicators have a latitude-dependent distribution at these times, a fact largely consistent with the palaeomagnetically determined position

150

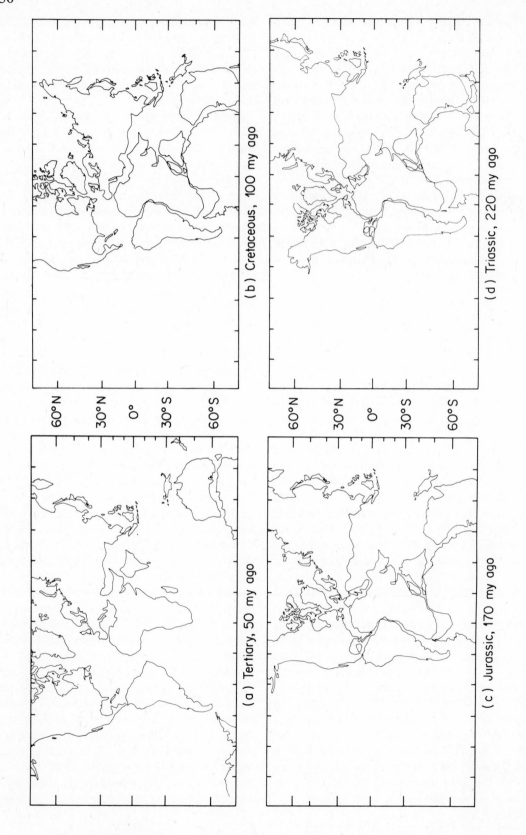

(a) Tertiary, 50 my ago

(b) Cretaceous, 100 my ago

(c) Jurassic, 170 my ago

(d) Triassic, 220 my ago

151

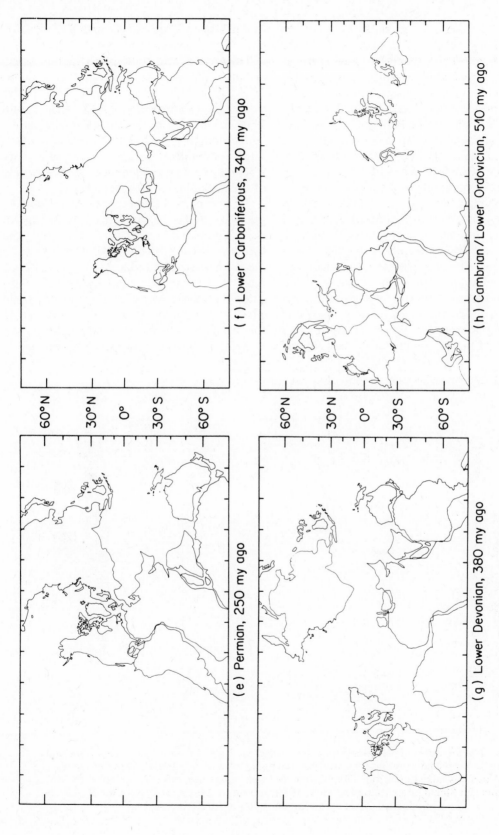

(f) Lower Carboniferous, 340 my ago

(h) Cambrian/Lower Ordovician, 510 my ago

(e) Permian, 250 my ago

(g) Lower Devonian, 380 my ago

Fig. 9.3. Continental drift from the Cambrian to the Tertiary illustrating the formation and break-up of Pangaea. (Maps based on Smith, Briden and Drewry, 1973; reproduced by permission of The Palaeontological Association)

of the continents. In particular, the tillites occur in high southern latitudes, evaporites and red beds are mostly in low latitudes between 35°N and 35°S, whilst coal measures occur predominantly in high latitudes, except for a minor coal belt that developed in humid tropical latitudes in the Permian.

Little happened to Pangaea for at least 50–70 my through the Permian and Triassic except for a small northward drift. The first continental fragments broke away from Pangaea in the late Triassic–early Jurassic (North America from West Africa and Europe) and the last in the early Tertiary (Greenland from Norway, and Australia from Antarctica), i.e. it took about 150 my to fragment the supercontinent completely. The study of this break-up and subsequent dispersal of the continents takes into account the oceanic magnetic anomaly patterns, and therefore the relative positions of the continents for the last 200 my

of earth history are known with a reasonable degree of confidence.

The first rifting took place in the North Atlantic and western Tethyan region about 200–180 my ago, whereas South America did not begin to break away from Africa until 130–120 my ago in the early Cretaceous, by which time the North Atlantic ocean was half open (for details of the great variety of geological and geophysical evidence for the opening of the Atlantic, see Chapter 14).

Antarctica began to separate from India by 100–105 my ago, as evidenced by the age of the Rajamahal traps in India, and the Cenomanian age of the earliest marine sediments in Southeast India (McElhinny, 1970). Analysis of magnetic anomaly patterns indicates that there was a period of very rapid sea-floor spreading from 110–80 my ago in the Pacific and Atlantic oceans (Larson and Pitman, 1972) and between 75–55 my ago in

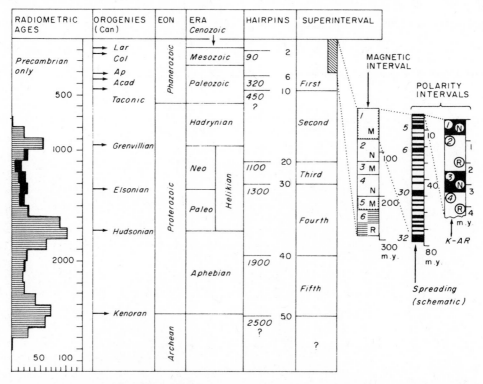

Fig. 9.4. Summary of palaeomagnetic time scales: the scale of Polarity Intervals (N denotes mostly normal polarity and R mostly reversed polarity), and the scale of Magnetic Intervals (M denotes mixed polarity). The radiometric ages, orogenies and eras are given for the Canadian Shield (after Irving and Park, 1972; reproduced by permission of the Research Council of Canada and *Canadian Journal of Earth Sciences*)

the Indian Ocean (McKenzie and Sclater, 1971). The latter motion assisted the separation of India from Antarctica, and finally Australia broke away from Antarctica about 45 my ago in the Eocene (Le Pichon, 1968). By the mid-Tertiary Pangaea was totally fragmented and the consistent parts well on their way to their present positions.

Hairpins and Superintervals through Time

Apparent polar-wander paths typically make several sharp bends or loops which describe major changes in the horizontal direction of movement (relative to the pole) of the lithospheric plate (or other tectonic unit). The turning points are called *hairpins*, the polar paths between them *tracks*, and the period of time relative to each track a *super-interval* (Irving and Park, 1972).

Fig. 8.1 shows the polar path and Fig. 9.4 a summary of palaeomagnetic time scales from the Archaean to the present relative to the Canadian Shield. Although this chapter is concerned with palaeomagnetic polar movements in the Phanerozoic, it is obviously convenient to consider here the full time scale. There are 5 tracks and 8 hairpins, labelled 2, 6, 10, 20, etc. Hairpins 2, 6 and 10 correspond in time with the Laramide–Columbian, the Appalachian–Acadian and Taconic orogenies of North America. The four hairpins 20, 30, 40 and 50 roughly coincide with the four main Precambrian isotopic closure periods reflecting late uplift following orogenesis—Grenvillian, Elsonian, Hudsonian and Kenoran.

The Phanerozoic hairpins 2 and 6 coincide with Magnetic Intervals 2 (the 'normal' one in the late Mesozoic) and 6 (the 'reversed' one in the late Palaeozoic). Irving and Park emphasize that parts of the polar path are tentative and that hairpins 10 and 50 are the most poorly defined, but nevertheless future revision will probably not change the fundamental picture and implied relationships.

Chapter 10

Palaeoclimatology and the Fossil Record

Palaeomagnetic data tell us the relative positions of the continents in the past; the drift of these continents across the palaeolatitudes was responsible for, and so can be double-checked by, the deposition of climatically sensitive sediments such as carbonates, evaporites, desert sand dunes and red beds in warm and/or arid climates not far from the palaeoequator, and the formation of tillites in glacial climates in the palaeopolar regions (Opdyke, 1962; Briden and Irving, 1964; Briden, 1970b; Stehli, 1973; Drewry, Ramsey and Smith, 1974).

However, one should be aware of the limitations of these palaeoclimatic indicators (Meyerhoff and Teichert, 1971). Sedimentary palaeodistribution does not *prove* the theory of continental drift. But, making the basic assumption that continental drift did occur we can predict that certain sets of circumstances must have existed which controlled different types of palaeoenvironments. The probable existence of *regular* sediment–latitude patterns in the past, as we shall see in this chapter, does provide some confirmation of the continental drift theory.

Whilst the continents were adrift during the Phanerozoic, life was evolving first in the sea and later on the land and in the air. The development of life was clearly affected by continental drift. The fragmentation and reassembly of the continents and their latitudinal movements affected a large number of life-sensitive phenomena, such as circulation of the oceans, supply of nutrients, climatic changes, formation and destruction of favourable habitats, transgressions and regressions,

etc. In the second part of this chapter we shall review these variables and then follow the fossil record through the history of the Phanerozoic.

Carbonates and Reefs

Most modern carbonates and reefs form in warm water (>20°C) within 30° of the equator, and most Palaeozoic–Mesozoic limestones, dolomites and organic (coral) reefs fall predominantly within 30° of the palaeoequator (Figs. 10.1, 10.3, 10.4) (Briden and Irving, 1974). Carbonate deposition and reef growth are favoured by a high-energy environment in the shallow-water wave-agitated zone on the windward side of low banks or islands such as off the Bahama Banks, Persian Gulf and the Great Barrier Reef, NW Australia. An increase in water temperature decreases the solubility of calcium carbonate, thus carbonate deposition is favoured by shallow warm water. As the solubility constants of calcium carbonate are unlikely to have changed with time, it is probable that both thick and/or pure carbonate sediments and reefs throughout the Phanerozoic are a good indication of tropical–subtropical low latitudes, although in periods with no ice-caps, such as the Cretaceous, warm seas and carbonate deposition may have extended to higher latitudes.

Simple calcareous reefs, constructed by stromatolitic algae, flourished in the Proterozoic but more complex types evolved in the early Palaeozoic due to the contribution of organisms such as sponges, archaeocyathids,

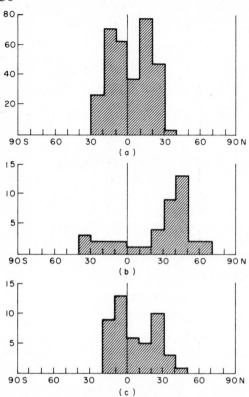

Fig. 10.1 Latitudinal histogram for organic reefs. (a) Present latitude of modern reefs. (b) Present latitude of fossil reefs. (c) Palaeolatitude of fossil reefs. (After Briden and Irving, 1964; reproduced by permission of Wiley–Interscience Publ.)

bryozoans, corals, crinoids, stromatoporoids, brachiopods, etc. They formed sporadically in the Cambrian and Ordovician but they underwent a spectacular development in the Silurian, especially in North America and western Europe. This burst in carbonate reef construction depended not only on the presence of a multitude of suitable organisms but also on the immense shallow-water sea that extended over the cratons in mid–late Silurian time in the aftermath of the Taconic orogeny. An excellent example of a late Silurian barrier reef is in the Michigan Basin where it is associated with both red beds and evaporites. The history of the tropical organic reefs throughout the Phanerozoic is well reviewed by Newell (1971).

Evaporites

The principal salts precipitated as evaporites are anhydrite ($CaSO_4$) and gypsum ($CaSO_4 2H_2O$), halite (NaCl), and sylvine (KCl)—this sequence indicates progressively more severe evaporative conditions (Stehli, 1973). The only requirement for evaporite deposition is that evaporation be sufficient to cause a significant concentration of brine. However, this requirement is favoured by a restricted set of circumstances such as limited circulation of sea water in a basin bordering the sea or in an enclosed basin with a high temperature and low rainfall. The limited inflow of marine water is often caused by a carbonate barrier reef and thus there is a common carbonate reef–evaporite association.

These conditions are met today in arid climates but not near the equator. It is well known that there is a strong excess of evaporation over precipitation in the present-day oceans in two zones between 5°S and 35°S and between 15°N and 40°N, whilst there is a reverse relationship between 5°S and 15°N (Stehli, 1973). Thus modern evaporites have a bimodal distribution with maxima in the subtropical high-pressure zones within 50° of latitude of the equator and a minimum in the low-pressure equatorial zone between 10°N and 10°S. Most fossil evaporites, at least as far back as the Permian and possibly from the beginning of the Phanerozoic, have a similar bimodal distribution with respect to their relevant palaeoequators (Fig. 10.3) (W. A. Gordon, 1975). The essential constancy of the evaporite regime provides a cogent argument that the atmospheric and oceanic circulations and the distributions of deserts and marine salinity have changed little throughout Phanerozoic time. Also the better correlation of ancient evaporites with their respective palaeolatitudes than with the present-day latitudes supports the lateral movement of continents during the Phanerozoic—in contrast to the fixist view of Meyerhoff (1970).

The earliest known evaporites are of late Precambrian age but these are rare. They generally increased in number with time through-

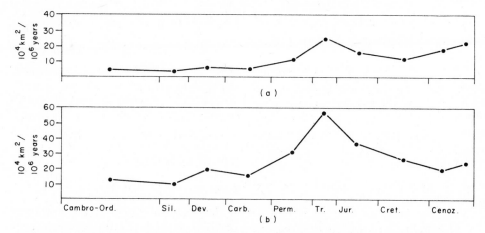

Fig. 10.2. Rate of accumulation of ancient evaporites through time. Curve A is based on the estimated surviving evaporite record and B represents an adjustment of curve A to compensate for the incompleteness of the record (after Gordon, 1975)

out the Phanerozoic reaching a peak in the Permian–early Jurassic (Fig. 10.2) when the continents were assembled into the supercontinent of Pangaea. The earlier and later periods of lower evaporite accumulation correspond to the major episodes of continental dispersal (Gordon, 1975).

In many places evaporite sequences range in thickness up to 300–500 m and three of the thickest older onland deposits are in the Michigan Basin (1000 m, late Silurian), the Paradox Basin, Utah (1330 m, Pennsylvanian), and in SE New Mexico (1500 m). They reach 7 km in thickness in the Red Sea (late Tertiary). For a review of evaporite formation in marginal basins see Chapter 14.

Red Beds

Red beds are arkoses, sandstones, shales and conglomerates which contain haematite as a visible pigmenting agent. The main requirements for their formation are an oxidizing environment and an adequate supply of ferric oxide.

There has, however, been much debate on the origin of the red coloration, with two main hypotheses:

1. Lateritic soils formed in humid uplands provide eroded iron-rich material which is transported and sedimented in drier lowlands (van Houten, 1968). However, as Seyfert and Sirkin (1973) pointed out, lateritic soils lack feldspar and thus red arkoses cannot be derived directly from them. Another reason against the laterite source area is the fact that many Phanerozoic red beds are closely associated with evaporite deposits.

2. The red pigment is formed *in situ* by oxygenated pore waters during intrastratal weathering and diagenesis of iron-bearing detrital minerals (T. R. Walker, 1967). According to Seyfert and Sirkin (1973) a hot arid climate is needed in which feldspars are relatively stable for limonite to be dehydrated to haematite, and Norris (1969) suggested that tropical and subtropical and even very dry zones are suitable for dune reddening, because of the ability of dune sand to absorb and retain rainfall. However T. R. Walker (1974) presented evidence suggesting that red beds may form diagenetically under warm moist (tropical) as well as hot dry (desert) conditions. Although he concluded that red beds *per se* are not reliable indicators of the climate, either in the source area or in the depositional basin, it is significant that the late Cenozoic–Quaternary sediments in his two areas of study (Baja California and Puerto Rico) are less than 30° from the present equator. For our purposes it is sufficient to know that red beds are forming at low latitudes today, which is relevant to the

fact that most major Phanerozoic red beds are situated within 30° of their Palaeoequators (Fig. 10.3).

Red beds formed in particular in Devonian–Trias times when extensive land areas resulted from coalescence of the continents. The construction of Pangaea gave rise to the ideal conditions for the formation of evaporites and red beds. In fact, the association with evaporites provides perhaps the most compelling argument for the formation of red beds in arid to semi-arid environments; excellent examples are the Permian red beds, interstratified with halite, in a belt from Texas to North Dakota, and the Lower Trias red beds associated with gypsum deposits in Arizona and Utah.

There is an interesting aspect of dune-bedded desert sandstones which is worth considering. Wind-blown sand dunes in present day deserts occur mostly between latitudes 18° and 40° (Briden and Irving, 1964). The orientation of the dunes is a reflection of the wind direction and thus much effort has been applied to working out the palaeowind directions from ancient cross-bedded aeolian sandstones (Poole, 1964; Runcorn, 1964), which rocks formed most extensively from the Permian to the Jurassic, probably in arid coastal areas or in inland arid desert basins.

The circulation of the present atmosphere is dominated by an easterly wind belt in low latitudes and a westerly one in intermediate latitudes, and since the sand dunes are deposited in low latitudes they might provide evidence of these easterly palaeowinds. Indeed the Triassic Bunter Sandstones of Britain and the Permo-Pennsylvanian Sandstones of western USA do show easterly wind directions, and Runcorn (1964) concluded that the consistency in the wind directions argued strongly in favour of them being planetary trade winds. However Stehli (1973) pointed out a major objection to such a conclusion. Detailed studies in modern deserts by Sharpe (1964) and Cooper (1967) showed that it was the occasional stronger or local wind, rather than the prevailing wind, that controlled the internal stratigraphy of dunes; from this Stehli (1973) concluded that fossil sand

dunes could not be used to determine reliably ancient planetary wind patterns.

The close association between and distribution of many Palaeozoic carbonate sequences, organic calcareous reefs, evaporites and red beds provides strong support for the approximate position of the relative palaeoequators suggested by palaeomagnetic data.

Coal

The accumulation and degradation of vegetation eventually leads under appropriate conditions to the formation of coal. Conditions most favourable for coal formation are warm and humid climate and shallow water; the presence of suitable vegetation is essential.

The earliest coal formed from algae in the early–mid-Proterozoic, in N Michigan and SW Greenland. The first land-plants appeared in the late Silurian, but it was not until the early Devonian that forests existed (although by then trees were at least 10 m high). The first land-plant-derived coal beds are of late Devonian age, in northern Russia and Bear Island near Spitzbergen; but the optimum conditions for coal formation appeared in the Carboniferous and Permian when extensive marine transgressions, caused by continental drift and closure of the oceans, gave rise to vast swamps on several continents. Coal beds continued to form on a smaller scale throughout the Mesozoic.

The syntheses of Briden and Irving (1964), Meyerhoff (1970) and Meyerhoff and Teichert (1971) showed that Devonian to Miocene coal beds tend to be concentrated in belts lying polewards of the contemporary evaporites (Fig. 10.3). Stehli (1973) remarked that this distribution is just what one would expect from the fact that these belts are characterized more by precipitation than evaporation. Such symmetrical evaporite–coal belts can be recognized from the Carboniferous to the Cretaceous. One might well ask then why more coal did not form in the high precipitation belt near the equator; the answer may be supplied by the observation that decomposition of vegetation in such tropical, wet belts is so rapid

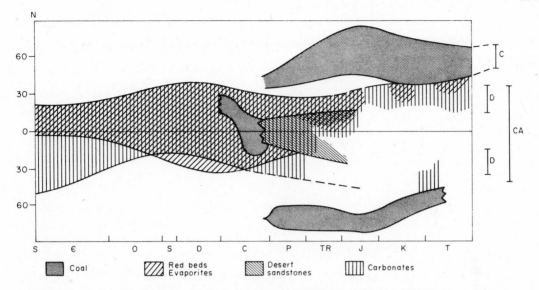

Fig. 10.3. Generalized plot of variation against palaeolatitude of major palaeoclimatic indicators with time. On the right the range of present-day counterparts is shown. D: desert sands; CA: shallow marine carbonates; C: peat. (After Briden and Irving, 1964; reproduced by permission of Wiley–Interscience Publ.)

that insufficient is left over for significant coal formation.

Fossil tree rings tell an interesting story (J. M. Schopf, 1973). Permo-Carboniferous tree trunks from low latitudes lack entirely or have thin growth rings, suggesting an absence of distinctive seasonal changes in humidity/temperature and a rapid growth in a subtropical environment. However in the late Permian many thick coal beds, associated with tillites in Gondwanaland, were formed within 5–30° of the south pole and they contain tree trunks with prominent growth rings implying slow growth in an environment comparable with that in present day muskeq swamps in Siberia (Seyfert and Sirkin, 1973), or in coniferous forests in temperate latitudes (Chaloner and Creber, 1973).

The distribution of coal beds seems to be a particularly useful palaeoclimatic indication of palaeolatitude, especially if the coals are taken in conjunction with the contemporary equatorward evaporites.

Oil in the Tethyan Seaway

In a stimulating analysis (which forms the basis of the following account), Irving, North and Couillard (1974a) identified four factors that controlled the generation and preservation of oil during the Phanerozoic: palaeoclimate, especially temperature, supply of mineral nutrients, plate tectonic factors that controlled initial basin formation and, later, the preservation of the basins.

For the Phanerozoic as a whole, over 80% of all oil occurs in reservoir and source rocks whose latitude at the time of deposition was less than 30°, and over 60% was less than 10°. Cenozoic oil is uniformly distributed with respect to palaeolatitude, Mesozoic oil shows a marked maximum at the equator, and in the Palaeozoic most oil is within 30° of the palaeoequator (the Cenozoic and Mesozoic estimates include the disproportionately large Persian Gulf deposits). Of the world's known oil, 60% was generated in 5% of Phanerozoic time, between 110 and 80 my ago (Albian to Turonian, inclusive).

The basic reason for the oil generation was that dispersal of the continents in the 110–80 my interval gave rise to the following set of favourable circumstances:

1. A narrow Tethyan seaway developed (see Fig. 10.4) between Laurasia and Gondwanaland within 30° of the equator (Smith, Briden

160

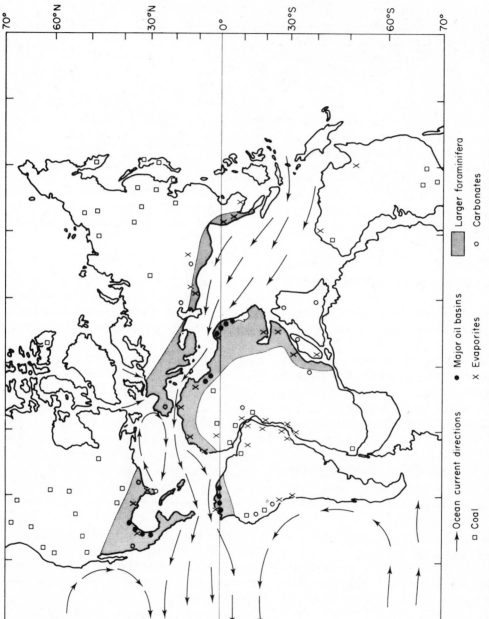

Fig. 10.4. The distribution of the continents in the Cretaceous, 100 ± 10 my BP (after Smith, Briden and Drewry, 1973; reproduced by permission of the Palaeontological Assoc.) showing the major Tethyan seaway. Oil basins after Irving, North and Couillard (1974) (reproduced by permission of *Canadian Journal of Earth Sciences*), foraminifera after Dilley (1973), ocean currents from Luyendyk, Forsyth and Phillips (1972) and sediments after Seyfert and Sirkin (1973)

→ Ocean current directions   ● Major oil basins   [▓] Larger foraminifera

□ Coal   × Evaporites   ○ Carbonates

and Drewry, 1973; Briden and coworkers, 1974) through which there was a strong westerly flow of wind and ocean currents (Luyendyk, Forsyth and Phillips, 1972).

2. There was very rapid sea-floor spreading in the Atlantic and Pacific oceans (Larson and Pitman, 1972). Fast-spreading ridges displaced ocean waters onto continental platforms (Valentine and Moores, 1970) giving rise to the greatest of all post-Devonian marine transgressions (Pitman and Hayes, 1973).

3. A dramatic increase took place in the number and variety of organisms, especially plankton, and the cosmopolitan distributions of the early Mesozoic were being replaced by a greater provinciality.

4. About 100 ± 20 my ago there was a peak in the intrusion of carbonatites, kimberlites, alkalic and ultrabasic rocks (MacIntyre, 1971), this peak reflecting a major renewal of mantle convection which was, in turn, responsible for the formation of the Tethyan seaway and for an increase in the supply of mineral nutrients in the oceans facilitating the rapid development of life.

5. According to oxygen isotope ratios in calcareous fossils, the temperature of ocean waters reached about 21°C—the highest level since the early Mesozoic (Emiliani, 1966).

Many oil basins were obliterated by later plate movements if, for example, they lay near leading plate edges or at points of continent–continent collision. The Gulf of Mexico was protected by the northeastward movement of the Antillean arc and the Persian Gulf by the rapid northward movement of the Indian plate.

Glaciation: Ordovician–Permian

There is well-documented evidence of early Palaeozoic glacial activity in the Saharan region (Fairbridge, 1970, 1971) which was close to the Ordovician south pole as determined palaeomagnetically. Other likely occurrences of this age are in Brazil, South West Africa, Nova Scotia, Spain, France and Scotland (Macduff Group of the Dalradian); all these lie within 30° of the pole (and 30° of latitude is

near the limit of most drifting Quaternary ice). Harland (1972) has reviewed most of the Ordovician localities.

In the late Palaeozoic the climate in the southern hemisphere cooled to such an extent that extensive permanent ice sheets developed across the supercontinent of Gondwanaland. It is the fragmentation and dispersal of the remains of these ice sheets that is one of the key palaeoclimatic indicators of later continental drift (Fig. 13.1).

As a reference source on these glacial features the reader should note the following main papers:

South America (Frakes and Crowell, 1969)
Southern Africa (Frakes and Crowell, 1970)
Antarctica (Frakes, Matthews and Crowell, 1971)
Southern Australia (Hamilton and Krinsley, 1967)
India (A. J. Smith, 1963)

A useful general, but anti-drift, commentary on Permo-Carboniferous glacial deposits is given by Meyerhoff and Teichert (1971).

The record of glaciation through the Palaeozoic can be correlated with movement of the Gondwanaland continents across the south pole (Crowell and Frakes, 1970). In the Ordovician the Sahara region was widely glaciated when North Africa lay near the pole. By Silurian to Lower Devonian time central South America arrived over the pole causing minor glacial activity there. According to Smith, Briden and Drewry (1973) the pole was still situated in this region in the Lower Carboniferous when alpine glacial deposits were formed. Eastern Antarctica had moved over the pole in the late Carboniferous when the maximum glacial activity gave rise to several ice sheets scattered across South America, southern Africa and Antarctica. By the Permian the continents had moved westward so that Australia and Antarctica were glaciated, but ice centres had diminished or disappeared in Africa and South America. For over 200 my the Gondwana supercontinent was affected by glaciation, so what events brought it to an end before the Trias? As Crowell and Frakes (1970) point out, the glaciation was not 'turned off' by the simple movement of the supercontinent away from the pole, because the south pole lay just off the edge of Antarctica from the Trias through to the Cretaceous (Smith, Briden and Drewry, 1973), long after the end of the glaciation. Crowell and Frakes (1970) suggest the following explanation: during the late Palaeozoic the Gondwana–Laurasia continent extended in a meridional direction from the south pole to northern high latitudes and this caused the oceanic and atmospheric circulation to be forced into meridional patterns around the world. The initial separation of Laurasia from Gondwanaland and the birth of the North Atlantic ocean enabled the air–ocean circulation to break into a more latitudinal pattern (that prevailed through the Mesozoic into the Tertiary) which prevented the growth and spread of further ice sheets.

Factors affecting the Distribution and Diversity of Species

We saw in the last section how drift of the continents in the past influenced the formation of various climatically sensitive sediments. The 'new global tectonics' have beneficially affected palaeontology, like most other branches of the earth sciences, by providing a geographical framework for the understanding of fossil fauna and flora. Here we shall see what factors controlled the distribution of life forms and were responsible for the radiation and extinction of species at certain critical times in the Phanerozoic.

Continental drift was not the only influence on the distribution, diversity and extinction of past life forms (Hallam, 1967); other factors include changes in continental area, sea level, seawater salinity, supply of nutrients to the oceans and, possibly, reversals of the earth's magnetic field. Let us look at these in turn as possible regulators of the fossil record.

If change of latitude is consequent upon continental drift, better or worse climatic conditions will create more or less favourable environments for organisms. Tolerance to temperature fluctuations is probably as important as availability of food resources, but climatic zonation is a major cause of latitudinal provincialization. Diversity usually increases towards the equator (Valentine, 1973b).

Plate tectonics enables continents to break-up and separate and eventually collide, thus changing the area of faunal provinces and the barriers to faunal migrations. This is well illustrated by the change from provincial to cosmopolitan biotic provinces in the Palaeozoic as Iapetus or the proto-Atlantic ocean closed during the formation of the Caledonian–Appalachian fold belt. But the destruction of ocean basins leading to extinction of marine species may be a boon to the development of new land plants and animals; such was the case at the inception of the Pangaea landmass in the Devonian.

In pointing out that physical isolation produces its own adaptive radiation and that continental fragmentation therefore tends to increase biotic diversity, Kurtén (1969) noted that reptile radiation took about 200 my from the late Carboniferous to the Mesozoic when for a long period the continents were contiguous, whereas the mammals evolving often on isolated continents radiated in only 100 my in the late Cretaceous and Cenozoic—in fact the major advance took only about 10–20 my from the beginning of the Palaeocene to the Eocene.

The diversity of many marine fauna can be related to major fluctuations in sea level throughout the Phanerozoic time scale which were once related to periods of orogenesis but are currently related to the pattern of fragmentation and reassembly of the continents (Fig. 10.5) (Valentine and Moores, 1970, 1972; Hays and Pitman, 1973). The plate tectonic mechanism is thought to control the world's marine transgressions and regressions in the following way. During periods of ocean-floor spreading the emergent active ridges have such a considerable volume that they form topographic highs with the result that oceanic waters are displaced to cause a transgression of the continents. Conversely, if spreading stops, the ridges subside (and their volume is more than that of the average trench at a subduction zone) so causing an increase in

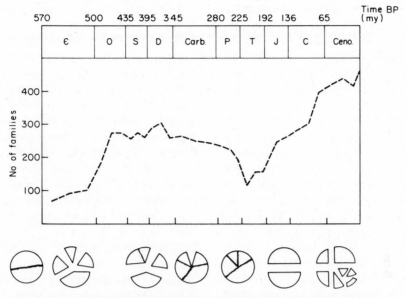

Fig. 10.5. A correlation of levels of faunal diversity with the patterns of continental assembly and fragmentation (indicated diagramatically) throughout the Phanerozoic (after Valentine and Moores, 1970; reproduced by permission of Macmillan Press)

the volume of the ocean basin, emergence of the continents and, therefore, a marine regression. The volume of present ridges is about $2 \cdot 5 \times 10^8$ km^3, enough to affect the ocean level by about 0·5 km around all the continents (Valentine and Moores, 1970). Thus continental assembly favours regression and fragmentation transgression, possible examples being:

1. Regression—formation of late Proterozoic supercontinent; Caledonian–Appalachian and Hercynian suturing and formation of Permo-Trias Pangaea.

2. Transgression—fragmentation of late Proterozoic supercontinent; periods of break-up of Pangaea, especially in the early Jurassic and mid-Cretaceous.

Between 110 and 80 my ago (Aptian to Coniacian in the mid-Cretaceous) the South Atlantic ridge formed when Africa parted from South America. This was a period of particularly rapid sea-floor spreading in both the Atlantic and Pacific oceans (Larson and Pitman, 1972) resulting in a spectacular marine transgression that almost doubled the area of continental shelves. When the spread-

ing rates returned to normal, near those of the present, about 60–50 my ago in the Palaeocene there was a corresponding major regression.

On transgression the continental shelf is covered with water and available habitats are considerably enlarged, increasing the total diversity of marine fauna. Alternatively a rapid fall in sea level eliminates most shore-beach and shallow-water habitats and so increases competition for space and nutrients, decreases faunal diversity and causes some extinctions (e.g. Newell, 1967). Kauffman (1972) found a positive correlation between these variables amongst North American Cretaceous molluscs and Axelrod (1972) suggested that the evolution of terrestrial floras was related to sea level changes.

Fluctuations in the salinity of ocean waters affect marine diversity; salinity reductions result from removal of salts from the water into evaporite deposits. The evaporites that formed in the Permo-Trias, in rifts and proto-oceans during the initial stages of continental separation (see Chapter 14), are commonly very large (up to 4000 km long, 600 km wide,

7 km thick, Kinsman, 1975a). The withdrawal of so much NaCl into evaporite sequences must have appreciably lowered the salinity of the world's oceans (Fisher, 1964). Nakazawa and Runnegar (1973) therefore considered that changes in seawater salinity were responsible for the crises in molluscan bivalves in Japan at the Permo-Trias boundary.

The supply of nutrients to the oceans, a fundamental factor regulating faunal diversity, may be controlled in two ways. Firstly, formation of a supercontinent leads to an increase in the seasonality in shelf waters and so to an increased nutrient supply. Conversely, if a supercontinent is broken up, the climates of the smaller continental masses would be more equable with less distinct seasonality and vertical mixing, so leaving fewer nutrients in shelf waters (Valentine and Moores, 1970). Secondly, sea-floor spreading releases mineral nutrients into the oceans. About 112 my ago this release was increased by a factor of two in proportion to the increase in sea-floor spreading rate (Larson and Pittman, 1972), one of many possible reasons for the generation of the mid-Cretaceous oil deposits (Fig. 10.4) (Irving and coworkers, 1974a).

Tappan (1968) and Tappan and Loeblich (1971) suggested that a reduction in phytoplankton productivity leads to nutrient depletion and a decrease in microflora, and Flessa and Imbrie (1973) made a comparison of their calculated rates of taxonomic change with variations in atmospheric oxygen by Tappan based on estimates of phytoplankton abundance. Although there is an overall correspondence, the only close correlation is between oxygen levels and the terrestrial and marine changes at the Cretaceous–Tertiary boundary. The idea that oxygen levels may be related to biotic changes was supported by McAlester (1970) who demonstrated that the high oxygen consumption rate of living taxa corresponds remarkably well to the high extinction rates of their fossil equivalents.

Now we come to the problematical question of the possible connection between reversals in the earth's magnetic field and faunal extinctions (J. F. Simpson, 1966; Black, 1971). Before speculating about causes, we should consider the fact that there is increasing evidence of a correlation between:

1. Periods of transgression and constant polarity, and of regression with mixed polarity. For example, during the Cretaceous, a period of widespread transgression, there was a long interval of normal polarity (Helsley and Steiner, 1968, 1969), whilst the early Triassic was a time of mixed polarity and regression (Burek, 1967; Helsley, 1969), as was the Cenozoic (Heirtzler and Hayes, 1967). As stated earlier, regressions are associated with decreases in marine faunal diversity and extinction, and transgressions with faunal explosions.

2. Geomagnetic polarity reversals and major faunal extinctions (Crain, 1971). The most impressive results are by Hays (1971) who, using deep-sea cores, found that during the last 2·5 my out of eight species of radiolaria (single-celled marine micro-organisms) that became extinct, six disappeared at times coincident with polarity reversals. Further back in time correlations are more speculative, but it is worth noting that roughly one-third of all living species, including the dinosaurs, became extinct at the end of the Cretaceous coinciding with a change from a long period of normal polarity (Helsley and Steiner, 1968, 1969) at the Campanian–Maestrichtian boundary (Keating, Helsley and Passagno, 1975) to one with many mixed polarity intervals (McElhinny, 1971). Likewise at the end of the Permian 225 my ago mass extinctions coincided with the resumption of reversal activity. In their reappraisal of the subject Flessa and Imbrie (1973) concluded that because the relationships are so close 'the geomagnetic polarity reversal hypothesis continues to be a strong candidate for the role of a major factor triggering past biotic changes'.

The question remains, what could be the causal link between these relationships? A variety of mechanisms have been suggested, such as climatic changes (Harrison and Prospero, 1974), increased doses of cosmic ray radiation related to the sun's position in the galaxy (Hatfield and Camp, 1970), or during periods of polarity reversal when the shielding effects of the geomagnetic field would be lowered (Uffen, 1963), direct magnetic effects on growth (Hays, 1971) and solar proton irradiation damaging the ozone screen (Reid and coworkers, 1976).

The Phanerozoic Fossil Record

A correlation by Valentine and Moores (1970) of species' diversity (based on benthonic shelf families of nine major invertebrate phyla) throughout the Phanerozoic with the patterns of continental fragmentation and assembly is shown in Fig. 10.5. In general terms, the graph illustrates that faunal diversity was lowest when there was a single supercontinent in the early Cambrian and early Trias, was highest when this was broken into several smaller continental plates in the mid–late Palaeozoic and late Mesozoic–Cenozoic, and that there has been an overall increase in

diversity over the last 600 my. The times of low diversity (depressions in the graph) were periods of significant faunal extinction. Although there has been an overall trend towards specialization throughout the Phanerozoic (Raup, 1972), the major rediversification in the Mesozoic–Cenozoic is thought to be due to continental fragmentation and drift leading to higher provinciality and to climatic zonations caused by continental configurations giving rise to high latitudinal temperature gradients (Valentine and Moores, 1970, 1972; Valentine, 1973a,b).

Raup (1972) make the criticism that the diversity levels in Fig. 10.5 have a sampling bias because there are less late Permian-Triassic sedimentary rocks preserved than there are rocks of other ages. During periods of regression, as in the Permo-Trias, continental shelves tend to be eroded and so one would

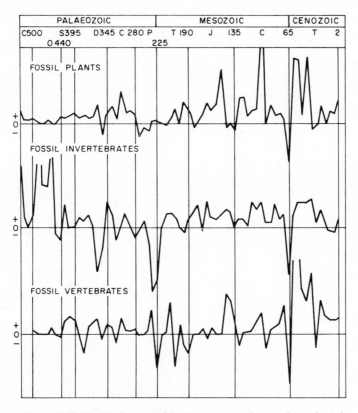

Fig. 10.6. Extinctions and appearances of new faunal and floral groups at stratigraphic boundaries during the Phanerozoic (modified after House, from *Understanding the Earth*; Artemis Press)

expect to find a relatively lower number of preserved shelf species in rocks of that age.

We shall now take a more detailed look at the faunal radiations and extinctions during the Phanerozoic (summarized in Fig. 10.6).

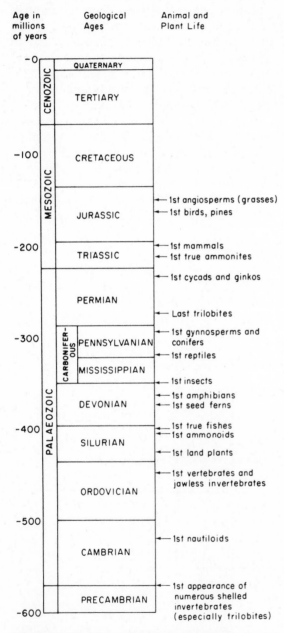

Fig. 10.7. The approximate time of appearance of major groups of animals and plants in the Phanerozoic (modified after Tarling and Tarling, 1971; reproduced by permission of G. Bell and Sons Ltd.)

The time of appearance of major faunal and floral groups is indicated in Fig. 10.7.

Early Palaeozoic

The Proterozoic supercontinent (Piper, 1974) had begun to break-up in the late Precambrian and by the late Cambrian the triangular European plate was surrounded by ocean (to close up and give rise to the Caledonian–Hercynian and Uralian belts), and North America had moved away from the southern continents which remained in a coherent group. The equator lay across North America, just to the north of Europe and across Siberia and Antarctica and warm shallow continental shelves (caused by a Cambrian transgression) produced abundant limestones (especially in North America) and reefs with calcareous algae and archaeocyathids along the presumed Cambrian tropical belt (Cowrie, 1971); the groups with calcareous hard parts were especially able to undergo rapid diversification. The overall climate seems to have improved considerably since the widespread glacial conditions of the late Precambrian. The combination of the climatic amelioration and the spread of shallow warm shelf seas probably greatly assisted the sudden diversification of life (House, 1971b).

The Precambrian–Cambrian boundary records the time when organisms developed skeletons allowing their preservation as fossils, and when there was an explosive evolution in marine life. The following nine phyla of marine invertebrates appeared for the first time in the early Cambrian (House, 1971b): Protozoa, Coelenterata (jelly fish, anemones), Archaeocyatha (sponges), Porifera, Bryozoa, Mollusca (gastropods, lamellibranchs, cephalopods), Brachiopoda, Arthropoda (trilobites) and Echinodermata (cystoids). Over 900 lower Cambrian fossil species are known. So rapid was the evolutionary explosion that all the remaining major groups of invertebrate fauna became evident by the middle Ordovician; in other words, it took about 120 my for the complete invertebrate range of phyla to evolve. During the Cambrian the rise of the trilobites was impressive (although two-thirds of the families were to disappear by the end of the Cambrian), and of the ammonites and graptolites. The Ordovician, a time of extensive sea floor spreading, was a major period of diversification (House, 1967)—the brachiopods, graptolites, ammonites, bryozoans, nautiloids and ostracods expanded considerably, corals and stromatoporoids became worldwide, bottom-dwelling graptolites evolved into planktonic types and the first fish appeared.

During the Silurian the marine transgression achieved its maximum extent giving rise to widespread reef limestones. A wide shallow-water sea covered much of Asia, Europe and North America and there were therefore no barriers to the migration of shelf faunas which by the Silurian had a cosmopolitan character (Cocks and McKerrow, 1973). Also, this was the first time that attempts were made to colonize fresh water leading to the appearance of vascular land plants and true fish; these were the earliest indications of the great changes that were to come in the late Palaeozoic.

Late Palaeozoic

In the period 395–225 my ago the continents were drifting together to form Pangaea; this involved closure of the pre-Caledonian ocean (Iapetus) by the early Devonian (it began closing by the mid-Ordovician), of the pre-Hercynian ocean by the late Carboniferous, and of the pre-Uralian ocean by the Permian. The effect on life forms was firstly to destroy the habitats of many marine invertebrates, causing the reduction in number of some and the extinction of others, and, secondly, to provide a suitable terrestrial environment for the evolutionary advance of plants and vertebrates.

Transgressions near the lower/middle Devonian boundary led to the establishment of widespread equable carbonate regimes in tropical latitudes, enabling corals and stromatoporoids to diversify rapidly, joining algae as reef-forming complexes, and to become worldwide. In the early Upper Devonian, however, the reef complexes became progressively attenuated, resulting in the virtual extinction of the reef corals and stromatoporoids; and, once the protection of the platform edge carbonates was lost, organisms living on the shelf (especially suspension-feeding brachiopods) were so adversely affected that some of them became extinct (House, 1975).

Trilobites declined considerably in the Devonian, many groups dying out leaving only a few to continue into the Carboniferous and Permian. Graptolites also declined, surviving only until the end of the Carboniferous.

There was some degree of provinciality of faunas in the early Devonian (Boucot, Berry and Johnson, 1968; House, 1971a) but by the mid-Devonian continental masses had moved together to produce a semicontinuous landmass. One result of this convergence was the formation of the 'Old Red Sandstone Continent' extending across North America and NW Europe with the first widespread terrestrial sediments of the Phanerozoic. This tectonic evolution stimulated the rapid development in the Devonian of fish (freshwater types with lungs) and vascular plants (the first forests appeared in the middle Devonian), and the appearance of amphibians, winged insects and terrestrial vertebrates. Fig. 10.8 shows that the distribution of late Palaeozoic labrinthodont amphibia and of Triassic reptiles are all grouped systematically across the relevant palaeoequator; they have a random distribution with respect to present latitudes.

Major transgressions in the Carboniferous led to the widespread deposition in low latitudes of shallow-water limestones with diversified corals and crinoids, and Coal Measures with seed ferns and giant cockroaches (10 cm long) and dragonflies (0·75 m wing span). A high latitude equivalent of the Coal Measure flora developed across Gondwanaland with the well-known *Glossopteris* plants associated with glacial deposits (Plumstead, 1973).

Similarly there were great advances on land during the Carboniferous. Amphibians flourished and gave rise to the reptiles, and forests of trees such as *Lepidodendron* and *Sigillaria* reached 30 m in height.

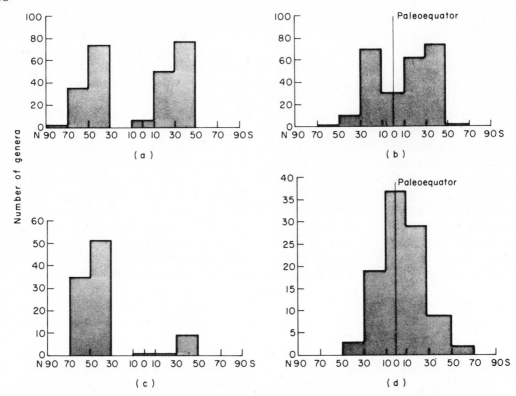

Fig. 10.8. Diversity of Triassic reptiles plotted against (a) Present-day latitudes and (b) Palaeolatitudes; and diversity of late Palaeozoic amphibia against (c) Present latitudes and (d) Palaeolatitudes (after Brown, 1968; reproduced by permission of *Australian Journal of Science*)

Regression began at the end of the Carboniferous and continued in the Permian. Although fusulinids (foraminifera) and echinoids were still diversifying, rugose and tabulate corals declined as reef builders, their place being taken by calcareous sponges and bryozoans. Permian brachiopod distribution enables the palaeoequator and geosutures to be defined (Waterhouse and Bonham-Carter, 1975). Amphibians declined as the reptiles flourished. The distribution of the terrestrial faunas of the late Carboniferous and early Permian supports the palaeomagnetic data, indicating that North America and Europe were contiguous at that time (Milner and Panchen, 1973). Conifers proliferated and in the insect world bugs, beetles and cicadas appeared.

By the end of the Permian all the continents were sutured together and uplift of the land following the several periods of orogenesis was responsible for the withdrawal of seas from the Pangaea landmass. The result of these tectonic circumstances is that the Permo-Trias boundary at the end of the Palaeozoic represents the greatest period of extinction in the fossil record (Rhodes, 1967) (Figs. 10.5, 10.6). Marine invertebrates were especially affected; trilobites, blastoids, fusulinid foraminifera, eurypterids, rugose corals, three-quarters of the bryozoan groups and several of the cephalopods, sponges, brachiopods and echinoderms, and many families of molluscs, all became extinct. Plant life was not seriously affected but vertebrates were. The agnatha vertebrates died out by the mid-Permian and between the Permian and the Triassic some 75% of amphibian and over 80% of reptile families disappeared.

Mesozoic

In the 30–40 my period between the mid-Permian and the mid-Trias the only new tectonic development was that the Pangaea land-mass drifted northwards about 30° (Briden and coworkers, 1974). In the late Triassic rift structures formed between North America and Gondwanaland representing the incipient stages of break-up of Pangaea. The other continental masses separated during the Jurassic and Cretaceous. These tectonic conditions allowed widespread transgressions and the formation of extensive continental shelves along the trailing edges of dispersing plates, favourable habitats for the rediversification of many shallow-water invertebrates; the vertebrates continued to evolve and diversify on the separating continents.

In the Triassic, faunas were largely cosmopolitan and only a few major provinces can be differentiated. At this time the belemnoids, oysters, complex-sutured ammonoids, crustaceans and echinoderms became important. Amongst the reptiles dinosaurs evolved on the land, plesiosaurs and ichthyosaurs in the sea and gliding pterosaurs in the air. Mammals evolved from the therapsid reptiles in the late Triassic. Land floras flourished with important palm-like cycads, horsetails, conifers, ferns and ginkgoes.

During the Jurassic those faunas that survived from the Palaeozoic or were established in the Triassic continued to diversify. Ammonoids and foraminifera flourished, as did corals and sponges in carbonate reefs on shallow shelves. The frog (an amphibian), coccoliths (planktonic plants), angiosperms (flowering plants) and the first bird (*Archaeopteryx*, with feathers developed from reptile scales) appeared. Terrestrial reptiles, such as dinosaurs, were prominent, the pterosaur reptiles learnt to fly, but mammals were slow to evolve. An important point is that by contrast with the cosmopolitan terrestrial faunas of the Triassic (Charig, 1971; C. B. Cox, 1973), by late Middle Jurassic times a Boreal Province among the ammonites and belemnites became distinguishable from a Tethyan Province in the northern hemisphere (Bassov and coworkers, 1972; Hallam, 1973). Separation of the faunal provinces reached a peak during the late Jurassic and early Cretaceous, coinciding with a period of extensive regres-sion, with the result that cephalopod correlations became impossible (Casey, 1971). In a similar way plant distribution shows a latitudinal–climatic zonation into a broad equatorial belt separating two east–west polar belts from the early Triassic to the early Cretaceous (Barnard, 1973).

In the Cretaceous Tethys was closing and the North Atlantic ocean was opening. Some Gondwana continents began to separate but about 100 my ago the Gondwana landmass lay in the same latitudinal position as in the Jurassic, 70 my before (Fig. 9.3). There were no faunal–floral crises at the end of the Jurassic so that the life forms established in the Triassic and Jurassic continued to develop. Planktonic globigerinas, coccoliths and diatoms became so abundant as to form foramineral oozes, especially along the Tethyan seaway (Fig. 10.4) (Dilley, 1973), and calcareous sponges, jellyfish, corals, bryozoans, pelecypod molluscs, belemnoid cephalopods, echinoids and crinoids all flourished in Cretaceous seas. Salamander amphibians emerged and terrestrial, marine and flying reptiles attained a peak of development. Most spectacular was the radiation of the angiosperms which overtook the gymnosperms during this period. According to Colbert (1973) reptiles retained a cosmopolitan character throughout the Mesozoic because, in spite of the fact that continents were drifting, there was sufficient connection between blocks to allow the dinosaurs and their contemporaries to wander far and wide.

At the end of the Cretaceous, 65 my ago, about one-third of all living species became extinct, including the ammonites, belemnites, plants, some bryozoa, bivalve molluscs, echinoids and planktonic foraminifera, and most of the terrestrial, marine and flying reptiles (e.g. dinosaurs, ichthyosaurs and pterosaurs); of the reptiles only the crocodiles, snakes, turtles and lizards survived into the Tertiary. The Cretaceous–Tertiary boundary was a time of change in two other respects. Firstly, after the long transgressive period during the Cretaceous, there was a regression at the Maestrichtian–Danian boundary; and secondly, the normal geomagnetic polarity era of the Cretaceous moved into one of mixed polarity in the Tertiary. Possible causes to account for such relationships were reviewed in the last section.

Tertiary

The continuing faunal and floral rediversifications during this period were a result partly of the increasing longitudinal provinciality created by fragmentation and separation of drifting continents, and partly of the widening latitudinal spread of the continents causing cooling of shelf waters in high latitudes and of a consequent latitudinal climatic zonation into separate provinces with individual diversity patterns culminating in major glaciation from mid-Miocene onwards in Antarctica (Valentine and Moores, 1970).

The most important development in the history of life in the Tertiary was the extraordinarily rapid rise in about 10–20 my of the mammals which took over the environments vacated by the reptiles on the land and in the sea and air. By the Eocene primates, elephants, rhinoceroses, pigs, rodents, horses, sea cows, porpoises, whales and bats, as well as most orders of modern birds and many families of modern plants, had all appeared.

The evolution of the vertebrates (fish, amphibians, reptiles and mammals) throughout the late Mesozoic and Cenozoic was strongly affected by the palaeogeography of the continents. Gondwanaland was more or less isolated from Laurasia in the late Mesozoic and this separation facilitated the development of two fairly distinct faunas (Cracraft, 1973).

Many foraminifera such as the globigerinids and nummulites evolved rapidly becoming abundant during the Tertiary. Gastropods, pelecypods and reef corals were the predominant invertebrates.

Chapter 11

Caledonian–Appalachian Fold Belt

The evolution of the continents during the Palaeozoic involved several phases of orogenesis. At present there is much debate about the way the orogenic belts formed by processes of sea-floor spreading and continental drift. So far the 'drifters' have accumulated an impressive body of evidence, but the 'non-drifters' (if there are any?) have been largely silent.

Within the North American/Eurasian continent there were five phases of orogeny during the Palaeozoic:
1. Caledonian: Upper Silurian-Lower Devonian (Britain, Scandinavia, E Greenland)
2. Acadian: Middle Devonian (Appalachians)
3. Hercynian: Upper Carboniferous (Europe)
4. Alleghanian: Upper Carboniferous or Lower Permian (Appalachians)
5. Uralian: Permian (Europe–Siberia)

In this chapter we shall make a special study of the Caledonian–Appalachian belt, which extends from Britain to the eastern USA, as this is better documented than any other Palaeozoic fold belt. Important evidence in this belt in favour of a former proto-Atlantic ocean includes differences between European and American fauna, and ophiolite complexes which are remnants of that oceanic crust.

Although 'proto-Atlantic' has commonly been applied to the Eur-American ocean, the term 'Iapetus' seems to be coming into favour (proto-Atlantic is not strictly correct) and will be used below.

The final part of this chapter will review the mineral deposits in these Palaeozoic fold belts—some regional metal zonations provide corroboration for the plate tectonic models.

The Iapetus (proto-Atlantic) Ocean

Application of the plate tectonic theory to the formation of the Caledonide–Appalachian fold belt depends on the opening of an ocean in early Palaeozoic time and its closure during the late Palaeozoic (J. T. Wilson, 1966; Harland, 1967). Let us look at the evidence for movement.

The Opening Stage

There is a lot of information in favour of the existence of an ocean along what is now the middle of the fold belt (Dewey, 1969; Bird and Dewey, 1970; McKerrow and Ziegler, 1972a):
1. Faunal province data suggest that there was increasing separation of the American and European species during the Cambrian and Lower Ordovician.
2. As would be expected, there are no Cambrian, but only Ordovician and Silurian oceanic sediments preserved along the site of the former ocean; the earlier sediments were destroyed during subduction.
3. The facies types and evolution of the early sediments are remarkably similar to those of Mesozoic–Cenozoic Cordilleran-type geosynclinal belts, viz. miogeoclinical clastic wedge followed by carbonate platform and pelitic-clastic continental rise.
4. The late Precambrian–Lower Palaeozoic structural history on either side of the central

'oceanic' belt was markedly different (McKerrow and Ziegler, 1972a).

5. There are slices of oceanic crust that were locally thrust on land as tectonic klippe (ophiolite complexes) which retain evidence of the nature of the early oceanic volcanics and sediments.

The Closing Stage

The following evidence is used to suggest that the Iapetus Ocean was contracting during the Silurian and was closed by the Middle Devonian (S. Turner, 1970):

1. The earlier distinctive American and European faunal provinces began to decline in the mid-Caradocian. Late Ordovician faunas show increasing evidence of mixing and, by Silurian times, the provinces are indistinguishable.

2. The formation of Benioff zones, and thus the contraction of the ocean, is indicated by the accumulation in the trench of scraped-up oceanic sediments and volcanics (the argille scagliose facies of the Dunnage mélange), the formation of high-pressure glaucophane-bearing assemblages and oceanic crust segments near the trench (e.g. Ballantrae ophiolite complex), and the intrusion and extrusion of calc-alkaline plutonic and volcanic rocks on the continental margins.

3. The gradual elimination of the ocean gave way in places to the onset of non-marine conditions in the Middle Silurian (Wenlock—the Midland Valley of Scotland). By the uppermost Silurian (Downtownian) fish were restricted to brackish environments (McKerrow and Ziegler, 1972b) and in the Devonian extensive Old Red Sandstone formed under desert (continental) conditions.

Now we shall look in some detail at the key evidence for the existence of that Iapetus Ocean: faunal provinces and ophiolite complexes.

Faunal Provinces

Faunal provinces are marine regions inhabited by a characteristic association of organisms, and bounded by barriers preventing the spread and mixing of the characterizing species. It was once generally thought that the faunal differences arose from separation by intervening landmasses; however the plate tectonic concept provides us with a firmer basis for reconstruction of the continents, confirmed independently by palaeomagnetic data for example, and thus it can now be shown with some degree of sophistication that the barriers were tectonically controlled by continental movements (Valentine and Moores, 1970, 1972; Sylvester-Bradley, 1971b). Considerable provinciality arises when the continents are fragmented and widely scattered, providing topographical barriers such as landmasses and deep oceans, together with climatic barriers when the continents stretch over great latitudes. In particular it was the formation of an ocean that created a barrier to migration of shallow-water organisms. Lowest provinciality and highest cosmopolitan distribution is associated with a supercontinent or Pangaea (Valentine, 1973a).

We shall now consider the faunal province data which suggest that the American and European continents were widely separated by the early Ordovician and were close together in the Silurian. Further consumption of oceanic crust caused by continental drift gave rise to an eventual Pangaea by the Permo-Carboniferous.

It is important to realize that it was not necessary for the ocean to be very wide for it to act as a barrier to migrating species. According to A. Williams (1969) the ocean currents flowed parallel to the North Atlantic continental margins in the Ordovician and therefore quite a narrow ocean would have sufficed to prevent migration and faunal mixing (McKerrow and Ziegler, 1972a).

Trilobites provide the main evidence for two Lower Cambrian provinces (Palmer, 1969; Cowie, 1971; Sdzuy, 1972):

1. Olenellid, divided into:

    a) Acado–Baltic (or Atlantic/European) Province.

    b) Pacific (or American) Province.

2. Redlichiid, in Asia–Australia–North Africa.

This faunal provinciality was, however, relatively weak (T. Fletcher in Bird and Dewey, 1970).

During the middle and late Cambrian seven trilobite provinces can be identified (Palmer, 1969); in regions with unrestricted access to the open ocean there were three provinces and in restricted regions four.

The greatest degree of provinciality amongst Palaeozoic organisms reflecting the maximum separation of the continents occurred in the Lower Ordovician when there were:

2 graptolite provinces (Bulman, 1971; Skevington, 1973)
4 trilobite provinces (Whittington and Hughes, 1972, 1973; Sadler, 1974)
5 brachiopod provinces (A. Williams, 1973)
2 crinoid provinces (Scotese, 1974)

In each of these the American (Pacific) and Acado–Baltic (Atlantic) provinces are prominent (Fig. 11.1).

There was, however, a drastic change in palaeofaunal boundaries by the latest Ordovi-cian (Ashgillian) caused by a decline in provinciality. The faunal evidence is considered by many palaeontologists to be consistent with subduction throughout the mid–late Ordovician, resulting in contraction of the Iapetus ocean and the mixing of the previously independent fauna (A. Williams, 1976). The brachiopod data suggest that the beginning of faunal exchange, following on the initial continental collision, took place in the mid-Caradocian (A. Williams, 1973). Continents continued to get closer during the Silurian with the result that there was a cosmopolitan or worldwide distribution of trilobites, graptolites, brachiopods, coelenterates, nautiloids, and crinoids; only ostracods show some sign of limitation to provinces (Holland, 1971). McKerrow and Ziegler (1972b) have discussed the faunal evidence for the distribution of Silurian continents and oceans. The Baltic Shield collided with the Canadian Shield in the late Silurian or early Devonian to produce the Caledonian Orogeny, but evidence from freshwater fish faunas suggests that not until the Middle Devonian was there a

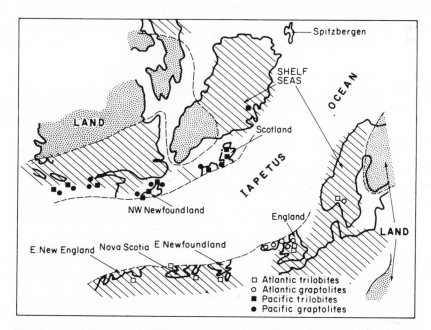

Fig. 11.1. Pacific- and Atlantic-type tribolites and graptolites in the shallow shelf seas bordering the Iapetus ocean in Cambrian-early Ordovician time. The Iapetus ocean is of unknown width. (Adapted from Cowie, 1974; reproduced by permission of J. W. Cowie)

land connection between the Canadian–Baltic Shield and Gondwanaland (Romer, 1966, in McKerrow and Ziegler, 1972a). According to House (1971a) in early Devonian times the Old World faunal province (N America, Europe, Asia) was still separated from the Austral province (S Africa, S America, Antarctica); likewise, Boucot, Berry and Johnson (1968) recognized a high degree of provinciality amongst early Devonian brachiopods in three world provinces. However, by the Middle Devonian these continental blocks were joined, thereafter leaving a semicontinuous landmass composed of all the continents (proto-Pangaea). It can be no coincidence that this increase in land area in the Devonian took place at the same time as the explosive evolution of terrestrial vertebrates and plants which became similar the world over (see further in Chapter 10). During the Permo-Carboniferous the gradual elimination of shallow sea environments gave rise, by the end of the Permian, to extinction of many marine faunas (e.g. fusulinids, blastoids, rugose corals, eurypterids, trilobites) and to the marked decline of others (e.g. foraminifera, bryozoans, brachiopods, ammonoids ostracodes) (Rhodes, 1967). Finally, in the Upper Carboniferous (late Westphalian) central Europe collided with the Baltic Shield giving rise to the climactic deformation of the Hercynian Orogeny—the formation of Pangaea was completed by the end of the Palaeozoic.

Ophiolite Complexes

Ophiolite complexes are stratigraphic units ideally comprising the upward sequence serpentinite and peridotite, gabbro, 'sheeted' basic dyke zone, pillow-bearing basic volcanic rocks, chert, pelagic limestone and argillite. Nowadays these are widely regarded as being slices of oceanic crust/mantle but there is a current tendency to label or misidentify some deformed mafic–ultramafic complexes as ophiolites. In the identification of Palaeozoic ophiolites much depends on the analogy with examples in the Alpine fold belt, such as Troodos in Cyprus and Vourinos in Greece, and in turn with present day oceanic crust. But

amongst the many complexes in the Appalachian–Caledonian belt are some key examples, well preserved and documented, which leave little doubt that they are of oceanic crust/mantle derivation: the several bodies in Newfoundland (Fig. 11.2) and the Ballantrae Complex in Scotland. In fact the Bay of Islands Complex in Newfoundland is as good an example of transported oceanic crust/mantle as the Troodos Complex.

Williams and Smyth (1973) stated that the analogy between the Newfoundland ophiolite suites and modern oceanic crust/mantle is based on the following considerations:
1. Similarities in the gross physical characteristics of the suite with geophysical models of oceanic crust and mantle.
2. Transported on-land ophiolite is rooted in oceanic lithosphere in Papua, New Guinea.
3. Strong lithologic similarities between the ophiolite suite and rocks of Macquarie Ridge (including sheeted dykes) where exposed and mapped at Macquarie Island.
4. Lithological and chemical similarities of oceanic tholeiites and pillow lavas of ophiolite suites.
5. A model involving sea-floor spreading is the only reasonable explanation for the extensive development of sheeted dykes like those in ophiolite complexes.
6. High pressure mineralogy of certain Alpine peridotites requiring mantle depths for conditions of crystallization.
7. Preponderance of metamorphic tectonites in the ultramafic fraction of ophiolites that display textures like those experimentally produced at conditions representative of the mantle.
8. Similar metamorphic mineral assemblages in oceanic rocks at mid-ocean ridges compared with those in ophiolites and similar vertical metamorphic variations related to geothermal gradient and depth of burial.

Newfoundland

In western Newfoundland there are four ophiolite complexes (Fig. 11.2): Bay of Islands, Baie Verte, Betts Cove–Tilt Cove

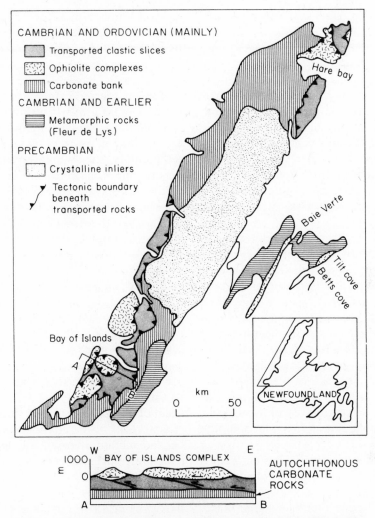

Fig. 11.2. The four ophiolite complexes of Newfoundland. The section shows the Bay of Islands Complex transported over a clastic wedge, itself thrust westwards over an autochthonous foreland. (Redrawn after Williams and Smyth, 1973; reproduced by permission of *American Journal of Science*)

and Hare Bay. The following general account is taken from the reviews of Church and Stevens (1971), Dewey and Bird (1971) and H. Williams (1971).

Most of the successions consist of layered plutonic rocks underlain by metamorphosed and deformed basic rocks and overlain by little-deformed pillow lavas and sediments. The general succession, which may be up to 10 km thick, is as follows:

Top

Mafic pillow lavas with interlayered sediments (chert and greywacke)

Meta-volcanic rocks cut by diabase dykes

Massive gabbros, diorites, quartz diorites and trondhjemites

Banded gabbros

Peridotites (lherzolites and harzburgites), dunites, pyroxenites

Garnetiferous amphibolites

Meta-volcanic greenschists and minor meta-sediments (pelitic–psammitic schists)

Bottom

Zircon ages of the trondhjemites by Mattinson (1975) indicate formation of the Bay of Islands complex about 508 my ago (late Cambrian) and the Betts Cove complex about 463 my ago (mid-Ordovician). The difference in age supports the idea that each ophiolite belt is the product of a separate spreading episode in independent narrow newly-rifted ocean basins.

The complexes occur in subhorizontal allochthonous thrust slices derived from the east and emplaced in the Ordovician. They are mostly underlain by sedimentary rocks belonging to transported clastic wedges (Fig. 11.2). Although greenschists and amphibolites form the base of the mafic–ultramafic successions in the Hare Bay and eastern Bay of Islands Complexes, peridotites, gabbros, volcanics and quartz diorites locally form the base of the western Bay of Islands Complex; in other words, the structural base of the tectonic slices has cut obliquely across the ophiolite succession.

The ultramafic rocks in the Bay of Islands and Hare Bay bodies are characterized by the presence of high-pressure mineral assemblages. Garnetiferous lherzolites and harzburgites contain aluminous ortho- and clino-pyroxenes, kaersutite and titaniferous phlogopite—a mineral assemblage indicative of high temperatures ($c.$ 1200°C) and pressures ($c.$ 25 kbs) of formation in the upper mantle (Malpas, 1974). Such high pressure mineralogy provides one of the most compelling indications that these mafic–ultramafic complexes did not form by *in situ* gravity accumulation in stratiform intrusions emplaced into continental crust.

The basal contact of the successions is either a thrust surface or a tectonic mélange, locally up to 70 m thick. Below part of the Bay of Islands Complex there is $c.$ 30 m of sheared serpentinite underlain by $c.$ 17 m of serpentinite mélange (consisting of serpentinite and gabbro boulders in a finely-comminuted serpentinite matrix) which, in turn, is underlain by another 17 m of sedimentary mélange containing sandstone and serpentine boulders in a shaly matrix. These mélange zones are a clear indication that the ophiolite complexes are sitting on major thrust surfaces.

Sometimes the high temperature peridotites are situated on thrust planes directly on top of sedimentary clastics and the fact that there is no sign of normal contact metamorphism is a further indication that the complexes must have been transported. But more interesting is the fact that there is a high temperature basal aureole of amphibolite and greenschist below the Hare Bay and Bay of Islands Complexes. Such aureoles below Alpine-type ophiolite suites have long proved difficult to account for, but Williams and Smyth (1973) have provided a satisfactory and interesting explanation. The contact aureoles are very odd as they have regional metamorphic/tectonic mineral fabrics (such as lineated schistosity and rotated garnet porphyroblasts). The proposed plate tectonic model envisages a subhorizontal sheet of oceanic crust being thrust over supracrustal rocks at the continental margin which became metamorphosed and deformed by the overriding thrust sheets. Subsequent to this the aureole rocks became welded against the basal peridotites, and the ophiolite slice moved along a lower structural base so that the aureole was included with it as a structurally underpinned slab. Tectonic transport then took place by cold gravity sliding giving rise to the characteristic mélange zone, and the aureole rocks moved with the ophiolite slice for at least 80–105 km. These aureole rocks provide us not only with a unique type of dynamic 'contact metamorphism', but also with evidence of how the ophiolite slices were transported to their present position.

Ballantrae

The Ballantrae ophiolite is described by Church and Gayer (1973) who compared it with the Newfoundland examples. They pointed out the following critical features:

1. The body is underlain by a foliated schistose amphibolite analogous to that below the Bay of Islands Complex.

2. The basal ultramafics contain lherzolites, amphibole spinel-bearing eclogitic rocks and high alumina clinopyroxenes—all features suggestive of upper mantle origin.

3. A zone of sheeted diabase dykes is overlain by pillow lavas.

4. Overlying the ophiolite is an olistostrome (wildflysch) unit containing fragments of glaucophane schist, chert, volcanic pillows, gabbro, pyroxenite and serpentinite, etc.; the silty shale matrix to the fragments has a highly sheared texture. The whole unit is regarded as being comparable to the tectonic mélange zones associated with the Bay of Islands Complex.

The Ballantrae Complex contains all the main components of a typical ophiolite succession and thus Church and Gayer (1973) regard it as a slice of pre-Arenig oceanic lithosphere overlain by a mid-Arenig wildflysch unit and younger arc-type volcanics indicating some late subduction. Using stable trace element distributions Wilkinson and Cann (1974) distinguished three different groups of basaltic rocks: hot-spot basalts, island arc low-K tholeiites and possible ocean floor basalts.

Current opinion on the Newfoundland–Ballantrae ophiolites is that the basal high pressure ultramafics represent upper mantle, the gabbros and 'sheeted' diabases layer 3 of the oceanic crust, and the pillow lavas, cherts and greywackes layers 2 and 1 (see further in Chapter 15). This sequence was emplaced by thrusting in the early Ordovician during an early phase in the closure of the Iapetus ocean within the European–North American continental plate.

But one must not think there is no controversy amongst the proponents of plate tectonics. As Church and Gayer (1973) pointed out, the Appalachian–Caledonian ophiolite complexes have been interpreted in several ways:

1. Serpentine intrusions into trench sediments and volcanics (Dewey, 1969; Bird and Dewey, 1970).

2. Layered complexes related to island arc volcanism (Bird and Dewey, 1970).

3. Folded and faulted overthrusted or obducted sheets of oceanic lithosphere (Church and Stevens, 1971).

4. Upthrust wedges of oceanic lithosphere representing inner margins of trenches (Fitton and Hughes, 1970; Dewey, 1971).

5. Autochthonous remnants of small ocean basins (Dewey, 1971).

Anglesey

In his 1969 paper Dewey suggested that a late Precambrian southward-dipping subduction zone was sited in the area of Anglesey on the basis of the following evidence. In the Mona complex there are flysch-type sediments, tectonic mélange, spilitic pillow lavas, chert, jasper, gabbro, serpentinized ultramafic rocks (dunite, harzburgite, pyroxenite, lherzolite) and lawsonite–pumpellyite glaucophane schists. These rocks do not occur in a single stratigraphic succession but are scattered throughout the island and there is no sheeted dyke complex. Nevertheless their fairly close association 'may' add up to a fragmented ophiolite zone (Wood, 1974). The lateral change from high pressure blueschists to low pressure sillimanite gneisses might be indicative of a paired metamorphic belt developed across an arc–trench system. Thorpe (1972) showed that the geochemical features (Y-La–Ce, Ti–Zr–Y, Cr–Ni–Ti) of the glaucophane schists are most similar to modern ocean floor basalts compared with basalts from other environments. Maltman (1975) pointed out that field relations do not support a low-angle thrust environment for the ophiolite suite; rather the presence of an aureole around the ultramafic rocks suggests the subvertical ascent of magmas, perhaps at the inflection of a subducting plate.

Palaeomagnetism and the Position of the Continents

One of the main limitations of the palaeomagnetic technique is that it defines the palaeolatitude, but not the palaeolongitude, of a particular continent. In other words, a particular continent may have a variety of positions relative to another at a certain palaeolatitude.

When considering post-Trias times other evidence come to the aid of those attempting to reassemble the continents. After the break-up of Pangaea new oceanic crust was formed between the drifting continents and much of that ocean floor is extant, and the continental plates have not drastically changed their shape in that time. Therefore, the ocean floor spreading data and the best-fit of the continental edges are used to assist in the reconstruction of post-Trias continental masses (Smith, Briden and Drewry, 1973).

However pre-Permian continental reassemblies pose a problem. Firstly it is not even known how many plates there were at this time, and then there are no accurate methods for reorientating plates separated by orogenic belts. This is made the more difficult because it is becoming increasingly apparent that the formation of some early Palaeozoic fold belts was ensialic (Hurley, 1972). Finally, there are no Palaeozoic ocean floors preserved. However, a first approximation of the continental outlines can be obtained by dividing Pangaea along the line of those pre-Permian fold belts that formed by continental collision.

In spite of these problems there is palaeomagnetic evidence that continental drift did occur in the Palaeozoic. Hailwood and Tarling (1973) and McElhinny and Briden (1971) found that rapid polar shifts relative to various continental masses took place particularly in the Ordovician, but that the magnitudes and timing of these shifts varied for different continents; thus continental drift probably took place. As concluded by Creer (1973) 'the wide distribution of palaeomagnetic poles as plotted on the map of Pangaea is not consistent with the existence of that single landmass in the Ordovician'. It seems to be difficult to date the actual time of closure of the Iapetus ocean. McElhinny (1973) suggests that North America and Europe came together in the Silurian–

Devonian/Carboniferous; according to Creer (1970a) and Phillips and Forsyth (1972) these continents were in contact from the Devonian, whilst Hailwood and Tarling (1973) concluded that final closure of the ocean occurred in the Carboniferous. Briden, Morris and Piper (1973) calculated on palaeomagnetic evidence that there has been 1000 ($\pm$800) km of closure across the British Caledonides since early Ordovician times.

The following example demonstrates how palaeomagnetic data can be usefully correlated with tectonic and palaeoclimatic data in the understanding of Palaeozoic fold belts. Palaeomagnetic orientations (Smith, Briden and Drewry, 1973), together with the distribution of sedimentary facies, reef complexes, red beds and carbonate sequences, indicate that the North American–Greenland continent went through a rotation of about 90° during the Palaeozoic. In the Cambrian the equator extended almost north–south across the continent, but by the Carboniferous it had an almost east–west trend. This, of course, affected the trend of palaeoclimatic zones, but it may also account for the fact that the continent is flanked on all sides by fold belts with similar Palaeozoic craton-directed folding and thrusting (Hatcher, 1974).

Fold Belt	Deformation Time
Appalachian	Ordovician–Silurian
	Devonian
	Early Carboniferous–Permian
Ouachita	Ordovician–late Carboniferous
Western Cordillera	Devonian–early Carboniferous
	Permo-Trias
Innuitian	Middle Devonian–early
	Carboniferous
East Greenland	Silurian–Devonian

Hatcher (1974) proposed that counterclockwise rotation coupled with northward to northeastward translation of the continent could account for the simultaneous tectonic histories of these fold belts. The early main deformation in the Appalachian–East Greenland, Innuitian and Ouachita fold belts was produced by the main rotation phase in the early to middle Palaeozoic, whilst the late Palaeozoic deformation in the Cordillera was caused by translational movement as rotation stopped when Iapetus was completely closed.

The British Caledonides

The first major paper describing the evolution of the British Caledonides according to plate tectonic theory was by Dewey (1969). The following summary is based on this model with some later modifications.

Subdivisions

The fold belt is divisible into three zones (Fig. 11.3). (For 11 zones, see Dewey, 1974.)

Fig. 11.3. Structural subdivisions of the British Caledonides with position of the Ballantrae and Anglesey ophiolite complexes

1. NORTHERN ZONE. This comprises:
   a) The unmetamorphosed late Precambrian Torridonian sandstones.
   b) Cambrian–Ordovician limestones and quartzites (unmetamorphosed) forming a thin, shallow-water shelf sequence now preserved only west of the Moine Thrust.
   c) The Moinian–Dalradian series of late Precambrian to Cambrian age occurring east of the Moine Thrust. This is a thick sequence of psammites, pelites, semi-pelites, marbles and basic volcanics which were deformed in recumbent folds and highly metamorphosed in the late Cambrian to middle Ordovician with a climax in the early Ordovician.
2. CENTRAL ZONE. This consists of a thick Ordovician volcanic pile and an Ordovician to Silurian sequence of shales, cherts and greywacke turbidites which underwent only low degrees of metamorphism and deformation in the late Silurian or Devonian (mostly in the Lake District and southern Uplands).
3. SOUTHERN ZONE. This consists of Precambrian rocks in the Mona Complex of Anglesey (including serpentines, cherts, pillow lavas and glaucophane schists) and the Church Stretton area (Uriconian volcanics and Longmyndian sediments), and a Cambrian to Silurian succession which was deformed and recrystallized to a low grade in late Silurian or Devonian times. The zone can be divided into two:
1. In North Wales there is a deep-water 20 km thick monotonous sequence of shales and turbidites with prominent volcanics in the Ordovician and few unconformities.
2. On the Midland Shelf (Church Stretton area) there is a thin sequence of shallow-water shales and silty mudstones with three major unconformities, indicating transgressions and regressions in a platform environment.

Tectonic Evolution

Firstly, a synopsis (Fig. 11.4). The sedimentary framework was established in an ocean spreading stage during the late Precambrian–early Ordovician. The ocean development was symmetrical with thin shelf, or miogeosynclinal, deposits bordering on the continents in the extreme north and south (Durness Group of limestones and quartzites and Church Stretton shales and mudstones), followed inwards by thick accumulations in deeper water on

180

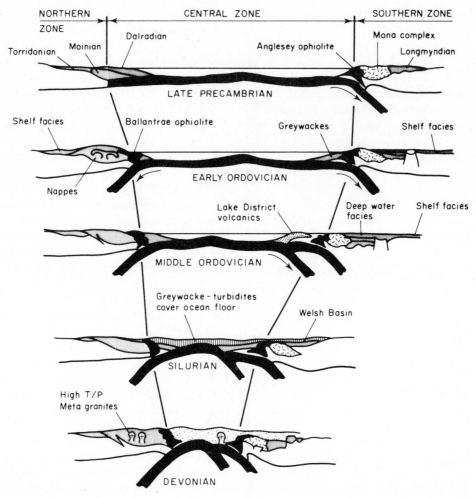

Fig. 11.4. A plate tectonic model to explain the evolution of the British Caledonides (redrawn and modified after Dewey, 1969; reproduced by permission of Macmillan Press)

each continental rise (Moine–Dalradian in the north and North Wales in the south). In the central zone the Ordovician ocean floor was covered with a thin veneer of pelagic deep-water cherts and shales, but in the Silurian during advanced stages of closure it was covered with a thick deposit of terrigenous turbidites and siltstones derived from the nearby landmasses.

Evidence of subduction lies in:

1. The late Precambrian glaucophane rocks and ophiolite suite in Anglesey on the south side, and in the early Ordovician Ballantrae Complex on the north.

2. Arc-type volcanics of Ordovician age in the Ballantrae, Lake District and North Wales areas.

Advanced movement on the Benioff zones was responsible for rise of geothermal gradients with production of regional metamorphism and for contraction and deformation of the sedimentary wedges along each continental rise. Early stages of deformation and metamorphism (sillimanite grade) in the Scottish Highlands reached a peak in the Lower Ordovician between 480 and 510 my. The two continents finally collided in early Devonian times when the last granites were intruded. After this, Caledonian orogenic activy ceased, the fold belt undergoing final uplift and erosion.

Now let us look at the evolution of the Caledonian belt step by step, from the late Precambrian to the Devonian.

Late Precambrian In the southern zone the Wentnor Series, consisting of red sandstones, siltstones and conglomerates, was deposited as a 5 km thick clastic wedge on the continental shelf; it may have been a post-orogenic molasse sequence following the deformation of the Mona Complex (Wright, 1969). The Charnian Group of water-lain tuffs and pyroclastics, and conglomerates, quartzites and shales were laid down in Leicestershire in a NW-trending aulacogen according to Dewey and Kidd (1974).

We can consider the late Precambrian supracrustal rocks of the northern zone in two parts:

The Torridonian This consists of a 9 km thick succession of predominantly red arkoses and cross-bedded sandstones which lie unconformably on the western foreland of the Caledonian mobile belt (A. D. Stewart, 1975). Two groups are distinguished: the Stoer Group, that locally infills valleys in the Lewisian land surface, and the Torridon Group which overlies it unconformably; their sedimentation occurred 995 ± 17 my ago (Moorbath, 1969). According to Dewey (1969) the Torridonian accumulated as a sedimentary wedge on the northern continental terrace, but Dewey and Kidd (1974) suggest that, together with its north–south facies and thickness changes, it was deposited in an aulacogen in the early rifted continental margin.

The Moinian/Mid-Dalradian The Moinian comprises a rather monotonous succession about 8 km thick of metamorphosed arenaceous and argillaceous rocks with minor impure dolomites (Johnstone, 1975). Terms commonly used to describe the present rocks are psammites or psammitic granulites, semi-

pelites and pelitic schists. Various sedimentary structures (current bedding, ripple marks, slump folds) show that Moinian sediments were deposited in shallow water to form thick, locally pebbly, sandstones and thinner mudstones and shales. The rocks have been overprinted by the Caledonian 'orogeny'. Several phases of folding are recognizable, including early recumbent folds, and large-scale interference patterns are well developed. Extensive chronostratigraphic correlations are made difficult by the lack of well-defined stratigraphic markers in a monotonous succession, by the lack of response of granitic gneisses to changes in metamorphic grade and by the complication of the folding. Metamorphic zonation is analysed by Winchester (1974).

It has long been thought that the Moinian is an easterly metamorphosed equivalent of the Torridonian and, more specifically, that a 750 my old isotropic age from the Moines represents a Morarian event that may correlate with the intra-Torridonian unconformity (e.g. Dewey and Kidd, 1974). However the recent Rb/Sr isochron age of 1050 ± 46 my from the Moines must date an early period of metamorphism and, accordingly, the Moine sedimentation probably occurred in the period 1250–1050 my which is before the sedimentation of the Stoer Group (Brook, Brewer and Powell, 1976). The Moinian may be more equatable with the Grenvillian of Canada.

The Lower and early Middle Dalradian sediments are of Precambrian age (Harris and Pitcher, 1975). The Precambrian–Cambrian boundary occurs somewhere within the Middle Dalradian. The major glacial tillite horizon that occurs intermittently over a distance of at least 600 km through Scotland to Ireland forms the base of the Middle Dalradian (Spencer, 1971). The Precambrian sediments consist largely of mature quartzites, pelites and limestones deposited in shallow water.

The Moine–Dalradian sequence has a total thickness of about 20 km. Dewey (1969, 1974) and Dewey and Pankhurst (1970) interpreted this sequence as a miogeoclinal continental rise deposit (Fig. 11.4) that derived its terrigenous material from the continent to the northwest, but Phillips, Stillman

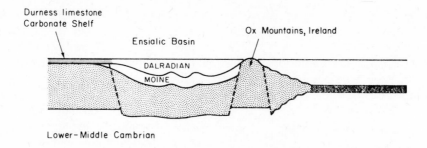

Fig. 11.5. The Moine–Dalradian sequence deposited in an ensialic basin (redrawn after Phillips, Stillman and Murphy, 1976; reproduced by permission of The Geological Society of London)

and Murphy (1976) conclude that it was laid down in an ensialic basin (Fig. 11.5) because clastic debris had a southerly derivation (e.g. from the Ox Mountains in Ireland). The latter conclusion is supported by the discovery that Carboniferous agglomerates in the Scottish Midland Valley have brought up fragments of granulite facies gneisses from a pre-Palaeozoic basement (Upton and coworkers, 1976).

Cambrian In this period, when the Iapetus ocean was expanding, there was sediment deposition on the two continental margins (northern and southern zones), but none in the central oceanic zone.

The Durness Group of limestones and quartzites and the Church Stretton quartzites were laid down as thin shallow-water shelf facies on the two continental margins. At the same time great thicknesses of terrigenous sediments were deposited in deeper water as flysch wedges on each continental rise. In the north the Upper Dalradian (partly Cambrian age on the basis of the agnostid trilobite, *Pagetia*, in the Leny Limestone) consists mostly of greywacke-like rocks and in the south (North Wales) a 5 km thick pile of shales and turbidites was deposited. But also in North Wales there are important Cambrian welded ignimbrites, agglomerates and tuffs (Arvonian Volcanic Series) and intrusive pot-ash granites (Coedana, Sarn and Twt Hill) with ages of 580–610 my. D. S. Wood (1974) interpreted both acid volcanics and granites as recycled upper crustal material by melting of the Monian rocks during early stages of sub-duction.

Finally, one should note the formation of an uplifted block, the Irish Sea horst, on which no Cambrian sediments were laid down.

Ordovician The most complicated phase in the evolution of the British Caledonides was in the Ordovician; there was sedimentation, subduction, obduction, volcanism, deformation, metamorphism and uplift.

The faunal evidence indicates that the maximum separation of the continents was in the early Ordovician. In Arenig times the obduction of the Ballantrae ophiolite complex and the extrusion of the overlying arc-type volcanics show that a Benioff zone had developed on the north side of the ocean. Movement on this consuming plate was intense, as the main deformation and metamorphism of the Scottish Highlands took place during a relatively short episode from 510–480 my—late granites have a Rb–Sr whole-rock isochron of 460 my (Pankhurst, 1974). Collapse of the continental rise and deformation of the Moine–Dalradian Series was diachronous in Ireland and Scotland from the early to late Arenig (Dewey and Pankhurst, 1970). This deformation was polyphase and marked by the early development of recumbent folds several kilometres across (the Iltay and Ballachulish nappes) prior to the onset of the metamorphism. There is some indication that a paired metamorphic belt (Miyashiro, 1967) formed in Arenig times:

high-pressure/low-temperature glaucophane schists in the trench at Ballantrae and high-temperature/low-pressure migmatites on the continental side of the plate margin in the northeast Grampians (Dewey, 1971).

On the south side of the ocean there was extensive volcanic activity in two areas:

1. North Wales. Most of the Llanvirn–Llandeilo succession in the Dolgelley–Arenig area and the Caradoc succession in the Snowdon area consists of andesites, ignimbrites and tuffs, with prominent gold and copper mineralization suggesting that the former Anglesey subduction zone was still operating—probably an active Ordovician island arc (D. S. Wood, 1974).

2. The Lake District. Here great volumes of andesite and thin tholeiitic basalt were extruded in late Llanvirn to mid-Caradoc times, suggesting the formation of a new subduction zone. If one compares these andesites and tholeiites derived from a shallow source with the Ordovician alkaline basalts in the Welsh border areas derived from a deep-seated origin, the subduction zone must have dipped to the south (Fig. 11.8). (Fitton and Hughes, 1970.)

In the mid-Ordovician the northern zone was undergoing uplift and erosion. The occurrence of detrital glaucophane in the Llandeilo Glen App conglomerates (E. K. Walton, 1956) shows that the Ballantrae ophiolite complex was exposed to erosion by this time, and detrital andalusite in Caradocian sediments indicates that at least 3 km of uplift had exposed the higher level nappes with Buchan assemblages in northeast Scotland (Dewey and Pankhurst, 1970).

The result of this erosion was increased deposition of terrigenous sediment in a flysch wedge on the north side of the trench—there are at least 6 km of mid-upper Ordovician greywackes, conglomerates, sandstones and limestones with trilobites and brachiopods in the Girvan area. Meanwhile, on the oceanic crust on the south side of the trench, only a thin 50 m sequence of pelagic cherts and graptolite-bearing shales accumulated. With increasing erosion the flysch wedges infilled the trench and encroached progressively southwards onto the oceanic plate during the Caradocian and early Llandoverian times when 8 km of turbidites were deposited (Dewey, 1971).

The Final Stages In the Silurian there was further sedimentation, but no volcanic activity indicating movement on the subduction zones had ceased. By mid-Silurian times the continents were approaching each other and thick turbidite–flysch sediments covered the intervening ocean; up to 3 km of turbidites were laid down in fairly deep water in the Girvan area (Cocks and Toghill, 1973). These thinned southwards with the result that only 1 km of limestones and shales (Upper Llandovery, Wenlock, Ludlow) with corals, brachiopods and trilobites was deposited on the continental shelf in the Midlands of England. Uplift, erosion and thermal instability continued through the late Silurian when the continents finally collided giving rise to deformation and metamorphism. The timing of collision seems to have been progressively later towards the southwest. Phillips, Stillman and Murphy (1976) accordingly advocate an oblique collision model, the unstable triple point migrating southwestwards across the British Isles from the late Ordovician to the late Silurian. Collision finally reached eastern Canada in the middle Devonian (Mitchell and McKerrow, 1975). The presence of Lower Old Red Sandstone andesites and of granite plutons (380–400 my) in Scotland suggests that some subduction may have continued in the early Devonian. Steady state conditions were not reached until middle Devonian times; the Iapetus ocean had opened and closed to give rise to a new fold belt that took 500 my to form.

The Appalachians

The Appalachian fold belt extends along the eastern side of North America from Newfoundland to Alabama. It has a history ranging from the late Cambrian to the Carboniferous and the main tectonic features can be matched with those of the British Caledonides (Dewey and Kay, 1968; Dewey, 1969). It is a classic fold belt as the original concept of the formation of a geosyncline was worked out here by Hall and Dana in the last century and it has become a key model for the evolution of a pre-Mesozoic fold belt by plate tectonic mechanisms (Bird and Dewey, 1970; H. Williams and Stevens, 1974; Williams, Kennedy and Neale, 1974).

Subdivisions

It is divisible into the following zones (Fig. 11.6):

1. The Eastern (Avalon) Platform in SE Newfoundland continues southwestwards along strike as the Cape Breton Platform and the Boston Platform, after which it passes under a cover of younger rocks of the coastal plain.

In late Precambrian time a 6 km thick sequence of volcanics and terrigenous clastics accumulated in this belt—early greywackes followed by arkoses, then quartzites. These sediments were faulted into graben and horst blocks, weakly metamophosed in places and intruded by granodiorite plutons (Holyrood gr.–Newfoundland; Dedham gr.– SE Massachusetts) in latest Precambrian time. This belt formed a broad epicontinental platform during the early Palaeozoic: basal Cambrian sandstones and conglomerates unconformably overlie Precambrian rocks and are followed by Cambrian–early Ordovician sediments—shales, algal limestones, and oolitic iron ores. Trilobites have European affinities.

2. The Central Mobile Belt extends from central Newfoundland through the Maritime Provinces to New England. Features of this belt include:

    a) Cambrian to Silurian sediments with prominent early Ordovician and Silurian volcanics.

    b) Climactic deformation during the late Silurian or Devonian associated with low-grade metamorphism in the northeast (Newfoundland) and mostly high-grade to the southwest.

    c) An abundance of acidic plutons of mid–Devonian age (Acadian orogeny) that intrude the older supracrustals or their metamorphosed equivalents.

3. The Piedmont of the southern Appalachians has an inner mobilized zone of high-grade schists and gneisses and is equivalent to the Fleur de Lys Group of Newfoundland.

4. The Blue Ridge Province extending from N Georgia to NW Newfoundland. This contains intermittent central cores of high-grade Precambrian—Great Northern Peninsula in Newfoundland, Green Mountains of Vermont, and the gneiss domes near New York,

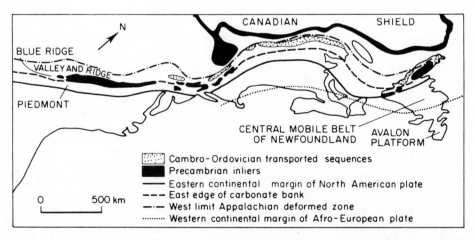

Fig. 11.6. The zonation of the Appalachian fold belt showing the position of major geological boundaries (modified Williams and Stevens, 1974; reproduced by permission of Springer-Verlag, Berlin)

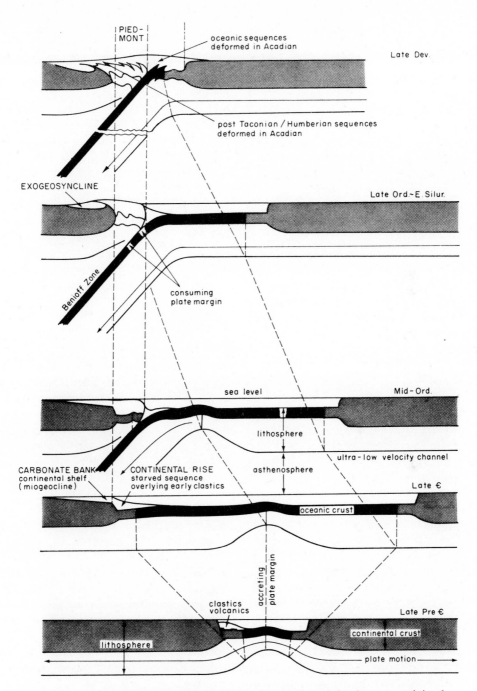

Fig. 11.7. A plate tectonic model based on westerly subduction to explain the evolution of the Appalachian orogenic belt (modified after Bird and Dewey, 1970; reproduced by permission of The Geological Society of America)

Philadelphia and Baltimore. These form a basement to autochthonous Cambro-Ordovician miogeoclinal deposits (with Pacific-type trilobites and graptolites) with a Middle–Upper Ordovician main deformation and variable weak to moderate metamorphism. Important in this belt are thrust sheets of Cambro-Ordovician sequences including the allochthonous ophiolite slices of NW Newfoundland which were transported westward during the early Ordovician.

5. The Valley and Ridge Province which extends northeastwards to the Lomond Zone in NW Newfoundland. This consists of a Cambrian (carbonate bank) to Carboniferous sedimentary sequence which has been weakly folded and thrust and little metamorphosed. The thrusts dip to the southeast and major overriding was to the northwest.

Further northwest lies the undeformed foreland of the orogenic belt in the form of the Appalachian Plateau with flat Devonian–Carboniferous sediments at the surface in the south, and of more deeply eroded Precambrian Shield in the north. Fig. 11.7 gives a tectonic model based on westerly subduction to explain the formation of the mainland Appalachians—this illustrates the following account.

Tectonic Evolution

Late Precambrian The evolution of the 'geosyncline' started in the late Precambrian; according to the plate tectonic theory the earliest sediments mark the initial opening of the Iapetus ocean.

On the northwest side of the belt there is an 8 km thick sequence of Eocambrian clastics (quartzites, arkoses, siltstones, sandstones) in the Great Smokey Mountains of Tennessee and North Carolina, and on the southeast side of the belt there is a similar 6 km sequence in the Avalon Peninsula of Newfoundland (Rodgers, 1970). These terrigenous sediments were derived by erosion of nearby mountainous Precambrian landmasses and they were deposited on miogeoclinal shelves as prograding continental terrace wedges, the detritus being derived consistently from the cratons. The sediments are locally associated with tholeiitic volcanic rocks and mafic dykes.

Cambrian–Early Ordovician During this period, when the ocean went through its main 'opening' stage, there were two sedimentary environments:

1. A shallow-water marine carbonate-rich sequence developed on a continental shelf. During the Cambrian, basal sands followed by carbonates encroached landwards in a gradual marine transgression until, during the late Cambrian and early Ordovician, a shallow-water carbonate bank up to 3 km thick extended on the western margin of the fold belt, from Texas in the south to Newfoundland in the north; its continuation can also be traced in NW Scotland and E Greenland (Swett and Smit, 1972). Rodgers (1970) compared this carbonate bank with the modern carbonate platform in the Bahamas off eastern USA (Dietz, Holden and Sproll, 1970). The carbonate sequence contains brachiopods (thus referred to as the shelly facies), and contains worm burrows, oolites, stromatolitic algal structures, desiccation cracks and flake conglomerates, indicative of an intertidal to subtidal environment (Swett and Smit, 1972).

2. A deep-water pelitic to clastic sequence developed during the Cambrian and early Ordovician on the continental rise to the east of the carbonates on the continental shelf (see Fig. 11.7). This is the Piedmont zone, now mostly metamorphosed and deformed, consisting of greywackes, pyrite- and graptolite-bearing black shales (thus referred to as the graptolite facies), and andesitic volcanics. The sediments coarsen and thicken eastwards reaching 15 km in thickness; the volcanic activity reached a climax in the early-middle Ordovician. The presence of pyrite in the shales, of turbidite structures in the greywackes, and the association with andesites suggests deposition in relatively stagnant deep troughs bordering volcanic islands—an environment comparable with that in the vicinity of Quaternary island arcs in the East Indies. For these reasons the pelitic–clastic–volcanic sequence is referred to as a eugeosynclinal facies.

A final important rock type must be considered. During the subduction, sediments and volcanics were scraped up off the oceanic floor and dumped in the active trench—seen today as the Dunnage mélange in Newfoundland which resembles the argille scagliose of the Appenines and the Franciscan mélange of California.

The boundary between the carbonate platform and the pelitic–clastic trough defines an important tectonic zone. It is marked by carbonate blocks in the eugeosynclinal deposits at several stratigraphic levels within the Appalachians, ranging from Middle Cambrian to Middle Ordovician—a classic example is the Cow Head Breccia in Newfoundland. These are regarded as submarine landslide breccias, the blocks having slid from the carbonate bank into the deeper water to the east; Rodgers (1970) pointed out that the steep to vertical eastern edge of the present-day Bahama Banks may be an analogous environment. He also suggested that the platform–trough boundary marks the true edge of the Cambro-Ordovician North American sialic continent, the crust to the east of it being oceanic, just as it is today off the Bahama Banks.

Middle–Late Ordovician An important orogenic phase (Taconic) took place in this period in the Northern Appalachians (Zen, 1972), and particularly in the Blue Ridge and Piedmont of the central and southern Appalachians (Hatcher, 1972).

In the early Middle Ordovician the previous tectonic and sedimentary pattern underwent drastic changes. Fragmentation and partial foundering of the carbonate shelf was caused by block faulting and at the same time a major uplift, or geanticline, developed (e.g. the Green Mountain dome) within the eugeosynclinal basin. Sediments derived from it were transported westward and deposited as an unconformable syntectonic flysch and then as black-mantling shale sequences on the subsiding carbonate rocks of the shelf. Continued uplift of the geanticline and subsidence of the former shelf led to gravity sliding of sediments in allochthonous slices into the subsiding miogeosyncline which now had the character of an exogeosyncline. Emplacement of the allochthons took place in the Middle Ordovician, during the main phase of the Taconic Orogeny in the northern Appalachians. Uplift of the geanticline also resulted in oceanward transport of flysch wedges out over the Dunnage mélange in the trench.

Continued compression in the Taconide Zone, caused by interaction between the oceanic and continental plates, resulted in further northwest thrusting and nappe transport (but now of crystalline rocks including Precambrian basement gneisses), and also in folding and regional metamorphism. This was the late Ordovician final phase of the Taconic Orogeny.

Late-orogenic clastic sediments (shales, sandstones and conglomerates) were deposited in eastern Pennsylvania during final stages of uplift.

Early–Mid-Silurian In the early Silurian post-orogenic alluvial clastics were laid down in the exogeosyncline. Sandstones coarsen and conglomerates increase eastwards showing that they were derived from the uplifted terrain of the eugeosyncline: they gradually transgressed further to the east as the relief of the mountainous area was reduced.

By Middle Silurian time there was an immense shallow-water sea over much of the North American craton in which carbonates with algal reefs and evaporites accumulated, particularly in the region of the Michigan Basin.

Late Silurian–Early Devonian In the central mobile belt in central Newfoundland, southeast Quebec and central Maine there was a minor late Silurian–early Devonian orogenic phase expressed as folding, metamorphism and granitic intrusions dated at 400 my. This took place in the eugeosyncline, the uplift of which gave rise to an eastward-thickening wedge of red beds in the exogeosyncline.

Middle–Late Devonian The Acadian orogeny occurred in this period when sediments and volcanic rocks of the eugeosyncline were folded and metamorphosed up to sillimanite grade in New England. In Newfoundland there was widespread intrusion of granitoid plutons. Uplift and erosion of these deep-seated rocks gave rise to the enormous volume (3 km thick) of clastics laid down in the exogeosyncline of this period. The molasse sediments even spread westwards to form an extensive subaerial alluvial plain, the Catskill delta, where continental arkosic red sandstones, conglomerates and shales were deposited in an arid climate about 20° from the equator.

Carboniferous–Permian Continued uplift and erosion of the mountain range produced a great thickness of coarse clastics in the exogeosyncline during the Carboniferous and early Permian. Locally extensive coal measures developed in shallow-water swamps during the Upper Carboniferous (Pennsylvanian). In the Permo-Carboniferous the Alleghenian orogeny caused folding, thrusting, metamorphism and granite intrusions in both halves of the southern Appalachian geosyncline. Continent-directed thrusting and folding was prominent in the Valley and Ridge Province giving rise to, for example, the Cumberland overthrust block and the Blue Ridge thrust sheet with a maximum horizontal displacement of 125 km.

The Alleghenian orogeny and its subsequent uplift terminated all deposition in the geosyncline: the formation of the Appalachian fold belt was complete.

Tectonic Summary

At this point it is worthwhile to review the development of the Appalachian fold belt in terms of plate tectonic theory. According to this viewpoint a geosyncline develops as a paired shelf and rise sequence along an inactive Atlantic-type continent–ocean boundary. It progresses into a Cordilleran-type belt with the formation of a Benioff zone, the collapse and deformation of the shelf (carbonate facies)–rise (clastic-volcanic facies) sequence, and the rise of calc-alkaline magmas generated by partial melting of the downgoing slab. Further oceanic lithosphere consumption leads to continental collision and formation of an Alpine-type fold belt.

The main features of the Appalachian fold belt may be considered in two stages of development. Firstly those associated with the early rifting and construction of the continental margins during the period of sea-floor spreading and, secondly, those associated with the destruction of the continental margins during the stage of ocean closure and eventual continent–continent collision (Williams and Stevens, 1974).

The first stage has certain characteristics. The deposition of thick late Precambrian–early Cambrian clastic sequences, comprising mostly quartzites and turbidites, in miogeoclinal wedges unconformably on crystalline basement—they are similar to modern continental shelf sediments. Tholeiitic volcanic rocks and mafic dykes are closely associated with the clastic sediments; they formed in connection with the continental rifting and separation. An eastward-thickening carbonate bank of early Cambrian–middle Ordovician age was deposited in shallow water conformably on the early clastic rocks, and its eastern edge defines the approximate margin of the American continent in early Palaeozoic times.

The second stage in development of the Appalachian fold belt is represented by the following features. The earliest indications of the destruction of the continental margin are shown by the break-up of the carbonate bank into horst–graben fault blocks, and the unconformable deposition of flysch wedges (up to 3,000 m thick and and derived from the east) of black shales, siltstones and greywackes upon the carbonates. Above the autochthonous flysch are allochthonous slices of clastic sequences derived from the continental margin overlain by transported ophiolite complexes representing sections of oceanic crust–mantle. The structural position in the stacking order indicates that the highest slices (i.e. the ophiolites) travelled farthest from the east (G. E. Williams, 1975). The formation and activity of subduction zones is expressed by the extrusion of abundant early–middle Ordovician calc-alkaline volcanic rocks interpreted as fossil island arcs.

According to many models (Dewey, 1969; Bird and Dewey, 1970; Dewey and Bird, 1971; Hatcher, 1972) observed geological relationships are ascribed to westward subduction in the Appalachians. However recent data summarized by Williams and Stevens (1974) suggest the presence in Newfoundland of an easterly-dipping subduction zone:

1. The calc-alkaline volcanic rocks decrease in thickness westwards implying lack of volcanism west of the continental margin;

2. The potash content of granitic rocks increases eastwards (Strong and coworkers, 1974);

3. The systematic variation of mineral deposits (Strong, 1974b);

4. Seismic-refraction profiles suggest the presence of southeasterly inclined rock layers

of contrasting velocity (Sheridan and Drake, 1968).

Fig. 11.8 gives a summary of the many subduction models for the Appalachians and British Caledonides.

Closure of the Iapetus Ocean in the northern Appalachians during the Taconic orogeny culminated in the final stages of continental compression between North America and Europe giving rise to the climactic deformation of the Acadian orogeny in the Devonian. In the southern Appalachians the ocean was finally closed during the Alleghenian

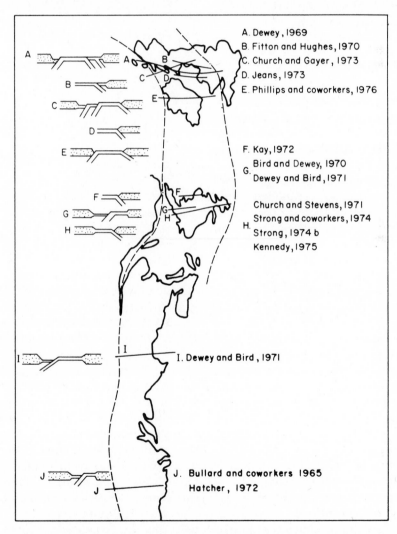

Fig. 11.8. Summary of plate tectonic models suggested for different sections of the Appalachian–British Caledonides fold belt (partly after Strong and coworkers, 1974; reproduced by permission of Macmillan Press)

(Variscan) orogeny in Permo-Carboniferous times when North America collided with Africa (McKerrow and Ziegler, 1972a).

In spite of local variations and problems of interpretation the Appalachian fold belt clearly went through a sequence of tectono-sedimentary–magmatic events that are remarkably similar to those which took place along active Mesozoic–Tertiary plate margins and which are preserved in the Cordilleran (Chapter 17) and Alpine (Chapter 18) fold belts. The current evidence therefore very strongly supports the proposal first outlined by J. T. Wilson (1966) that the Appalachian fold belt formed by the opening and closure of an ocean.

Mineralization

The question that concerns us here is: Do the mineral deposits in the Appalachian Caledonide fold belt have a systematic regional distribution and, if so, is it similar to that in Mesozoic–Tertiary fold belts? Such a correspondence in metallogenic provinces would provide us with additional evidence that the fold belt was situated at lithospheric plate junctions.

The first problem we come up against is the scarcity of mineral deposits of this age. From his comprehensive survey Ridge (1972) calculated that Palaeozoic mineral deposits account for only 9% of all the Phanerozoic deposits of the western hemisphere. To explain this, he suggested that the Palaeozoic was a period of less intense development of ore deposits than were the Mesozoic and Tertiary eras; but Sawkins (1972) thought the reason was the deeper erosion level of the older fold belts. No doubt both were contributing factors.

Newfoundland

Strong (1974a,b) gives a review of the Newfoundland mineral deposits, in relation to possible plate tectonic environments. The general features of the deposits are similar to those of 'modern' accreting and consuming plate margins. The evolutionary development of the ores is summarized in Fig. 11.9.

Southern Appalachians

Across the Appalachian fold belt there is a systematic progression in the ore types and this correlates well with the main tectonic belts. Gabelman (1968) outlined the distribution of the metal zones and Strong (1974b) related them to an easterly-dipping subduction zone under an Andean-type continental margin to the east (Fig. 11.10).

Passing eastwards from the foreland with Mississippi Valley-type Pb–Zn deposits there is the following zonation. The Blue Ridge province contains largely Cu–Fe deposits, followed to the southeast by a narrow belt with vein-type Au along the border of the Piedmont which contains, in its main part, polymetallic Cu–Fe–Zn–Pb–Au–Ag as well as some Sn deposits. In the Piedmont there are no economic gold occurrences today although this was the main source of gold in the USA in the first half of the nineteenth century.

This zonal distribution of metals is particularly interesting as it compares favourably with those extending across the Mesozoic–Tertiary circum-Pacific island arcs and Cordillera:

Southern Appalachians:
    Fe Cu Au Cu Fe Pb Zn Au Ag Sn
      Strong (1974b)
Cordillera of W America:
    Fe Cu Mo Au Pb Zn Ag Sn Mo
      Sillitoe (1972a) and others
Pacific Island Arcs:
    Cu Sn W F Ba
      Mitchell and Garson (1972)
    Au Cu Mo Au Pb Zn Sn W Sb Hg
      Zonenshain and coworkers (1974)

British Caledonides

In the southern British Caledonides metal occurrences can be related to the three subduction zones passing through Ballantrae, the Lake District and Anglesey. Dunham (1953) and Wheatley (1971) illustrated the distribution of volcanogenic ores, mostly of Ordovician age, but their possible relationship with the subduction zones is unpublished.

Whilst Pb and Zn are common throughout most the region, Cu occurs with them as a

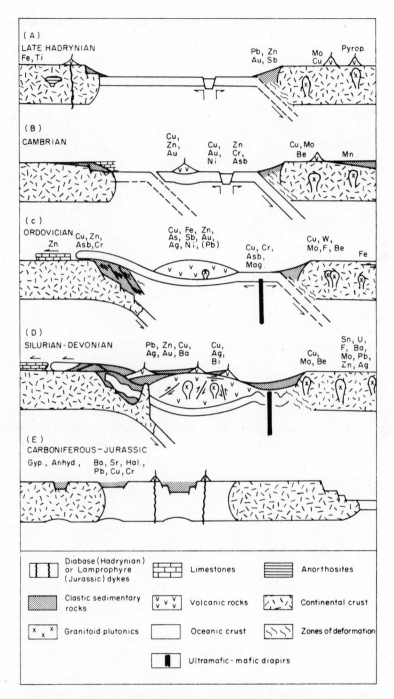

Fig. 11.9. Plate tectonic model to explain the zonation of mineral deposits of the Newfoundland Appalachians (after Swinden and Strong, 1976; reproduced by permission of The Geological Association of Canada)

192

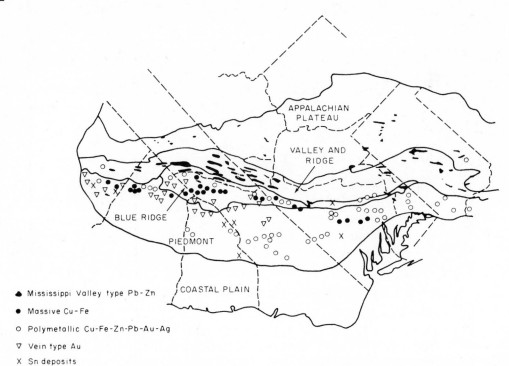

Fig. 11.10. Palaeozoic mineral zonation of the southern Appalachians (after Strong, 1974b; reproduced by permission of the Association of the Institute of Mining Engineers)

prominent ore only in the two belts immediately overlying the two southerly-dipping subduction zones. In these belts Au occurs in southeast Ireland and in the Harlech Dome of Wales, minor Ag in Parys Mountain, Anglesey, and Fe oxide (magnetite) in the Avoca area of southeast Ireland. Further to the south there is a belt with Pb, Zn and minor Cu and Ba, followed to the south by Mo, F, Ba, Pb and traces of W (scheelite) in the Mountsorrel Granodiorite in Leicestershire (R. J. King, 1973).

Chapter 12

The Hercynian Fold Belt

The Hercynian of northwest Europe is a segment of an extensive fold belt extending from the Gulf of Mexico to eastern Europe, and includes the Ouachita belt, the Alleghanian part of the southern Appalachians and the Mauritanides of northwest Africa. The fold belt formed between the Devonian and Permian in eastern North America as a late expression of the Appalachian fold belt, whereas in Europe it developed as a separate entity, physically apart from the Caledonian fold belt to the north. The Uralides of eastern Europe emerged at approximately the same time.

For a variety of reasons knowledge of the belt is scanty and tectonic models to explain its origin have been slow to advance. In the first place, the general poor degree of exposure makes correlation difficult, the Ouachita belt is extensively covered by Mesozoic–Cenozoic sedimentary rocks, the Alleghanian was superimposed on and is not easily separated from the Appalachian events, the Hercynian of southern Europe is heavily overprinted by the Alpine belt and, not least of all, publications on the subject are in many languages.

Remember that whilst the initial stages in the evolution of the Hercynian were taking place in Europe during the early Devonian, so also were the final stages in the evolution of the Caledonides in Scotland. This means that about 380–400 my ago marine sediments were being deposited only a few hundred kilometres south of Scotland where andesites were being extruded, granites intruded and continental sediments deposited. In the Appalachians, however, all this activity continued along the same belt.

We shall first consider the main components of the belt and afterwards the tectonic models proposed to explain its origin.

Components of the Belt

Fig. 12.1 shows the distribution of the three main zones of the belt as subdivided by Dewey and Burke (1973).

Zone 1 was a foreland in the Devonian during continental sandstone deposition, a shelf in early Carboniferous times, and the site of discontinuous external (paralic) coal basins in the late Carboniferous. This zone was never affected by Hercynian 'orogeny'.

Zone 2 was a region of mixed marine–nonmarine sedimentation in the early Devonian, of basic volcanism in the Givetian (mid-Devonian), of lutite deposition in the mid-Devonian to early Carboniferous and of northward-moving flysch wedges in the Namurian (mid-Carboniferous).

Zone 3, which includes the Moldanubian of Central Europe and the Piedmont of the Appalachians, is characterized by the presence of many windows of Precambrian basement, such as the Bohemian Massif, Black Forest, Vosges, and Massif Central, of high temperature metamorphism and, locally, of remobilization of the basement with formation of migmatites and reworked gneisses (e.g. Massif Central, Bohemian Massif and southern Brittany). In the latest Carboniferous (Stephanian) the zone was the site of intermontane coal/clastic basins and potassic ignimbrite volcanism.

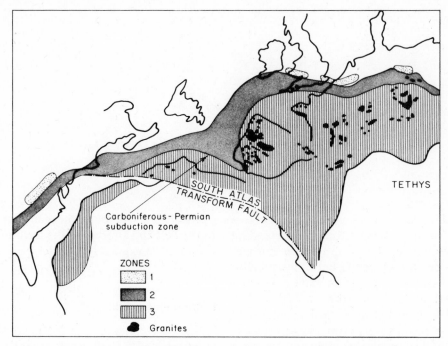

Fig. 12.1. Major zones of the Hercynian–Mauritanide–southern Appalachian fold belt on a Permo-Triassic reconstruction of the continents across the central and north Atlantic. For details of zones see text. (Redrawn after Dewey and Burke, 1973)

Deformation Phases

There were several periods of deformation during the formation of the Hercynian belt but they were inhomogeneously distributed, being confined to certain major zones. It is not easy to be specific about the effects of the phases as there is some disagreement in the literature about their presence or importance in different areas. The three principal phases are as follows:

The Bretonic phase deformed part of the southern border of the geosyncline (Brinkmann, 1969; Read and Watson, 1975) but, as noted later, the rocks in Cornwall much further to the north were also deformed at the end of the Devonian. This orogenic phase had the effect of severely reducing in width the Hercynian geosyncline into a narrow foredeep (Zeigler, 1975).

The Sudetic orogeny transformed the foredeep and remainder of the original

			Deformation Phase	Age my
Upper Carb.	Silesian or Pennsylvanian	Stephanian		
			Asturic	290–295
		Westphalian		
		Namurian		
			Sudetic	325
Lower Carb.	Dinantian or Mississippian	Viséan		
		Tournaisian		
			Bretonic	345
Devonian				

geosyncline into an uplifted Hercynian mountain range, leaving only a narrow trough to the north (Pfeiffer, 1971). According to Dewey and Burke (1973) 'the main deformation in zone 2 occurred during the Sudetic phase in mid-Carboniferous time'. But rocks in southern Cornwall were deformed in the Bretonic phase at the end of the Devonian (Dearman, 1971) and those in the Ardennes, the Rhenische Schiefergebirge and southwest England underwent deformation in the late Carboniferous Asturic phase (Fitch and Miller, 1964; Read and Watson, 1975).

The Asturic phase caused décollement folding of the foredeep and final consolidation of the mountain chain (Ziegler, 1975), commonly ascribed to the last stages of continental collision. The Asturic earth movements are said by Johnson (1973) to affect particularly the northern shelf area of zones 1 and 2 rather than zone 3, but according to Dewey and Burke (1973) zone 3 is characterized by Asturic age deformation. As Ager (1975) points out, the name comes from the Asturias in northern Spain where the deformation phase is developed in a more spectacular way than anywhere else in Europe.

To the non-specialist these discrepancies in the timing of the orogenic phases throughout the fold belt are confusing. The deformation phases are particularly important as they are currently taken to indicate the timing of plate collision.

Sedimentation and Volcanism

Uplift on the site of the Caledonian orogenic belt in the Devonian gave rise to an erosional source area in Spitzbergen, East Greenland, western Norway and northeastern Scotland which has been termed the 'Old Red landmass' (Friend, 1969) because it was an area of continental red sandstone deposition in inter- and extra-montane basins. To the south of this landmass there was a downwarp or basin in which accumulated thick marine sediments, and along an intermediate zone these were intercalated with non-marine sediments.

The Devonian period began with a transgression. The early Devonian rocks in zone 2 belong to a continental facies, clastic detritus being derived from the landmass to the northwest; greywackes, sandstones, orthoquartzites and shales predominate. A shallow-water marine facies evolved in the mid-Devonian with conspicuous reef limestones. There was widespread extrusion of albite-rich soda-keratophyres and spilitic lavas and tuffs in the Rhenisch Schiefergebirge in the mid-Devonian (Lehmann, 1952), and of spilitic lavas in north Cornwall and south Devon from the mid-Devonian to the early Carboniferous (Floyd, 1972a).

Zone 2 passes northwards in southwest England into zone 1 by an increase in the proportion of continental (sandstones, arkoses and conglomerates) at the expense of marine (shales and limestones) sediments. The Old Red Sandstones of Wales are typical of zone 1.

Sedimentation in zone 3 was controlled by the distribution of basement uplifts. Generally sediments are thinner than in zone 2 and clastics are less important than limestones and marls.

Following a regression at the end of the Devonian the Carboniferous period began with a transgression that may have been caused by growth of a new mid-oceanic ridge in the Tethyan region (Johnson, 1973). The sea advanced across the Old Red Sandstone plain transforming zone 1 into a shallow-water carbonate shelf; by late Carboniferous times only a few islands, such as St George's Land in Wales, stood above sea level. There is a facies change from the shelf limestones via a marginal reef facies to the deeper water, partly detrital, sediments (sandstones, cherts, limestones and black shales) of the Culm trench of southwest England (Ramsbottom, 1970) with which are associated pillow lavas and the Lizard ophiolite complex.

The Culm facies was also developed in zone 2 in southwest Ireland, Belgium and the Ruhr in the Rheno-Hercynian zone. In central and western Europe there was a major phase of tholeiitic lava extrusion in the mid-Lower Carboniferous. The Sudetic deformation in the mid-Carboniferous was associated with granitic intrusion and with acid-to-intermediate explosive volcanism of Cordilleran type (Nicholas, 1972), giving rise to intermontane basins in which Westphalian continental sediments and coal deposits were localized. The felsic volcanic and plutonic activity continued in the late Carboniferous through the Asturic phase into the early Permian by which time the formation of the Hercynian was complete, except for localized post-orogenic intrusion of lamprophyres and extrusion of acid ignimbrites.

As will be seen later in this chapter, there is currently a wide divergence of opinion on how the movements of continental plates caused the production of the Hercynian fold belt. Not until a particular tectonic model becomes widely acceptable will it be possible to give not just a descriptive summary of the sedimentation and volcanism as above, but rather a genitic account related to the opening and closing of different segments of oceanic crust.

Granites, Migmatites and Metamorphism

The Moldanubian zone 3 is characterized by the presence of much pre-Hercynian basement, an abundance of granites and migmatites, and a low-pressure/high-temperature facies series of regional metamorphism. This zone extends from the Bohemian Massif, through the Black Forest, the Vosges, the Massif Central, southern Brittany and the Iberian peninsula to Mauritania in NW Africa (Fig. 12.1). (Details are summarized by Zwart, 1967a).

The basement rocks include early Proterozoic gneisses in the Channel Islands as well as late Proterozoic granulites, kyanite gneisses and eclogites in the Bohemian Massif and Cabo Ortegal, NW Spain.

Fig. 12.1 shows the large number of granites in this zone which range from mesozonal to high-level types. The deeper seated granites are associated with migmatites and probably formed by remobilization of gneissic basement rocks.

The Hercynian regional metamorphism was superimposed directly on a Proterozoic metamorphism—Caledonian effects are largely absent. Typical results of the Hercynian metamorphism are schists and gneisses containing andalusite, cordierite, sillimanite

and in places staurolite. Zwart (1967a) concluded that the Hercynian low pressure metamorphism took place at depths up to 20 km with a thermal gradient of about 150°C/km and a pressure range of 1·5–5 kb. The depth of erosion lies mostly between 5 and 15 km and so the uplift of the belt amounts to an average of 1 mm per 10–25 years.

Mineralization

During the formation of the Hercynian fold belt there evolved several important types of mineral deposit (Fig. 12.2) including lead–zinc in shelf limestones and tin–tungsten in association with post-orogenic granitic plutons. Although both types are known from earlier times, these Hercynian examples represent a peak of development reached late in earth history. The following classification of Hercynian 'granite' mineralization in Europe is by Badham (1967b).

Hercynian (*c.* 370–230 my)	*Saxonic* (*c.* 180 my)
1. Sn–W, Au–Qu, Mo	1. Ba–Fe–Qu–(Cu Pb Zn) sulphides
2. Qu–(Cu Pb Zn Fe) sulphides	2. (Ni Co Fe) arsenides–Ag–Bi
3. U–Qu–calcite	3. Ag–CO_3–Hg sulphosalts
4. Ag–Sb–CO_3	

The Hercynian mineralization is associated spatially and genetically with the granites, the Saxonic mineralization is located in fault systems near but less dependent on the plutons.

Pb–Zn in limestones

Galena–sphalerite deposits formed in Carboniferous carbonate sediments in the shelf facies of zone 1 in the Tristate field of the USA (Missouri–Kansas–Oklahoma) and the Pennine field of England (Fig. 12.2). They are commonly referred to as the 'Mississippi Valley-type ores', whose modes of origin are much debated (e.g. J. S. Brown, 1970).

The deposits typically occur as lenses conformable with bedding and as infillings of discordant fractures often associated with major fault sets in limestones or dolomitized limestones. The common association with bioher-

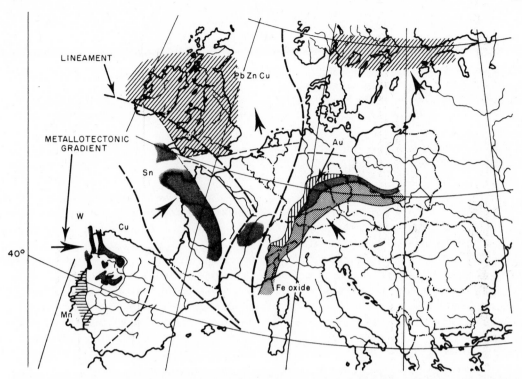

Fig. 12.2. Hercynian mineralization in western Europe (after Gabelman, 1976; reproduced by permission of The Geological Association of Canada)

mal reefs substantiates a generally accepted shallow-water shoreline or shelf environment, a direct result of the widespread Carboniferous transgression. Barytes and fluorite are characteristic accessories and the composition of fluid inclusions indicates that the metals were precipitated from liquids of high salinity comparable to brines.

A widely held view is that the metals were transported by connate, marine, evaporitic brines and deposited epigenetically in suitable host rocks (e.g. Dunham, 1966; Roedder, 1967; Jackson and Beales, 1967). Consequently, although the ores occur in Carboniferous limestones, they may have been introduced at a later date—those in the North Pennines are considered by Deans (1950) to be late Permian and those in the South Pennines by Ford (1969) to be late Triassic in age.

But to date there has been little attempt to relate the occurrence of these ores to plate tectonics. There may be a connection between the Pennine ores and aulacogens. The movement of heavy metals and connate waters

through evaporites in the Red Sea is well known (e.g. Degens and Ross, 1969). Evaporites commonly form in the early rifting stage of the break-up of continents, being found in marginal basins on the coasts of opened oceans or in failed rift arms (aulacogens) (see Kinsman, 1975a and b, and Chapter 14). In the present context the Permo-Triassic failed rifts or graben (with their evaporites) of the North Sea and the Cheshire graben are relevant, because they flank the Pennines which constituted an intervening ridge or area of uplift at the time (Whiteman and coworkers, 1975a,b). The Benue Trough in Africa is an excellent example of a fossil aulacogen (Burke and Dewey, 1973a) that contains Cretaceous Pb–Zn ores and present day brine seepage (Farrington, 1952), the mineralization having most likely formed from geothermal heavy metal-bearing brines (Grant, 1971). Ford (1969) suggested that the South Pennine ores may have formed from metal-bearing brines that migrated up-dip from the lower lying North Sea basins. The faults and fractures that assisted the fluid transport and finally localized the ores in the Pennine field should be seen, therefore, as structures related to contemporaneous North Sea graben and so to the abortive attempt to fragment this intracontinental shelf at the time that the North Atlantic rift system was initiated. Viewed in this way the Hercynian-age Mississippi Valley-type Pb–Zn mineralization may be structurally related to proto-plate tectonics.

Sn, W, etc. in Granites

Tin–tungsten mineralization associated with Carboniferous post-orogenic granite plutons is well known in Europe (Fig. 12.2), i.e. Cornwall, Brittany, Massif Central, Germany and the Iberian peninsula (Schuiling, 1967; Gabelman and Krusiewski, 1972), and is present but less common in the Appalachians (e.g. Dagger, 1972). There is a great range of elements divisible as follows:

Hypothermal Sn W As Cu
Mesothermal U Ni Co Ag Pb Zn
Epithermal Fe Sb

The minerals concerned occur mostly as fracture fillings in the country rocks surrounding the granitic 'emanative' centres. In Cornwall the most deep-seated tin-bearing zone is overlain by a wider copper zone which, in turn, is overlain by a more extensive lead–zinc zone (Hosking, 1966).

The mineralization has not been easy to fit into the plate tectonic models applied to the Hercynian belt. It was suggested by Mitchell (1974) that the Cornish granites and associated mineralization are the result of a continent–continent collision and are thus comparable with tin granites in the Himalayas. But most authors relate the granites to an Andean-type plate boundary. In the model of Johnson (1973) the European granites are situated in the Sudetic Cordilleran belt at the leading edge of the mobile central European plate; ideally their K_2O contents should increase southwards. In contrast, in the models of Nicholas (1972), Floyd (1972b) and Badham and Halls (1975), the subduction zone dips northwards from southern Europe

and accordingly the potash values should increase in that direction, as is suggested to be the case by Badham and Halls (1975). Bromley (1975) has argued that the 700 km distance from southern Europe to Cornwall is rather excessive, but this is not so in comparison with the 3000 km from the Japanese trench westwards to the tin granites on the Asian mainland (Zonenshain and coworkers, 1974). As Badham and Halls (1975) point out, such a distance is feasible with either a slow rate of, or shallow-dipping subduction (cf. Uyeda and Miyashiro, 1974).

Models for Tectonic Evolution

Recently there has been a proliferation of papers offering explanations, mostly with respect to possible plate tectonics, for the evolution of the Hercynian fold belt; they are especially interesting as they differ so widely. We shall consider here the main arguments for these models of which there are two principal types (Fig. 12.3).

1. Nicholas (1972) proposed that it was an Andean-type mountain belt. He based this on the following features: lack of a eugeosyncline, formation on a continental basement, clastic volcano-sedimentary formations in shallow basins, chiefly andesitic volcanicity of explosive type, granites followed by rhyolitic volcanism, and weak metamorphism. The Hercynian continent–ocean boundary is postulated to lie in the northern part of the Alpine fold belt and is thus difficult to define.

A somewhat similar model by Floyd (1972b) envisages a Benioff zone somewhere in the Alpine region, dipping northward under a continental crustal plate, in and on which the Hercynian belt developed. He has two key pieces of evidence. Firstly, the spilitic pillow lavas (mid-Devonian–early Carboniferous) of Cornwall and Devon have the composition of continental alkaline basalts, whereas Devonian spilites from the Harz in Germany are continental and oceanic tholeiites. This northward increase in alkalinity is thought to be analogous to the continentward increase at recent continental–oceanic plate boundaries (Kuno, 1966). Secondly, the unusually high

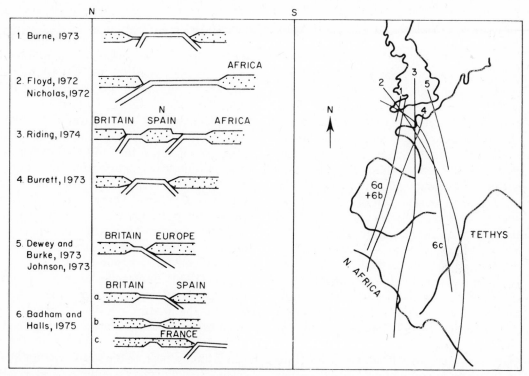

Fig. 12.3. Summary of plate tectonic models suggested for the Hercynian mobile belt of NW Europe. All sections have similar N–S orientations. Sections 6a,b,c refer to a time sequence, 400 my, 340 my and 300 my ago respectively

contents of Li, Rb, Cs, F, Cl, B, Sn, U and Pb in the Cornubian granites demonstrate their highly evolved nature. A variant of this Andean-type model was applied to the Iberian Variscan by Bard and coworkers (1973).

Badham and Halls (1975) also envisage, like Floyd, a subduction zone that dips northwards from southern Europe. Evidence cited in favour of this includes a northward increase in K_2O in upper Palaeozoic volcanic rocks from southern to central France (quoted from an unpublished thesis by C. Boyer, 1974), a probable northward increase in the potassium content of the Carboniferous post-orogenic granites from southern Europe to Cornwall, and typical arc and back-arc sequences and associated ore deposits in western Germany (Amstutz, Zimmermann and Schot, 1971) and Bohemia (Pouba, 1970). The obviously broad Sn–W–K granite zone across Europe is ascribed to the low rate and shallow dip of the subduction.

According to Riding (1974) the deformation in zone 3 in Europe was caused by the northward subduction of the Tethys oceanic plate, after which Africa collided with the Euramerican continental plate giving rise, in particular, to the Alleghanian deformation in the southern Appalachians.

2. Several authors consider that zone 2 incorporates a major suture formed as a result of the removal of an oceanic plate during collision between the two continental plates of northern and southern Europe (Burrett, 1972; Laurent, 1972; McKerrow and Ziegler, 1972a; Burne, 1973; Dewey and Burke, 1973; Johnson, 1973; Mitchell, 1974).

Although there are several minor points of departure amongst these papers, the model is based particularly on the following lines of evidence:

a) The 350 my old Lizard Complex of serpentines, gabbros and basic dykes has long been regarded as an ophiolite complex

(Thayer, 1969). The identification of the complex as a fragment of ancient oceanic crust gains much credence from the recent discovery of the remains of a sheeted dyke complex in the upper gabbro zone (Bromley, 1975). In the Culm trench there is a pillow lava–black shale–radiolarian chert assemblage suggestive of an oceanic environment. Associated with the ophiolite complex is a Middle–Upper Devonian succession of thick flysch-like turbidites with volcanic fragments (Gramscatho Beds), breccia conglomerates full of exotic blocks (wild flysch–Veryan Series and Gidley Well Beds), spilitic volcanics, cherts, black shales and the Meneage crush zone—a combination that might be expected in a deep sea trench in the vicinity of a subduction zone. This wide assortment of evidence is strongly in favour of a major suture reflecting a Carboniferous Benioff zone through this part of SW England.

Note, however, that Badham and Halls (1975) whilst accepting the Lizard Complex as a fragment of oceanic crust do not conclude that its presence must indicate the existence in Cornwall of a former subduction zone. They suggest that lateral migration between microplates may cause obduction without subduction. Other objections are, firstly, that ophiolite complexes may be transported a long way from their oceanic provenance and so their present position need not necessarily indicate the existence in that spot of a former oceanic crust; and, secondly, it is being increasingly realized that ophiolitic complexes are less likely to be a remnant of a major ocean than of a back-arc marginal ocean, on-land thrusting of which need not involve subduction.

b) If the suture is the site of a vanished ocean, the two bordering continental plates might well have significantly different sedimentary histories.

North of the paratectonic zone, with its suture and flysch wedges, etc., there was a wide shelf with a shoreline near Scotland. In the early Carboniferous the sea to the south transgressed widely over the Old Red Sandstone continent to the north causing widespread shelf carbonate deposition. In the Upper Carboniferous there were discontinuous coal basins between isolated projecting islands, essentially a continental facies paralic (marine) zone bordering a foreland to the north. To the south of the suture in the Lower Carboniferous there was also extensive shallow-water sedimentation with limestones and reefs, but in a zone which after the mid-Carboniferous deformation was characterized by a limnic (freshwater) facies of intermontane clastic/coal basins. There is thus a clear distinction between the northern paralic and southern limnic facies of sedimentation, marked in particular by the contrasting types of coal basins (Johnson, 1973).

Burne (1973) quotes the example of SW England in Westphalian time when there was a stable mature continental margin to the north bordered by the paralic sequences of the Bideford Group and a relatively deep water limnic basin to the south, exemplified by the Bude Formation composed largely of submarine fan deposits with fresh or brackish water fossils.

However Ager (1975) disputes the existence of the former ocean stating that in rocks of all ages concerned there is a single broad faunal province across the whole of Europe.

c) There should also be a marked tectonic difference between the two plates. The northern zone is largely undeformed; the main deformation occurs in the narrow paratectonic zone 2 just north of the suture (Culm Trench) in pre-Westphalian time (Sudetic Phase), whilst in the southern zone 3 it was during the Westphalian (Asturic Phase) according to Dewey and Burke (1973). As the southern zone was developed on a Precambrian (Cadomian) basement, the deformation locally caused remobilization giving rise to low-pressure metamorphism, granites, migmatites and mantled gneiss domes (e.g. Massif Central, Bohemian Massif) (cf. Zwart, 1967a). The southern zone, and locally the suture zone (Cornwall), was intruded by potassic tin-bearing granites (Brittany, Erzgebirge, N Portugal).

Johnson (1973) has illustrated the possible palaeogeographic–tectonic environment at this time. The Hercynian developed on the site of a mid-European sea which had been in existence since the Ordovician (Whittington and Hughes, 1972). Closure of this sea by northward movement of the southern Europe plate produced the Hercynian continental collision-type fold belt and, concurrently, the Tethys ocean opened on the south side of the southern European plate.

By now it will be clear what enormous disagreements there are between the plate tectonic protagonists for the Hercynian belt. This is well brought home if we ask the question how many subduction zones were there and which way did they dip? Replies include the following (Fig. 12.3).

	Number	Dip direction
Floyd (1972b) Nicholas (1972)	One	north
Dewey and Burke (1973)	One	south
Riding (1974)	Two	both north
Burrett (1972) Burne (1973) Badham and Halls (1975)	Two	south and north

Nevertheless, although it may be frustrating not to be given a definitive picture, it is very instructive for the student of geology to follow out the reasoning behind these tectonic models. The disparate views concerned illustrate rather well a current problem in the earth sciences: having sorted out with a considerable degree of satisfaction the broad evolution of the Mesozoic–Cenozoic fold belts in plate tectonic terms, it is not so easy to unravel the origin of pre-Mesozoic belts.

But not everyone believes that 'the new global tectonics' is applicable to the Hercynian orogen. Krebs and Wachendorf (1973) considered the central European area from the Vosges to the Bohemian Massif (belonging to the southern plate and underlain largely by pre-Hercynian sialic basement, according to Dewey and Burke, 1973), and found no signs of a Hercynian oceanic crust or subduction zone, or of a southern continental margin as required by Nicholas (1972). They accordingly erected an intracontinental model for the formation of the fold belt involving little horizontal shortening; it would seem unfortunate, however, that their study was limited to that southern plate which, according to any of the plate tectonic theories, should not contain a subduction zone etc., but rather should consist of remobilized older basement, just as Krebs and Wachendorf found.

Chapter 13

Pangaea: Late Carboniferous–Early Jurassic

All the continents had drifted together to form a continuous landmass by the late Carboniferous—the final stage of fusion is marked by the post-Westphalian Asturic folding in the Variscides and the Allegheny folding in the Appalachians. This Pangaea remained intact for about 150 my, until the early Jurassic. The formation of such a supercontinent gave rise to a new set of tectonic, climatic and biogeographic conditions, very different from those in the Cambrian–Devonian when continents were drifting and oceans opening and closing. Thus it is not surprising that different types of sediments accumulated, fauna and flora flourished, and a vast new glacial epoch evolved in this period. We shall look at some of the geological evidence that supports the concept of such a Pangaea, including some of the pre-separation geology, because if the continents formed a single plate, their geology prior to fragmentation should be broadly similar.

There are three historic landmarks in the evolution of ideas on the formation of Pangaea. Observing the distribution of the Glossopteris flora and glacial sediments in the southern hemisphere, Suess proposed in 1885 the existence of a supercontinent called Gondwanaland incorporating India, Africa and Madagascar and he later included South America and Australia. Realizing the problem of migration of the plant species, many geologists then thought that the intervening oceanic areas between the present continents had sunk or that land bridges had existed between them; but in 1929 Wegener proposed the revolutionary idea that a former supercontinent had broken up and the individual continents had drifted to their present position. In 1937 du Toit produced ten lines of geological correlation supporting juxtaposition of South America and Africa and he proposed the existence of two supercontinents, Laurasia and Gondwanaland, separated by a seaway, the Tethys, which prevented mixing of the northern and southern flora.

Du Toit also produced a reconstruction of Gondwanaland based on general morphological fit and geological correlations. Using the 500 fathom isobath Bullard, Everett and Smith (1965) made the first computerized reassembly of the continents around the Atlantic and a fit of the southern continents was made by Smith and Hallam in 1970. The similarity between the continental arrangements of du Toit and Smith and Hallam is remarkable. Whilst the mathematical models give an air of precision, it should be borne in mind that there is much disagreement about certain features. Does Madagascar lie against E or SE Africa? Does India lie opposite W Australia or Antarctica?

Now let us consider some of the geophysical and geological evidence that is consistent with the geometrical arrangement of a late Palaeozoic–early Mesozoic supercontinent.

Palaeomagnetic Data

As considered briefly in Chapter 9, by determining palaeomagnetically the pole positions and the palaeolatitudes for sedimentary and igneous rocks in different continents and by comparing their polar-wander curves, it is possible to get a fix on the relative positions of the continents for any given period. Such palaeomagnetic data indicate that the continents of Gondwanaland and Euramerica came together to form a first supercontinent in the Silurian (Fig. 9.2), and then in Permian–Trias times Asia combined with Europe to form the Laurasia plate; and so Pangaea was formed. All the continents were clustered closely together in this way from the late Carboniferous to the early Jurassic (Creer, 1970a; McElhinny and Luck, 1970). The palaeomagnetic maps of Smith, Briden and Drewry (1973) show that the continents underwent very little change in position for over a quarter of Phanerozoic time (Fig. 9.3).

Gondwanaland was intact for 450 my from the early Cambrian to the mid-Cretaceous, when it fragmented (Veevers, Jones and Talent, 1971). It drifted and rotated as a supercontinent until the Trias (220 my

ago), after which its position relative to the south pole did not change until the mid-Cretaceous (100 my ago)—see Fig. 9.3.

Gondwanaland Glaciation

This subject has been dealt with in Chapter 10 on palaeoclimatic indicators. Fig. 13.1 shows the area of Gondwanaland affected by the ice sheets. The close grouping of the glaciated areas of the individual continents provides one of the clearest indications of subsequent continental drift. For an account of the palaeoclimatology of the Permo-Trias period, see Robinson (1973).

Glossopteris Flora

The Glossopteris land plants flourished in the areas south of the Tethys in the Lower Gondwana Formations. The first species evolved during the later stages of the Carboniferous or Permian glaciation in some

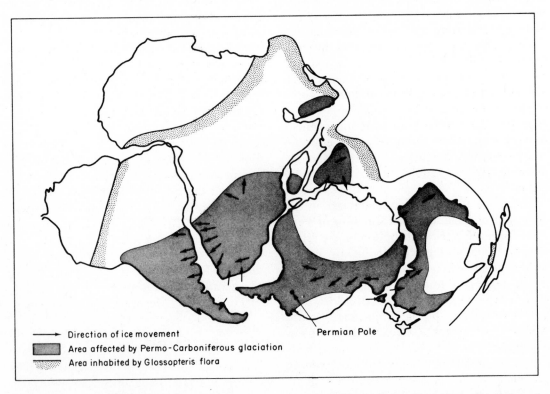

Direction of ice movement

Area affected by Permo-Carboniferous glaciation

Area inhabited by Glossopteris flora

Permian Pole

Fig. 13.1. A reconstruction of Gondwanaland in Permo-Trias times according to Smith and Hallam (1970) showing the areas affected by glaciation (mostly after Crowell and Frakes, 1970 and Seyfert and Sirkin, 1973) and inhabited by Glossopteris flora (after Plumstead, 1973)

areas, and soon afterwards in others (Plumstead, 1973); their distribution is larger than that of the glacial deposits (Fig. 13.1); clearly the stimuli provided by the cold climate favoured the growth of this unique flora. Plumstead (1973) suggested that the cause of this plant evolution may have been cosmic radiation, which is known to be greater in polar regions and which did not affect the northern coal plants of Laurasia; she also pointed out that the special climatic conditions favourable for development of the flora could not have operated independently in each of the (at present) widely separated continents, but could have done in a supercontinent dominated by enormous ice sheets.

Tetrapods

Terrestrial vertebrates obviously require a land connection in order to wander from one continent to another, and since the idea of land bridges is no longer tenable, contiguity of the continents gives us a viable model. According to Cox (1973) the Triassic is the only period during which tetrapods show clearly that land connections existed between every one of today's continents. There seems little dispute that the tetrapod evidence favours easy communication between the continents of Laurasia and Gondwanaland during the formation of Pangaea (Hallam, 1967; Colbert, 1973); a result of this free access is that the fauna is comparatively homogeneous or cosmopolitan throughout the world (Charig, 1971). Only a small barrier existed between Upper Carboniferous to Lower Permian aquatic tetrapod faunas of North America and Europe (there was no barrier for the terrestrial types) and this may have been a range of hills formed by the Caledonian–Appalachian fold belt (Milner and Panchen, 1973).

Probably the most cogent evidence for continental drift from terrestrial vertebrates comes from the early Permian reptile, Mesosaurus, which only occurs in the Gondwana Formations of South America and South Africa. It has a peculiar distinctive structure and no close relatives elsewhere. Although it was aquatic, it was a reptile adapted to limited swimming, probably in shallow freshwater, and so one cannot conceive that it swam 5000 km across the Atlantic; the most likely answer is that the two landmasses were joined.

It is difficult to quantify the palaeontological data in favour of Pangaea and subsequent continental drift, but a notable attempt by Brown (1968) produced a significant result. It is well known that the maximum degree of diversity of present day fauna occurs in the tropics and that the diversity decreases polewards. Brown plotted the number of genera of late Palaeozoic amphibia and Triassic reptiles against the present day latitudes of their localities (Fig. 10.8A and C) and found no correlation. But when the number of genera were plotted against their palaeolatitude (Fig. 10.8B and D), they have a symmetrical distribution. Also the diversity of invertebrates decreased to a major low in Permo-Trias times coinciding with the period of maximum assembly of the continental fragments (Fig. 10.5).

Matching of Age Provinces

Late Precambrian and Palaeozoic fold belts tend to wrap around older undisturbed Precambrian cratons (Fig. 1.1). It is often not difficult to define fairly accurately the boundaries of these cratons with the adjacent reworked or accreted belts, several of which were fragmented by Mesozoic–Tertiary continental drift. The matching of the pieces across facing continents provides a useful means of regrouping the continents and of demonstrating drift.

A general reassembly of these cratons, on different continents in a pre-drift reconstruction, was made by, amongst others, Hurley and Rand (1969) and Hurley (1970). Those that fit well together are:
1. The pre-2000 my cratons between South America and Africa (Fig. 13.2); in particular that between Guyana and West Africa is well documented, both structurally and radiometrically (Hurley, 1972).
2. The pre-2500 my craton that extends across the Davis Straits between West Greenland and Labrador (Fig. 1.2).

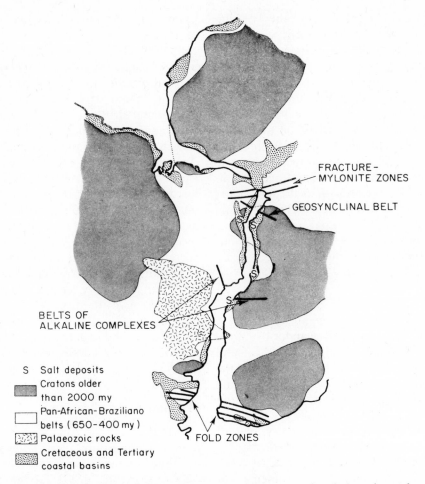

Fig. 13.2. Correlation of cratons, structures etc. between South America and Africa (compiled from Asmus and Ponte, 1973; Nairn and Stehli, 1973; Neill, 1973)

Many of the internal major features of the Precambrian fold belts can be followed across facing continents. For example, there is the line of mid-Proterozoic anorthosites across both Laurasia and Gondwanaland which, in the former, are accompanied by rapakivi granites, alkaline complexes, and acid volcanics (Chapter 6), and the line of early Proterozoic banded iron formations crossing every continent (Goodwin, 1973b).

Linear Phanerozoic fold belts provide an excellent means of checking the reassembly of the continents suggested by other criteria. There are two important regions: Gondwanaland and the North Atlantic part of Laurasia.

Du Toit (1937) recognized that the Palaeozoic fold belts of the southern hemisphere form an integrated pattern when the supercontinent is reconstructed. A map of Gondwanaland (Fig. 13.3) shows that the Palaeozoic and Mesozoic fold belts extend from South America across Antarctica to Australia (Engel and Kelm, 1972). When NW Europe is joined to North America on a pre-Trias fit, three fold belts fall into alignment: Grenville–Gothide, Caledonian–Taconic, and Acadian–Hercynian (Wynne-Edwards and Hasan, 1970).

If the outline of the Phanerozoic fold belts can be followed across the continents, so too should their internal geological features. For

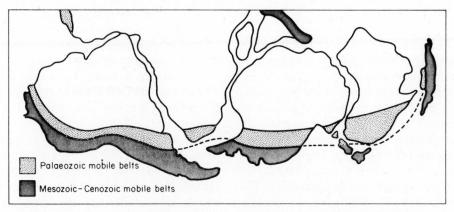

Fig. 13.3 The supercontinent of Gondwanaland and its peripheral mobile belts

example, in the Palaeozoic belts there is the carbonate platform extending from East Greenland, through Scotland, to eastern USA (Swett and Smit, 1972), and there are many other stratigraphic correlations between the British Caledonides and the Appalachians (see Chapter 11). There is a remarkably close match of geological features between South America and Africa (see Fig. 13.2 for details), and a close correlation between volcanics, coal deposits and tillites across all the continents of Gondwanaland between the Upper Carboniferous and the Lower Jurassic (Figs. 13.4, 14.7).

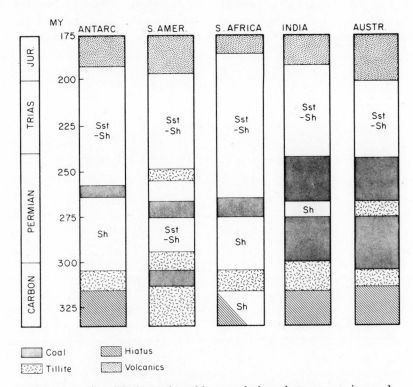

Fig. 13.4. Simplified stratigraphic correlations between major rock formations across the Gondwanaland continents in Pangaea times. SST–SH = sandstone–shale (adapted from Doumani and Long, 1962)

Metallogenic Provinces

If the Precambrian cratons and Phanerozoic fold belts can be used to assist the reassembly of continents separated by drift, so too can the metallogenic provinces they contain.

Schuiling (1967) made a general correlation of tin belts on both sides of the Atlantic. He suggested that there is a N–S coastal belt along the juxtaposed margins of South America (Brazil) and West Africa (South West Africa to the Benue Trough), but Petrascheck (1973) realigned these to cross from one continent to the other in a NE–SW direction. Nevertheless, the reader is not told to which tectonic belts these ore provinces belong; in either case Petrascheck's tin belt does not correlate with the tectonic belts and older cratons of Clifford (1971). The tin mostly occurs with Li, Be, W, Nb and Ta in pegmatites and granites (Hunter, 1973) in the highly metamorphosed parts of the Pan-African belts (Clifford, 1971).

Petrascheck (1973) suggests there is an Upper Proterozoic (600–1000 my) pegmatite province in SE Africa, S India, Madagascar and W Australia, characterized by Nb, Ta, Be and Zr—that is assuming Madagascar lies in its southern preferred position and India lies against W Australia. On this morphological fit of the continents the gold deposits in the Archaean greenstone belts in Rhodesia, Barberton, the Dharwars, the Pilbara and Kalgoorlie regions also form a coherent group.

When the continents were aggregated to form Pangaea in Devonian–Trias times conditions were ideal for the deposition of continental red sandstones, siltstones and argillites which were hosts for uranium, vanadium and copper mineralization. Copper deposits formed mainly during the Devonian–Trias and to a lesser extent in the Jurassic (they are often associated with anhydrite beds and contain important subsidiary cobalt—Jacobsen, 1975), whilst the uranium–vanadium ores reached their peak of development in the Triassic and Jurassic. The Colorado Plateau region of western USA is the principal area in the world for such ores. These clastic sedimentary copper deposits mostly formed in shallow-water lacustrine or lagoonal continental depressions or basins.

Chapter 14

The Break-up of Pangaea: Mesozoic–Cainozoic

There is good evidence for believing that Pangaea broke up into the present landmasses largely during the Mesozoic. The first indications of rifting actually came in the Permian (Oslo Graben), and of continental separation in the Trias; but the main separation stage was during the Jurassic and Cretaceous, with some late rifting continuing into the Tertiary and Quaternary (Red Sea and East African Rift Valley). The total time span of this fragmentation process was at least 225 my. In this chapter we shall look at the whole gamut of geological evidence for this mega-fragmentation, which was the incipient stage in the great drift of the continents which dominated the most recent geological history of our planet.

Triple Junctions, Aulacogens, Domes and Rifts

A new approach to the study of continental fragmentation came from the discovery of aulacogens by Shatski (1961) and this idea evolved via Salop and Scheinmann (1969) into the dome–triple junction–aulacogen–rift concept of Burke and Dewey (1973a). These sorts of structure apparently began to appear about 2000 my ago when stable continental platforms had evolved. Prominent examples are found in Proterozoic and Palaeozoic terrains, but the development of this kind of structure was most notable during the Mesozoic and Tertiary when Pangaea fragmented. All the various stages in this break-up are in evidence today on or around the continents, but it was

not a simple progression with time (some of the best preserved early stages in the rifting process are of Tertiary age). It will be convenient for us to study the full time range of these structures, as together they provide us with the complete break-up story.

Fig. 14.1 shows the development sequence from early domes, triple junction rift networks (rrr junction) based on deep-seated axial dykes, followed by various modifications. Spreading may take place on any of the three arms. If it develops on all three, then three plates are formed, but if only on two, then the formation of two plates leaves an abandoned rift or aulacogen. This is defined as a linear trough extending from a continental margin or geosyncline into a foreland platform or continent.

For examples of this kind of structure look at Fig. 14.2 which shows, firstly, the Cretaceous triple junction joining South America and Africa—two arms separated to form the South Atlantic, the third was abandoned and remains as the Benue Trough (Burke and co-workers, 1971; Grant, 1971; Nwachukwu, 1972; Olade, 1975)—and, secondly, the Tertiary triple junction centred around the Afar Depression—two arms have just begun to separate (the Red Sea and the Gulf of Aden), whilst the third (the Ethiopian Rift) failed to move.

But let us start at the beginning of the development sequence and look at the crustal doming, warping and flexure.

As long ago as 1939 Cloos realized the importance of doming to rifting when he

210

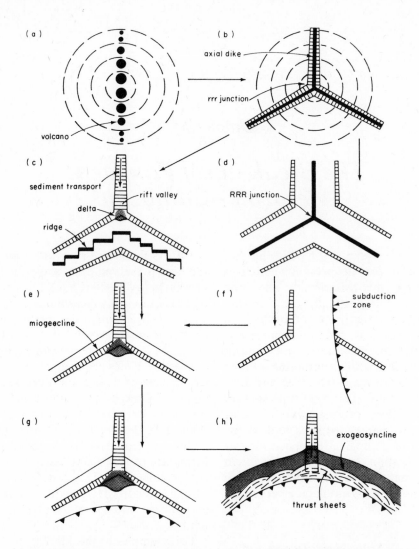

Fig. 14.1. Schematic inception and evolution of plume-generated triple
junctions. (a) Uplift develops over plume with crestal alkaline volcanoes
(e.g. Ahaggar). (b) Three rift valleys develop meeting at an rrr junction (e.g.
Nakuru). (c) Two rift arms develop into a single accreting plate margin
(ridge) and continental separation ensues, leaving the third rift arm (failed
arm) as a graben down which a major river may flow and at the mouth of
which a major delta may develop (e.g. Limpopo). (d) Three rift arms
develop into accreting plate margins meeting at an RRR junction (e.g. early
Cretaceous history of Atlantic Ocean/Benue Trough relationship). (e)
Atlantic-type continental margin evolves with growth of delta at mouth of a
failed arm and miogeoclines (e.g. Mississippi). (f) One arm of RRR system
of D begins to close by marginal subduction; if ocean is sufficiently wide, a
chain of calc-alkaline volcanoes will develop along its margin; any sediments in the closing arm will be deformed (e.g. Lower Benue Trough). (g)
Atlantic-type continental margin with miogeoclines and failed rift arms
approaches a subduction zone. (h) Continental margin collides with subduction zone, collisional orogeny ensues, sediment transport in the failed
arm reverses polarity, and failed arm is preserved as an aulacogen striking
at a high angle into an orogenic belt (e.g. Athapuscow). (After Burke and
Dewey, 1973a)

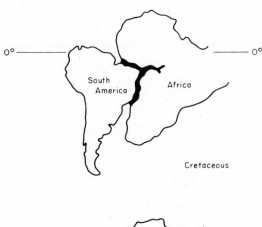

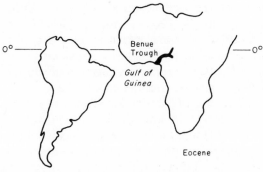

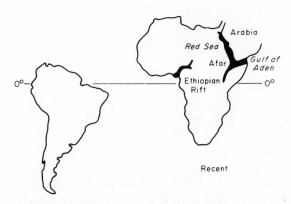

Fig. 14.2. Maps showing origin of Benue Trough as rift arm abandoned during continental rift separation of Africa and South America in Cretaceous time and, similarly, Ethiopian Rift valley during separation of Africa and Arabia in last 25 my (data from Smith, Briden and Drewry, 1973). Plume-induced three-armed rift systems are believed to have radiated during Cretaceous time from what is now the mouth of the Benue Trough and the Afar region. Rifting is believed to mark times when continent was stationary with respect to deep mantle plumes. (After Hoffman, Dewey and Burke, 1974; reproduced by permission of the Society of Economic Palaeontologists and Mineralogists)

experimentally demonstrated their relationship with respect to the Rhine and Red Sea rifts. But in recent years most information on this subject has come from Africa (Bailey, 1964; Gass, 1970a; Le Bas, 1971); a general account of the uplift, rifting and break-up from the Jurassic to the Tertiary is given by Burke and Whiteman (1973). The domal uplifts (Fig. 14.3) are usually about 1 km high, 100 km wide and 200–300 km long, and are characterized by negative gravity anomalies reaching about 100 mgal and by alkaline vulcanicity.

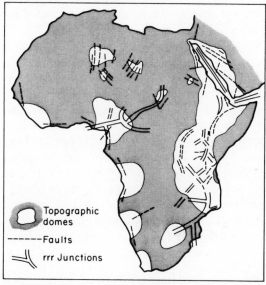

Fig. 14.3. Areas of late Mesozoic and early Tertiary domal uplift in Africa (after Gass, 1972b [reproduced by permission of the Geological Department, University of Newcastle-upon-Tyne], developed from Le Bas, 1971, and Gass, 1972a) with related triple junctions (after Burke and Whiteman, 1973; reproduced by permission of the Academic Press; London, New York and San Francisco)

There is a linear crustal arch related to the rift between South America and Africa. The Palaeozoic–Mesozoic sedimentary basins of Brazil (the Parana basin and the Sao Fransisco–Parnaíba–Salitra group) and of the Congo and Kalahari are separated from the Atlantic by a coastal basement strip 300–600 m wide, which marks a linear Cretaceous arch, the precursor of the protorift zone (Neill, 1973).

The next stage in development was the formation of three rifts which ideally were symmetrically orientated at 120° to each other. Burke and Whiteman (1973) recognized ten such triple junctions in Africa (Fig. 14.3) and Burke and Dewey (1973a) 45 throughout the world.

The oldest rift systems concerned with the break-up of Pangaea are the Midland Valley of Scotland, with its Carboniferous alkalic basalts and Permian tholeiites, the Oslo graben with its Permian alkaline rock suite and the troughs of the North Sea. These rifts did not reach a spreading stage and they represent the first attempt at mega-fracturing of Pangaea soon after its coalescence.

As Gondwanaland began to break up into separate continents during the Jurassic and Cretaceous (200–130 my ago), there was extensive intracontinental rifting in eastern Africa (Fig. 14.2) associated with widespread igneous activity. The rift valleys are located across several domes: a Jurassic one in Tanzania and Tertiary domes in Somalia, Ethiopia, Kenya and the Red Sea (Gass, 1970a). There were two peaks in rift formation: Jurassic–Cretaceous (180–130 my ago) and since the start of the Neogene (25 my ago). The intervening 100 my saw rift development but no rift inceptions (Burke and Whiteman, 1973).

The Rhine graben is one of the world's best known rift valleys. It is of Tertiary age and is located over a dome forming the southern arm of a triple junction centred on Frankfurt (Burke and Dewey, 1973a). It is associated with alkaline volcanics, high heat flow, recent faults and earthquakes (Mueller, 1970) and has undergone 4·8 km spreading in the last 45 my (Illies, 1970). Also, part of the East African rift valley has opened up 10·0 km since the Miocene, that is, 12–30 my ago (Khan, 1975). These are interesting examples of rift systems still undergoing mild tectonic activity.

Burke and Whiteman (1973) point out that whether or not the rifts developed by spreading to give rise to separate plates depended on whether the evolving world-wide plate system could accommodate the new structures. This is illustrated by the fact that African rifts striking close to east–west remained inactive, whilst several rifts that were aligned roughly north–south underwent spreading because their poles of plate motion were close to the Earth's spin axis (Le Pichon, 1968). For example, Madagascar was separated from the African mainland by spreading on two rift arms with nearly polar trends.

Structural Control of Rifting

The questions we need to ask ourselves here are: 'Do new rifts tend to follow old structures?' and 'Why have continents rifted where they have?' In other words, to what extent do former mobile belts/sutures/lineaments exert a structural control on the location of younger rift systems and continental margins?

There must be doubt that a fracture pattern the length (5500 km) and breadth (1000 km) of the Rift System of eastern Africa could be due simply to pre-existing lineaments in the basement; it must have formed in response to a definite stress pattern of its own (Le Bas, 1971; McConnell, 1972), and there are several places where the rift valleys cut obliquely across Precambrian mobile belts to demonstrate this.

Nevertheless, there are so many places where young rifts or continental margins follow the grain of old fold belts that local structural control may be suspected. Three examples will suffice to illustrate this relationship:
1. The Rift System of eastern Africa commonly parallels the trend lines of Precambrian mobile belts (e.g. the Mozambique) and in places follows the border of older cratons (e.g. Tanzania Shield, Zambian Block), a conformability distance of at least 4000 km (McConnell, 1972).
2. The join between South America and Africa between the Niger delta and Walvis Bay for the most part follows the grain of the Pan-African/Braziliano mobile belts for at least 2400 km (Fyfe and Leonardos, 1973; Nairn and Stehli, 1973).
3. The North Atlantic ocean opened in the Mesozoic along roughly the suture of the

proto-Atlantic that closed in the Palaeozoic (J. T. Wilson, 1966). More recent geophysical data of Ballard and Uchupi (1975) suggest the presence in the Gulf of Maine, off the eastern coast of North America, of a 250 km long linear Carboniferous–Permian basin that developed as part of a major right-lateral shear zone caused by the sliding past each other of the two continental masses in the final stages of closure of the Iapetus ocean. The structural weakness created by this basin probably controlled the subsequent major crustal rifting in the Trias in this area.

There is a general consensus amongst the above authors that old sutures or vertical mobile belts are planes of weakness extending deep into the lithosphere. They are subject to easy rejuvenation and are more susceptible to fissure than normal continental lithosphere with discordant trends. The high thermal gradients associated with the old mobile belts become the site of renewed thermal activity, a rift generates giving rise to a spreading ridge, and thus the continent is split and plate separation takes place.

According to the concept of J. T. Wilson (1965) transform faults, which control offsets of mid-oceanic ridges and are matched by offsets of the rifted continental margins, were initiated by lines of weakness in the original single continental mass. But such fractures within the continents are not well documented. Good examples are demonstrated by Garson and Krs (1976) in the Red Sea where a great many transform faults extend offshore into ENE-trending transverse fractures in Precambrian rocks. The direction of sea-floor spreading was clearly guided by the original continental fracture patterns. Similar transverse structures controlled the initial splitting of the continent in the Afar rift of Ethiopia (Barberi and coworkers, 1974).

The Timing of Break-up

It takes a long time to fragment a supercontinent. The break-up of Pangaea began in the early Jurassic, 180 my ago, and the last continent was not separated until the early Tertiary (45 my ago). In other words some continents broke away long after others: the separation of North America was very advanced, but Gondwanaland fragmentation was rather late (Pitman and Talwani, 1972; Smith, Briden and Drewry, 1973).

The sort of criteria used to date the time of separation of the landmasses are geologically meagre in quantity, and subject to some uncertainty. Nevertheless they do demonstrate distinctive environmental changes that can be dated by isotopic or fossil methods. The most common types are basic dykes, alkaline complexes often associated with faults, the first appearance of marine sediments along continental margins after a long period of continental sedimentation or lack of it, and affinities and differences between marine fauna on facing coasts suggesting contiguity or separation of the continents (Smith and Hallam, 1970).

The first landmasses to break apart were North America and Africa/South America, separation beginning about 180 and 200 my ago according to the following evidence (for a discussion of the early history of the North Atlantic, see Noltimier, 1974).

1. The onset of rifting is expressed by the formation of fault-bounded late Triassic basins with continental sediments and volcanics of the Newark Group on the eastern coast of the USA (Klein, 1969); these sediments indicate that insufficient separation had taken place by this time to allow ingress of marine waters. The intrusion of the Palisade sill at c. 190 my ago (Dallmeyer, 1975) and the Holyoke and Deerfield lava flows between 191 ± 6 my and 202 ± 10 my ago (data quoted in Dewey and coworkers, 1973) immediately preceded the start of active lateral spreading. The intrusion of coast-parallel dolerite dykes in Liberia between 173 and and 192 my ago probably took place during the incipient stage of separation (Dalrymple, Grommé and White, 1975) at about the time the Freetown Igneous Complex was intruded, 180 my ago, on the coast of Sierra Leone (Briden, Henthorn and Rex, 1971).

2. Palaeontological evidence of the age of sediments overlying layer 2 basalts allows the determination of the sea floor spreading rate

which, if extrapolated westwards, gives an age for the American continental margin of 190–200 my (Phillips and Forsyth, 1972).

3. Off the continental shelves of North America and North Africa there are 'quiet zones' in which the linear magnetic anomalies are weak or absent (Heintzler and Hayes, 1967). The 400 km wide western zone is twice the width of the eastern. They are regarded as congruent isochrons (Pitman and Talwani, 1972). The western zone probably formed in the late Triassic period of constant normal polarity 190–165 my ago, the edge of the continental shelf having an age of about 200 my (McElhinny and Burek, 1971; Phillips and Forsyth, 1972). This is in agreement with the time of onset of active lateral drift estimated by Pitman, Talwani and Heirtzler (1971) and Pitman and Talwani (1972). In order to account for the asymmetry of the quiet zones Noltimier (1974) suggested that the axis of the active spreading ridge shifted to the east by several hundred kilometres 160 my ago after

this time (McKenzie and Hussainy, 1968). Geophysical work on the South Atlantic oceanic crust indicates that the spreading ridge axis formed by 127 my ago in earliest Cretaceous time (Ladd, Dickson and Pitman, 1973; Larson and Ladd, 1973). During the Aptian (middle Cretaceous—112 my) sea water flooded into part of the graben and extensive evaporite deposits accumulated which are preserved on both the Brazilian and West African coasts (Burke, 1975) (Fig. 13.2); palaeomagnetic poles on lavas from SW Africa and Brazil suggest that the South Atlantic 'had not opened appreciably by 112 my BP' (Gidskehaug, Creer and Mitchell, 1975). The detailed ammonite biostratigraphy of Reyment (1969) documents the oscillatory inundations (transgressions and regressions) of sea water into the narrow gulf during the Albian (106 my), Cenomanian (100 my) and Turonian (94 my). Faunal evidence suggests that the final break occurred in the early Turonian (Reyment and Tait, 1972):

Successions	Age	Environment
Marine limestones and sandstones	Albian (106 my) and younger (Upper Cretaceous)	Freely circulating sea-water in transgressions and regressions—comparable to the Red Sea
Evaporites (up to 2 km thick)	Aptian (112 my) (mid-Cretaceous)	Inflow of sea-water restricted by local barriers—comparable to the Afar Depression and the Red Sea
Non-marine clastic sandstones and shales	Pre-Aptian (Upper Jurassic–Lower Cretaceous)	Continental lagoons, alluvial plains, flood plain deposition on levelled basement surface

which spreading continued symmetrically until the present.

There is more detail known about the timing of the break between South America and West Africa than between any other two landmasses; the sediments and fauna in marginal basins on the facing coasts are comparable and ideal for erecting a chronology (Reyment and Tait, 1972). The following sequence of events took place.

The presence of identical non-marine ostracods of late Jurassic to early Cretaceous age in Brazil and West Africa shows that the two continents must have been contiguous at

There is a controversy regarding the timing of break-up of Gondwanaland. The discrepancy between the post-late Carboniferous polar-wander paths for the South American-African block and Australia suggests that the fragmentation of Gondwana started in Permo-Carboniferous or early Permian times (Valencio, 1974). Geological evidence indicating that eastern Gondwanaland (Antarctica, Australia, India and Madagascar) must have separated from Africa–Arabia by mid-Jurassic times includes:

1. the age of the earliest marine sediments in North Kenya and North Madagascar (late

Triassic/early Jurassic) and in Tanzania and South Madagascar (early/mid-Jurassic),

2. the faunal similarities between early Upper Jurassic sediments in South Africa and northwest Madagascar, and

3. the early/mid-Jurassic age of the Karroo dolerites (McElhinney, 1973). The palaeogeographic maps of Dietz and Holden (1970) and Seyfert and Sirkin (1973) support this age of separation. However, the 100 ± 10 my Cretaceous map of Smith, Briden and Drewry (1973) and Briden and coworkers (1974) shows Gondwanaland as still intact, and according to Hallam (1967) the distribution of terrestrial vertebrates and neritic molluscs suggests that land connections were in existence between most segments of Gondwanaland until well into the Cretaceous.

Marginal Basins

In the initial stages of break-up fractures and faults formed and down-faulted basins and structural depressions were infilled with continental sediments. With increasing separation of the continental 'plates' more and more marine water invaded the basins, until narrow seaways were produced which eventually gave rise to the present oceans. Today this course of events is recorded in marginal basins along the trailing edge of the continents in which the progression from non-marine to marine sequences is characteristic within sedimentary piles that may reach up to 16–17 km in thickness (Kinsman, 1975b).

Along the Atlantic margin of the USA there are several fault-controlled linear basins infilled with up to 8 km of Late Trias non-marine sediments and some volcanics (Klein, 1969). The sediments include sandstones, shales, and arkosic red beds with mudcracks and rainpit impressions suggestive of subaèrial deposition in flood plain depressions, whilst the remains of dinosaur footprints, and fossil freshwater fish confirms a non-marine environment. Basaltic sills and intrusives (such as the famous Palisade Sill of New York) in these sequences result from magma injection into early rifts along the continental margin. Seismic reflection profiles have delineated three comparable major Triassic crustal rifts off the eastern coast of North America in the Gulf of Maine (Ballard and Uchupi, 1975). Also Triassic to early Jurassic red beds and evaporites are recorded on the continental margin off Nova Scotia and Newfoundland (Jansa and Wade, 1975).

The continental margin of Tanzania is characterized by normal faults with throws of up to 10 km (Kent, 1973). These are associated with a Permo-Trias (Karroo) block-and-trough structure with continental sediments in the troughs. In early Jurassic times marine oolites and limestones were deposited in a shallow water shelf environment, but in the middle Jurassic there was a widespread marine transgression that dates the early age of the Indian Ocean.

The continental break-up and the development of Mesozoic marginal basins around southern Africa are reviewed by Dingle (1973) and Dingle and Scrutton (1974). The first rifting took place 180 my BP along the eastern side of southern Africa (separation from Antarctica and formation of proto SW Indian Ocean), and between 130 and 125 my BP the western side separated from South America. In the period from the late Permian to the late Cretaceous southern Africa moved from a mid-continental high latitude to an ocean-dominated middle latitude position; these movements can be traced in comparable facies changes etc. along the present coasts on the west side of Africa and the east side of South America.

The features outlined in this and the last section show that marginal sedimentary basins, formed along the plate accretion boundaries and preserved along the present continental coasts, provide key evidence for the sequence of events and timing of the early stages of continental break-up.

Evaporites

Evaporite deposits often form, given certain limitations, in the early rifting stage of the fragmentation of continents and of the formation of proto-oceans.

The following evaporites occur in rifts, proto-oceans or on the margins of separated continents:

1. The North Sea troughs contain Permian evaporites (Whiteman and coworkers, 1975a, b; Ziegler, 1975) and the Cheshire graben has Triassic (Lower Keuper) evaporites (Evans and coworkers, 1968). The trilete trough patterns with their triple junctions are seen as failed rift arms that have undergone no active spreading, sited on an inner continental margin and related to the initial rifting of the North Atlantic.

2. The Red Sea is a 20 my old proto-ocean with evaporites up to 100 km wide and 7 km thick overlying new oceanic crust (Fig. 14.4a) (Kinsman, 1975a). There are also continental-based 1 km thick Pleistocene–Pliocene evaporites in the marginal graben of the Danakil Depression of Ethiopia (Hutchinson and Engels, 1970).

3. Late Permian to late Trias evaporites (up to 1·5 km thick in Greece and 1 km in Italy) border the western proto-Tethys (see Chapter 18).

4. There are evaporites in the nearshore or onland mid-Carboniferous to Triassic graben of the Canadian Maritimes, the Appalachian Piedmont and in NW Africa (Schenk, 1969; Rona, 1970). Upper Triassic to mid-Jurassic evaporites also occur off the North Atlantic continental margins (Pautot, Auzende and le Pichon, 1970; Emery and coworkers, 1975), and they originally extended from the Gulf of Mexico via the Old Bahama Channel to the coast of Senegal (Aymé, 1965; Burke, 1975).

5. Aptian evaporites up to 2 km thick occur on the continental margins (Fig. 13.2) of Brazil (Wardlaw and Nicholls, 1972) and West Africa (Belmonte, Hirtz and Wenger, 1965; Baumgartner and van Andel, 1971; Pautot, Auzende and le Pichon, 1973) and they extend for 250 km offshore (Roberts, 1975). Let us look more closely at these as they have

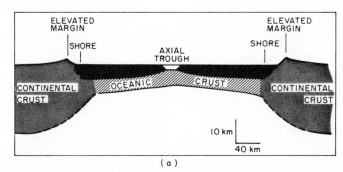

(a)

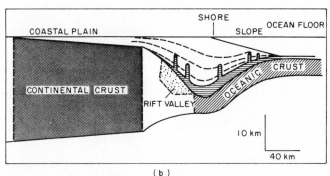

(b)

Fig. 14.4. (a) Proto-ocean evaporites in the southern Red Sea. The upper 3–4 km of evaporites are proved by drilling, the lower section from seismic data. (b) Proto-ocean evaporites deeply buried at the base of a trailing continental margin sediment prism. Similar sections seen in Mississippi Delta and northern Gulf of Mexico. (After Kinsman, 1975a; reproduced by permission of Macmillan Press)

distinctive chemical similarities, unusual in other evaporites of the world, which lend support to the idea that they were once contiguous (Wardlaw and Nicholls, 1972).

a) Tachyhydrite, a rare $CaCl_2$-bearing evaporite mineral occurs in sequences at least 100 m thick and on both coasts it tends to overlie halite and carnallite and underlie sylvite.

b) Beds of $CaCO_3$ and $CaSO_4$ normally underlie salt sequences, but they are lacking at the base of these evaporites and present at their top.

c) Comparable ranges of concentration of Ba, Rb and Sr were found in equivalent minerals and there is a general similarity in the relative abundance of B, F, Mn, Ca, Zn, Ba and Pb in carnallite.

These evaporites also provide some clue to the question of the movement direction of marine waters into the opening seaway, because it has been suggested that the fracture between South America and Africa opened from north to south by Kennedy (1965), but from south to north by Belmonte, Hirtz and Wenger (1965) and Dietz and Holden (1970). In an evaporite basin the more soluble phases tend to form in the closed end, whereas the relatively insoluble phases predominate at the open end. Wardlow and Nicholls (1972) found that the more soluble phases were concentrated in the north, suggesting that this was the closed end of the seaway. This evidence is consistent with the proposal of Grant (1971) that the fracture opened by E–W movement along the Gulf of Guinea transform faults and that the seaway opened by marine ingression from the south.

The above examples show that evaporites occur in rift systems, marginal basins, down the continental slope and on oceanic crust, in all cases having formed during the early stages of continental rifting. Kinsman (1975a) calculated that evaporite deposition occurs during the initial 10–20 my of spreading and in proto-oceans the masses may reach up to 4000 km long, 600 km wide and 7 km thick. Along trailing continental margins such evaporites are likely to be buried at the base of miogeoclinal sedimentary piles (Fig. 14.4b) (Kinsman, 1975b).

Besides structural control the development of evaporites depends on (1) the climatic palaeolatitudinal position, and (2) the correct balance between restriction and availability of saline ocean waters (Kinsman, 1975a; Burke, 1975).

1. As pointed out in Chapter 10, evaporites can only form where evaporation exceeds rainfall and the present day Hadley cells controlling atmospheric circulation limit the location of such conditions to two zones 5–35°S and 15–40°N; modern evaporites thus have a bimodal distribution, being absent in a zone close to the equator where equatorial rains are high. The distribution of evaporites and other climatically sensitive sediments by Seyfert and Sirkin (1973) and Drewry, Ramsey and Smith (1974) shows that since the Mesozoic the high pressure subtropical belts of net evaporation have roughly maintained their present width and position relative to their palaeoequator. It is therefore likely that since the Mesozoic, most evaporites formed within 35° (Kinsman, 1975a) or 50° (Gordon, 1975) of the equator. Thus evaporites could only form on the margins of fragmented continents that were located within these latitudinal limits.

2. There has to be a delicate balance between the influx of sea water and the total evaporation and this is best achieved by the presence of physical barriers. Burke (1975) interpreted the salt deposits as the products of evaporation of oceanic waters repeatedly spilled over structural sills into low latitude sub-sea level graben formed in the early stages of continental rupture.

The introduction of fresh water would be inimical to salt deposition. Early rifts are located on domal uplifts and so rivers at this stage generally flow down the slope of the domes. Later spreading is associated with subsidence of the the rift margins and a reversal of the drainage pattern, so preventing further salt formation.

Igneous Activity

Continental rifting was commonly associated with three types of igneous activity: intrusion of basic dykes and alkaline complexes, and extrusion of lavas. It is usually

218

possible to obtain an isotopic age on these rocks and they therefore provide another means of dating the break-up of continents.

We shall consider two aspects of these igneous rocks: their tectonic and age relationships.

Basic Dykes

During the early stages of opening of the North and Central Atlantic many parallel sets of dolerite dykes were intruded in late Triassic to early Jurassic times in the coastal regions of eastern North America, West Africa and northeastern South America (May, 1971) (Fig. 14.5). On a reassembly of the continents for the Triassic the dykes form radial alignments interpreted by May as being parallel to lines of tension in a principal stress trajectory field imposed on the continental margins by movements in the upper mantle at the onset of Atlantic sea floor spreading.

In the Skaergaard area of East Greenland there is classic evidence of the mechanism of continental break-up 60 my ago in the North Atlantic, namely a major coastal flexure and associated dyke swarm recognized by Wager and Deer (1938). The area was reinvestigated by Nielsen (1975) who found that the flexure corresponded to half of a graben structure. The sequence of events was as follows: the extrusion of lavas and the intrusion of a first generation of coast-parallel dykes (dipping inland at 40–50°), correlated with a pre-spreading updoming; flexure and collapse of the coast associated with much faulting; intrusion of a second generation of vertical coast-parallel dykes correlated with active spreading which opened up the North Atlantic.

The similarity of the East Greenland flexure to the Lebombo Monocline was recognized by Cloos (1939). Burke and Dewey (1973a) suggest that coastal flexures of this sort mark rifted arms of junctions that have developed to

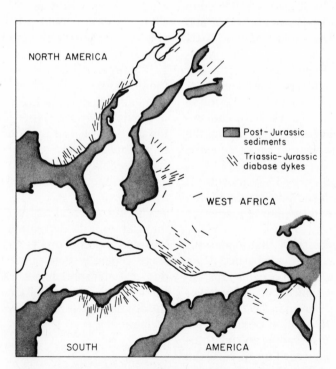

Fig. 14.5. Basic dyke swarms intruded during the incipient stages of opening of the Central Atlantic ocean in early Mesozoic times (redrawn after May, 1971; reproduced by permission of The Geological Society of America)

the ocean opening stage, although they are not often preserved along present coasts. Similar flexures are known on the Bombay coast, the east side of the Afar Depression and the north side of the Gulf of Aden. Other coast-parallel basaltic dykes which are a useful indication of early activity along rifted continental margins are the 162 my old dykes in SW Greenland (Watt, 1969) and the dykes parallel to the axis of the coastal flexure of Brazil (Neill, 1973).

There are thousands of tholeiitic dykes in the Karroo swarms of southern Africa (K. G. Cox, 1970; Vail, 1970). They have an interesting tectonic setting as they lie parallel to the arms of the lower Zambesi–lower Limpopo triple junctions of Burke and Dewey (1973a) and, in particular, parallel to the Lebombo Monocline which is situated along the eastern north–south arm (Fig. 14.6). This 700 km long monocline is a major crustal flexure near the margin of the continent with an enormous difference in structural level on either side— estimates vary from 9–13 km (K. G. Cox, 1970). The dykes were intruded into the zone of maximum extension or warping located along the rifted continental margin.

Tholeiitic Flood Basalts

There were enormous volcanic outpourings of predominantly quartz-tholeiitic lava in Gondwanaland during the Jurassic and Cretaceous (Fig. 14.7). The Jurassic lavas extend in a zone across the centre of Antarctica and no doubt reflect an abortive attempt at fragmentation, but the Cretaceous lavas are located along or near separated continental margins. As would be expected, most of the volcanicity was associated with faulting, some faults being slightly older and others slightly younger than the lavas.

What is remarkable is the scale of lava expulsion at this time. The Parana basalts of Brazil cover an area of 1,200,000 sq km which is larger than that covered by the Deccan Traps and the Columbia basalts put together (Herz, 1966). In places the Karroo lava pile reached 9 km in thickness and the lava field, probably covering 2,000,000 sq km, is the largest on earth (K. G. Cox, 1970, 1972).

There was continental distension and rifting in the western Tethys prior to the accreting ridge activity associated with the early Jurassic

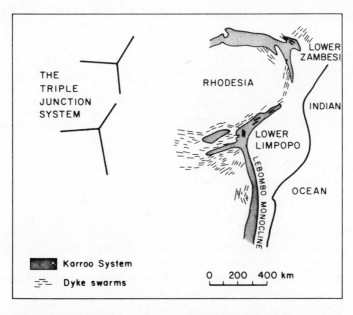

Fig. 14.6. Basic dyke swarms of Karroo (Jurassic) age emplaced parallel to triple junction boundaries near the downwarped rifted continental margin of southeast Africa (compiled from Vail, 1970; Burke and Dewey, 1973a)

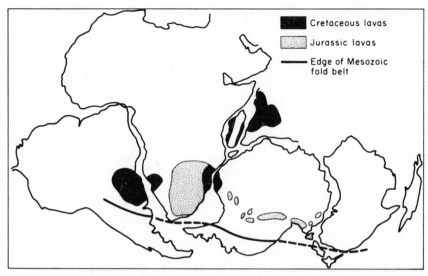

Fig. 14.7. The distribution of Mesozoic lavas plotted on a map of Gondwanaland. The Jurassic lavas broadly parallel the margin of and lie in the foreland of the Mesozoic fold belt, whilst the Cretaceous lavas are localized near separated continental margins. (Compiled from Cox, 1970; Smith and Hallam, 1970)

plate motions. Late Triassic flood basalts were extensively erupted in a region extending from Morocco to Greece, the Caucasus and Iran (Dewey and coworkers, 1973).

Alkaline Volcanic Rocks and Intrusive Complexes

The close association between continental rifting and alkaline magmatism is well known (Bailey, 1974; Goles, 1976). The East African rift system with its early Jurassic–Recent igneous activity is the most pertinent to consider (P. B. King, 1970; Le Bas, 1971; Baker and coworkers, 1971, 1972).

Of the volcanic rocks alkali basalts are the most common, especially in Ethiopia (Goles, 1975), although phonolites and trachytes are locally extensive, as in Kenya (Wright, 1970). Plutonic ring complexes, especially south of the Rungwe junction, contain syenites and nepheline syenites, which are deep-seated equivalents of the trachytic volcanoes, and the critical association of carbonatites (Barber, 1974) and nephelinites (King, Le Bas and Sutherland, 1972). The keynote of the rift magmatism is the abnormal enrichment of volatiles (and alkalis) expressed by the highly explosive and fragmental nature of the volcanism (Bailey, 1974). Rift volcanic rocks are also characterized by light REE enrichments and Sr^{87}/Sr^{86} ratios in the range $0 \cdot 703$ to more than $0 \cdot 710$ (Condie, 1976b).

The first appearance of three magmatic episodes are of paramount importance and correlate with critical tectonic phases:

3	Pliocene	Flood trachytes in rift floor	Main rifting
2	Upper and Middle	Flood phonolites and fissure eruptions	Main doming
1	Miocene	Regional nephelinites and carbonatites; alkali basalts	Early uplift

This sequence represents a progression from incipient melting of the upper mantle associated with early uplift, through extensive partial melting of the uppermost mantle and lower crust giving rise to phonolites, to pervasive partial melting of the lower crust due to lowering of the central rift segment into the heated zone causing trachyte outpourings (Bailey, 1974).

What factors have contributed to the association between alkaline magmatism, rifting and doming? According to Harris (1969, 1970) and Gass (1970a) the cause was thermal expansion of the upper mantle. In the uplifted domal areas with their high geothermal gradients, the downward increase in temperature was accompanied at mantle depths by thermal expansion, causing high density phases to transform to less dense types, thus producing uplift at the surface. The domes were split by rifts, ideally in a trilete pattern, caused by the downslope movement of the lithospheric plates off the uplifted area (Morgan, 1971).

Bailey (1974) considers that uplift was caused by the fact that less dense material was added to the upper mantle below the rift zone. From a gravity profile Sowerbutts (1969) suggested that there is a mass deficiency in the upper mantle below the rifts. The arching was a mechanical response to stresses transmitted through the rigid African plate and it gave rise to partial melting by decompression in the underlying zone of reduced pressure, into which volatiles, heat and alkalis were transported from the degassed surrounding mantle (Sowerbutts, 1969; Bailey, 1970, 1974). Under these conditions low density phlogopite, amphibole and carbonates were formed in the upper mantle, themselves causing further expansion and uplift, and providing parental material for the alkaline magmatism (see also Lloyd and Bailey, 1975). The advantage of this model is that it provides a means of concentrating the heat needed for the partial melting, a mechanism for the uplift and the chemical components for the alkaline magmatism.

The first eruptive rocks in the cycle, which we see in Ethiopia today, were extensive alkali flood basalts generated from melts at depths greater than 35 km (Green and Ringwood, 1967). More saturated basalts were then erupted in the rifts and generated in the order of 10 km depth and, finally, with incipient separation of the continental blocks, dykes were injected and new oceanic crust formed (volcanic islands in the Red Sea: Mallick and Cox, 1973) with the composition of oversaturated low-K tholeiitic basalt generated at very shallow depths of about 3–5 km (O'Hara, 1968). The sequence of igneous rocks related to this early rift development was controlled by the progressive rise of geoisotherms in the crust (Gass, 1972b).

Geophysical data support the concept that a diapir of relatively low density material is situated below parts of the rift system formed from some higher density material either by heating and expansion or by change in bulk chemical composition (Girdler and coworkers, 1969). Seismic data suggest that the East African rift system is a zone of shallow earthquakes (average focal depth of 20 km) similar to the Rhinegraben, Lake Baikal rift and the mid-Atlantic ridge. Contrary to early conclusions that the rifts are associated with negative Bouguer anomalies, recent gravity surveys indicate a more complicated picture. In Ethiopia (and the Red Sea) the anomaly is positive, in the Gregory rift there is a narrow positive within a broader negative, and to the south of Kenya there is only a negative which decreases in amplitude southwards, eventually disappearing in North Tanzania (Khan, 1975). In other words, in the Red Sea, where crustal separation has clearly taken place and dense material has intruded upwards, the anomalies are highest. A progressive southward decrease in anomaly intensity is correlated with an expected decrease in the quantity of upwelled material. A combination of seismic refraction (Griffiths and coworkers, 1971) and gravimetric data (Khan and Mansfield, 1971) suggest the presence below the Gregory rift of a body of material of low density, $3 \cdot 15$ g/cm^3 (P wave velocity c. $7 \cdot 5$ km/sec) from 20 to 60 km depth and 200 km wide. This is consistent with seismic results indicating that the crust below the rift is thinned to 20 km (from a normal 36 km for surrounding East Africa) and has an anomalously low sub-Moho velocity (Long, Sundaralingam and Maguire, 1973). There is a striking similarity between the model proposed for the structure below the Gregory rift based on the above data and that suggested for the mid-Atlantic ridge (Talwani and coworkers, 1965; Khan, 1975).

But some alkaline volcanics have a shallower source and a different structural control. The Jurassic rocks of the Jos plateau

(Jacobsen, MacLeod and Black, 1958; Bowden, 1970) and the Tertiary rocks of Tibesti (Vincent, 1970) are characterized by alkaline rhyolitic volcanics—evidence of substantial melting of continental crust—and are not associated with rifts, merely uplifts and faults. It is suggested by Burke and Whiteman (1973) that if the evolutionary path proceeds from uplift to alkaline volcanism with considerable partial melting of the crust, then this impedes, or halts, further structural development, i.e. rifts fail to form.

It is well known that peralkaline (soda-rich) subvolcanic ring complexes are associated with continental rift valleys. The rocks are silica-undersaturated and include feldspathic types (nepheline syenites, phonolites), nonfeldspathic types (ijolites, nephelinites) rich in nepheline and pyroxene, carbonatites, and fenites formed by *in situ* alkali metasomatism of the country rocks. The close association may be explained in three ways: by simple crustal extension; by the passing of mantle plumes under stationary continental lithospheric plates; or of continental lithosphere over stationary mantle plumes.

Fig. 13.2 shows the location of linear belts of alkaline complexes and carbonatites in Brazil and Angola; the lineaments may be failed arms of triple junctions formed in association with the opening of the South Atlantic (Neill, 1973) or they may be onland extensions of transform faults (Marsh, 1973).

Kimberlites

The majority of kimberlites are of Mesozoic–Cenozoic age and are confined to the interior and margins of stable continental cratons (Dawson, 1967, 1970); they rarely occur within the late Tertiary–Recent rift valleys of eastern Africa (Nixon, 1973). The intrusions took place during uplift or dilation of the cratonic areas along deep-seated fractures; some are located on domes, others in graben or transform lineaments. The 1200 km long belt of alkaline complexes in Angola (Fig. 13.2) contains 94 Middle Cretaceous kimberlite pipes, especially in the Lucapa graben near the Zaire border (Reis, 1972; Rodrigues,

1972). These intrusions lie on NE–SW trending lineaments that occur on small circles centred on the Cretaceous pole of rotation for the South Atlantic and which can be correlated with transform faults offsetting the mid-Atlantic Ridge (Marsh, 1973; Mitchell and Garson, 1976).

Early Mesozoic kimberlite dykes in SW Greenland were intruded during the early stages of continental rifting and separation from Canada (Andrews and Emeleus, 1975).

Age relationship between continental rifting and igneous activity

Dietz and Holden (1970) thought that these two phenomena were contemporaneous, but according to Vine and Hess (1971) and Scrutton (1973) the continental break-up usually occurred about 25 my after the beginnings of igneous action. From his review of the subject (Fig. 14.8) Scrutton (1973) demonstrated that the main period of igneous activity lasted on average about 30–40 my and that the break-up occurred at or just after, but not before, a peak in the activity. He proposed the following typical sequence of events from the rifting of a continent to the formation of a new ocean:

1. Arching of a continental area with precursory igneous activity.
2. Rifting and onset of more continuous igneous activity; first marine incursions.
3. Igneous activity reaches a peak.
4. Break-up occurs at, or soon after, the igneous peak.
5. Fully marine conditions established.
6. Igneous activity on the continental margins decreases as an active mid-ocean ridge develops.

Igneous rocks related to the continental break-up range in age from mid-Trias (200 my ago) to the present (Fig. 14.8). They had a diachronous development as different oceans opened at different times. We are fortunate in having every stage of igneous activity preserved today, from rifts that failed to spread, to proto-oceans like the Red Sea, to broad oceans with continental margin volcanics, dykes and plutonic complexes.

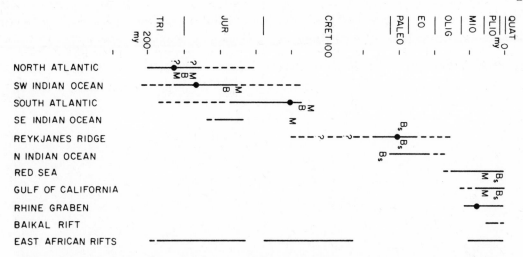

Fig. 14.8. A summary of data relevant to the timing of break-up of Pangaea. —●— duration of igneous activity with peak if known; – – – possible continuation of activity; B$_s$ time of break-up based on sea-floor spreading data; B approximate time of break-up based on projected sea-floor spreading data, first appearance of marine sediments and other criteria; M earliest marine sediments known. (After Scrutten, 1973; reproduced by permission of the *Geological Magazine*)

Mineralization

Some mineral deposits were formed during the break-up of the continents prior to their actual separation; some formed during early stages of spreading and today they are found in rift valleys and along the present continental margins.

Rifts with Pb, Zn, etc.

It has long been recognized that rift systems and their associated igneous rocks are of major significance in the localizing of ore deposits. One of the most well known is the Oslo Graben with its Permian magmatic suite. This structure probably represents one of the first abortive attempts at the break-up of Pangaea. According to Vokes (1973) the mineral deposits comprise oxides of Fe and Mn and sulphides of Pb, Zn, Cu, Mo, Bi and Ag.

A similar group of mineral deposits (Au, Ag, Pb, Zn, Cu, Mo, Sn and Ba) is associated with the Late Miocene–Late Pliocene Rio Grande rift system in New Mexico. According to Kutina (1972) the mineralization is con-- centrated in the graben, even though some is older than the rifting; some ore deposits occur in stocks of Laramide age (45–75 my),

although the rifting did not begin until about 20 my ago. Several deposits are located in the projected extensions of rift arms.

The Benue Trough in Nigeria has a 560 km long central belt of Pb–Zn mineralization (vein-fillings) developed about 80 my ago in the Upper Cretaceous–Santonian (Farrington, 1952; Nwachukwu, 1972). Earlier in this chapter the Benue Trough was described as a failed rift arm of a triple junction; the Pb–Zn mineralization establishes a link with the geothermal brines and associated metal concentrations enriched in Pb–Zn in the Red Sea rift which has undergone a small amount of opening. There is dilute brine seepage in the Benue Trough at present and isotopic data from Benue galenas indicate the lead was derived from low U/Pb sites, analogous to the dissolved salts in present day geothermal brines; thus it is likely that the Benue Pb–Zn mineralization was associated with geothermal heavy metal-bearing brines (Grant, 1971).

Sn in Alkaline Granites

Tin deposits occur in non-orogenic alkaline and peralkaline granite plutons of Jurassic to Tertiary age—for example, the Younger

Granites of Africa in Nigeria, the Tibesti area of Chad, at Mayo Darlé in Cameroun, in Damaraland, South West Africa, and also in Transbaikalia–Mongolia (Sillitoe, 1974). These tin-bearing complexes typically form in linear belts associated with domes and their formation ceased before the inception of rifting.

Chapter 15

Plate Tectonics and Sea Floor Spreading

After the break-up of Pangaea individual or groups of continents began to drift apart. Throughout the Mesozoic and Cenozoic they were dispersed in different directions until they took up their present positions. They did this because the intervening ocean floor began, and continued, to grow. Since the early nineteen sixties intensive exploration of the ocean floor in the form of magnetic and seismic surveys, dredging, coring and deep-sea drilling (and isotopic dating of the samples) has enabled the growth history of the oceans to be worked out in surprising detail and, as a result, a time scale for the drift of the continental plates.

In this chapter we shall look at several aspects of this new oceanic crust (in particular its spreading history and structure) because its evolution strongly affected that of the continents.

The Plate Mosaic

The theory of global plate tectonics evolved from that of sea-floor spreading, the main advance coming from the field of seismology. Using data from 29,000 earthquakes between 1961 and 1967 Barazangi and Dorman (1969) outlined the seismic belts of the world. Fig. 15.1 shows the dip direction of the subduction zones towards the areas of intermediate and deep focus earthquakes; shallow focus earthquakes are not shown. The seismic belts are narrow, continuous and never transect each other, the interiors of blocks are relatively stable, and the seismic zones follow the mid-oceanic ridges, island arcs and Meso-

zoic–Tertiary mountain belts, the divergent margins having less activity than convergent ones.

Morgan (1968) and Le Pichon (1968) outlined six major and about fourteen minor lithospheric plates bordered by tectonic belts (rises, trenches and transform faults), and Isacks, Oliver and Sykes (1968, reviewed by Sykes, Oliver and Isacks, 1970) made a detailed study of the relations between seismology, the plates and their movements. By this time the seismic data were combined with the ocean floor magnetic data and the hypothesis of global plate tectonics was on a sure footing.

Although some are solely oceanic, most of the plates consist of continental and oceanic parts. The reason for this combination within one plate is that new oceanic has been attached to old continental crust by sea-floor spreading. Some continent–ocean boundaries occur within plates and are seismically stable. The margins of the present Atlantic and Indian Oceans are seismically inactive and thus contrast with the active margins of the Pacific Ocean (Fig. 15.1).

Morgan (1968) adapted J. T. Wilson's (1965) transform fault concept to a spherical earth. The tensional mid-ocean ridges are generally oriented radially to a pole and the pure shear transform fractures follow orthogonal small circles; thus the pole of spreading lies roughly perpendicular to the transform faults. The rate of spreading, therefore, increases progressively away from the poles where it is zero to a maximum at the equator of rotation. Following these principles Le Pichon (1968)

225

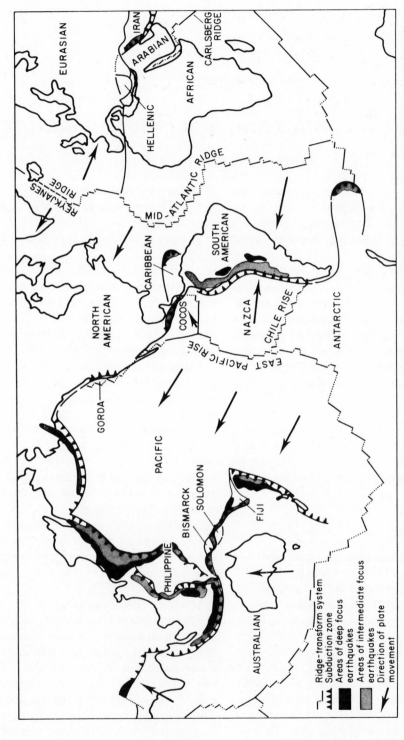

Fig. 15.1. The major lithospheric plates (each named) of the world and their accreting, transform and converging boundaries (partly after Dewey, 1972). Areas of deep and intermediate focus earthquakes (based on Barazangi and Dorman, 1969) and movement directions of plates (after Le Pichon, 1968)

established the poles, rates and direction of the six major lithospheric plate movements.

The increasing rate of spreading with distance from the pole of rotation is reflected in the progressively greater distance between particular magnetic anomalies and the ridge axis. Similarly, the rate of plate convergence at subduction zones increases away from the pole of rotation (Dewey, 1972). A corollary of this is that the intensity of development of fold belts, including the generation of calc-alkaline magmas, should be at a maximum near the equator of rotation.

Broad Structure of the Plates

Fig. 15.2 shows that the outermost part of the earth is made up of a number of layers. The plates consist of the rigid lithosphere, which includes both the continental or oceanic crust, and the underlying denser layer of mantle down to the top of the asthenosphere. The continental crust averages 40 km in thickness and increases to more than 70 km under the Andes (D. E. James, 1971) and 80 km under the Himalayas, while the oceanic crust (layers 1, 2, 3) averages 7 km. The lithosphere is 100–150 km thick under the continents and 70–80 km under the oceans; it is characterized by appreciable strength or rigidity with resistance to shearing stress and thus undergoes little significant internal deformation during lateral drift. The lithospheric plates ride or drift over the asthenosphere which extends to about 700 km depth, the topmost part having no significant strength and is thus plastic and capable of internal flow with lower S and P wave velocities than the overlying or underlying mantle.

The low velocity zone is widely thought to consist of partially molten upper mantle and is regarded as the source region for magmas that intrude the continental crust. Lambert and Wyllie (1968) and Kushiro (1968) proposed that traces of water locked in amphibole may assist partial melting in the low velocity zone. The exact composition of the low velocity zone is unknown—under the continents it may consist of garnet peridotite and amphibole or eclogite and amphibole, and under the oceans spinel peridotite and amphibole or eclogite and amphibole (Wyllie, 1971).

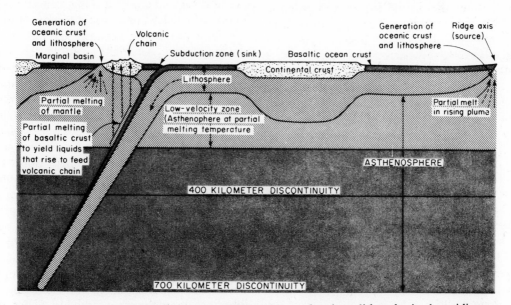

Fig. 15.2. Cross-section of the upper mantle and crust showing a lithospheric plate riding on the asthenosphere. The continent is embedded in the plate and moves with it. (After Dewey, 1972; reproduced by permission of *Scientific American*)

Types of Plate Boundary

The plates are bordered by three types of seismically and tectonically active boundaries: divergent or spreading zones, transform faults, and convergent or subduction zones.

Divergent Zones

These are mid-ocean ridges, radial to poles of rotation; they are narrow seismic belts, totalling 80,000 km in length, characterized by very shallow earthquakes (Fig. 15.1). The seismic activity is typically far smaller than at major convergent zones. According to Sykes, Oliver and Isacks (1970), the largest earthquake known in the ocean-ridge system has about one eighth the energy of the largest shock in an arc. The seismic data are consistent with a thin lithosphere at the ridges subject to tensional stresses and lack of a shear component. The ridges are also zones of high heat flow (up to ten times the average) and active tholeiitic magmatism responsible for generation of new oceanic lithosphere.

Topographically the ridges stand up above the surrounding ocean floor as 500–1000 km wide and 2–3 km high mountain chains. Bott (1971) notes that in the Indian and Atlantic oceans where they have slow spreading rates (1·5 and 1·0–1·5 cm/yr, respectively) the ridges are relatively narrow and rugged with a median rift along their crest. In contrast, the fast-spreading East Pacific Rise (4·4 cm/yr) is much wider, less rugged and has no prominent median rift.

Transform Faults

Although transcurrent faults were long known from the continents, it was not until 1952 that extensive fracture zones were recognized across the ocean floor—in the Mendocino area of the Pacific (Menard and Dietz, 1952). Vacquier in 1959 demonstrated that the zones displaced magnetic anomaly patterns and J. T. Wilson (1965) called them transform faults.

Transform faults are pure shear boundaries parallel to the movement direction vector between two plates and they lie on small circles normal to poles of rotation. Along them there is no generation or destruction of lithosphere. The only part of the fault that is tectonically active is that lying between two ridges, and this is characterized by shallow focus earthquakes and strike–slip movement along a sub-vertical plane. Many transforms were only located by the offset of the oceanic ridge seismic belts.

The motions of transforms, sinistral or dextral, are opposite to those across transcurrent faults on the continents. There are twelve main types (J. T. Wilson, 1973): six dextral and six sinistral, named according to their terminations, e.g. ridge–ridge, ridge–concave arc, ridge–convex arc, etc. The six possible dextral transforms are illustrated in Fig. 15.3.

Around the present-day Pacific Ocean there are several on-land transforms, the Philippine Fault, the Alpine Fault in New Zealand, and the Atacama Fault in Chile, but the most well known is the San Andreas Fault in California.

The San Andreas is a dextral ridge–ridge fault with both ends in the sea. Magnetic anomalies in the NE Pacific indicate that there has been a constant rate of movement of 6 cm/year parallel to the fault and 1900 km of lateral displacement since Oligocene times—30 my ago (McKenzie and Morgan, 1969). The fault developed because the westward-moving North American continental plate flanked by a subduction zone overrode the Pacific plate with its oceanic ridge (East Pacific Rise). There was an interesting, but complicated, sequence of events. In the early stages when continental and oceanic plates approached each other, the latter was subducted in a coast-parallel trench. Soon the oceanic rise abutted against the continental margin and the former trench deposits became attached to the advancing continental plate (they later became the California Coast Ranges). The continental margin rode over the crest of the ridge and moved onto the westerly oceanic plate. At this stage the continental and oceanic plates were moving at different rates in the same direction. So far, the continental plate had managed to advance due to shear at the subducting edge, but now that the two plates were advancing in the same

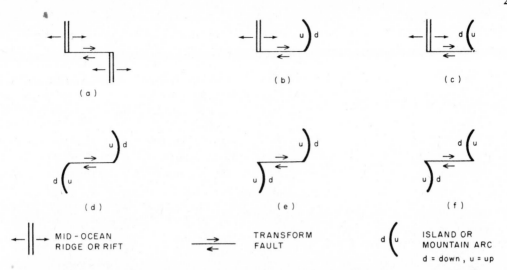

Fig. 15.3. The six possible types of dextral transform faults. (a) Ridge to ridge. (b) Ridge to concave arc. (c) Ridge to convex arc. (d) Concave arc to concave arc. (e) Concave arc to convex arc. (f) Convex arc to convex arc. Note that the direction of motion in (a) is the reverse of that required to offset the ridge. (After L. J. T. Wilson, 1973; reproduced by permission of J. Wiley)

direction subduction was no longer possible. The need for shear movements at the leading continental edge was then accommodated by the formation of the San Andreas Fault, acting as a transform of the overridden ridge (Jacobs, Russell and Wilson, 1974).

Convergent Boundaries

These occur where two plates have come together and one has overridden the other. Four types of collision are possible (Dewey and Bird, 1970a,b), between:
1. Oceanic/oceanic plates along island arcs (see Chapter 16)
2. Oceanic/continental plates along island arcs (see Chapter 16)
3. Oceanic/continental plates along continental margins (see Chapter 17)
4. Continental/continental plates (see Chapter 18)
On collision one plate becomes bent and is thrust down a subduction zone into the mantle where it is melted. Where continental and oceanic plates converge, the latter is subducted due to its higher density. The convergent boundaries are the regions of maximum shortening and highest shear motion on the earth's surface, and thus it is not surprising that of all seismic belts they are the widest, have the maximum activity and the largest earthquakes (Sykes, Oliver and Isacks, 1970). They are characterized by oceanic trenches, shallow and deep earthquakes; the deep foci locate the position of the subducting surface that dips at an average of 45° and extends to depths of up to 700 km (Isacks, Oliver and Sykes, 1968).

Dewey and Burke (1974), having analysed continental margins involved in plate separation and collision, showed that ideally they should be irregularly shaped, i.e. not straight, because the initial fracturing and break-up developed along interlinked triple junctions tends to give rise to a series of projections and embayments (Fig. 15.4). Upon progressive plate convergence the impinging projections become points of high strain, ophiolite obduction and basement reactivation, and within embayments (which tend to be zones of lower strain) remnants of the former ocean may be preserved. The difference in timing between early and late collision zones causes diachronous development of structures such as basement nappes and flysch fans within the orogenic belt. This idea has been applied to

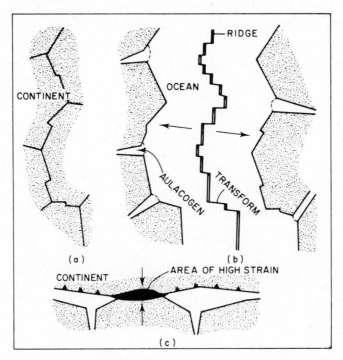

Fig. 15.4. Idealized plate boundaries during the opening and closing of oceans. The triple point configuration in (a) leads to an irregular plate margin with projections and embayments in (b). In (c) a projection impinges against another colliding continent causing high stress and possible ophiolite obduction and basement reactivation. In the embayments former oceanic crust may be preserved. (Redrawn after Dewey and Burke, 1974; reproduced by permission of The Geological Society of America)

the structural variations in the Appalachian–Caledonian belt by Dewey and Kidd (1974) and Phillips, Stillman and Murphy (1976).

Oceanic Magnetics and Sea Floor Spreading

Landmarks in the evolution of ideas on the structure of the oceanic crust are well known:
1. Hess (1962), based on a 1960 report to the US Office of Naval Research, and Dietz (1961) proposed that the oceans evolved by spreading of the ocean floor due to upwelling of new material at mid-oceanic ridges.
2. R. G. Mason (1958), Mason and Raff (1961), and Vacquier, Raff and Warren (1961) discovered magnetic anomaly strips in the floor of the Pacific Ocean off California. The origin of these variations in magnetic intensity was at the time unknown.

3. Morley (1963) in an abortive paper (see Wyllie, 1976), Vine and Matthews (1963) and Morley and Larochelle (1964), using the discovery that the earth's magnetic field had reversed direction several times in the past, proposed that the alternating magnetic anomalies were caused by these periodic reversals.

A good summary of the development of these ideas is by Wyllie (1971), and a review paper on sea-floor spreading is by Vine and Hess (1970).

When lavas, containing magnetic particles such as oxides of titanium and iron, crystallize below their Curie point, just below 500°C, in the presence of an external magnetic field, the particles align themselves in the direction of the prevailing magnetic field of the earth. The magnetic anomalies, found by trailing a magnetometer from a research ship across the oceans, reflect variations in the intensity of the earth's magnetic field. By subtracting the regional geomagnetic field of the earth the local magnetic deviations are obtained, locked into the upper 0·5 to 2 km of the basaltic layer of the oceanic crust. The anomalous field due to a normally magnetized strip reinforces the earth's field and gives rise to a positive anomaly and conversely for reversedly magnetized strips. The sort of magnetic anomaly pattern discovered in the ocean floor is shown in Fig. 15.5, where it can be seen that the axis of the Reykyanes Ridge pattern points directly

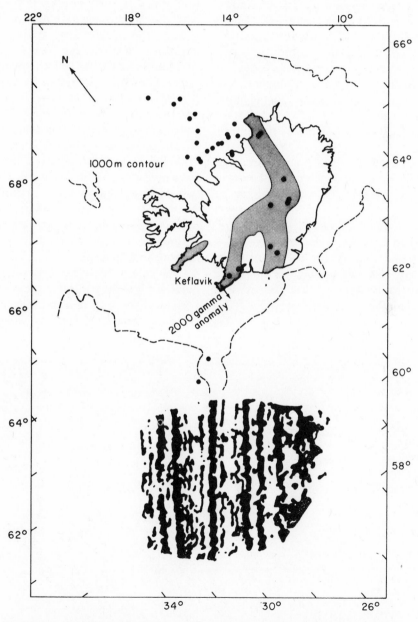

Fig. 15.5. Positive magnetic anomalies (black) across the Reykjanes Ridge (see Fig. 15.1) pointing towards the rift structure on Iceland (after Heirtzler, Le Pichon and Baron, 1966; reproduced by permission of Pergamon Press)

towards the active graben that extends across Iceland. There are similar patterns throughout most of the world's oceans (Larson and Pitman, 1972a) (Fig. 15.6). The remarkable feature of such a pattern is its almost perfect symmetry about a central positive anomaly (labelled no. 1) parallel to the mid-oceanic ridge. The positive anomalies are numbered outwards on each side of the ridge anomaly.

The significance of these anomalies was not realized until Vine and Matthews proposed that they reflected strips of oceanic lava which had been magnetized in normal (as today) or reversed directions. Positive anomalies, like no. 1, have a normal polarity and negative ones a reversed polarity. Two important results of the Vine–Matthews hypothesis were:

1. It confirmed the sea-floor spreading hypothesis of Hess. If the symmetrical pattern was caused by the emplacement of magma at the ridge axis, and older reversally magnetized lava was carried further away from the ridge, then the ocean floor must be accreting.

2. It added the time factor to the process. The anomaly patterns enabled a magnetic stratigraphy to be erected for the oceans, comparable to the fossil stratigraphy of the continents.

The application of K/Ar dating first to magnetized lavas from the continents and then to the oceanic lavas gives rise to two further results:

3. The erection of a time scale for the stratigraphy of normal and reversed magnetic strips. The present normal polarity goes back about 700,000 years—this is the longest duration of any polarity interval in the last 4·5 my and the shortest is 20,000 years. Beyond 4·5 my the radiometric dating method is imprecise; nevertheless, by assuming that the spreading rate for the last 4·5 my was constant for earlier time, the age of earlier anomalies can be worked out—that is, back to at least 180 my ago in the early Jurassic (Fig. 15.7). Extrapolation of the time scale enables isochrons to be drawn on the ocean floor up to 2000 km from the ridge axis. Such isochrons have been drawn right across the North Atlantic. Because new material at ridge crests in different oceans was similarly affected by polarity reversals, it is possible to correlate anomalies of specific number and age from one ocean to the other.

4. The rate of sea-floor spreading can be calculated by comparing the age of an anomaly with its distance from its mid-oceanic ridge. In spite of variations, such as wide anomalies

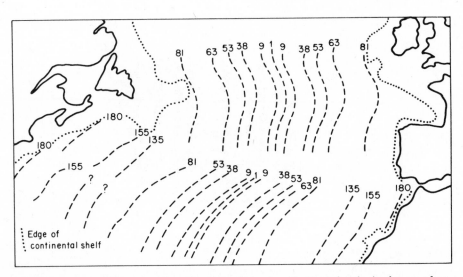

Fig. 15.6. Magnetic anomalies (numbered) in the North Atlantic (redrawn after Pitman and Talwani, 1972; reproduced by permission of The Geological Society of America)

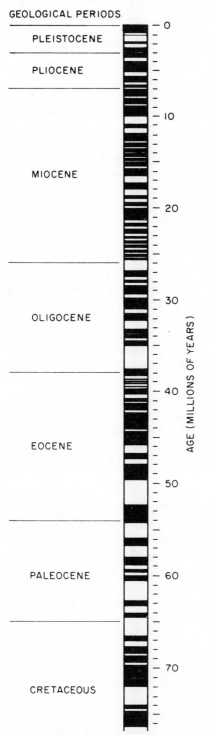

GEOLOGICAL PERIODS

PLEISTOCENE

PLIOCENE

MIOCENE

OLIGOCENE

EOCENE

PALEOCENE

CRETACEOUS

AGE (MILLIONS OF YEARS)

0

10

20

30

40

50

60

70

Fig. 15.7. Chronology of geomagnetic field reversals extrapolated from sea-floor spreading data. Periods of normal and reversed polarity are in black and white respectively (after Heirtzler, 1968; reproduced by permission of *Scientific American*)

reflecting a longer duration of polarity than short ones and the spreading rate decreasing towards the spreading poles, it is clear that the rate of spreading varies from ocean to ocean (the *spreading rates* quoted are valid for one side of a ridge and are half the *separation rate*), e.g.

East Pacific Rise	5–6 cm/yr
Pacific–Antarctic Rise	4
S Indian Ocean/N Pacific	3
S Atlantic	1·5
N Atlantic/Red Sea	1

One effect of the variation in spreading rate is that it changes the thickness of the accreting volcanic layer: as the rate increases, the thickness decreases. In other words, a fast rate of spreading does not give much time for volcanic material to accumulate, and a thin layer results, and vice versa.

Finally, there was an interesting suggestion by Nisbet and Pearce (1973) that the TiO_2 content of oceanic basalts may be used as an indication of the spreading rate. They found that the lowest contents (1·02%) occurred in basalts from the Gulf of Aden with the slowest spreading rate of $1·0 \text{ cm yr}^{-1}$; data from 14 other sites defined a crude linear positive correlation. In a more comprehensive treatment Sugisaki (1976) established that the K_2O content of oceanic basalts increases as the spreading rate of plate movement decreases: 0·13% at 2·9 cm/yr to 0·22% at 1·0 cm/yr. This correlation is presumably due to the fact that the depth of origin of magmas is dependent on the spreading rate.

The net result of the information outlined so far not only provides compelling evidence for sea-floor spreading and therefore the drift of continents, but also enables us to follow the history of the oceans since the early Jurassic. As Fig. 15.6 shows, the evolutionary pattern of the North Atlantic is complete from side to side. Whilst it has so far been necessary in this chapter to take a backward step through time—from anomaly 1 to 180—it is now possible to reverse the procedure and follow the opening of the oceans, and thus the position of the continents, from the break-up of Pangaea to the present.

Stratigraphy of the Oceanic Crust—Layers 1, 2, 3

There are two ways to find out how the oceanic crust varies with depth: firstly by sampling on the surface and by drilling, and secondly by seismic surveys. With this information we can find out about the age, thickness, composition and physical properties of the layers that make up this new oceanic floor.

Sampling has difficulties. Much dredging has been done but it is not always sure which layers the samples come from, and drilling is expensive, particularly for penetration to the deeper layers. But sampling has revealed that the top layer is sediment and that this is underlain by basalt.

The seismic methods obviously have greater potential in this field and we are fortunate here that the rather simple structure of the oceanic crust (as shown by magnetics, for example) is suitable for a method, the main assumption of which is that the crust must be formed of uniform layers with different seismic velocities.

Whilst there are still some minor disagreements, the many investigations have shown that the crust is composed of three, or perhaps four, uniform layers.

Layer 1 consists of unconsolidated sediments with low seismic velocities of 1·5–3·4 km/sec. One of the remarkable features about the sediments is their variation in thickness across the oceans. It was no surprise to find great thicknesses of terrigenous sediment near the continental margins (they reach 7 km off Argentina—Zambrano and Urien, 1970), but the almost complete absence of any sediment over the crests of mid-oceanic ridges was unexpected (Ewing and Ewing, 1967, 1970; Van Andel and Bowin, 1968). It is now established that there is a general increase in thickness and age of sediments away from the mid-oceanic ridges.

What types of sediments are there? Important are oozes enriched in the four groups of planktonic micro-organisms, viz. foraminifera, radiolaria, diatoms and calcareous nannoplankton (Saito and Funnell, 1970). The use of foraminifera to subdivide stratigraphy of

Cretaceous to Recent sediments (about 140 my) has reached a high degree of sophistication. There are also zeolite clays enriched in palygorskite, montmorillonite and phillipsite. Ferromanganese coatings and nodules are locally important on the ocean floor.

In places the sediments have an extremely regular stratigraphy. In the North Atlantic there are three marker horizons which can be picked up on seismic profiles. The most prominent, horizon A, was earlier thought to be turbidite material consisting of slumped reef debris (Saito, Burkle and Ewing, 1966), but more recently it has been suggested that it may be a layer of Eocene chert—lithified siliceous ooze (Ewing, Windisch and Ewing, 1970).

Layer 2 is 1 to 4 km thick, has a seismic velocity of 5·0 km/sec, consists of pillow-bearing tholeiitic basalts probably with much pyroclastic material, and outcrops extensively on the sediment-free oceanic ridges. It decreases in thickness from the ridge crests towards the oceanic basins. Most ocean floor basalts are olivine tholeiites, calcic (11·44%), aluminous (16·49%), potash-poor (<0·3%) with extremely low contents of Rb, Sr, Ba, U, Th, Pb and Zr, and high ratios of Si/K (200), Na/K (>10) and K/Rb (>1000), a low Th/U ratio (<2) and flat REE distribution patterns. Their Sr^{87}/Sr^{86} ratio is between 0·702 and 0·704 (Engel and Engel, 1970; Peterman and Hedge, 1971). They originated by partial melting at shallow mantle depths.

The basalts are often weakly metamorphosed, belonging now to the zeolite and greenschist/amphibolite facies. Most of the original igneous textures and chemical compositions are preserved, except for introduction of H_2O and Na_2O (spilitic types) and decrease in CaO (Miyashiro, 1973b). The zeolite facies rocks occur in normal oceanic basins, but the greenschist- to amphibolite-grade rocks only occur at mid-oceanic ridges where there is the highest heat flow. Low temperature alteration is also common, increasing progressively away from the ridges (Christensen and Salisbury, 1972).

There are some minor acidic, more highly differentiated volcanics in layer 2; they are confined to the vicinity of the mid-oceanic ridges and thus are associated with higher than normal heat flow (McBirney and Gass, 1967).

Layer 3 is 5–6 km thick and has seismic velocities of 6·4–7·0 km/sec but is less easy to define. Although Hess (1962) thought it was a homogeneous layer of 70% serpentinized mantle peridotite, it is now thought it may consist of more than one layer, for which there are several petrological models. But first it should be noted that it is absent from the crest of the mid-Atlantic ridge although present at the crest of the East Pacific Rise; this variation may be due to different rates of sea-floor spreading. It increases in thickness progressively from the mid-oceanic ridges and yet the total thickness of layers 2 and 3 tends to be constant across the oceans (Miyashiro, Shido and Ewing, 1970). Christensen (1970) suggested that it consisted uniformly of amphibolite or hornblende gabbro, but Cann (1970b, 1974) proposed that it consisted of an intense basic dyke swarm overlying gabbros with some serpentinite diapirs intruded from the mantle (Fig. 15.8). Cann's crustal succession—sediments, volcanics, dykes, gabbros, serpentinites—is similar to that found in obducted ophiolite complexes such as the Troodos complex in Cyprus (Gass, 1968) and the Vourinos complex in Greece (Moores, 1969), and it has been widely accepted in subsequent literature (e.g. Vine and Moores, 1972).

Peridotites, serpentinites and minor anorthosites occur on some mid-oceanic ridges (Engel and Fisher, 1969, 1975; Miyashiro, Shido and Ewing, 1970; Aumento and Loubert, 1971). The ultramafics include the following types:

Dunites: ol
Harzburgites: ol, opx
Lherzolites: ol, opx, cpx
Wehrlites: ol, cpx

The association of these rocks with gabbros with distinctive cumulate textures has led to the conclusion that in places oceanic ridges contain major layered basic–ultrabasic complexes in which gravity differentiation was prominent (Melson and Thompson, 1970; Engel and Fisher, 1975).

The Life Cycle of Ocean Basins

If we look at all the ocean basins in the world, we see that they are in different stages

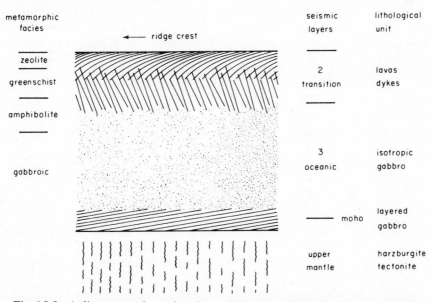

Fig. 15.8. A diagrammatic section through the oceanic crust below the sediments of layer 1. Note the asymmetry of the structure and the direction towards the ridge crest (modified after Cann, 1974; reproduced by permission of Blackwell Scientific Publ. Ltd.)

of growth or decline. They can therefore be arranged in an order representing the life cycle of a typical ocean basin (Table 15.1) (J. T. Wilson, 1973).

The East African Rift System may be considered to be in an embryonic stage of development; it is an arm of a triple junction which has failed to open or lags behind the other two arms, the Red Sea and the Gulf of Aden, which represent the incipient stage of growth of an ocean. The Atlantic Ocean began to open in the early Jurassic and is still growing in an advanced stage of maturity. The Indian Ocean floor has begun to be subducted below Java on its eastern side, whilst the Pacific is being subducted on both sides. Although it is still the largest ocean, the Pacific is smaller than when it surrounded Pangaea in the late Carboniferous to early Jurassic and J. T. Wilson (1973) regards it as now being in the first stage of decline. In the Mediterranean region today there are few remnants of the former Tethys Ocean that existed between Laurasia and Gondwanaland (see Chapter 18); the Indus fault line in the Himalayas represents a cryptic suture between the colliding Indian and Asian landmasses.

Table 15.1 shows that ocean basins in their different stages of growth and decline are associated with different types of sedimentary and igneous rocks. For a detailed account of the life cycle of an ocean basin, the reader is referred to Jacobs, Russell and Wilson (1974).

Sedimentary Evolution from Rifts to Oceans

The tectonic evolution from a rift to a Red Sea-type proto-ocean to a wide mature ocean basin naturally creates different sedimentary environments in each stage of growth. By considering these changes in order we can erect the sedimentary history of an evolving ocean; the following stages of development can be distinguished (Schneider, 1972; Dickinson, 1974a).

Rift Valley

The East African Rift Valley provides a modern, and the Triassic basins of eastern North America (Klein, 1969) an ancient, example of this early stage.

The rift valley is typically located across a dome and thus a two-way drainage pattern is set up. Regional drainage moves away from the rifts down the flanks of the domes and local drainage moves from the rift valley walls to the bottom of the rifts giving rise to fresh water lakes in which, under suitable climatic conditions, non-marine evaporites may form as in the Afar Depression. In the North American Triassic basins there are granitic conglomerates derived from the nearby fault walls.

The rift valley stage is thus characterized by locally-derived terrigenous river and lake sediments together with non-marine evaporites.

Table 15.1. Stages in the life-cycle of ocean basins and their properties (after J. T. Wilson, 1973; reproduced by permission of J. Wiley)

Stage	Example	Mountains	Motions	Sediments	Igneous rocks
1 Embryonic	East African rift valleys	Block uplifts	Uplift	Negligible	Tholeiitic flood basalts, alkalic basalt centres
2. Young	Red Sea and Gulf of Aden	Block uplifts	Uplift and spreading	Small shelves, evaporites	Tholeiitic sea floor, basaltic islands
3. Mature	Atlantic Ocean	Mid-ocean ridges	Spreading	Great shelves (miogeosynclinal type)	Tholeiitic ocean-floor, alkali basalt islands
4. Declining	Pacific Ocean	Island arcs	Compression	Island arcs (eugeosynclinal type)	Andesitic volcanics, granodiorite-gneiss plutonics
5. Terminal	Mediterranean Sea	Young mountains	Compression and uplift	Evaporites, red beds, clastic wedges	Andesitic volcanics, granodiorite-gneiss plutonics
6. Relic scar (geosuture)	Indus line, Himalayas	Young mountains	Compression and uplift	Red beds	Negligible

Proto-oceanic Gulf

This stage forms as a result of two developments. Firstly, new oceanic crust is added to the rift valley floor due to spreading of the continental borders and, secondly, the floor of the basin subsides relative to the continental walls and the thermally expanded and heightened ridge axis. This is followed by restricted inflow of sea water from a nearby ocean. Four types of sediments may be laid down in this stage:

1. If the latitude (<30° from the equator) and water temperature (>20°C) are suitable, carbonate reefs form in shallow water as on the present margins of the Red Sea. Vogt and Ostenso (1967) calculated that during seafloor spreading continental margins sink at rates of 1–4 cm per 1000 years. Reef growth could keep pace with such slow subsidence and therefore long periods of growth may be expected.

2. Marine evaporite deposits up to 7 km thick form by evaporation of trapped sea water in narrow basins with restricted circulation (Burke, 1975; Kinsman, 1975a). They may be preserved on facing continental margins belonging to oceans like the Atlantic that have evolved via the Red Sea stage (Pautot and coworkers, 1970, 1973) (see further in Chapter 14).

3. Circulation of oxygenated deep-ocean currents is restricted in the Red Sea stage and this gives rise to anoxygenic reducing bottom conditions. Sapropelitic carbonaceous muds and euxinic clays are found in the Red Sea today and in Lower Cretaceous sediments in the North Atlantic.

4. Sediments from the geothermal brine basins in the Red Sea (Fig. 19.15) are enriched in metallic sulphides (Cu, Pb, Zn), iron minerals and Mn-bearing minerals (Degens and Ross, 1969).

Narrow Ocean

The growth of oceans can be divided into narrow and open stages (Dickinson, 1974a). In the early stage the elevated mid-oceanic rise may divide the ocean into two halves with separate depositional histories. Basal clastic terrigenous sediments typically form a seaward-thickening miogeoclinal wedge at the periphery of the continental shelf off eastern USA (Dietz and Holden, 1966) and fossil examples have been suggested in the Cordilleran (J. H. Stewart, 1972), Appalachian (Rodgers, 1970) and Caledonian (Dewey, 1969) fold belts and the Coronation Geosyncline—quartzite phase (Hoffman, 1973).

In the central parts of the narrow ocean calcareous sediments with micro-organisms such as radiolaria, diatoms and foraminifera may be deposited on the rise crest and flanks that lie above the carbonate compensation depth; below that depth pelagic red clay (lutite) or siliceous sediment (chert) may accumulate on the lower flanks and in deeper basins (Berger and Winterer, 1974).

Open Ocean

Three groups of sediments may be deposited along the rifted continental margin of a mature wide ocean (Dickinson, 1974a) (see Fig. 15.9).

1. The basal clastic wedge of the continental terrace is succeeded by a carbonate platform, a modern example of which is the Great Bahamas Bank off eastern USA (Rodgers, 1970). Upon decoupling of the continental–oceanic lithosphere, collapse of the carbonate bank gives rise to a basin–seamount submarine topography (Jenkyns, 1970b) and succeeding deeper water shales starved of clastics.

2. Turbidity currents pass down the continental slope and deposit terrigenous sediments (flysch) on the continental rise.

3. Locally immense thicknesses of clastic sediments may be deposited along rifted continental margins in the form of major river deltas giving rise to progradational continental embankments. Thicknesses of up to 17·5 km are possible if subsidence takes place caused by flexure and isostatic compensation of the underlying lithosphere (Walcott, 1972).

The sedimentation that takes place during the growth of oceans, as described above, is followed within the plate tectonic scheme of events by sedimentation within the three types

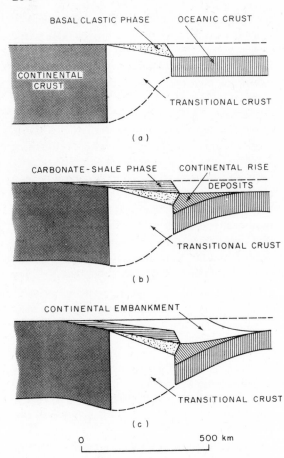

BASAL CLASTIC PHASE   OCEANIC CRUST

CONTINENTAL CRUST

TRANSITIONAL CRUST

(a)

CARBONATE-SHALE PHASE   CONTINENTAL RISE DEPOSITS

TRANSITIONAL CRUST

(b)

CONTINENTAL EMBANKMENT

TRANSITIONAL CRUST

(c)

0   500 km

Fig. 15.9. Idealized diagrams to illustrate the successive depositional phases in the evolution of a rifted continental margin at the interface with a mature ocean. See text for details (after Dickinson, 1974a; reproduced by permission of Soc. Econ. Pal. Mineral.)

of fold belt caused by plate collision (reviewed in Chapters 16, 17 and 18).

Oceanic Islands

There are probably at least 7000 submarine volcanoes higher than 1 km in the Pacific and Atlantic (Engel and Engel, 1970), but only a few stand above sea level as islands for ready examination. Nevertheless, these islands provide one of the most important clues to magma generation in post-early Mesozoic times. Only a few islands are not volcanic, e.g. Seychelles (granitic), St Paul Rocks (mantle ultramafics) and Macquarie (uplifted oceanic crust).

The oceanic islands developed after the oceanic crust on which they sit. It is generally held that the lavas of layer 2 are tholeiitic basalts; the lower parts of the volcanics, mostly submarine, are tholeiitic (Engel, Engel and Havens, 1965); and the upper parts, commonly seen as subaerial islands, consist of alkali basalts enriched in Na and K and their subordinate differentiates (Borley, 1974)—andesine and oligoclase andesite (hawaiite and mugearite), trachyte and phonolite. In other words, the alkaline rocks represent the later stages of evolution. The main demonstration of this stratigraphic relationship was made by Aumento (1968) from rocks dredged from part of the mid-Atlantic ridge; he showed that the basalts changed from tholeiitic with normative olivine to alkaline with normative nepheline with increasing height in the volcanic sequence. Large volcanoes, such as Hawaii, may ideally consist of tholeiite capped by alkali lavas. Alkali-rich basalt comprises 85–99% by volume of the accessible part of most islands and large submarine volcanoes (Engel and Engel, 1970), but the hypothesis that these all pass downwards into less alkaline, and eventually tholeiitic, types must remain conjectural until the lower parts of these volcanoes are sampled.

In 1963 Wilson considered that the islands may have drifted by sea floor spreading and therefore suggested that those furthest from a mid-oceanic ridge should be the oldest. McBirney and Gass (1967) and Middlemost (1973) disagreed with this idea; Engel and Engel (1970) thought it was 'true, at least in the Pacific'; and Baker (1973) concluded that the relationship remained debatable, particularly since quoted ages are only from the upper exposed parts of the volcanoes and thus give no indication of the age of the start of activity. The important question that remains here is whether the volcanoes are the same age as the floor on which they rest.

Most interesting is the fact that the chemical composition of oceanic island lavas may vary systematically with distance from the mid-ocean ridge (McBirney and Gass, 1967). For example, in the South Atlantic there are three groups of volcanic islands (Baker, 1973):

mildly alkaline or transitional (to tholeiitic) basalts occur on islands close to the mid-Atlantic ridge (Ascension, Bouvet), more alkaline types are off, but not far removed from, the ridge (Tristan da Cunha, St Helena, Gough), and extremely undersaturated basanitic lavas occur on islands furthest from the ridge (Cape Verde, Fernando de Noronha). These differences are accentuated in the late differentiates. Similarly, in the Pacific (McBirney and Williams, 1969) the Galapagos and Easter Islands near the East Pacific rise crest have tholeiitic basalts and siliceous differentiates and so contrast with the more distant Tahiti and San Felix islands with basanites and phonolites.

Fig. 15.10 shows that silica saturation decreases with increasing distance from the mid-Atlantic ridge; the most siliceous tholeiitic lavas occur on the ridge crest where the heat flow is highest. However, Borley (1974) produced data that disputed this relationship

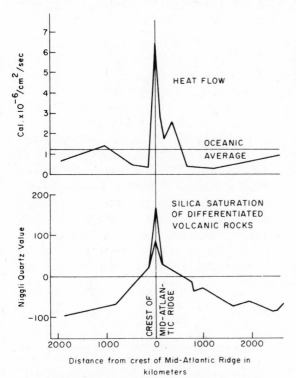

Fig. 15.10. Comparison of heat-flow profiles and silica saturation curve of differentiated rocks across the Mid-Atlantic Ridge (after McBirney and Gass, 1967; reproduced by permission of North-Holland)

(rocks from San Miguel in the Azores and Tristan, 480 km from the ridge, are much more undersaturated than those from Gough, 625 km from the ridge crest) and proposed caution in the interpretation of these undersaturation trends.

Based on the work of McBirney and Gass, Harris (1969) suggested that the basalt magma type and its equilibration depth and degree of fractionation are directly dependent on the geothermal heat-flow environment. Baker (1973) applied this scheme to the South Atlantic oceanic islands and found that near the high heat-flow ridge magmas do not fractionate until they reach shallow levels and tholeiitic lavas are erupted. Away from the ridge, in areas with lower heat flow, equilibration takes place at greater depth and erupted lavas are more evolved and undersaturated.

Green and Ringwood (1967) produced experimental data on the segregation depths of different basalt types. Using this data Baker (1973) suggested that in the South Atlantic the depth of magma segregation and equilibration increased with increasing distance from the mid-oceanic ridge. Thus the mildly alkaline to tholeiitic basalts on Bouvet and Ascension near the ridge crest were generated at depths of less than 35 km, and the strongly alkaline basalts on Trinidade and Cape Verde Islands, furthest from the ridge, were derived from depths of 100 km or more.

Mineralization at Plate Accretion Boundaries

In Chapter 19 we shall review the mineralization in ophiolite complexes, widely regarded as on-land fragments of oceanic crust; we shall restrict ourselves here to four mineralization environments within the present growing oceans:

  geothermal brines in the Salton and Red Seas
  sediments of layer 1 near active oceanic ridges (East Pacific Rise)
  volcanics of layer 2
  Iceland

There are considerable concentrations of a variety of metals in the brines and sediments, and some are of economic potential, but all these examples are of great importance since they establish a direct connection between sea floor spreading and ore genesis.

Geothermal Brines

The Red Sea metalliferous brines and muds occur in three main depressions (12×5 km; $4 \times 2 \cdot 5$ km; $3 \times 0 \cdot 66$ km) along the median valley (Fig. 19.15) (Degens and Ross, 1969). Different layers of the muds are enriched in metallic sulphides–sphalerite and lesser pyrite, galena, chalcopyrite and marcasite; iron minerals such as goethite, haematite, iron montmorillonite and siderite; and Mn-bearing minerals like rhodochrosite. Metal contents reach 85% Fe_2O_3, 21% ZnO, $5 \cdot 7\%$ Mn_2O_3, 4% CuO and $0 \cdot 8\% PbO$.

The brines are very rich in Na and it is widely held that they are of meteoric origin, the high salinity (up to 255,000 ppm) being due to percolation through halite-rich evaporite deposits which are common in the Red Sea.

The Salton Sea brines are noted for their high metal concentrations, in particular Cu, Ag, Au, Fe, Mn, B, Zn and Pb (Skinner and coworkers, 1967).

Sediments near Ocean Ridges

Layer 1 sediments on the flanks of mid-oceanic ridges, in particular the East Pacific Rise, contain considerable quantities of Fe, Mn, Pb, Zn, Ni, Cu and Ba (Boström and Peterson, 1966, 1969; Cronan, 1976). Other enriched elements are U, V, Ag, Sn and Ti (Horowitz, 1970; Boström and Fisher, 1971). The close spatial relationship of these metalliferous sediments with the volcanics of the rift

zones has led to the common view that the metals were precipitated from hydrothermal exhalations of volcanic origin. But, more precisely, how? It is possible that circulating heated sea water picked up the metals when percolating through layer 2 basalts (Corliss, 1971). Alternatively such metalliferous ferromanganoan clays may be the result of reaction of the sea water with the erupting basalts at the ridge (Bonatti and Joensuu, 1966). There are similar metal-rich sediments of Tertiary age in the Pacific and Atlantic, well distant from the mid-ocean ridges, thought to have been carried to their present positions by sea floor spreading (Cronan and coworkers, 1972; Boström and coworkers, 1972). For comparison with umbers on Cyprus see Chapter 18.

Layer 2 Basalts

Dredged basalts from the oceanic ridges have been found to contain copper-bearing sulphides as spherules in vesicles (Moore and Calk, 1971) and in veins (Dmitriev, Barsukov and Udintsen, 1971).

Iceland

The basalt lava flows of SE Iceland are intruded by Tertiary granophyres with sulphides of Cu, Pb, Zn and molybdenite, and by explosion breccia rhyolitic pipes with chalcopyrite (Jankovic, 1972). The sulphides are enriched in Te, Ag, Bi and Cd. δS^{34} analyses of sulphides exhibit near-zero per mil values of $+0 \cdot 7$ to $-1 \cdot 2\%$, indicative of primordial sulphur from the upper mantle. This example from Iceland, situated on the mid-Atlantic ridge, is particularly important as it demonstrates unequivocally that base metal mineralization can be derived from the upper mantle.

Chapter 16

Island Arcs

An island arc is a tectonic belt of high seismic activity characterized by a high heat flow arc with active volcanoes bordered by a submarine trench (Mitchell and Reading, 1971; P. J. Coleman, 1973, 1975; Sugimura and Uyeda, 1973). It forms where a plate of oceanic lithosphere collides with, and is subducted beneath, another oceanic or a continental plate along a Benioff or subduction zone which extends to about 700 km depth and is the focus of shallow (0–80 km) and deep earthquakes. Partial melting of the downgoing slab at 150–200 km depth gives birth to magmas that rise and are extruded in volcanoes located 150–200 km from the trench axis and 80–150 km vertically above the Benioff zone. Other important features of island arcs are sedimentary rocks including, in particular, turbidites, tectonic mélanges on the inner wall of trenches, paired metamorphic belts with high-pressure glaucophane-bearing types, plutonic intrusions (granodioritic) in the deeper roots of the magmatic arcs, Kuroko-type mineral deposits, and a systematic chemical variation of volcanics, plutonics and mineral deposits across the arcs.

In this chapter we shall review some of the main features of island arcs which are important belts of continental construction.

Location and Age

The active arcs of today are situated predominantly on the west side of the Pacific from the Aleutians in the north to the Kermadec–Tonga arc in the south (Fig. 16.1); important also are the Scotia and Antillean (Caribbean) arcs in the Atlantic. Most arcs are intra-oceanic (Marianas, Scotia), some lie close to continental margins (Japan), and the Aleutian arc passes laterally into a Cordilleran-type fold belt. Thus, as Mitchell and Reading (1971) pointed out, there may be transitions from suboceanic to subcontinental lithospheric junctions.

Many of the arcs are active today—there are over 200 Quaternary volcanoes in Japan and the world's earthquakes are largely confined to these areas. The most recent volcanic arcs have been active during the last 10 my since the early Pliocene. According to Spencer (1974) the following were developed primarily in the Tertiary: Macquarie Ridge, Tonga–Kermadec, Fiji, New Hebrides, Solomons, Marianas, Aleutians and the Lesser Antilles. The formation of the Puerto Rico and Cuba arcs was completed in Cretaceous to Palaeogene times.

There is evidence that some arcs were constructed on older subducting plate boundaries (Mitchell and Bell, 1970). The history of the East Indies arcs ranges from the Mesozoic to Recent, whilst in Japan the presence of glaucophane-bearing Ordovician, Carboniferous, Permian–Jurassic, and Jurassic-Cretaceous successions, each with contemporaneous low-pressure metamorphic belts, is indicative of a long history of lithospheric subduction beneath this arc system (Kimura, 1973), and tectonic activity continues today (Miyashiro, 1973a) (Table 16.1).

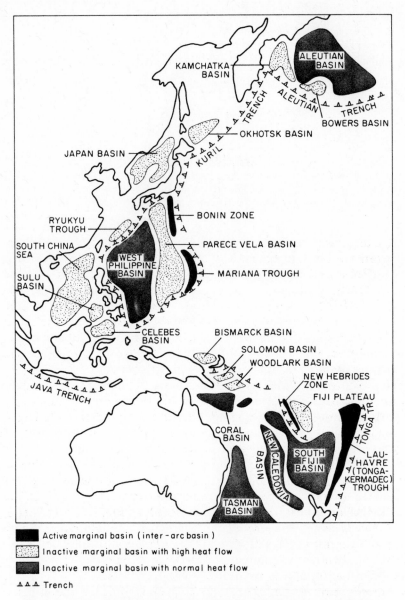

Active marginal basin (inter-arc basin)

Inactive marginal basin with high heat flow

Inactive marginal basin with normal heat flow

▲▲▲ Trench

Fig. 16.1. Distribution of trenches and, in particular, marginal basins in the western Pacific showing the threefold classification based on increasing age, depth and crustal heat flow (based on Karig, 1974; reproduced by permission of Annual Reviews Inc.)

Subduction Zones

Seismic foci show that subduction zones dip from a trench towards an arc at an average of 45°, the seismic activity being concentrated on the upper surface of the downgoing slab of lithosphere, thus defining the 'seismic plane' or Benioff Zone which may be up to 50 km wide. Fig. 16.2 shows a vertical section through the northern Isu–Bonin arc.

Earthquakes vary from shallow to deep. Focal mechanisms for shallow earthquakes below the seaward slope of the trench indicate the presence of extensional stresses normal to the trench axis in the sediments and crust in the upper 75–100 km of the lithospheric slab

Table 16.1. Chronology of regional metamorphic events in the Honshu arc of Japan (after Miyashiro, 1973a; reproduced by permission of the American Journal of Science).

Event No.	Geologic age	High-pressure metamorphic area (radiometric age)	Broadly simultaneously metamorphosed area of low-pressure type with granites (radiometric age)
6	Miocene to the present	Not exposed yet (?)	Northeast Japan (3–25 my)
5	Jurassic to Cretaceous	Sanbagawa belt (110 my); Part of Shimanto terrane	Ryoke belt (c. 105 my); Abukuma Plateau (100–120 my)
4	Permian to Jurassic	Sangun belt of western Honshu (160 my)	Hida IV (180–240 my)
3	Carboniferous	Circum-Hida zone including Omi area (320 my)	Hida III (320 my) Maizuru zone (330 my)
2	Ordovician	Kiyama (450 my)	Hida II (500 my, Cambro-Ordovician); Kurosegawa zone (430 my)
1	Precambrian		Hida I (1640 my)

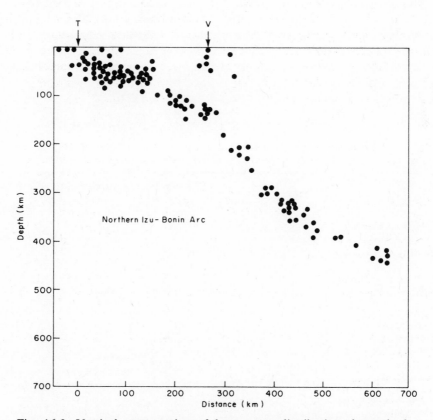

Fig. 16.2. Vertical cross-section of hypocentre distributions beneath the northern Isu–Bonin arc. T = trench, V = recently active volcanic chain. Distance measured from trench axis. (After Condie, 1976b; reproduced by permission of Pergamon Press)

(Stauder, 1968). A tensional zone exists where the layer bends sharply to plunge beneath the arc trench. The most active shallow seismic zone lies mainly in the crust on the landward side of the trench, where the slip vector of earthquake foci lies on the plane of shallow dip of the arc and normal to the trench, indicating underthrusting of one arc slab beneath the other. The focal mechanisms of intermediate-to-deep earthquakes indicate that the maximum compression axis is usually parallel to the dip of the seismic zone and that the downgoing slab at depth is under compression and meeting resistance (Sykes, Oliver and Isacks, 1970; Isacks and Molnar, 1971).

There is a close relationship between the amount of seismic activity, depth, seismic velocities, depth of trench, and the rate of arc convergence. Earthquake activity in the upper 100–200 km decreases progressively with depth. This decrease correlates with a decrease in seismic velocities, and viscosity and variations in partial melting temperatures and ductile/shear deformation ratios. There is a positive correlation between the depth of seismic activity and the slip rate (cm/yr) of plate convergence: the deepest earthquakes

occur in arcs with the greatest rate of under-thrusting (Isacks, Oliver and Sykes, 1968; Sykes, Oliver and Isacks, 1970). Table 16.2 from Miyashiro (1972) lists data from some well-known arcs and shows that there is clear correlation between the depth of the trench and earthquake depth and rate of plate motion; the petrological information on this table will be discussed later.

Fig. 16.3 shows the relationship between the structure of a typical arc system and gravity anomaly profiles. There is a gravity minimum reaching about -200 mgal on the inner side of the trench, and a positive anomaly of the same amplitude on the island side reaching a peak in the arc–trench gap. The positive anomaly is widely ascribed to the higher density of the cool downthrust lithospheric slab, whilst the negative is attributed to the accumulation and subsequent underthrusting (especially below the inner slope) of sediments in the trench (Hatherton, 1974). Fig. 16.3 also shows the typical heat-flow values across arc systems (for theoretical studies of the expected patterns induced by the subducted lithospheric slab, see Minear and Toksöz, 1970; Oxburgh and Turcotte, 1970).

Table 16.2. Correlation between the activity and volcanic rock series in island arcs; see text for discussion (after Miyashiro, 1972; reproduced by permission of the American Journal of Science)

Group	Arc and trench	Rate of plate convergence (cm/yr)	Maximum depth of earthquakes (km)	Maximum depth of the trench (km)	Volcanic rock series
	Tonga	9	700	11	Th
I	Izu-Bonin	9	600	11	Th + (C) + (A)
	Northeast Japan	9	600	11	Th + C + (A)
	Kurile-kamchatka	8	600	10	Th + C + A
	Aleutian	6	300	8	(Th) + C + A
	Indonesia	5–6	600	7	(Th) + C + A
II	Ryukyu	?	300	7	(Th) + C
	North Island (New Zealand)	Slow (3?)	300	4	C + A
	Hellenic (Aegean)	Slow (3?)	200	4	(Th) + C + A
III	Calabrian (Sicily)	Very slow (2?)	300	Buried	A
	Macquarie	Very slow	100	Shallow	(Th) + (C) + A

Note: The activity of arcs decreases in the order: group I → II → III. Th = tholeiitic series, C = calc-alkali series, A = alkali series. The rocks of the series shown in parentheses ( ) are not typical of the series and are very small in quantity. The rates of plate convergence are mainly after Le Pichon (1968).

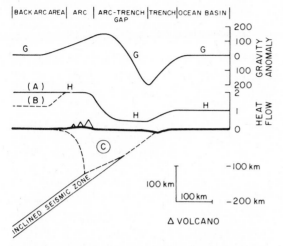

Fig. 16.3. Sketch showing characteristic geophysical features of arc systems in profile. Heavy line, with volcanoes, is topographic–bathymetric surface. Line G is gravity anomaly profile. Line H is heat-flow profile; in retroarc area line A is the observed pattern for marginal seas, line B is the inferred pattern for thrust belts of foreland basin behind fringing arcs. Triangular space C, bounded within dashed lines, is region of diffuse shallow seismicity beneath main arc and arc trench gap. The inclined seismic zone may vary in dip from 20° to nearly vertical, with 30°–60° dips most common, and is undulatory rather than strictly planar. Vertical distance from volcanoes to hypocentres of intermediate depth tends to lie within the range 100–175 km. (From Dickinson, 1973b; reproduced by permission of the University of Western Australia Press)

Types of Arcs and their Structure

There are several types of island arc. Some are clearly intra-oceanic (Tonga, Solomons, New Hebrides, Macquarie Ridge, Scotia, Marianas and Lesser Antilles); some are separated from sialic continent by narrow semi-oceanic basins (Japan, Kuril, Banda, Andaman islands and Sulawesi); some pass laterally into a Cordilleran-type fold belt (Aleutian); and some are built against continental crust (Sumatra–Java). Thus there are transitional types between intra-oceanic arcs and Andean-type continental margins in which the volcanic belt is an integral part of the continental landmass (Mitchell and Reading, 1969, 1971; Dickinson, 1974b). Fig. 16.4 illustrates diagrammatically a cross-section through a typical intra-oceanic arc.

There is a systematic variation in structural zones as one passes across the strike of an island arc (Dickinson, 1971a,b). The trench is typically 50–100 km wide and may be up to 11 km deep (Mariana), and contains pelagic and turbidite sediments, a tectonic mélange of thrusted oceanic and terrigenous sediments and layer 1–3 ophiolites, an inactive arc–trench gap 50–250 km wide with diverse sediments, and a magmatic arc 50–100 km wide composed of high-level volcanics and deep-level plutonic intrusions. The back-arc or

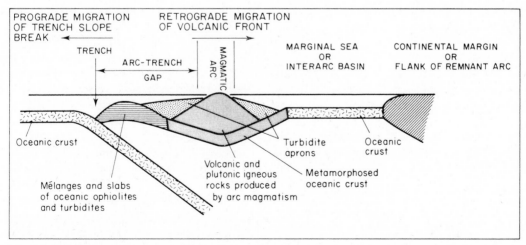

Fig. 16.4. Diagrammatic transverse crustal section through an idealized intra-oceanic island arc (modified and redrawn after Dickinson, 1974b). Dual migration of trench and arc (after Dickinson, 1973a; reproduced by permission of the Society of Economic Paleontologists and Mineralogists)

inter-arc basin (often called a marginal sea) is sited on thin oceanic crust which may have grown by back-arc spreading (Karig, 1970, 1971, 1972; Packham and Falvey, 1971).

There are two types of intra-oceanic arc:

Those with thin sub-arc crust, e.g. 15–25 km total thickness in the Kuril Islands and 15 km in the Marianas and Tonga–Kermadec arcs; these have no pre-arc sialic basement.

Those with thick sub-arc crust. The Japanese arcs have a crustal thickness of 30–35 km and are underlain by a layer with seismic velocities typical of continental crust (Rikitake and coworkers, 1968). This may contain a large component of intrusive plutonic arc rocks (Mitchell and Reading, 1971) and/or a segment of detached continental basement (Matsuda and Uyeda, 1971).

There is an evolutionary sequence amongst island arcs which is related to their type of volcanism (Baker, 1968). Young arcs such as South Sandwich and Tonga are composed largely of basalt and basaltic andesite, but more evolved arcs, like Japan, Kamchatka, Aleutians, Indonesia and the Lesser Antilles, consist predominantly of andesite.

The widths of arc–trench gaps in arc systems are proportional to the past duration of magmatic activity in the main arc (Dickinson, 1973a). This growth is ascribed to prograde or outward migration of the trench and subduction zone away from the arc–trench gap plus an inward migration of the magmatic arc axis away from this region (Fig. 16.4).

The dips of the downgoing lithospheric plates beneath island arcs are variable. In general, individual subduction zones tend to dip more steeply with depth. Luyendyk (1970) suggests that the slower the rate of consumption of a plate the steeper will be its angle of entry. Because the downgoing lithospheric plate is heavier than the asthenosphere, it tends to sink passively to an equilibrium at a near-vertical position. With time, therefore, the dip at depth increases.

Volcanism

Volcanic rocks make up the most important part of island arcs. The smaller islands give somewhat isolated information of the most recent volcanic activity, but the larger islands have a long history, even back to the Ordovician in the Honshu arc of Japan (Miyashiro, 1973a). This incidentally suggests that the islands have grown by progressive addition and accumulation of volcanic debris from many successive arc systems. But it also means that one has to be careful to take account of the precise age of the volcanic successions when considering their chemical variation across the arcs.

There are three series of volcanic rocks in island arcs (Kuno, 1966; Miyashiro, 1972, 1973a; Jakes and White, 1972):

1. Tholeiitic Series, including tholeiitic basalt, icelandite (andesite), and some dacite. SiO_2 contents range from 48–63% by wt. The rocks contain groundmass augite, pigeonite and sometimes orthopyroxene but no hornblende or biotite, and show little or no increase in SiO_2 but considerable increase in total FeO with crystallization.

2. Calc-alkali Series with high alumina basalt and abundant intermediate rocks (andesite and dacite) with some rhyolite. SiO_2 contents are in the range 52–70% by wt. They contain orthopyroxene but no pigeonite in the groundmass and may have hornblende and biotite but show little iron-enrichment during fractional crystallization.

3. Alkali Series, divisible into:

   a) the sodic group with alkali olivine basalt, hawaiite and mugearite (alkali andesites), trachyte and alkali rhyolite;

   b) the shoshonite group including shoshonite (with K_2O/Na_2O ratios near unity), latite and leucite-bearing rocks. These are undersaturated with respect to SiO_2 and typically contain alkali feldspars and feldspathoids.

Ringwood (1974) summarizes (Table 16.3) the geochemical trends of island arc volcanism indicated by Jakes and Gill (1970), Fitton (1971) and Jakes and White (1972). The first stage is often characterized by the tholeiitic series possessing low abundances of incompatible elements (e.g. K, U, Ba, REE) and unfractionated (relative to chondrites) rare earth patterns. On the other hand, the more

Table 16.3. Characteristics of island arc tholeiites and calc-alkaline suites (after Ringwood, 1974; reproduced by permission of Geological Society of London)

	Island arc tholeiites	Calc-alkaline suite
Examples	Mariana, Tonga, Izu, S Sandwich arcs	West Indies, Indonesia, Aleutians, Japan, Kamchatka, New Zealand
Stage	Early	Late
Dominant magma	Basalt, basaltic andesite	Andesite
SiO_2-mode	55%	60%
Fractionation	Tholeiitic	Calc-alkaline
Cr, Ni, Mg, Ti	Low	Low
K/Na	Low	High
Rare earths	Unfractionated, chondritic	Strongly fractionated Light REE enriched
Incompatible elements e.g. Ba, La, U, Th, Zr, Ta.	$10–30 \times$ chondritic	$30–100 \times$ chondritic

evolved stages of volcanism are characterized by the calc-alkali series with higher abundances of incompatible elements for a given SiO_2 content and with fractionated rare earth patterns.

In mature island arcs such as Japan and Kamchatka there is a progression from the tholeiitic to calc-alkali and alkali series from the oceanic to the continental side (Fig. 16.5) (Kuno, 1966; Jakes and White, 1969). In these and some less mature island arcs (i.e.

Indonesia and the Kuril Islands) the most characteristic variation is an increase in the alkali content towards the continent in island arc tholeiites and calc-alkaline suites; that is, for rocks with a similar SiO_2 content an increase in K_2O, $Na_2O + K_2O$ and K/Na ratio (Dickinson, 1968; Hatherton and Dickinson, 1969). Fig. 16.6 shows the positive correlation between the potassium contents of lavas plotted against the depth of the volcanoes to the Benioff zones.

The variation in composition of the eruptive products of the primary magmas correlates closely with the pattern of isobaths of mantle

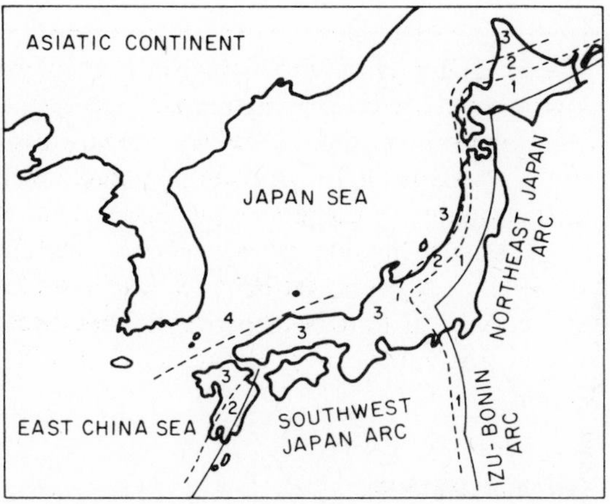

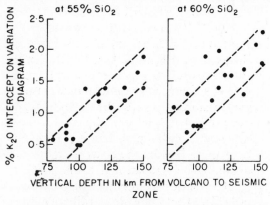

Fig. 16.5. The progressive change in composition of Quaternary volcanic rocks in Japan across the arc. 1. Tholeiitic and calc-alkali rocks. 2. Calc-alkali rocks. 3. Calc-alkali and sodic alkali rocks. 4. Sodic alkali rocks. (After Miyashiro, 1972; reproduced by permission of the *American Journal of Science*)

Fig. 16.6. Plot of potash levels in selected arc lavas versus the depths from corresponding volcanoes to the Benioff zone beneath the arcs (from W. R. Dickinson, *J. Geophys. Res.*, **73**, 2261–2269, 1968, copyrighted by American Geophysical Union)

earthquake foci. This compositional zonality may be explained by the difference in the depth of magma generation. Thus the primary magma composition is most siliceous and least alkaline at the volcanic front where the depth of magma generation is the shallowest (Sugimura and Uyeda, 1973).

There is also a close correlation between the three chemical series and the degree of tectonic activity of different types of arcs. Table 16.2 shows that in the youngest (least mature) group 1 arcs, with the deepest earthquakes and highest rate of plate convergence, the tholeiitic series is typical and the calc-alkali and alkali series only occur in the mature stages of development. In more mature group 2 arcs the calc-alkali series is typical, the alkali series present but unimportant, and tholeiitic volcanics uncommon. In group 3 arcs, with the slowest rate of plate convergence and shallowest earthquakes, alkali volcanics are common and the other two series atypical. The systematic abundance variation in elements, elemental ratios and isotopes across a typical arc are as follows (J. Gill, 1974):

these granodiorites decreases from the oceanic to the continental side of the arcs, just like the Quaternary volcanics (Taneda, 1965, quoted in Miyashiro, 1973a).

In intra-oceanic arcs the earliest volcanics were probably submarine but they soon built up large strato-volcanoes with the result that many of the youngest arcs consist of chains of volcanic islands (Kuril, Marianas, Tonga–Kermadec). Uplift and erosion has given rise to larger islands which contain a long history of older arc complexes (Japan).

The petrological evolution of island arcs is reviewed by Ringwood (1974). There are two stages of development, as summarized in Fig. 16.7:

1. Amphibolite from the subducted oceanic crust is dehydrated at 70–100 km depth. The water produced causes partial melting in the upper mantle wedge above the Benioff zone, so giving rise to magmas that differentiate to produce the early tholeiitic series.
2. With further subduction of oceanic crust to 100–150 km depth serpentinite is dehydrated. The quartz eclogite oceanic crust is partially

---

continent $\longrightarrow$ *ocean*

Fe (max), heavy REE, Y, K/Rb, Na/K, Sr^{87}/Sr^{86}

K, Rb, Ba, Cs, P, Pb, U, Th, light REE, Th/U, Rb/Sr, La/Yb

$\longleftarrow$

In island arcs separated from continental crust by marginal basins there seems to be a correlation between the composition of erupted lava and the type of crust beneath the enclosed seas (P. E. Baker, 1972). When the crust is closer to the oceanic type (like the Bering Sea behind the Aleutian arc) the lavas are gradational towards the tholeiitic type. On the other hand, where the back-arc crust has more continental affinities (as in the Indonesian region), the lavas are of more continental type, gradational towards the calc-alkaline series common along continental margin fold belts like the Cordillera of western America.

Older, more eroded, arcs have some granitic rocks exposed; they account for 15% of the surface area of Japan. The calcium content of

melted giving rise to rhyo-dacite magmas which react with overlying mantle pyrolite to form pyroxenite, diapirs of which rise upwards and partially melt to produce magmas which fractionate by eclogite crystallization at 80–150 km depth and amphibole crystallization at 30–100 km thus creating the calc-alkaline series. Note that both processes operated at high water vapour pressures (see also T. H. Green, 1973).

Sedimentation

Sediments are an important component of island arcs. Within the plate tectonic framework they are deposited in four areas: the trench, trench–arc gap, arc and marginal basins.

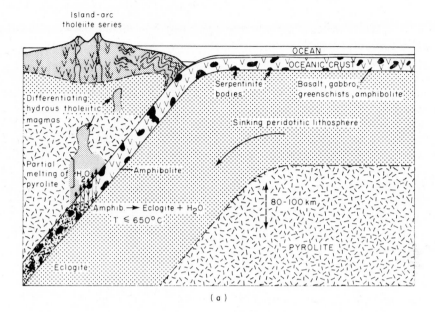

(a)

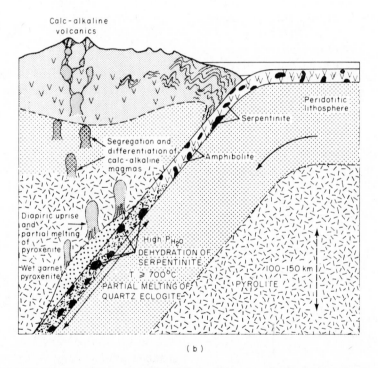

(b)

Fig. 16.7. Early and late stages in the petrological evolution of an island arc. (a) Involves dehydration of subducted amphibolite, introduction of water into the overlying wedge and generation of the island arc tholeiite igneous series. (b) Involves partial melting of subducted oceanic crust and reaction of liquids with mantle above Benioff zone leading to diapiric uprise and formation of calc-alkaline magmas. (After Ringwood, 1974; reproduced by permission of the Geological Society of London)

The Trench

According to plate tectonic theory the trenches should contain pelagic oceanic sediments (and thrusted slices of ophiolite) off-scraped into the mouth of the subduction zone with an accumulation rate of 1 km/my as estimated by Oxburgh and Turcotte (1971). According to Bogdanov (1973) the average thickness of sediment (mostly greywacke turbidites) in the recent trenches of the western Pacific ranges from 1–3 km, but Scholl and Marlow (1974) found the island arc trenches along the western Pacific to be largely devoid of sediment and concluded from this that if there were any offscraped sediments they must have been transported down the subduction zone (see further in Chapter 17).

But there may be another reason for the remarkable paucity of some trench sediments. According to Heezen and Hollister (1971) the South Sandwich trench lacks sediments but does contain ripple and scour marks as well as crag-and-tail structures which led them to the conclusion that the Antarctic bottom currents have swept the trench sediments into the Argentine basin.

The Trench–Arc Gap

Sediments deposited in forearc basins within trench–arc gaps are flat lying and mostly undeformed and unmetamorphosed. Some modern gaps are submarine (central Aleutians, Tonga–Kermadec) and some partly submarine and partly subaerial (Japan). The following types of sediment are possible (Dickinson, 1974a,b):
1. Fluviatile–deltaic–shoreline complexes of subsiding coasts.
2. Shallowing marine sediments on unstable shelves.
3. Slope-covering and trough-filling turbidites.

The sediments are mostly derived from the nearby arc volcanoes (volcaniclastic) or by erosion of uplifted basement metamorphic and plutonic rocks. Examples include the late Mesozoic Great Valley sequence of California (Bailey, Blake and Jones, 1970) and the late Palaeozoic–early Mesozoic western marginal facies of New Zealand (Landis and Bishop, 1972).

The Arc

In intra-oceanic arcs the volcanoes are commonly submarine and the sediments mostly of volcanic derivation. According to Dickinson (1974b) three sedimentary facies types may be expected in intra-arc basins:
1. A central facies near the eruptive centres of pyroclastics with andesites and pillow lavas, bounded by a proximal facies such as biogenic reefs on the flanks of partly submerged volcanoes;
2. A dispersal facies as aprons and blankets of clastic sediment;
3. A basinal facies of volcaniclastic turbidites, marine tuffs and submarine ash flows.

Examples of intra-arc sediments include late Tertiary marine beds on the Tonga Ridge (Karig, 1970), late Tertiary marine sediments capped with lacustrine deposits in Honshu, Japan (Matsuda, Nakamura and Sugimura, 1967), late Cenozoic volcaniclastic marine beds (10 km thick) in the Kuril arc (Markhinin, 1968), and similar rocks in Fiji (Dickinson, 1967), the early Miocene Matanui Group in the New Hebridean arc (Mitchell, 1970), and the late Cretaceous–early Tertiary beds of the Greater Antillean island arc (Donnelly, 1964).

The Marginal Basin

There is no simple pattern of sedimentation in marginal basins adjacent to continents except for large inputs of terrigenous material deposited by rivers, but in basins remote from such terrigenous debris there is a regular pattern of sedimentation dominated by volcanic input (Karig and Moore, 1975). Against the volcanic chain accumulate volcaniclastic aprons several kilometres thick, beyond the distal ends of which is deposited either pelagic brown clay rich in montmorillonite, glass and phenocrysts, or calcareous biogenous ooze.

Within the back-arc area Dickinson (1974a) distinguishes *inter-arc basins*, floored by oceanic crust which receive turbidite wedges of volcaniclastic beds shed from the main arc, from *retro-arc basins*, floored by continental basement that contain fluvial, deltaic and marine beds up to 5 km thick mostly derived from the uplifted area in the fold–thrust belt behind the arc.

Paired Metamorphic Belts

Island arcs, particularly around the Pacific, are characterized by paired metamorphic belts of similar age but different type (Miyashiro, 1961). Typically an inner low-pressure belt

with andalusite and associated granitic rocks lies in and below the magmatic arc, and an outer high-pressure belt with glaucophane and associated ophiolites and serpentinites occupies the outer part of the arc–trench gap near the inner wall of the trench. Thus the relative position of the paired belts gives an indication of the dip direction of the subduction zone. A long-standing anomaly in this respect has been the pair on Hokkaido Island in Japan (Fig. 16.8) where the high P belt lies on the continental side of the low P belt, suggesting the presence of a 'reversed arc' with

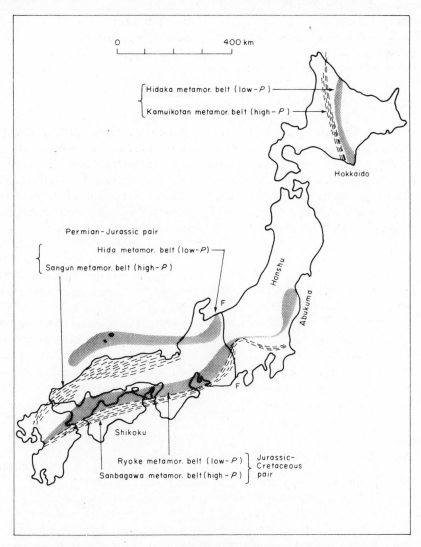

Fig. 16.8. Three pairs of regional metamorphic belts in Japan (after Miyashiro, 1973a; reproduced by permission of the *American Journal of Science*)

a Benioff zone dipping eastwards to the ocean in the opposite direction to the other Japanese and west Pacific arcs. This anomaly was resolved by Okada (1974) who demonstrated that these belts belong not to each other but to two further belts, giving two paired belts each sited on westerly-dipping subduction zones.

Parts of the most recent (Miocene–Present) paired belts are not yet exposed (e.g. Honshu arc, Japan) and thus the best developed pairs are in Mesozoic–Cenozoic arcs. Probably the best documented examples occur in Japan (Fig. 16.8) (Miyashiro, 1973a). There are 14 paired belts around the Pacific (Fig. 19.9) including the following:
1. New Zealand—Jurassic/Cretaceous (Landis and Coombs, 1967)
2. California—late Jurassic/late Cretaceous (Hamilton, 1969)
3. Washington State—Permian/Trias (Misch, 1966)
4. Chile—late Palaeozoic (González-Bonorino, 1971).
As these examples show, they occur in both intra-oceanic and continental margin arcs, the older paired belts occurring in the larger islands and the continental margins which have undergone considerable uplift and erosion.

The high-pressure/low-temperature belts are caused in general terms by the tectonic underthrusting in the trench and a resultant very low geothermal gradient, whilst the low-pressure/high-temperature belt is associated with the steep geothermal gradient in the vicinity of the magmatic arc caused by high heat flow and ascent of new magmas through the crust. Miyashiro (1974) estimated the following geothermal gradients:

Low-pressure metamorphism: $>25°C\ km^{-1}$
High-pressure
metamorphism: $\sim10°C\ km^{-1}$, or less.

The low P/T type is characterized by andalusite and sillimanite, whilst cordierite is common and pyralspite garnet is rare; the high P/T type typically has glaucophane, jadeite and lawsonite. Both types range from greenschist to amphibolite facies and in Japan the progressive mineral changes through these facies for both types are well documented (Banno, 1964; Katada, 1965; Miyashiro, 1958). Ernst (1965) described the mineral parageneses in the Franciscan assemblage of California and comparative studies of the Japanese and Californian rocks were made by Ernst and Seki (1967) and Ernst and co-workers (1970).

The formation of the paired metamorphic belts is reflected in the present heat-flow distribution calculated by Uyeda and Vacquier (1968) across the Japanese Island Arc. Near the trenches heat-flow values are below $1\ \mu$ cal. $cm^{-2}\ sec^{-1}$, and the values increase towards the inner side of the arc where they reach $2·5\ \mu$ cal $cm^{-2}\ sec^{-1}$ in the Okhotsk and Japan marginal seas.

Mélanges

Mélanges are tectonically chaotic rock units that are most important products of plate collision. There are two types of mélange, tectonic and sedimentary (or olistostrome); both are found on the outer side of the trench–arc gap near the landward side of the trench and both are related to the underthrusting tectonics at convergent plate margins.

Olistostromes are sedimentary stratigraphic units of limited scale which contain numerous exotic and native, largely sedimentary, blocks, usually in a pelitic matrix. They formed by subaqueous gravity sliding in a topographic depression such as on the side of a trench—they are bedded mud slide deposits (Marchetti, 1957; Abbate, Bortolotti and Passerini, 1970). The Dunnage mélange of Newfoundland (Horne, 1969; Kay, 1972) may be a Lower Palaeozoic olistostrome formed in a trench related to subduction of the Iapetus ocean (Dewey, 1969) and the Gwna mélange of Anglesey, Wales, may be an olistostrome associated with a late Precambrian subduction zone (Wood, 1974).

Tectonic mélanges are rock bodies composed of tectonically-mixed exotic blocks in a sheared shaly matrix; they may reach enormous size, some blocks ranging up to several hundred metres in length (Hsü, 1968). The Franciscan mélange of California is probably the best known. It contains blocks of ophiolites from a mid-oceanic ridge, pelagic cherts and radiolarites from the ocean floor, deepwater turbidite greywackes poured into a deep-sea trench, and eclogites and glaucophane schists formed by high P/T metamorphism. These rocks from widely diverse origins are thought to have been intermixed tectonically in the mouth of a late Mesozoic subduction zone (Hamilton, 1969; Hsü, 1971a; Blake and Jones, 1974). The Franciscan mélange occurs at a continent–ocean plate boundary which will be discussed in Chapter 17, but parts of this Cordilleran fold belt did form from old trench material belonging to an island arc system (Hsü, 1971a).

Other examples of tectonic trench mélanges preserved as remnants in continent–continent collision fold belts are in the Swiss Alps (Bearth, 1967), the Apennines (Bailey and McCallien, 1963), the Ankara mélange in Turkey (Bailey and McCallien, 1950), and the Exotic Block zone of the Himalayas (Gansser, 1964).

One problem outlined by Hsü (1974) is that if sedimentary olistostromes are sheared by later deformation they may be difficult to distinguish from tectonic mélanges, particularly if an olistostrome is incorporated into a mélange. For example, the argille scagliose of the Apennines is interpreted by Maxwell (1959) as an olistostrome, but Hsü (1974) considers that it is a tectonic mélange containing allochthonous slabs of olistostrome beds.

Most blueschist–mélange terrains exhibit a clear decrease in depositional and metamorphic age proceeding towards lower grade rocks situated successively farther from the major suture zone (Ernst, 1975). The higher grade rocks are more intensively deformed than the lower grade or unmetamorphosed rocks disposed nearer to the coast. The tectonic model in Fig. 16.9 shows how higher grade, more deeply subducted, older, landward mélange units tend to be thrust ocean-

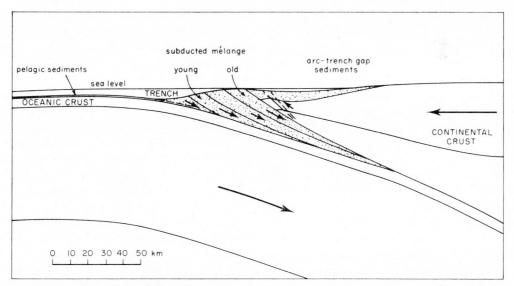

Fig. 16.9. Interpretive cross-section of a convergent plate junction showing successive imbrications of younger material within basal portions of the trench mélange. As underplating continues, the inclination of the plate junction apparently decreases and moves seaward away from the accreting stable plate margin. Underplating results in the elevation, exposure and erosion of older, once deeply subducted and recrystallized portions of the terrain. Low-grade metamorphic portions of the underplated younger rocks are exposed seaward, whereas the coeval higher grade, more deeply subducted sections generally are exposed only subsequent to the erosional stripping of the overlying older rocks. (After Ernst, 1975; reproduced by permission of Dowden, Hutchinson and Ross)

ward over the more weakly recrystallized, nearer-surface, younger, seaward mélanges. In other words, underplating results in the elevation and exposure of older, once-deeply subducted portions of the trench mélange. This style of successive underthrusting is well demonstrated in the Aleutian (Grow, 1973) and the Lesser Antilles arc (Westbrook, Bott and Peacock, 1973). The accumulation of the underplated mélanges and buoyant rebounded tectonic slices along the front of the nonsubducted plate should cause a decrease in the angle of dip and an episodic seaward stepping of the subduction zone (Ernst, 1975).

Marginal Basins

The term 'marginal basin' or 'sea' is used to refer to two different types of geological environment; that described in Chapter 14 formed near a plate accretion boundary and is found today along the coasts of facing continents, such as Brazil and West Africa. The type described here lies behind island arc systems and is best seen along the western side of the Pacific, from the Aleutian Basin to the Tasman Sea (Fig. 16.1) (Karig, 1974). These basins are underlain by oceanic crust and are situated between the main magmatic arc and the corresponding continental margin (e.g. Japan Basin) (Kaseno, 1972; Hilde and Wageman, 1973) or another island arc–trench system (e.g. Philippine Basin and Fiji Plateau).

The first attempt to interpret marginal basins in terms of modern plate tectonic theory was by Karig (1970) who explained the origin of the Lau–Havre trough west of the Tonga–Kermadec trench by extensional rifting. In 1971 Karig, Packham and Falvey, and Matsuda and Uyeda all proposed from indirect evidence that marginal basins formed by a type of sea-floor spreading which caused the opening of the area behind the arcs. The extensional origin for marginal basins has gained wide acceptance in recent years, in contrast to the early idea of Umbgrove, Beloussov and others which postulated an origin through the subsidence of continental crust.

Karig (1971, 1974) divided marginal basins into actively-spreading and older (mostly early to late Tertiary) inactive types (Fig. 16.1). Active basins have a very high heat flow and occur in five arc systems (Tonga, Kermadec, Mariana, Bonin and New Hebrides). The remaining basins are inactive and are subdivided into two groups: an older one with normal heat flow (e.g. South Fiji, West Philippine and Aleutian), and a younger with high heat flow that has yet to cool back to normal (e.g. South China, Kamchatka, Okhotsk, Japan and Parece Vela).

The type and origin of volcanic material in active marginal basins is not clear. Pillow basalts occur in the axial zone of the Mariana and Lau basins (Hawkins, 1974) but the axial zone has no well-defined central rift or axial symmetry. Volcanism may be related to tilted fault blocks, suggested by basin morphology of linear ridges and troughs, but data on this subject are rare (Karig, 1974).

Sedimentation rates in active marginal basins are very high but there is a remarkable paucity of sediment. Accumulation rates of volcaniclastic aprons extending from the volcanic chain are in excess of 100 m/my (Deep Sea Drilling Project Reports quoted in Karig, 1974). Montmorillonitic clays, fine-grained volcanic detritus, foraminiferal and calcareous nannofossil oozes are the predominant sediments.

Inactive marginal basins have a basement morphology similar to the active basins, but older basins have a thick veneer of sediment that ranges from 100 m of pelagic mud in those protected from continental sources (e.g. Philippine Sea) to several kilometres of terrigenous turbidites in those flanked by a continent (e.g. Bering Sea).

The depth to basement in marginal basins increases with age from about 2·5 km in active to more than 5·5 km in older basins. This increases in depth with age, accompanied by a fall in crustal heat values from more than 3 HFU to normal, is similar to that recorded in the main oceans (Karig, 1974). Anderson (1975) shows that the active Mariana Basin has a central zone of high heat flow (4·50 HFU to 8·50 HFU at a fracture zone intersection) corresponding to an axial topographic high and a flank zone of low heat flow (0·00 to 0·90 HFU). These results suggest that the axial high with its high thermal anomaly is a narrow zone of magmatic intrusions which may be the centre of sea-floor spreading activity. Thus marginal basins may be generated by a spreading mechanism not fundamentally different from that operating in mid-oceanic ridges.

However marginal basins do not have well-developed symmetrical linear magnetic anomaly patterns like those in the main ocean basins. Linear anomalies do exist but they are largely incoherent or have amplitudes only half of those generated at mid-oceanic ridges. The reasons for the different magnetic characteristics of marginal basins and main oceans are not clear; after consideration of several possibilities Karig (1974) suggests that the most reasonable cause is a type of crustal extension in marginal basins that is different from the passive spreading at mid-oceanic ridges.

There is some analogy between the extensional marginal basins in island arcs and a mature continental system such as New Zealand, Kamchatka, Sumatra, Java and central America, where the tectonic position of the back-arc basin is occupied by a volcanic rift zone with prominent welded and non-welded silicic tuffs. The Basin and Range Province of the western USA may be such an ensialic back-arc basin (Scholz, Barazangi and Sbar, 1971). In island arc systems such as the Philippine Sea this early volcanism occurs as dacite and rhyolite (Karig and Glassley, 1970). Active diapirism above the Benioff zone may provide the excess heat required for partial melting in the lower crust to account for the generation of such silicic magmas in early stages of extension.

Further results of back-arc spreading may be a decrease in dip of the Benioff Zone (Karig, 1971) and the cuspate curvature of the arc system (Matsuda and Uyeda, 1971).

Mineralization

Across island arcs there is a succession of mineral deposits that correlates with the variation of rock types and geological structure; much of the mineralization is characteristic of, or indeed unique to, consuming plate boundaries of the island arc type. There is considerable complexity because of the superimposition of arc systems, and thus to understand the metallogeny one may have to separate in a mature arc the successive developments from the late Palaeozoic to the Quaternary. The following metallogenic epochs are recognized:

Japan	West Pacific in general
(Tatsumi, Sekine and Kanehira, 1970)	(Zonenshain and coworkers, 1974)
1. Permian–Carboniferous	Early Mesozoic (Trias–mid-Jurassic)
2. Late Jurassic–early Cretaceous	Late Mesozoic (Late Jurassic–early Cretaceous)
3. Cretaceous–Palaeogene Tertiary	Late Cretaceous–early Palaeogene
4. Neogene Tertiary	—
5. Quaternary	—

Firstly, the volcanic ores. According to Tatsumi, Sekine and Kanehira (1970) the mineralizations in the geosynclinal stages are simple, consisting of three types: Cu–pyrite, Cu–Pb–Zn and Fe–Mn. On the other hand, the late- to post-orogenic mineralizations are characteristically polymetallic. Some kinds of mineralizations are associated with certain types of volcanic activity, for example, the Cu–pyrite and Fe–Mn occur with basic submarine volcanics, while Cu–Pb–Zn deposits are related to acid submarine volcanics. None of these three ore types is closely associated.

There are two principal types of sulphide ore developed during the calc-alkaline stage of volcanism (Fig. 16.10).

1. BESSHI-TYPE ORES In the early main stage of arc development Besshi-type stratiform sulphide deposits are associated with andesitic or basaltic, largely pyroclastic, volcanics and deep-water sediments such as shales and mudstones. There are three varieties: compact pyrite, chalcopyrite, sphalerite; banded sulphides and silicates; copper-rich ore with chalcopyrite and minor pyrite. They all tend to have high Pb, Zn, Ag and Ba contents (Sillitoe, 1972b). Examples

256

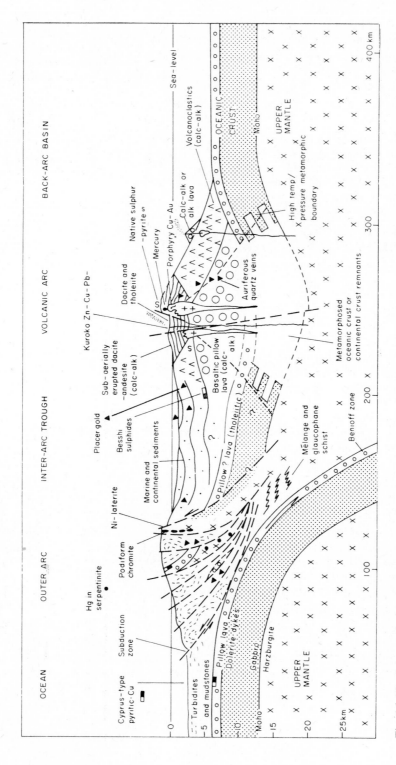

Fig. 16.10. Schematic cross-section through an island arc with a well-developed outer arc showing principal types of mineralization (after Mitchell and Garson, 1976; reproduced by permission of *Minerals, Science and Engineering*)

occur in the high-temperature Sanbagawa, Hida and Abukuma metamorphic belts of Japan, whilst a lower Palaeozoic deposit occurs near Bathurst, NSW, Australia (Stanton, 1955).

2. KUROKO-TYPE ORES

In the late stages of arc development Kuroko-type stratiform sulphide mineralization is associated with dacitic-rhyolitic lava domes (Saito, 1974) or, more rarely, andesitic pyroclastic rocks together with shallow-water near-shore marine sediments. The main ores are enriched in Zn–Pb–Cu–Ag and are typically associated with barites (Matsukuma and Horikoshi, 1970). They are probably volcanic exhalative deposits formed during the last stages of arc volcanism (Horikoshi, 1969). Their type locality is in the Miocene Green Tuff belt of Japan in which the ores were emplaced in shallow water up to 500 m deep within a time interval of 0·2 my (Ueno, 1975). In their review of the Kuroko ores, Lambert and Saito (1974) show that they lie between 130 and 155 km above the Benioff zone.

Manganese deposits are commonly present in island arcs associated with andesite or basaltic volcanics and reef limestones (Stanton, 1972). They occur in Recent arcs of the SW Pacific and in the Tertiary arcs of Japan, Indonesia and the West Indies (Evans, 1976).

Mercury deposits (quicksilver and cinnabar) occur within some volcanoes in Japan, the Philippines and New Zealand and there are Quaternary native sulphur–pyrite deposits related to volcanoes on Honshu in Japan.

Gold deposits occur in three magmatic environments in island arcs (Mitchell and Garson, 1976). Gold is associated with quartz diorite plutons on the Solomon Islands, gold-quartz veins occur in meta-andesites and volcaniclastic rocks in New Zealand, and auriferous sulphides and tellurides occur in veins in andesites in Fiji.

In the western Pacific the metallogenic belts are related to westerly-dipping subduction zones (Zonenshain and coworkers, 1974). Here there is complete zonation of mineral deposits: Cr, Ni and Pt occur on the inner side of the trench, presumably situated in tectonic slices of ophiolites, the arc–trench gap is devoid of magmatic rocks and mineralization, and the main volcanic–plutonic arc is characterized by the calc-alkaline series which passes westwards into the alkaline series. West of the granite front four types of granitoids are distinguished:
1. Granite–granodiorite (Au–Mo mineralization).
2. Diorite–monzonite (polymetallic ores with Sn and sometimes Au).
3. Lithium–fluorine granites (Sn–W and rare metals).
4. Alkaline granites and volcanics (rare-earth, niobium–zircon mineralization, sometimes with Hg and Sb).
The Cu–Pb–Zn mineralization of the Kuroko-type is situated in the centre of the calc-alkaline series of the magmatic arc.

Tertiary porphyry copper deposits occur in some island arcs—Bougainville in the Solomon Islands, Taiwan, Burma, Ryoke in Japan and Puerto Rico (see further under porphyry copper ores in Chapter 17).

Marginal seas or back-arc basins may involve extension and a small degree of sea-floor spreading with resultant emplacement of diapiric ultramafics and possible Cr deposits. These marginal basin 'ophiolites' may be favourable sites for Cr because there appears to be a correlation between Cr and high oxygen/water pressure (Shiraki, 1966; Challis, 1971). Subduction of oceanic lithosphere would involve hydration of the upper mantle (Green and Ringwood, 1967) which might give rise to a high oxidation state in magmas generated in the extensional zone below marginal basins.

Chapter 17

Continental Margin Orogenic Belts—The American Cordillera

Continental margin orogenic belts such as the Cordilleran–Andean belts of western America evolve along tectonically–seismically active margins where an oceanic lithospheric plate is consumed beneath a continental plate; they are convergent plate junctions adjacent to subduction zones. Their main features include a trench with turbidite-type sediments, high heat flow and regularly arranged metamorphic zones, calc-alkaline volcanics (especially of the andesite–rhyolite type), monzonitic–granodioritic plutons and batholiths, molasse-type successor basins, and extensive sulphide mineralization including the well-known porphyry copper deposits.

In this chapter we shall review the main features of these belts and consider some of the complexities of their evolution.

Tectonic Evolution

Analysis of the general structure of the Cordilleran orogenic belt in North and South America is complicated by two main factors.
1. The belt has a history ranging from the late Precambrian to the Quaternary, and so one must distinguish between tectonic belts developed at different times, e.g. in the Palaeozoic and Mesozoic.
2. Any one 'orogenic phase' has a history ranging from the formation of early basins (oceanic or continental) with sedimentation and volcanism to later episodes of magmatism, metamorphism, deformation and uplift. Rast (1969) emphasized the need to distinguish between the different tectonic stages in the evolution of an orogenic belt, and Rutland (1973a) applied these distinctions to the Cordilleran belt as a whole.

Thus, if one looks at the structure of any one section of the Cordillera today, one sees a bewildering complexity caused, firstly, by the juxtaposition of rock units of vastly differing age and, secondly, by the late epeirogenesis which has caused differential uplift and erosion and consequently variable preservation of individual sub-belts. To overcome these difficulties, the first need is to consider separately the Palaeozoic and Mesozoic–Tertiary histories (Fig. 17.1) and then within these to distinguish the supracrustal from the tectono-magmatic evolution.

It is widely held that the present Pacific coastline has been the site of a major continent–ocean boundary for the last thousand million years. Decoupling probably started in the late Proterozoic, and in the Phanerozoic there has been semicontinuous eastward subduction of the oceanic lithospheric plate and concomitant deformation and intrusion of the leading edge of the continental plate.

The most simple cross-sections are seen in southern Canada and central Andes. The western United States is complicated by the presence of the Basin and Range Province, the Colorado and Columbia Plateaux, and the San Andreas transform fault.

According to Rutland (1973a) there were two distinct breaks in tectonic development of the Cordillera, and so three evolutionary periods can be distinguished: late Proterozoic

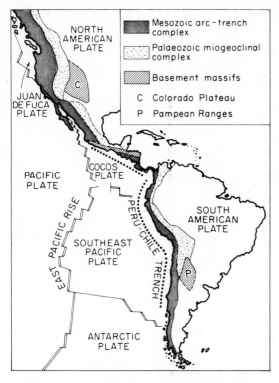

Fig. 17.1. Major stratotectonic subdivisions of the Cordilleran orogenic belt (after Rutland, 1973a; reproduced by permission of *American Journal of Science*)

to late Palaeozoic, Mesozoic, and Cainozoic.

Let us now consider the tectonic development of these periods in the North American Cordillera (summarized in Fig. 17.2).

Late Proterozoic to late Palaeozoic

In North America the Belt–Purcell Supergroup (1400–900 my old) was overlain by the Windermere rocks 800–600 my ago (see Fig. 7.1 and Chapter 7). These strata consist predominantly of shallow-water clastic sediments which pass westwards (in the Purcell) into deep-water turbidites and eastwards thin progressively onto the craton. They are interpreted as a miogeoclinal wedge formed on a continental shelf–slope–rise prior to the main decoupling of the oceanic–continental plates in the early Palaeozoic (J. H. Stewart, 1972; Wheeler and Gabrielse, 1972). This environment may have been interrupted by a short-

lived subduction giving rise to the Racklan–East Kootenay orogenies.

During the earliest Palaeozoic the Cordilleran belt continued to be the site of a continental terrace miogeocline. In North America from the Cambrian and until the late Devonian (Churkin, 1974), or mid-Triassic according to Wheeler and Gabrielse (1972), a carbonate–orthoquartzite bank extended from Alaska to California; it formed a broad continental shelf marginal to the landmass analogous to the Bahama Banks off Florida today.

The decoupling of this oceanic–continental margin was diachronous along the Cordillera, evidence of subduction and formation of island arc-type rocks appearing earlier in Alaska (Eberlein and Churkin, 1970) than further south. Beginning in the late Cambrian and continuing until the Middle Devonian a volcanic–greywacke suite with affinities to modern island arc deposits appeared west of the carbonate shelf (Fig. 17.2), providing evidence of oceanic plate consumption during this period. The lavas range from basaltic to rhyolitic and include pillow breccias, aquagene tuffs, volcaniclastic sediments and welded rhyolitic tuffs (Brew, 1968).

From the Ordovician to the Lower Devonian a series of interconnected extensional back-arc basins developed between the carbonate bank and the arc to the west (Fig. 17.3) (Churkin, 1974). They contain graptolitic shales and cherts and were regions of high heat flow as indicated by the presence of barytes deposits: similar barytes mineralization occurs on the East Pacific Rise (Boström and Peterson, 1966, 1969) and in the Afar and Red Sea rifts (Bonatti and coworkers, 1972).

These marginal basins were closed by deformation in the late Devonian and Mississippian and the pelagic sediments were thrust eastwards onto the carbonate shelf. This was the Antler Orogeny that caused uplift and erosion of the belt and deposition of thick sequences of conglomerates and sandstones in unconformable clastic wedges (Churkin, 1974). The Devonian–Mississippian boundary thus marks an important change in tectonic development along the North American margin as this was

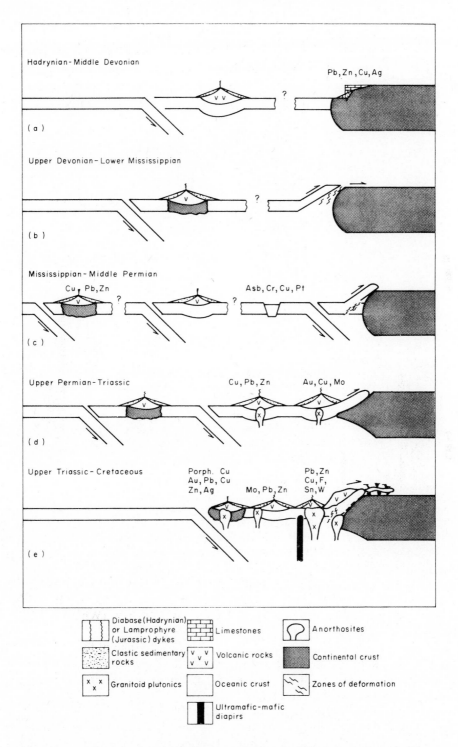

Fig. 17.2. Plate tectonic models for the development of the Canadian Cordillera and its mineralization (after Swinden and Strong, 1976; reproduced by permission of The Geological Association of Canada)

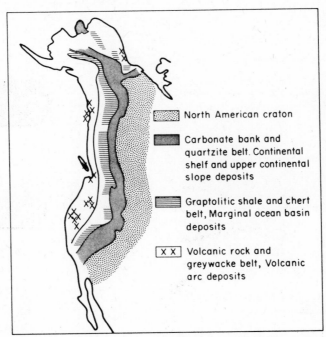

Fig. 17.3. Major lower Palaeozoic stratigraphic belts within the Cordilleran fold belt of North America (redrawn after Churkin, 1974; reproduced by permission of the Society of Economic Paleontologists and Mineralogists)

the first time that clastic detritus was shed eastward onto the craton (Monger, Souther and Gabrielse, 1972) and that oceanic and slope sedimentary rocks were thrust eastward across the continental margin (Burchfiel and Davis, 1972).

The formation of new back-arc basins and oceanic crust was initiated after early Mississippian time in what is now the Canadian Intermontane Belt and Omineca Crystalline Belt (Fig.17.4). The new oceanic rocks are basalts, ultramafics, cherts and locally gabbros. According to Churkin (1974) a series of extensional basins, apparently floored by new oceanic crust, developed in the Pennsylvanian and Permian periods. During the late Permian and early Triassic the basin rocks were again thrust eastwards onto the craton.

Thus, during the Palaeozoic there was a succession of tectonic events: formation (and migration) of volcanic arc complexes, back-arc basins and continent-ward thrusting—that is, closely analogous with the succession of events along the western Pacific arcs. This comparison lends support to the idea that plate tectonics operated in this part of the world in the Palaeozoic.

Mesozoic

In the Canadian Cordillera there is clearly defined evidence of an island arc or continental margin environment in Triassic times in the form of andesites, sub-volcanic granitic plutons and glaucophane-bearing blueschist metamorphism in the Atlin terrace and the Cascade fold belt (Wheeler and Gabrielse, 1972).

A major change in the world's plate movements took place in the early Jurassic marked by the beginning of fragmentation of Gondwanaland. During this Mesozoic cycle from 180–80 my ago the world's spreading systems developed, there was an ubiquitous marine transgression of continents and major plutonism took place in the Cordilleran orogenic belts. The end of the cycle in the late Cretaceous was defined by widespread erosion and

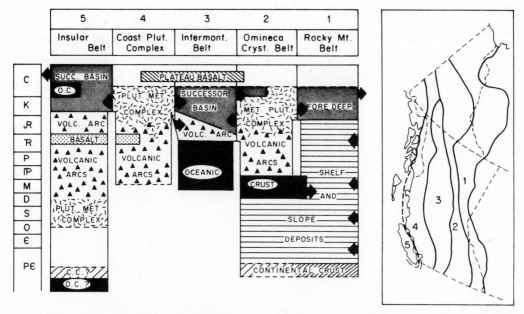

Fig. 17.4. Space and time distribution of lithological assemblages in the five main geological belts of the Canadian Cordillera. Spaces between the vertical columns indicate that no linkages are known between the belts. These linkages take the form of detritus shed from one belt to another, shown by arrows that indicate the direction of clastic movement. (Compiled from Monger, Souther and Gabrielse, 1972; reproduced by permission of *American Journal of Science*)

the cessation of major plutonic activity within the Cordilleran belts, and the beginning of a major continental regression.

In the North American Cordillera there were three 'orogenic episodes' within this cycle:

Columbian (latest Jurassic to early late Cretaceous)
Sevier (late Jurassic to late Cretaceous)
Nevadan (late Jurassic)

The tectono-plutonic activity occurred in a western eugeosynclinal belt and was similarly and synchronously developed along much of the eastern Pacific from Canada to the central Andes. There was a general tendency for plutonism to migrate eastwards towards the continental margin where it continued in the late Tertiary. This Mesozoic orogenic belt truncated the Palaeozoic–Proterozoic basement as shown in Fig. 17.5

The two most characteristic products of this activity were the andesite-dominated volcanics and the granitic plutons and batholiths which can reasonably be ascribed to processes above eastward-dipping subduction zones

(Fig. 17.2). During early stages of uplift of the Cordillera in the Mesozoic and early Tertiary clastic rocks were deposited in successor basins and easterly foredeeps and exogeosynclines.

In the late Mesozoic and early Tertiary deformation migrated eastward into the foreland, mainly by thrusting and overriding of the foredeeps. In the southern Rocky Mountains there was a 140 km shortening of sedimentary beds between the early Cretaceous (136–100 my ago) and the early Eocene (54–49 my ago)—a gross mean rate of 2 mm per year of crustal shortening (Wheeler and coworkers, 1974). The stacking of the thrust sheets caused 8 km of tectonic thickening of cover rocks above a passive but depressed basement (Wheeler and Gabrielse, 1972).

Cainozoic

About 80 my ago there was another dramatic change in tectonic pattern in the Cordilleran belt (Coney, 1971). This period marks the end of major plutonism in the Sierra

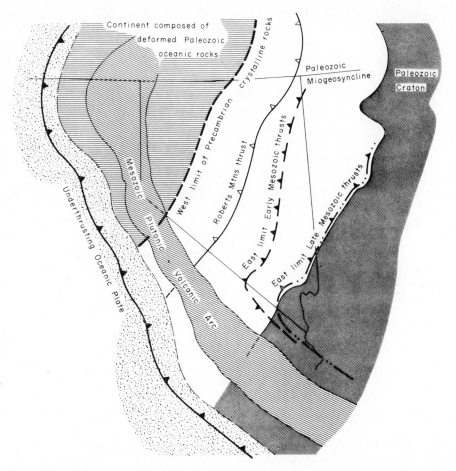

Fig. 17.5. Tectonic sketch map for the late Mesozoic showing truncation of Palaeozoic geosynclinal and deformational trends by a Mesozoic plutonic–volcanic arc of Andean type (from Burchfiel and Davis, 1972; reproduced by permission of *American Journal of Science*)

Nevada and of Franciscan deposition. Starting in the early Tertiary there was extensive erosion of the mountain belt and the planation surfaces were preceded and accompanied in the early to middle Tertiary by major volcanism. This significant volcanism–planation relationship is recognizable along much of the Cordillera from Canada to Chile and Bolivia. Following the eastward migration in the Mesozoic, tectonic activity during the Laramide orogeny continued in the early Tertiary to move towards the cratonic interior.

There was a further important period of tectonic reorganization about 40 my ago (middle to late Eocene) which marks the end of Laramide compressive deformation in the

western United States (Coney, 1971). Apart from the localized Cascadan orogeny in British Columbia, plutonism, volcanism and deformation ceased 40 my ago in much of North America (NW Canada to the Caribbean), after which there was only ignimbrite eruption in the Rocky Mountains and Great Basin.

To account for the two tectonic transitions Coney (1972) suggested that the differences in Cordilleran tectogenesis may be correlated with the variable motions of the North American plate which were due to the spreading history of the North Atlantic oceanic crust. The Nevadan and Sevier–Columbian deformations in the period 180–80 my were related to the northwesterly rotation of North America away from Africa, the Laramide deformation was a reflection of the southwesterly rotation of North America from Europe, and in the post-40 my period North American plate motion was reduced as indicated by Pacific submarine data. According to this model the Cordilleran–Andean orogeny developed along the leading edges of westward-moving continental plates actively driven by spreading in the Atlantic to the east.

An alternative reason for the change in tectonic character may be a decrease in dip of subduction zones (Lipman, Prostka and Christiansen, 1971, 1972). Rutland (1973a) proposed other possibilities: a more rapid rate of ocean floor spreading in the Pacific in the Mesozoic; a closer coupling of the Pacific ocean plate and the Cordilleran continental margins before compared with after 80 my ago.

Sedimentation

Like island arcs, Cordilleran–Andean fold belts have four sedimentation areas: the trench, trench–arc gap, main volcano–plutonic arc, and back-arc basin (see Fig. 16.4). However they also contain molasse-type sediments deposited in late successor basins. The following discourse on sedimentation is largely from the review by Dickinson (1974b).

The Trench

It is widely believed that pelagic sediments were scraped off the ocean floor and deposited tectonically in the deep-sea trenches and that major parts of the circum-Pacific fold belts consist of such uplifted trench deposits—that is, in the eugeosynclinal tectonic mélanges like the Franciscan Assemblage (Hsü, 1971a; Page, 1972a,b) and in the outer of the paired metamorphic belts (Ernst, 1970; Ernst and coworkers, 1970). If this were so, the modern trenches should contain vast sedimentary thicknesses a few tens of kilometres thick, i.e. of geosynclinal proportions.

However the studies of Hayes and Ewing (1970) and Scholl and Marlow (1974) suggest

otherwise. The trenches on the east side of the Pacific contain up to 2 km of sediments in two layers (upper terrigenous turbidites overlie pelagic sediments), while western trenches are either empty or at the most have 100–400 m of pelagic sediments; 55% of all trenches contain less than 400 m of sediment. The turbidite sediments are a special result of the late Cenozoic glaciation (the thickest present trench-fills around the Pacific are off glaciated coasts (von Huene, 1974)) and therefore earlier trenches could not be expected to have thick turbidites on account of the lack of glaciations in the Mesozoic, for example. Also, the modern eastern Pacific trenches are so narrow (50–100 km) and shallow that they could not accumulate more than about 2–3 km of sediment without overflow onto the adjacent ocean floor. And so, where are all those off-scraped oceanic sediments that should be in the trenches? Since the entire area of the Pacific oceanic crust has been renewed at least once since the early Cretaceous (Larson and Chase, 1972), millions of cubic kilometres of pelagic sediment have been transported into the trenches. In view of these results Scholl and Marlow (1974) were forced to two interesting conclusions: firstly, that most of the off-scraped oceanic sediments must have been transported down the subduction zone and, secondly, the enormous thickness, 15–30 km, of terrigenous and volcaniclastic sediments in older (e.g. Mesozoic) eastern Pacific eugeosynclines could not have been derived from trench deposits (as is usually assumed); they must have accumulated in more landward basins such as the arc–trench gap and the back-arc basin.

Sediments beneath the inner trench slope are often complexly folded by compression caused by underthrusting of the oceanic plate (Seely, Vail and Walton, 1974). There is an imbricate thrust zone below the inner trench slope on the Oregon continental margin where younger unconsolidated abyssal sediments have thrust beneath and thus elevated older sediments; the underthrusting produces rapid uplift of the lower continental slope at a rate of 1 km/my (Kulm and Fowler, 1974).

The Arc–Trench Gap

The arc–trench gap on continental margins is the depositional site of fluviatite, deltaic and pro-deltaic sediments belonging to progradational coastal plain complexes built upon subsiding shelves against the open sea.

Along the eastern Pacific Mesozoic–Tertiary arc–trench sequences have been uplifted and are locally preserved in the present mountain belt. Their original tectono-stratigraphic relationships can still be seen as they are ideally situated between the contemporaneous mélanges representing deformed oceanic and trench material and the coeval metavolcanic–plutonic rocks representing the magmatic arc belts. Probably the best known example is the Late Mesozoic Great Valley sequence of California, comprised largely of sandstones and shales, situated between the coastal Franciscan assemblage with its blueschists, eclogites and mélanges, and the Sierra Nevada Batholith (Bailey, Irwin and Jones, 1964; Dickinson and Rich, 1972). The Great Valley arc–trench depositional site lay across the margin of the continent, because on its west side it consists of a deep trough facies overlying an ophiolite complex, and on its east it includes 12–15 km of turbidites with longitudinal (coast-parallel) palaeocurrent structures and further to the east this passes into a thinner shelf facies (Dickinson, 1974b).

The Arc

Sediments within arc basins are largely of volcanic origin and ideally may consist of sub-aerial tuffs and volcaniclastic intermontane red beds in extensional graben-like troughs, together with unstable shelf facies and conglomerates. However, in continental margin fold belts like the Cordillera and the Andes that have undergone substantial uplift they are unlikely to be extensively preserved; here the underlying deeply eroded granitic plutons are best displayed.

Examples of preserved intra-arc sediments include the Upper Triassic Nicola Group in central British Columbia (Schau, 1970), early Jurassic sediments in western North America

(Stanley, Jordan and Dott, 1971), and continental red beds intercalated with Tertiary volcanics in the central Andes (D. E. James, 1971).

Back-Arc Basins

The back-arc area in continental-margin belts varies from the shallow seas with marine beds on the Sunda Shelf behind Sumatra and Java to the continental facies in the Amazon Basin behind the Andes. Ideally the latter may consist of piedmont or deltaic clastic wedges that thicken towards the magmatic belt.

According to Churkin (1974) extensional marginal ocean basins developed behind the frontal volcanic–greywacke arc in the Ordovician, Silurian and Lower Devonian from Alaska to California (Fig. 17.3). To the east was a carbonate–quartzite belt interpreted as a continental shelf–upper continental slope deposit. The marginal basins contain graptolitic shales, cherts and minor quartzites regarded as deep-sea pelagic sediments deposited on oceanic crust. The marginal basins were closed by deformation in the Late Devonian and Mississippian.

Later back-arc sedimentation took place in the Cordillera from Alaska to the Gulf of Mexico within a foreland basin in which marine Cretaceous beds grade westwards to a deltaic and coastal-plain facies and, further west, to piedmont aprons of clastics that lie against the thrust belt (Weimer, 1970).

Scholz, Barazangi and Sbar (1971) proposed the Basin and Range Province of the western United States is an ensialic back-arc basin.

Successor Basins

Successor basins (equivalent to the epieugeosynclines of Kay, 1951) are deeply subsiding troughs with predominantly clastic greywacke–arkosic sediments initiated by the uplift of a mountain belt and which overlie the partly eroded, deformed and intruded eugeosynclines (P. B. King, 1966).

During the Mesozoic and early Tertiary successor basins were developed in the Intermontane, Rocky Mountain and Insular belts of the

Canadian Cordillera (Fig. 17.4). In the Inter-montane Belt there are three successor basins (Eisbacher, 1974).

The development of the successor basins records the progressive continentalization of the crust in the mobile belt. Gradual uplift and erosion unroofed the adjacent crystalline complexes which contributed the clastic components of the younger basins.

Igneous Activity

Continental margin orogenic belts have a long and complicated history of igneous activity. For example, in the belts along the eastern side of the Pacific magmatism has a life span equal to the tectonic history extending from the late Precambrian to the Quaternary; it was a complex development dependent on the multiple stages of tectonic evolution of the belts.

Broadly the igneous products are of two types, volcanic and plutonic. Because these belts typically undergo extensive post-orogenic uplift, many (but not all) of the volcanic rocks formed in the main arc zone are eroded and the underlying granitic rocks in the arc core are exposed. Thus in the Cordillera of western North and South America both magmatic types are available for inspection today.

We shall now consider in turn the volcanic and granitic rocks within the eastern Pacific Cordillera.

Volcanic Rocks

Rocks of volcanic affinity are abundant, but where many of them formed depends on individual interpretation of the tectonic environment. The following types are given in an evolutionary sequence.

The earliest volcanics are late Precambrian tholeiitic basalts, near the bottom of the Belt Supergroup, which were considered by Souther (1972) to be related to the thinning and rifting of the crust during initial continental separation. Some late Proterozoic and early Palaeozoic basalts were extruded as flows into the continental wedge near the carbonate–shale boundary at the slope–rise

transition; examples occur along the east side of the Selwyn Basin in southern Canada (Wheeler and Gabrielse, 1972). According to Roberts (1972) volcanic rocks, clastics and chert accumulated on the oceanic crust in an offshore deep-water environment from the late Precambrian until the late Devonian; they form part of the western eugeosynclinal zone.

In late Palaeozoic times there was only a minor development of volcanic rocks. In Canada there are two groups (Wheeler and Gabrielse, 1972):

1. Tholeiitic basalts with ultrabasics and gabbros, possibly extruded on the ocean floor.
2. Basalts, andesites, and acid volcanics with abundant pyroclastics that formed in a volcanic arc environment. According to Churkin (1974) this eugeosynclinal arc extended from Alaska to California from the Ordovician to the Lower Devonian; volcanic products include abundant basaltic–andesitic breccias and tuffs, aquagene tuffs, pillow basalts and abundant broken pillow breccias and structureless submarine and subaerial lavas.

There are two groups of Mesozoic volcanics: meta-basalts in the ophiolite mélanges originally formed in an oceanic environment, and andesites with more silicic types and pyroclastics in the main arc.

1. OCEANIC OPHIOLITES Rocks formed in the oceanic crust typically include meta-ultramafics and serpentinites, meta-gabbros and basalts, sheeted dykes (diabases), cherts possibly with radiolaria, and pelagic limestones (Dickinson, 1971a, 1972; Cann, 1974). Because these are transported into a trench setting they may be accompanied by eclogites, tectonic mélanges, greywacke flysch turbidites and bluescists. This suite is thus a mixture of rocks emplaced partly at a mid-oceanic ridge and partly in a trench (Hsü, 1971a).

Important examples in the Cordillera of these oceanic rocks either with or without their trench accompaniments include:

a) The Franciscan mélange of California (Ernst, 1965; Hsü, 1967, 1968; Hsü and Ohrbom, 1969; Coleman and Lanphere, 1971).

b) The sequence underneath the western

part of the Great Valley sequence, California (Bailey, Blake and Jones, 1970). These may be Franciscan rocks thrust eastward beneath the contemporaneous Great Valley strata (Ernst, 1971).

c) The sequence underlying the Intermontane Belt of Canada—Carboniferous to Triassic (Monger, Souther and Gabrielse, 1972).

d) In the coastal belt of Peru and Chile (González-Bonarino and Aguirre, 1970).

2. MAIN-ARC ANDESITES, ETC. Mesozoic volcanics along the eastern Pacific margin are dominantly fragmental augite andesites, ignimbrites, pyroclastics and subordinate dacites and rhyolites (Dickinson, 1962). Andesitic volcanicity began in the Permian in the western United States (H. E. Wheeler, 1940), the Trias in northern Peru and the Lias in southern Peru (J. J. Wilson in Cobbing and Pitcher, 1972). It continued extensively during the Jurassic in the main arc zone that was strongly deformed and intruded by granitic batholiths in the Cretaceous (Dickinson, 1962).

Whereas island arc volcanism largely produces andesites, dacites and high-alumina basalts, volcanism in the continental margin orogenic belts related to subduction activity produces these plus abundant silicic types (Hamilton, 1969). The shallow granitic batholiths developed along the main arc axis were roofed by intermediate-to-silicic calc-alkaline volcanic fields (Hamilton and Myers, 1966). Fig. 17.6 gives a compilation of the ages of salic plutons and volcanics in the North American Cordillera in Mesozoic and Cenozoic time.

Great volumes of volcanic rocks, mostly andesites and ignimbrites, were extruded in the Cordillera during the Cenozoic, which was a period of important plate margin tectonic activity (Francis and Rundle, 1976). Considerable effort has been made in recent years to see if there is a relation between igneous activity and tectonic movements of the lithospheric plates along continental–oceanic and oceanic–oceanic margins. Much depends on whether major changes in the rate and direction of plate movements can be correlated over great distances. Those in favour include Dott

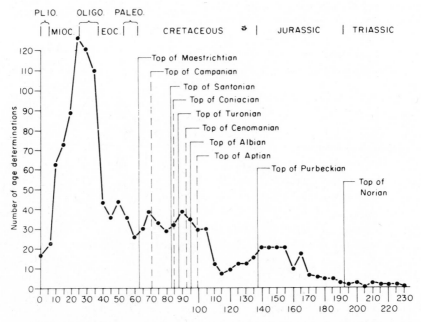

Fig. 17.6. Post-Palaeozoic dates on salic plutons and volcanics in the North American Cordillera between Alaska and Baja California (after Gilluly, 1973; reproduced by permission of Geological Society of America)

(1969), Coney (1971), Mitchell and Reading (1971), Larson and Pitman (1972) and Noble and coworkers (1974); Rodgers (1971) and Gilluly (1973) are against such correlations.

Noble and coworkers (1974) suggested that episodes of Cenozoic volcanism in the Andes of Peru could be correlated with tectonic movements of the oceanic plates (Fig. 17.7). The late Eocene pulse of igneous activity may be related to the abrupt change in the rate and direction of rotation of the Pacific plate deduced by Clague and Jarrard (1973). The Oligocene was a period of volcanic and tectonic quiescence. The reinception of tectonism and volcanism at the start of the Miocene may correlate with the increase 20–25 my ago in the rate of rotation of the Pacific plate from less than 0·5° per my to 1·3° per my (Clague and Jarrard, 1973). This change may also have been responsible for on-land events of this age in the western United States (Noble, 1972) and the Aleutian and Lesser Antilles arcs (Marlow and Scholl, 1973). The present-day rapid spreading of the East Pacific Rise began about 10 my ago (Larson and Pittman, 1972) and this timing coincides with

the inception of extensive middle and late Miocene and Pliocene igneous activity and middle Miocene deformation.

In the Canadian Cordillera there was an abrupt change from explosive eruption of calc-alkaline lavas associated with subvolcanic plutonic rocks in the Eocene to quiet effusion of alkali–olivine plateau basalts in late Miocene and early Pliocene times (Souther, 1970; Monger, Souther and Gabrielse, 1972). This may mark a change from calc-alkaline magma generation above an eastward-dipping subduction zone to derivation of alkali–olivine basalt directly from the mantle at depths of 300–400 km (Kuno, 1966), implying that oceanic plate consumption ceased prior to Miocene time and was substituted by simple shear motion along the Denali–Fairweather–Queen Charlotte transform fault that today occupies most of the Canadian part of the North American coast.

In the western United States a similar termination of the subduction system took place in late Cenozoic time and was replaced by an oblique coastal transform system and extensional faulting inland (Atwater, 1970). Volcanism associated with this faulting was dominantly basaltic, including alkali basalts and bimodal basalt–rhyolite suites (Christiansen and Lipman, 1972).

During Pleistocene and Recent times central British Columbia has been the site of nearly 150 alkali–olivine basaltic cinder cones and stratovolcanoes arranged in linear belts associated with north–south graben—this may be the first expression of an opening rift system along this non-subducting plate margin (Souther, 1970).

This review of the volcanic rocks preserved in the Cordillera demonstrates that they vary from early tholeiitic basalts formed in a continental rift–miogeoclinal environment, to oceanic tholeiitic basalts largely subducted but locally preserved as in the Franciscan mélange, to predominantly andesitic calc-alkaline lavas developed above eastward-dipping subduction zones, to alkali–olivine basalts in extensional rift environments.

At this point we shall consider the chemical variation of the volcanic rocks across the

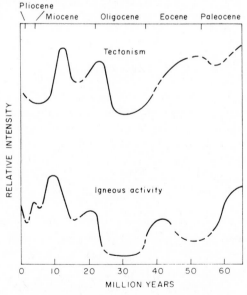

Fig. 17.7. Schematic diagram showing inferred relative intensities of igneous and tectonic activity in the Andes of Peru during the Cenozoic (after Nobel and coworkers, 1974; reproduced by permission of North-Holland)

continental margin orogenic belts. It is well known that across island arcs (see Chapter 16) the composition of lavas changes from tholeiitic to calc-alkaline and then alkaline and/or shoshonitic with increasing distance from the trench (Kuno, 1966; Dickinson, 1968; Jakes and White, 1972; Nielsen and Stoiber, 1973). Similar chemical variations take place in volcanic rocks across continental margin orogenic belts like the Cordillera, and these are usually expressed in terms of K_2O/SiO_2 and K_2O/Na_2O ratios.

Across the Andes of South Peru there is a zonal arrangement of late Cenozoic lava types according to their distance from the trench, but the unique feature here is the absence of the tholeiitic type (Lefèvre, 1973). The coastal lavas belong to an andesitic calc-alkaline suite which shows a progressive eastward increase in K_2O content and in K_2O/Na_2O ratio, towards shoshonitic lavas which are more potassic and titaniferous but within which the K_2O/Na_2O is approximately constant. The potash variation may be correlated with the inclination of the seismic zone down to depths of about 250 km (D. E. James, 1971).

In the western United States lower and middle Cenozoic calc-alkaline volcanic rocks similarly display increasing alkali contents towards the continental interior (Lipman, Prostka and Christiansen, 1971, 1972). In their 1971 paper these authors found there were two eastward-dipping zones defined by increasing K_2O values. Because these are similar to K_2O/depth plots for modern arc volcanics it was inferred that these marked the position of early and middle Cenozoic subduction zones. An eastward shift of the subduction systems may account for the distribution of related volcanic rocks in the western United States. But this dual subduction zone model was challenged by Gilluly (1971).

Granitic Batholiths

Granitic rocks form extensive batholiths and plutons along the Cordillera from Alaska to Chile. Prominent in North America are the Coast Range, Boulder–Idaho, Sierra Nevada and California batholiths (Fig. 17.8), and in

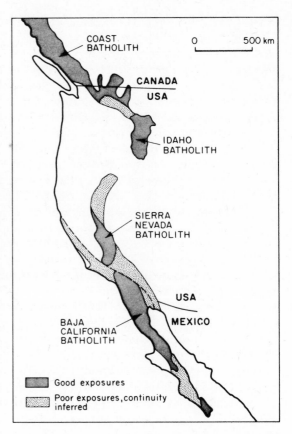

Fig. 17.8. Late Mesozoic batholiths of west–central North America. Magmatism culminated in the late Cretaceous giving rise to an almost continuous batholithic belt. (Redrawn after Hamilton, 1969; reproduced by permission of The Geological Society of America)

South America the Peruvian batholith. The largest are more than 1500 km long and 200 km wide. The batholiths have a composite structure—the Sierra Nevada consists of some 200 plutons separated by many smaller plutons, some as small as a few kilometres across. They contain a variety of rocks belonging to a consanguineous magma series with a regular basic-to-acid sequence of intrusions in roughly the following proportions (Cobbing and Pitcher, 1972): gabbro-diorite, 7–15%; tonalite, 50–58%; granodiorite and adamellite, 25–34%; granite, 1–4%.

The complexes were intruded into regionally metamorphosed and deformed sedimentary and volcanic rocks of the western eugeosyncline on which they have superimposed contact metamorphic aureoles, the mineralogy of which suggests emplacement at depths between 6 and 10 km. According to Hamilton and Myers (1967) the batholiths were emplaced beneath a cover of their own volcanic ejecta. They are regarded as the deep-seated equivalents of the arc volcanics, exposed today because of the major late uplift of the continental margin orogenic belts.

Since Lindgren made the suggestion in 1915, there has been much controversy about possible systemic variations in age and composition of the plutonic rocks from west to east across the Cordillera. Let us look at these two aspects in turn.

The granitic complexes vary in age from Triassic to Miocene with the main intrusive phase in the mid–late Cretaceous. The age problem is complicated because too many age determinations have been made by the K/Ar method. A recent re-evaluation by Evernden and Kistler (1970) showed that there were five periods of emplacement of the granitic rocks in the western USA, each lasting 10–15 my, and revised dates are given in Table 17.1 for the USA, Canada and Chile. There is clearly no simple progressive eastward decrease in age of the granitic rocks. The earliest are found on both east and west margins, but during the late, main and post-orogenic phases, from the mid-Cretaceous to the early Tertiary, there was an eastward migration of plutonic centres. However, according to Kistler (1974) the youngest Oligocene–Miocene rocks in the Cordillera of the western USA occur in both the east and the west.

Moore (1959) demonstrated that in western USA a line separates quartz-diorites to the west from largely granodiorites and quartz monzonites to the east. Systematic increases in K_2O, SiO_2, K_2O/Na_2O and K_2O/SiO_2 away from the continental margin are reported across the batholiths of the Sierra Nevada—but with no age correlation (Bateman and Dodge, 1970), southern California (Baird, Baird and Welday, 1974), northern Chile (Farrar and coworkers, 1970) and Peru (Giletti and Day, 1968). There are two contemporaneous magma series in the Boulder Batholith with the sodic to the west, based on K_2O/SiO_2 and Na_2O/SiO_2 variations (Tilling, 1973), but there is no systematic variation in $(K_2O + Na_2O)/SiO_2$ across the batholith in Baja California (Gastil, Phillips and Allison, 1975).

An interesting new development concerns the initial strontium isotope ratios which vary with geographic position but not with age from 0·703 to c. 0·709 (Kistler, 1974). The quartz-diorite line of Moore (1959) in California coincides essentially with the line of the initial Sr^{87}/Sr^{86} of 0·704. Rocks to the west, with a ratio of less than 0·704, are quartz-diorites and trondjhemites, whilst those with a ratio of between 0·704 and 0·706, to the east, are principally quartz-diorites and granodiorites. Quartz monzonites are developed only where the ratio exceeds 0·706.

Table 17.1 Phases of plutonic activity in the Cordillera of western USA, Canada and Chile (from Rutland, 1973a; reproduced by permission of American Journal of Science)

	Canada my	USA my	Chile my
Generative phase preserved on east margins	200–175	235	210–195
Early orogenic phase on east and west margins	170–160	180–160	190–180
Middle orogenic phases	143–110	148–132 121–104	144
Late main orogenic phases showing eastward migration: quartz-diorite and granodiorite	105–85	90–79	107–87
Post-orogenic phases showing further eastward migration: quartz monzonites	76–64 50–40	75–50 40–20 16–0	67–56 43–29 10–3

Kistler (1974) also states that there is a correspondence between the isotopic variations and other chemical parameters. Plutons in the western Sierra Nevada with initial Sr isotope values less than 0·704 are depleted in alkali and rare earth elements and are similar to oceanic tholeiites and alkaline basalts with regard to K, Rb, Sr, REE and Sr and Pb isotopes. To the east of the Sr 0·704 line plutons have more fractionated REE and K, Rb and Sr abundances greater than those in oceanic basalts. Kistler concludes that these chemical and isotope variations can only reflect a pattern of lateral chemical inhomogeneity that existed in the source regions of the granitic rocks since at least the late Precambrian. However, to account for the increase in potassium and other elements away from the convergent plate boundary Best (1975) suggests that hydrous melts derived from the descending oceanic crust migrate upwards into the overlying mantle wedge, thereby scavenging and zone-melting potassium etc. from peridotite during ascent. Progressively greater amounts of K are abstracted from increasingly thicker vertical sections of the mantle wedge.

Metamorphic Belts

Like island arcs, Cordilleran-type orogenic belts may have a sequence of metamorphic belts. Ernst (1971) recognized four metamorphic zones formed at successively greater depths:
1. Zeolitized rocks, especially with laumontite,
2. Pumpellyite-bearing rocks,
3. Greenschists and/or blueschists with glaucophane and lawsonite,
4. Albite amphibolites.
The first three types occur in the Franciscan terrain of western California (Fig. 17.9). The lowest grade laumontite-bearing rocks occur along the present coast, and there are inclusions of rutile-bearing amphibolites and eclogites largely in the inner two zones. The eastward progression towards higher pressure parageneses marks the direction of presumed lithospheric descent.

Similar blueschist rocks occur in many parts of the Cordillera, but their metamorphic zonations are not so well defined as in the Franciscan. There are important examples in Chile (González-Bonorino and Aguirre, 1970; González-Bonorino, 1971), Venezuela (Shagam, 1960), Guatemala (McBirney, Aoki and Bass, 1967), Mexico–Baja California (Cohen and coworkers, 1963), Klamath Mountains, California (Davis, 1968), Oregon (Swanson, 1969), Washington (Vance, 1968), British Columbia (Monger and Hutchinson, 1970), and Alaska (Forbes and coworkers, 1971).

Returning to the Franciscan section, which is the best documented, the predominant rocks are greywacke, microgreywacke, shale and serpentinized peridotite with minor mafic pillow lava and chert. But little stratigraphy is preserved as the rocks have been highly deformed into a tectonic mélange. The western zeolite facies zone was recrystallized at depths not greater than 5–10 km, but the mineral assemblages of the pumpellyite-, lawsonite-, and jadeite-bearing rocks must have been buried to depths of at least 35 km, equivalent to 9+ kb pressure (Ernst, 1965, 1971). Oxygen isotopic and phase equilibrium data suggest pressures in excess of 8 kb (Ernst and coworkers, 1970). These figures lead to the conclusion that the blueschist metamorphics were presumably dragged down into the throat of the subduction zone and buried beneath the overriding lithospheric plate, but subsequently rebounded isostatically to shallow depths. The leading edge of the overlying plate seems to experience upward drag near the suture (Ernst, 1971).

Interpretations of the site of deposition of the Franciscan rocks vary slightly. Following many earlier authors, Ernst (1970) concluded they were 'deposited in and adjacent to' an oceanic trench, whilst according to Hsü (1971) they are a mixed series: the ophiolites formed at a mid-oceanic ridge, the radiolarites were deposited on the oceanic plain, and the flysch turbidites were poured into a trench.

The Franciscan rocks pass eastwards into the contemporaneous Great Valley sequence, and the interrelationships between them have

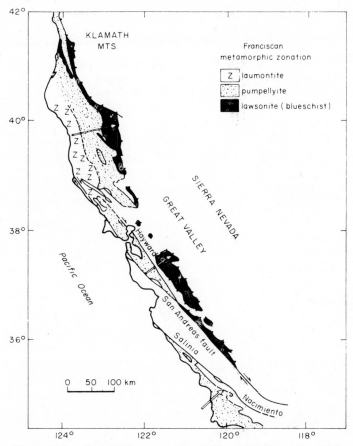

Fig. 17.9. Progressive metamorphic zonation in high-pressure/low-temperature assemblages in chiefly metaclastic rocks of the Franciscan Formation. Arrows indicate direction of presumed underflow (after Ernst, 1971, based largely on Bailey, Blake and Jones, 1970). (Reproduced by permission of *Contrib. Mineral. Petrol.*)

been the subject of much controversy (Taliaferro, 1943; Bailey, Irwin and Jones, 1964; Hsü, 1968). The Great Valley sequence comprises conglomerates, lithic sandstones, siltstones and shales, interpreted by Ernst (1970) to be a continental shelf–slope deposit. The junction between the ensimatic Franciscan and the miogeosynclinal Great Valley sequences is marked by a major fault considered to be the crustal expression of the late Mesozoic Benioff zone (Ernst, 1970) which marked the boundary between the Pacific and western North American lithospheric plates at that time (Ernst, 1971).

The main metamorphic event that gave rise to the blueschists is dated at 150 my (Coleman and Lanphere, 1971) and this marks the change from the geosynclinal to the orogenic stage in the Californian Cordillera. The range of K–Ar dates from the Franciscan as a whole is 150–70 my (Suppe and Armstrong, 1972), which is identical to the period of extensive magmatism in the batholithic belt (Rutland, 1973a). This is not surprising if the Klamath–Sierra Nevada–Salinia Complex is interpreted as the high temperature–low pressure half of a paired metamorphic belt sequence (Ernst, 1971).

Mineralization

The western Cordillera of North and South America is well endowed with a variety of mineral deposits and there is a systematic change in types across the mountain belts, which reflects the increasing depth of the underlying subduction zone away from the continental margin. The metals were largely emplaced as components of the calc-alkaline magmas that gave rise to the granitic batholiths and andesitic–rhyolitic lavas in the main arc zone.

The generalized sequence of metal provinces across the mountain belt in terms of the plate tectonic–subduction model was summarized by Sillitoe (1972a). From west to east the sequence is: Fe; Cu, Mo, Au; Pb, Zn, Ag; Sn, Mo; the metal provinces are aligned parallel to the axis of the belt. The porphyry copper deposits, often with molybdenum, are probably the most distinctive type.

However the development of the metal zones is variable with few sections across the Cordillera containing all types. The following table demonstrates the incomplete nature of many sections, but nevertheless the ones present do not depart significantly from the generalized sequence.

there are porphyry copper deposits of late Mesozoic and Pliocene–Quaternary (Gilleti and Day, 1968; Mitchell, 1973) and in Chile of Pliocene age (Quirt and coworkers, 1971); in the Cordillera as a whole they are largely restricted to the Cretaceous and early–middle Tertiary (Stanton, 1972), but in places there has been a repeated development of a particular metal. For example, in the Cu province of Chile copper deposits range in age from Jurassic to Pliocene (Sillitoe, 1972a) and in the Sn province of Bolivia there was tin deposition in late Triassic, Miocene, Pliocene and perhaps Pleistocene times (Turneaure, 1971). This recurring Sn mineralization has been ascribed to the periodic tapping of a persistent Sn anomaly in the upper mantle rather than to subduction zone activity (Clark and coworkers, 1976).

Porphyry copper deposits occur in some island arcs but are more common in continental margins (e.g. the Andes, the Rockies and Iran) due to the greater abundance there of suitable granitic host rocks. They are situated near the apex or around the margins of

Metals		Region	Reference
West	East		
Hg, Cu, Au, Ag, W, Pb, Mo		Western USA	Noble (1970)
Fe, Cu, Mo, Zn, Pb		British Columbia	A. S. Brown (1969)
Fe, Cu, Au, Pb, Zn, Ag, Cu, Au, Pb, Cu, Sn		Peru	Bellido, de Montreuil and Girard (1969)
Cu, Mo, Au, Pb, Zn		Ecuador and Columbia	Goosens (1969); Singewald (1950)
Cu, Ag, Pb		Mexico	Noble (1970)
Fe, Au, Cu, Pb, Zn, Ag		Mexico	Gabelman and Krusiewski (1968)
Cu, Cu–Pb–Zn–Ag, Sn–W–Ag–Au–Sb–Bi–Pb–Zn		Central Andes	Clark and coworkers (1976)

Wolfhard and Ney (1976) and Swinden and Strong (1976) emphasize the fact that the metallogeny in the Cordilleran belts is complicated because of the long succession of events from the late Proterozoic to the present (see Fig. 17.2). The coastal Fe occurs as an oxide but most of the remaining ores are sulphides. The formation of the metals in one traverse may be essentially contemporaneous. In Peru

small porphyritic, predominantly quartz monzonitic–granodioritic plutons intruded into or below andesitic or dacitic volcanics (Mitchell and Garson, 1972; Sawkins, 1972; Sillitoe, 1972c; Jacobsen, 1975). They were emplaced at only a few kilometres depth (Lowell and Guilbert, 1970), almost contemporaneously with their intrusive or extrusive host rocks—only two million years difference in Utah (Moore and Lanphere, 1971). Association with calderas and strike–slip faults suggests they were emplaced late in the development of the magmatic arc (Mitchell and Bell, 1973). Two types of porphyry copper deposit are defined by their respective molybdenum/gold contents (Hollister, 1975). Those richest in Mo pierced mainly a continental crust that was relatively thick and included at least Palaeozoic metamorphics (and often Precambrian basement, Fig. 17.10); they are well developed in the Cordillera of North and South America as well as in the Alpine belt from Yugoslavia to Pakistan, whereas those with appreciable Au intruded thick marine

volcanics and are prominent in the island arcs of the SW Pacific.

In 1973 Mitchell produced an interesting idea relating the type of metallogenic belt to the angle of dip of the Benioff zone. An increase in inclination may be related to a decrease in rate of lithospheric descent (Luyendyk, 1970). According to Mitchell Sn–W–Bi–Fe deposits and associated granites are emplaced above shallow-dipping Benioff zones, and porphyry copper–gold deposits in andesites and tonalites form above steeply dipping zones in island arcs and in silicic volcanic rocks above tin-bearing granites. To account for the metallogenic development of

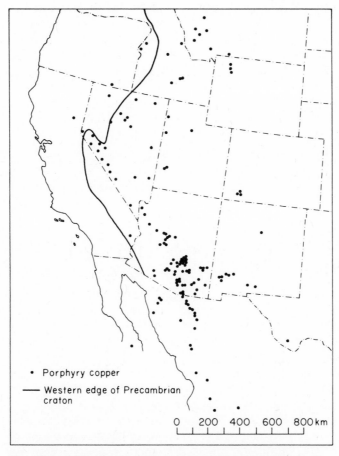

Fig. 17.10. Distribution of porphyry copper deposits in the Cordillera of the western USA relative to the edge of the Precambrian craton, to the west of which molybdenum is not a significant constituent in the deposits (after Hollister, 1975; reproduced by permission of *Minerals, Science and Engineering*)

some continental margin belts it may be necessary to consider a major change in dip of the Benioff zones. For example, in Peru the late Mesozoic porphyry copper-molybdenum associated with andesitic volcanoes and tonalitic plutons formed above a steeply-dipping subduction zone, but the Quaternary porphyry coppers in silicic volcanics and granitic plutons formed when the subduction zone had decreased in dip.

Chapter 18

The Alpine Fold Belt

In this chapter we shall consider the complex network of structures that make up the Alpine fold belt that extends from Gibraltar to the Middle East; further east it passes along strike into the Himalayan belt. The Alpine belt is far from simple as it consists of a multitude of small parts and, because of its accessibility, it is known in some detail. It began to form in the early Jurassic with the break-up of Pangaea and the movement of major plates, it reached its climactic deformation stage in the Tertiary, and tectonic activity still continues today, as is evidenced by the frequent earthquakes and volcanoes. In plate tectonic terms it is a continent–continent collision belt formed by interaction of the African and Arabian plates with the Eurasian plate. In the past there were several classical ideas to explain its formation, but we shall consider it in terms of the current plate tectonic model. This concerns not only the formation of the on-land structures, but also the disappearance of the Tethys Ocean.

The evolution of the Alpine belt is complicated since not only does it consist today of a complex mosaic of microplates, but also it evolved over a period of about 200 my by the continuous motion of a large number of plates and microplates (Fig. 18.1). There was thus a constantly evolving interconnecting network of mid-oceanic ridges, continental margins, island arcs, back-arc basins and transform faults. Unlike the Cordilleran continental margin orogenic belt, which has a semi-continuous trench system and a main arc axis that can be defined for stretches of thousands of kilometres, the Alpine belt has an extremely complex and variable structure because most of the individual mini-belts or zones have different tectono-metamorphic and stratigraphic histories, each phase being largely diachronous. Thus no one evolutionary sequence is applicable to all parts of the belt. For example, according to the view of Dewey and coworkers (1973), there were six periods of basalt formation, seven of ophiolite formation, three of ophiolite obduction, eleven of deformation and seven of high T/P metamorphism taking place diachronously in different areas.

Nevertheless, it is possible to outline a generalized sequence of events applicable to an idealized complete tectonic cycle with respect to the history of a single plate from its birth at an accreting ridge to its disappearance at a consuming plate margin, with the consequent appearance of new material at the convergent juncture. Evidence of individual segments of this history can be found in different parts of the belt. The following is a synopsis of the principal events that contributed to the development of this orogenic belt.

Synopsis

1. Fracturing of the pre-Alpine continental area of Pangaea and extrusion of flood basalts in Triassic times.
2. Formation of new oceanic crust and mantle between separating microplates, now preserved locally in obducted ophiolite complexes.
3. Formation of late Triassic carbonate platforms often sited on evaporite sequences

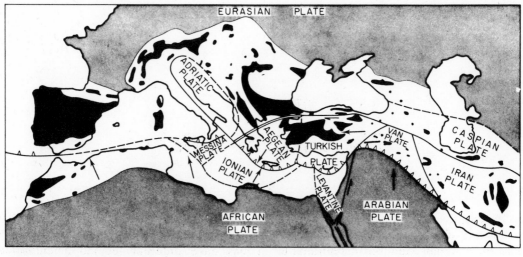

Consuming plate margin – triangles point in direction of dip
Accreting plate margin
Transform faults
Areas of pre‑Alpine continental basement
External forelands

Fig. 18.1. The present lithospheric plates and types of plate boundary in the Alpine System of Europe together with the distribution of pre-Alpine continental basement. Arrows refer to plate motions relative to the Eurasian plate. (Compiled and redrawn after Dewey and coworkers, 1973; reproduced by permission of The Geological Society of America)

along the rifted continental margins. Red beds border the continents.

4. Collapse of the carbonate platforms began in the early Jurassic and was followed by deposition of deeper water pelagic facies with radiolarites and shales.

5. At plate-consuming margins deposition of Cretaceous flysch by turbidity currents in trenches associated with nearby ophiolite-bearing mélanges and blueschist metamorphism. Acid volcanism in the arc environment, closure of marginal basins, thrusting and obduction.

6. Upon continent–continent collision (Eocene–Oligocene) overriding of nappes and thrust-sheets with diachronous 'synorogenic' flysch deposition. High temperature regional metamophism. Marginal basins continued to open in the Neogene in the western Mediterranean.

7. Late orogenic uplift and consequent erosion led to deposition of mostly late Tertiary, non-marine, clastic wedges (molasse) in foredeeps or exogeosynclines along the cratonic margins. Evaporites laid down in sev-

eral basins in late Miocene (Messinian). Continued tectonism in the Pliocene in some areas such as the Jura.

Rock Units

The rocks in the Alpine Belt will be dealt with in order of appearance, from the early to the late evaporites, after which the tectonic evolution of the mountain belt will be reviewed.

Late Triassic Evaporites

Evaporites started forming in the late Permian and reached a climax in the late Trias in the region bordering the Tethys Ocean (Fig. 18.2); they are now preserved throughout the present Mediterranean area (W. D. Gill, 1965). Salt layers are more than 1500 m thick in western Greece and 1000 m in Italy (references in de Jong, 1973).

The evaporites are one of four Upper Triassic sedimentary facies in the Mediterranean region (Bosellini and Hsü, 1973). On the bor-

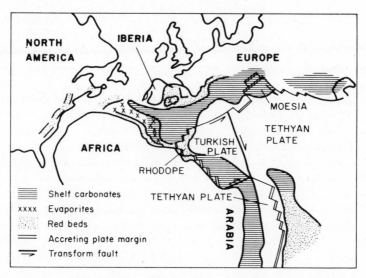

Fig. 18.2. Late Triassic palaeographic reconstruction and plate margins in the Alpine System (redrawn after Dewey and co-workers, 1973; reproduced by permission of The Geological Society of America)

dering continental areas extensive red beds were deposited in terrestrial semi-arid conditions. Evaporites may have formed in inland salt lakes and on continental sabkhas. Shallow-water reef and shelf carbonates formed in tidal flats and lagoons, and in deeper water there were cherty micritic radiolarian limestones. Desiccation of the salt basins took place when they became isolated from the Tethyan Ocean by the reef and shelf carbonates; the formation of the basins may have been controlled by early rifts and graben structures. Alternatively, the evaporites may have been deposited on oceanic crust in a proto-rift Red Sea environment (Hutchinson and Engels, 1972; Dewey and coworkers, 1973; Kinsman, 1975a), similar to that of the Aptian evaporites in the marginal basins along the coasts of eastern South America and West Africa which formed at an early stage of the separation between these continents.

The evaporites are also important tectonically because they later formed the décollement (detachment) zone that enabled the overlying Mesozoic sediments to be thrusted over their basement in many parts of the Alpine System such as the Jura, the Atlas Mountains, the northern Apennines and the Subalpine chains of the western Alps (W. D. Gill, 1965).

Carbonate Platforms

A miogeoclinal, shallow water, carbonate platform is a characteristic feature of a rifted continental margin in a near-equatorial region (Dewey and Bird, 1970b). The modern type example is the Bahama Bank off the east of North America (Dietz, Holden and Sproll, 1970), and a similar Cambro-Ordovician carbonate bank bordered the western margin of the Iapetus Ocean (Rodgers, 1970).

The growth of carbonate platforms in the Alpine belt began in the middle/late Trias and continued mostly to the early Lias. Subsidence appears to have kept apace with rapidly accumulating carbonates with the result that enormous thicknesses of carbonate deposits developed. Subsidence and sedimentation rates were about $100 \text{ mm}/10^3$ years (Garrison and Fischer, 1969). In the peri-Adriatic zone of the Dinarides exceptional carbonate formation continued from the early Triassic to the late Cretaceous and Tertiary giving rise to a sequence about 7 km thick (Herak and coworkers, 1970).

The growth (and disintegration) of a Sicilian carbonate platform is well documented by Jenkyns (1970b,c) and a review of the Mediterranean occurrences is given by Bernoulli and Jenkyns (1974).

The platform rocks consist of supratidal to shallow subtidal limestones and dolomites, containing stromatolitic, pelletal and oolitic varieties, commonly with gastropods and calcareous algae. Evidence that these sediments locally emerged to near or above sea level is given by intercalated coal seams and bauxites (d'Argenio, 1970). The platform rocks are also associated with carbonate reef deposits containing many sponges, corals, hydrozoans and calcareous algae.

The carbonate platforms throughout the Alpine–Mediterranean region began to break up in the early Jurassic. Neptunian dykes and sills, formed by fissure infilling, penetrate the platform sediments and are related to larger scale extensional tectonics expressed by block faulting that caused differential subsidence of the platforms and formation of a submarine seamount–basin topography (Bernoulli and Jenkyns, 1974) (Fig. 18.3a). This block faulting was associated with widespread volcanic activity locally associated with lead–zinc mineralization. On the seamounts carbonate sedimentation continued at depths of less than 200 m with formation of condensed sequences (starved of sediment), crinoidal calcarenites, iron pisolites, ferromanganese nodules, the well-known nodular 'ammonitico rosso' facies (Jenkyns, 1974), and locally algal stromatolites that grew in the photic zone. Very different deeper water sedimentation took place in the basins, giving rise to marls and radiolarites and a noticeable lack of pure carbonates. As the carbonate compensation depth increases with the depth of water (Fig. 18.3b) (Berger and Winterer, 1974), the presence of the Alpine radiolarites and sympathetic absence of carbonates in the basins suggests that the deposition depth there was of the order of 4·5 km (Garrison and Fischer, 1969). However, there is considerable disagreement about the depth of deposition on the seamounts in the early–middle Jurassic: less than 200 m (Bernoulli and Jenkyns, 1974) to 1000–

2400 m (Bosellini and Winterer, 1975). Fig. 18.3(a) shows a possible sequence of events in the collapse, fragmentation and subsidence of a carbonate platform situated on a rifted continental margin. As blocks subsided, deeper water sediments took the place of carbonates, and submarine topography became smoother, so that by the late Jurassic a deep-water facies was widespread.

It is interesting to compare these Alpine structures with modern equivalents (Bernoulli and Jenkyns, 1974). The carbonate platforms are similar to the Bahama Platform built on the rifted continental margin of North America. The pelagic sediments resemble platform-margin deposits surrounding carbonate platforms in the Bahamas. The rifted environment may also be compared with the Red Sea where early clastics pass into a carbonate–evaporite facies, where block faulting and Recent volcanics are common and where lead–zinc enrichments occur in the geothermal brines (Degens and Ross, 1969). There are several modern seamount terraces with depths, fauna and sediments similar to those on the early Jurassic Alpine seamounts (Jenkyns, 1971).

Ophiolites—Remnants of Oceanic Crust

The Alpine belt is the homeland of ophiolites, so-called by Steinmann (1926) for the serpentinite–spilite–chert trinity. Today the succession commonly found in ophiolite complexes (magnesian ultramafics, gabbros, dykes, pillow lavas and pelagic sediments such as radiolarian chert) is widely considered to represent a section through layers 3, 2, 1 of the oceanic mantle–crust formed at a spreading centre (Cann, 1970b, 1974; Vine and Moores, 1972) and emplaced tectonically as slabs during some form of plate collision into or onto continental crust (R. G. Coleman, 1971; Moores and McGregor, 1972; Moores, 1973; Moores and Jackson, 1974). Examples in the Alpine belt are common, especially in the eastern part (Fig. 18.4) and include the Troodos complex, Cyprus (Moores and Vine, 1971; Greenbaum, 1972; Gass and Smewing, 1973), the Oman (Reinhardt, 1969), the

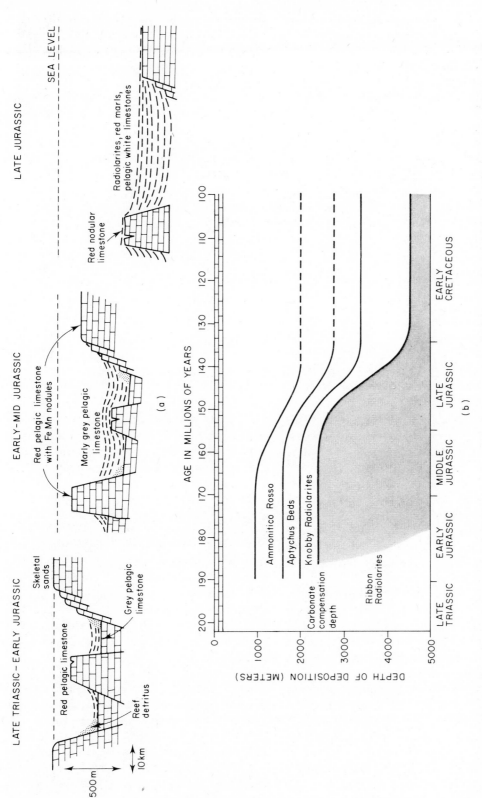

Fig. 18.3. (a) Diagrammatic palaeographic evolution of the collapse of the carbonate bank of the southern rifted continental margin of Tethys during Jurassic time (redrawn after Bernoulli and Jenkyns, 1974; reproduced by permission of the Society of Economic Paleontologists and Mineralogists). (b) Facies diagram for Mesozoic pelagic limestone and radiolarite sequences of the Tethyan region showing, in particular, the increase in the carbonate compensation depth with time (redrawn after Bosellini and Winterer, 1975; reproduced by permission of The Geological Society of America). (See text for discussion)

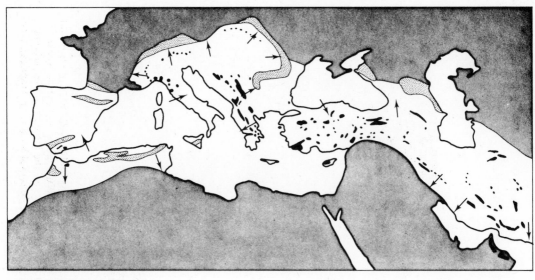

●   Ophiolite complexes
◄—   Direction of overthrusting
▨   External molasse troughs
▨   External forelands

Fig. 18.4. Distribution of ophiolite complexes and molasse basins along the foreland margin and direction of late overthrusting in the Alpine System (compiled from Dewey and coworkers, 1973; some ophiolites from De Jong, 1973; reproduced by permission of The Geological Society of America)

Othris Mountains, Greece (Menzies, 1976), the Vourinos complex, Greece (Moores, 1969), Masirah island, Oman (Moseley, 1969), and the northern Apennines (Bezzi and Picardo, 1971). According to Dewey and coworkers (1973) ophiolite complexes in the Alpine belt formed in the Triassic, the early and late Jurassic, the early, middle, and late Cretaceous and the Eocene, and there was ophiolite obduction in the early, middle, and late Cretaceous.

It is difficult to summarize the main features of many, or most, ophiolite complexes as so few are well described; a review of the better documented Troodos Massif is therefore given here as it contains the main characteristics of a typical ophiolite complex.

Fig. 18.5 shows a map and cross-section of the complex, which has the form of an elongate dome, and Fig. 18.6 lists its major rock units, correlated with the seismic layers of oceanic lithosphere.

The lowermost unit of the plutonic complex consists largely of foliated harzburgite. The ultramafic cumulate zone comprises dunite,

clinopyroxene dunite, poikilitic wehrlite and picrite, whilst the overlying zone above the seismic 'moho' contains metagabbro, olivine and pyroxene gabbro (Greenbaum, 1972). The granophyre, often called quartz-diorite, forms residual segregations in the gabbro and screens between basic dykes. The sheeted intrusive complex is made up of 90–100% dykes, mostly basaltic in composition, separated by thin screens of pillow lava (typically multiple dykes with chilled margins). They have moderate to high SiO_2 (50–60%) and low alkali contents (less than $0 \cdot 2\%$ K_2O). The lower pillow lavas characteristically contain 10–50% dykes with chilled margins; they are mainly oversaturated (45–65% SiO_2) basalts, often intensely silicified, with a K_2O peak around $0 \cdot 25\%$. Erosional conglomerates associated with sedimentary sulphide deposits (ochres) occur in depressions on the Lower Pillow lava surface (Constantinou and Govett, 1972). The Upper Pillow lavas are generally free of dykes, undersaturated (40–50% SiO_2), are often olivine-bearing basalts with low to high K_2O values (up to *c.* 2%) and contain

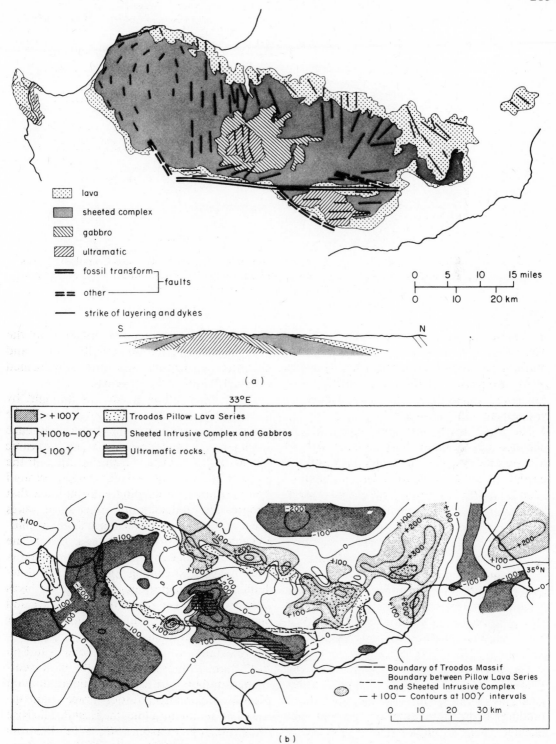

Fig. 18.5. The Troodos Complex, Cyprus. (a) Map and cross-section (after Moores and Vine, 1971; reproduced by permission of The Royal Society). (b) Magnetic anomaly map (after Vine, Poster and Gass, 1973; reproduced by permission of Macmillan Press) in which an appropriate regional field has been removed from the total field aero survey

284

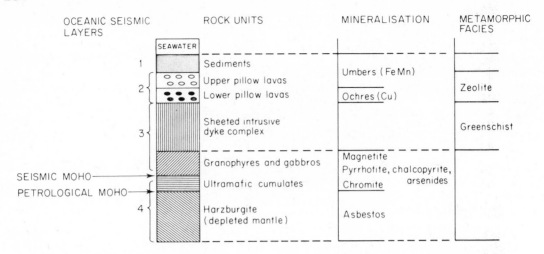

Fig. 18.6. The rock units, mineralization and metamorphic facies of the Troodos Complex, Cyprus (redrawn and modified after Searle, 1972; Gass and Smewing, 1973; reproduced by permission of Macmillan Press), correlated with oceanic seismic layers

more basic (picrite) and even ultrabasic types. The pelagic sediments consists of Fe–Mn rich mudstones—umbers (Robertson and Hudson, 1973; Robertson, 1975), radiolarites, volcanogenic (bentonitic) clay and sandstone (Robertson and Hudson, 1974).

The lowermost harzburgite is regarded as depleted upper mantle (Greenbaum, 1972); Moores and Vine (1971) include the dunite as part of the depleted mantle. The pyroxenite-gabbro may be a product of partial fusion segregated from parent mantle material and crystallized in cumulate intrusive bodies, the granophyres being their residual liquids (Moores and Vine, 1971). In contrast to stratiform intrusions the plutonic lithological sequence is not a time sequence: all products crystallized at the same time, but in different parts of the reservoir and under different P/T conditions and volatile fugacity (Greenbaum, 1972).

The Sheeted Dyke Complex and the Lower Pillow Lavas (with similar chemistry) are genetically related and are the only direct products of the spreading axis process. The Sheeted Dyke Complex has a greenschist facies metamorphic overprint and the Lower Pillow Lavas a zeolite facies, from which a thermal gradient of 150°C km^{-1} has been calculated (Gass and Smewing, 1973). Cann

(1970a) estimated that the gradient at the ridge axis would be about 500°C km^{-1} and thus the Cyprus rocks were probably formed within 100 km of the ridge axis.

The Upper Pillow Lavas are thought by Gass and Smewing to be unrelated to the Lower Lavas and are an off-axis sequence analogous to the seamounts and volcanic islands that are distant from the ridge in the modern ocean floor. Their undersaturated composition is in keeping with the idea that progressive undersaturation in silica takes place with increasing distance from a mid-oceanic rise (McBirney and Gass, 1967). The Upper Lavas have zeolite facies metamorphic phases, including phillipsite, a low-temperature mineral, suggesting that the lavas were extruded over a cold oceanic surface and they were then responsible for producing their own thermal gradient.

According to Robertson and Hudson (1973) and Robertson (1975) the basal sediments (umbers) are closely comparable with Recent ferromanganiferous, trace element-enriched sediments along the East Pacific Rise (Boström and Peterson, 1966, 1969; Boström and coworkers, 1974) and with their Tertiary 'rise' equivalents of the Atlantic (Boström and coworkers, 1972) and Pacific Oceans (Cronan and coworkers, 1972). Corliss (1971), Robertson and Hudson (1974) and Bonatti (1975) suggested that the metals (and REE enrichments—Robertson and Fleet, 1976) were picked up when circulating heated sea water percolated through hot layer 2 lavas; chloride complexes in reduced hydrothermal solutions were exhaled as thermal springs into aerobic sea water. Sulphide mineralization formed as chemical precipitates derived from metal-rich sulphur-bearing exhalations related to fumaroles (Johnson, 1972). Similar occurrences at mid-oceanic ridges are reviewed by Sillitoe (1972b). The preservation of overlying radiolarites may be associated with silica derived from the thermal springs (Robertson and Hudson, 1973, 1974). The cherts contain Campanian (Upper Cretaceous) radiolaria and this agrees with the 76+ my age for the igneous rocks of the complex (Vine, Poster and Gass, 1973).

There is a close correlation between the gravity (Gass and Masson-Smith, 1963) and magnetic (Vine, Poster and Gass, 1973) anomalies over the Troodos Massif. The positive magnetic anomalies coincide with the strongly and normally magnetized Lower Pillow Lavas (Fig. 18.5b). In the absence of reversely magnetized material, Vine, Poster and Gass (1973) suggest that the lavas may have formed in the 10 my period just before 75–80 my ago when the earth's magnetic field was of a single normal polarity. It is interesting to note that the adjoining eastern Mediterranean is magnetically undisturbed (Vogt and Higgs, 1969; Woodside and Bowin, 1970); it may have formed in the same mid-late Cretaceous, constant, normal polarity field as the Troodos Complex (Rabinowitz and Ryan, 1970; Vine, Poster and Gass, 1973).

On the basis of major oxide contents Miyashiro (1973b) is of the opinion that the Troodos Complex formed as a basaltic volcano in an island arc, but most people today would probably not agree with this interpretation. Smewing, Simonian and Gass (1975) discuss the relevant geochemical parameters and conclude that the Complex formed within a slowly spreading small back-arc marginal basin.

Alpine Flysch and Plate Collision

The term *flysch* has had problems of definition (Reading, 1972). For the present purpose flysch 'designates Cretaceous and Early Tertiary marine shaly formations in the Alps, characterized by the presence of regular intercalations of sandstone and/or impure limestone beds' (Hsü, 1970). Accompanying depositional types include olistoliths and wildflysch.

It has been conventional to consider Alpine sediments in terms of pre-Alpine evaporites, carbonates and pelagics, synorogenic flysch and post-orogenic molasse. In plate tectonic terms flysch sedimentation took place in a variety of environments created as a result of different types of plate movements; they are synorogenic because they owe their origins to the movements of microcontinents. The flysch has a distinctive palaeotectonic setting because it did not come into existence in the early 'pre-Alpine' stage of continental fragmentation, but later when the microcontinental interactions resulted in rising cordilleras, island arcs and coastal ranges, with consequent erosion and transport of clastic terrigenous debris that largely constitutes this facies (Hsü, 1972).

The mode of formation of Alpine flysch, long misunderstood because of its variable occurrence, must be seen in terms of complex plate tectonic settings. There are at least three principal flysch environments (Hsü, 1972):
1. Geophysical evidence suggests that the Balearic and Tyrrhenian basins in the western Mediterranean had an extensional origin and deep-sea drilling has confirmed a graben structure in the Balearic basin as well as the presence there, and in the Alboran and Valencia basins, of extensive turbidites of flysch type (Ryan and coworkers, 1970). In fact, seismic data indicate that these deep flysch-like basins in the western Mediterranean are floored by abyssal plains with Recent flysch-type sediments up to 1 km thick, many of which were transported there by bottom currents from the basin margins (Stanley, Gehin and Bartolini, 1970; Bartolini, Gehin and Stanley, 1972). Hsü and Schlanger (1971) suggest that the

Ultrahelvetic Flysch of Switzerland was deposited in extensional basins, possibly in a back-arc setting.

2. Recent flysch-like turbidites make up the submarine fan of the Nile delta. According to Mutti (1974) many Tertiary turbidite sequences in the Tethyan region were deposited in deep-sea fan environments, e.g. Upper Eocene–Oligocene and Miocene in the northern Apennines and Oligocene–Lower Miocene on Rhodes, and Whitaker (1974) has reviewed the many Mesozoic and Tertiary fan deposits in this region that were channelled down submarine canyons. The evidence of Picha (1974) shows that Eocene–Oligocene turbidites in the flysch trough of the western Carpathians were not supplied from 'internal' zones (uplifted cordilleras), as is commonly supposed, but were funnelled down submarine canyons in the continental shelf.

3. Flysch may be deposited in active trenches bordering island arcs if subduction rates are low or if the sediment supply rate is high (Dewey and coworkers, 1973). Quaternary compositionally-immature turbidites occur in the Hellenic trench (Ryan and coworkers, 1970) formed by underthrusting of the Aegean (Hellenic) Arc (sometimes called the Ionia Basin) by the African plate (McKenzie, 1970; Rabinowitz and Ryan, 1970; Comninakis and Papazachos, 1972). Modern flysch sedimentation in the Hellenic trench is reviewed by Stanley (1974).

The most extensive Alpine flysch deposits are of Cretaceous age and occur in the external zone bordering the Alpine Front and extending from the Swiss Alps eastwards to the Carpathians and the Balkans. Hsü (1972) suggests that the early Cretaceous flysch in Rumania was deposited in a trench where the Greco-Italian microcontinent overrode the Tethys. Because Bavarian flysch sedimentation in the East Alps took place continuously for about 50 my from the early Cretaceous to the Palaeocene–Eocene, it may be difficult to imagine its accumulation in a trench where underthrusting would be expected to remove it quickly within a single tectonic–subduction cycle of 3–4 my. To overcome this problem Hsü (1972) suggested that the relative motion between the plates at that time was largely lateral.

The classical model of Alpine flysch formation is in a compressional foredeep in front of a rising cordillera (Argand, 1916). If we reinterpret this in plate tectonic terms, the flysch is developed in a trench bordering an uplifted arc.

Late- to Post-orogenic Molasse

The accumulation of clastic wedges of the molasse facies in basins along cratonic margins (foredeeps or exogeosynclines) is a hallmark of the last stages of orogeny; it is initiated by uplift of the mountain belt following the last major tectonism. In the Alpine System the first molasse sediments were deposited before the last stages of nappe movement and the last were entirely post-tectonic (for distribution see Fig. 18.4). Van Houten (1974) gives a review of molasse deposits in the northern Alpine, Aquitaine and Ebro foredeeps.

The last main deformation phase in the European Alps began in the early Oligocene. By late Oligocene there was extensive uplift and transport of nappes (Ultrahelvetic and South Helvetic) and consequent erosion contributed 6 km of terrigenous detritus to the Oligocene–Miocene Molasse Basin of the Central Alps in Switzerland (Fig. 18.9) (Hsü and Schlanger, 1971). Some proximal molasse deposits were overridden by nappes between the late Oligocene and the late Miocene (Helvetic nappes), and post-tectonic molasse continued to accumulate until the early Pliocene. Commonly the autochthonous flysch sequences grade upward into molasse, which is dominated by a distinctive proximal non-marine facies consisting largely of alluvial fanconglomerates which may exceed 1 km in thickness (Van Houten, 1974).

During molasse accumulation sedimentation kept pace with subsidence with the result that a near sea level surface fluctuated between non-marine and marine conditions; molasse may thus be intercalated with, or grade laterally into, marine sequences with lignite, coal, freshwater limestones and evaporites.

Each of the molasse foredeeps reviewed by Van Houten (1974) underwent a distal axis migration with time, the northern Alpine axis moving 60 km between the Oligocene and Pliocene at an average rate of 2 mm/yr. The molasse accumulation lasted for 25–35 my with a preservation rate of between 150 and 400 m/my. The average vertical stripping for the three regions varied from 3–4 km, suggesting that a maximum of 7–10 km must have been locally eroded from source areas. From geochronologic and heat flow data, Clark and Jäger (1969) calculated an erosion rate of 0·4–1 mm/yr in the Central Alps (0·6 mm/yr according to Trümpy, 1973), from which they estimated that a maximum local stripping of 10–25 km took place during 25 my of molasse accumulation. The difference between this maximum value and that recorded by Van Houten reflects the proportion of eroded material that is not preserved in the molasse basins.

Late Miocene Evaporites

It is ironic that, having started its history with extensive evaporite deposition, the Tethys–Mediterranean should end the same way.

Deep-sea drilling has revealed the existence of substantial late Miocene (Messinian) evaporites beneath all the major basins of the Mediterranean, namely the Tyrrhenian, Ionian, Balearic and Levantine Basins (Hsü, Ryan and Cita, 1973; Kidd, 1976). Seismic reflection profiles show these 6 my old evaporites to be more than 1 km and sometimes up to 2 or 3 km in thickness. To account for this extraordinary occurrence these authors suggest that during desiccation of *deep* depressions in the area of the present-day Mediterranean (this is supported by the important Pontian regression recorded on land in the form of buried canyons, alluvial and terrestrial clastics in channel cuttings) eight to ten marine invasions took place via a giant waterfall at the Straits of Gibraltar, the last of which was probably related to rifting movement along the Azores–Gibraltar fracture zone. Such desiccation would have sub-

stantially lowered the sea level and subtracted considerable quantities of salt from the world's oceans. Compatible with this proposal is the fact that the carbonate compensation depth was suddenly much shallower in the late Miocene compared with the Pliocene in the Atlantic and even in the Pacific (Cita, 1971; Berger, 1972).

However a contrary viewpoint is given by Sonnenfeld (1974, 1975) who, pointing out that the Upper Miocene evaporites stretch from SE Spain to the Caspian Sea and from the Carpathian Foreland to the Yemen (via the Red Sea), suggests that they formed in a series of interconnected *shallow* basins which pre-concentrated, or locally diluted, circulating bottom currents. Rather than normal oceanic salt water from the Atlantic via the Straits of Gibraltar, he envisages a supply of brackish water from the Black and Caspian Seas; evidence for this is provided by the fact that the evaporites are everywhere in juxtaposition with freshwater deposits (an oceanic connection would have supplied normal marine salinities to these sediments) and bottom-dwelling faunas are endemic or brackish Paratethys types (and not oceanic).

Mineralization

There is a consistent and expected relationship between the kinds of Alpine mineral deposits, their host rocks, and their plate tectonic environment (Dixon and Pereira, 1974; Petraschek, 1976). The main types are as follows (see Fig. 18.6):
1. Those associated with early graben. Evans (1975) proposed the following examples.

Syngenetic lead–zinc deposits of the Bleiberg type (Fig. 18.7) occur in Middle Triassic limestones containing evaporites; barytes, celestine and anhydrite are found in the ores. This association may be analogous to the late Tertiary lead–zinc deposits in the Red Sea area (Dadet and coworkers, 1970; Anwar, El-Madhy and El-Dahhar, 1972) which probably formed in early graben. A modern expression of this activity may be in the metal-rich brines and sediments of the Salton Sea (California) and the Red Sea (for

288

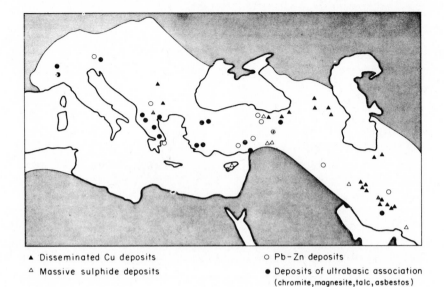

▲ Disseminated Cu deposits   ○ Pb–Zn deposits
△ Massive sulphide deposits   ● Deposits of ultrabasic association
                                        (chromite, magnesite, talc, asbestos)

Fig. 18.7. Important mineral deposits in the Alpine System (redrawn after Dixon and Pereira, 1974; reproduced by permission of Springer-Verlag, Berlin)

references see Degens and Ross, 1969; Backer and Schoell, 1972; Hackett and Bischoff, 1973; Bignell, 1975). Note that the Bleiberg-type limestone–lead–zinc association is equated with the Mississippi Valley-type ores (Mitchell and Garson, 1976).

2. Chromite in ophiolite ultramafics. Lenticular and lineated (podiform) chromite deposits occur in the lower serpentinized ultramafics of many ophiolite complexes in the Alpine belt (Fig. 18.7) (Thayer, 1969; Engin and Hirst, 1970; Peters and Kramers, 1974) which formed at oceanic spreading centres (Cann, 1970b, 1974) and were subsequently emplaced tectonically (obducted) into or onto continental crust during some form of plate collision (Moores and Jackson, 1974).

3. Massive cupriferous sulphide deposits (ochres) occur in many obducted ophiolite complexes, such as the Troodos Complex, Cyprus (Fig. 18.7) (Constaninou and Govett, 1972; Searle, 1972), at Küre, Turkey (Suffel and Hutchinson, 1973), and Ergani–Maden in Turkey (Griffitts, Albers and Öner, 1972). The Cu sulphides are thought to have formed in fumarolic exhalations (Constantinou and Govett, 1972) discharged into a highly reduc-

ing environment in depressions near the crest of a mid-oceanic ridge (Spooner and Fyfe, 1973). Late- to post-volcanic Fe–Mn sediments (umbers) (Robertson and Hudson, 1974; Robertson, 1975) are comparable with ferromanganoan concentrations from the East Pacific Rise (Boström and Peterson, 1969; D. Z. Piper, 1973; Boström and coworkers, 1974; Cronan, 1976). Mn-rich deposits in cherts and Fe–Cu–Zn sulphide deposits in basalts occur in Apennine ophiolites (Bonatti and coworkers, 1976); they formed as a result of mobilization of metals from basalt during circulation of thermal waters near the Mesozoic spreading centre, following the model of Bonatti (1975).

4. Porphyry copper deposits occurring in the eastern part of the Alpine belt in calc-alkaline plutons, largely in the granodiorite–quartz monzonite range associated either with (Lowell and Guilbert, 1970) or without (Hollister, 1975) molybdenum concentrations. They formed, typically within 0·5–2 km of the surface, either on continental margins or in island arcs in relation to palaeo-Benioff zones and plate collision (Mitchell and Garson, 1972; Jacobsen, 1975). One of the most prom-

inent is the Sar Chesmeh deposit in Iran (Bazin and Hübner, 1969) which lies in a belt of related copper occurrences over 160 km long (Dixon and Pereira, 1974).

Tectonic Evolution

Having considered the development of the principal rock units we shall now review the tectonic evolution of the Alpine belt.

Early Rifting

The earliest expression of the rupture of a continent is the formation of graben-like rift structures characterized by clastic sediments such as fanconglomerates, non-marine or saline lake deposits, and alkaline magmatism. Flood basalts are also related to early continental rifting.

The first magmatic activity belonging to the Alpine belt is represented by late Triassic flood basalts that are preserved in at least nine regions. In Morocco they occur in a clastic-filled graben, in Algeria and SE Spain they are associated with shallow-water hypersaline sediments, and in the Balearics, Carnics and Hellenides they were erupted in a shallow-water marine carbonate environment. In the Othris Mountains of Greece the inception of rifting in the Trias was accompanied by extrusion of low K_2O, nepheline-normative, light REE-enriched basalts (Menzies, 1976). In Sicily alkali trachytes were erupted in the Toarcian of the early Jurassic (Jenkyns, 1970a) and in the Venetian (southern) Alps there was rhyolitic–trachytic volcanism in the late Jurassic (Bernoulli and Peters, 1970). Dewey and coworkers (1973) suggest that all this volcanic activity was related to early episodes of continental distension and rifting prior to the development of an ocean in the regions concerned. Permian–mid-Triassic volcanics in the southern Alps (*sensu stricto*) may mark a graben-like zone of rifting (Evans, 1975). The early Jurassic Tethys seaways opened in places parallel to, and elsewhere transverse to, the Triassic rifts (Scandone, 1975).

A relevant question that may be asked at this stage is what was going on in the European platform when major plate movements were beginning to take place to the west and south between the late Trias and late Jurassic? In Britain during the Triassic a horst–graben topography controlled continental sedimentation (Audley-Charles, 1970). A similar tectonic regime continued to operate into the Jurassic when basement faults controlled depositional patterns during the formation of a north European epeiric sea (Sellwood and Hallam, 1974). The North Sea graben (which controlled much oil accumulation) formed in the Callovian (Middle Jurassic) as a series of failed arms (Whiteman and coworkers, 1975a,b; Ziegler, 1975). The formation of all these structures is a reflection of the extensional tectonics that characterize shelf regions that border rifted continental margins opening into new oceans (Bott, 1971; Hutchinson and Engels, 1972).

Oceanic Crust

The date of 180 my ago is the commonly accepted age for the initiation of continental separation between Africa and North America—for details see Chapter 14 (A. G. Smith, 1971; Dewey and coworkers, 1973). The date of 180 my is important with respect to the Alpine System as it is the age of collapse of the carbonate platforms which on stratigraphic grounds can be dated as late Pliensbachian–Toarcian (Lower Jurassic).

Here we must be careful to distinguish between the pre-Triassic Tethys oceanic plate and the oceanic material that was accreted in the period between the early Jurassic and Eocene. The Tethys plate that originally lay between Africa and Europe was largely consumed by inter-plate activity by the end of the Cretaceous, probably along the subduction zone of the Major Caucasus and the Pontide–Minor Caucasus. Possible remnants of the original Tethys plate may be found in the oceanic crust segments of the Black Sea and the South Caspian Sea. Relevant geophysical data for the three-layer crust in these areas are as follows:

	Sediment	? Basalt	? Gabbro	Reference
Black Sea				
Thickness	8 km	6 km	8 km	Neprochnov, 1968
Seismic velocity		4–5 km/sec	6–7 km/sec	
S Caspian Sea				
Thickness	4 km	6 km	18 km	Rezanov and
Seismic velocity	<4 km/sec	4·5–5·6 km/sec	>6·5 km/sec	Chamo, 1969

An important point is that the present Mediterranean is not a remnant of the western original Tethys (Le Pichon, 1968).

A prominent feature of the eastern Mediterranean basin, which probably has a mid–late Cretaceous ocean floor, is the Mediterranean Ridge, a topographic high lying south of, and parallel to, the Hellenic, Pliny and Strabo trenches. Geophysical data across these structures are summarized by Rabinowitz and Ryan (1970). The trenches are thought to be caused by northward under-thrusting of the Aegean Arc by the African plate (McKenzie, 1970) and the Troodos Complex on Cyprus may be an upthrust slab of the Mediterranean Ridge (Vogt and Higgs, 1969). Ryan and coworkers (1971) suggest that the Mediterranean Ridge is a flexural swell related to the Hellenic Trench and not a mid-oceanic ridge.

Plate and Microplate Movements

Fig. 18.1 show the main present-day micro-plates from the Atlantic to Persia and between the major Eurasian and African plates, all plates being defined by the seismicity along their margins (McKenzie, 1970). Each of these plates is in motion although there is no unidirectional sense. The African plate is gen-erally moving northwards; however the net effect against the Eurasian plate is not simply compressional, because it is complicated by the variable interplay of movements by the intervening microplates. For example, a result of the northward movement of the Arabian plate is that the Turkish plate is being wedged sideways to the west, and therefore the Aegean plate is advancing southwestwards against the Ionian plate.

So much for the present picture. The ques-tion now is were similar microplate move-ments reponsible for the generation of the individual fold belts within the Alpine belt from the early Jurassic to the present? Argand, in 1924, was the first to suggest from geological data that continental drift and mi-crocontinental rotations were the cause of the mega-structures in the Mediterranean area. More recently geophysical data have con-firmed this approach. Hsü (1971b) and A. G. Smith (1971) independently suggested three main stages of movement between Africa and Eurasia, which were dependent upon the rela-tive opening rates of the central and north Atlantic and which controlled all minor move-ments of intervening smaller plates. After the incipient rifting the main stages were as fol-lows:

Period	Time (my)	Motion of Africa relative to Europe
1. Middle Jurassic to Upper Cretaceous	165–80	Eastward
2. Upper Cretaceous to late Eocene	80–40	Westward
3. Late Eocene to present	40–0	Northward

The stage 1 'eastward' movement occurred because the central Atlantic had begun to open but Europe was still joined to North America. The second stage resulted from the fact that the North Atlantic was now opening at a faster rate than the central Atlantic. Stage 3 occurred when the opening rates of both parts of the Atlantic were similar.

The next major development of this approach to microplate dispersal and Alpine evolution was by Dewey and coworkers

(1973) who proposed eight stages of Tethyan history (after 180 my ago) related to eight major changes in Atlantic geometry and plate motion. Fig. 18.8 shows the evolving plate boundaries for three important stages in Alpine history and, in particular, the decrease in size of the Tethyan plate, a modern remnant of which may be found in the Caspian and Black Seas.

Whilst the relative movements of some of the older microplates may be subject to re-appraisal and discussion, the rotation of some of the younger microplates has been success-fully determined by palaeomagnetic data:

Spain (Van der Voo, 1969; Sylvester-Bradley, 1974)

Sardinia–Corsica (Zijderveld and cowor-kers, 1970a; Alvarez, 1973)

Italy–Dunarides block (Zijderveld and co-workers, 1970b; De Jong, 1973)

Italy (Channell and Tarling, 1975; Kloot-wijk and Van der Berg, 1975; Lowrie and Alvarez, 1975)

Turkey, southern and western Alps (Smith, 1971)

There were three phases of Betic–Alpine contraction caused by subduction from the Cretaceous to the Eocene (Boccaletti and Guazzone, 1974). By the Oligocene all Mesozoic oceanic crust in the western Mediterranean area had been consumed, so that there was a continuous Alpine orogenic belt. The continuing migration of arc–trench systems opened up new marginal basins in the Miocene and rotated the Sardinia–Corsica microplate (Alvarez, Cocozza and Wezel, 1974). According to Boccaletti and Guazzone (1974) the present western Mediterranean is formed from Neogene and Quaternary margi-nal basins.

Nappe–Thrust Tectonics

Whilst flysch was being deposited in Tethyan basins, compressional and tangential movements between microplates led to continent–continent collisions, terminal sutur-ing of oceans and nappe–thrust tectonics (Dewey and coworkers, 1973). Evidence of this type of tectonics is found throughout the

Alpine belt, but the most spectacular develop-ment, which will be considered here, was in the Swiss Alps and adjoining areas. A classic review of Alpine geology was published by Trümpy in 1960. Other important syntheses are:

Western Alps (Ramsay, 1963b; Debelmas and Lemoine, 1970)

Central Alps (Oberhauser, 1968; Oxburgh, 1968; Trümpy, 1973; Bernoulli and coworkers, 1974)

Eastern Alps (Clar, 1973; Oxburgh, 1974)

Jura Mountains (Laubscher, 1973)

In most areas there is evidence of a pre-Triassic, mostly crystalline basement and a cover of mostly marine, Mesozoic–Tertiary sediments; these are often separated by a décollement of Triassic evaporites. The cover rocks have been deformed into folds and nappes. There are two main types of nappes, namely with or without cores of crystalline basement, and they mainly formed by gravity tectonics (see papers in De Jong and Scholten, 1973). There was a long history of fold, nappe and thrust tectonics in the Alpine belt, ranging from the late Cretaceous in the eastern Alps and Carpathians to late Oligocene–Miocene in the Helvetic zone and Pliocene in the Jura.

Fig. 18.9 shows a tectonic cross-section across the Swiss Alps from the Jura Mountains to the northern Apennines. The following summary of the tectonic evolution of this belt is from Trümpy (1973) and Bernoulli and coworkers (1974).

The oceanic opening stage took place dur-ing the late Jurassic and early Cretaceous; ophiolites were emplaced between the late Jurassic and Albian during terminal closure of the oceans. The first compressional events occurred in the Middle–Upper Cretaceous (Albian–Maestrichtian) in the southern Pen-ninic and Lower Austro-Alps, when flysch sedimentation started in the Valais and Pie-mont troughs. The main deformation and metamorphic stage in the Alps (especially the eastern Alps and Penninic belt) was confined to the period between the late Upper Eocene and the late Lower Oligocene, at a time when flysch sedimentation was continuing in the as yet unfolded Helvetic zone. In the middle and

292

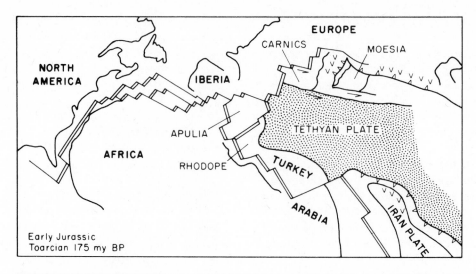

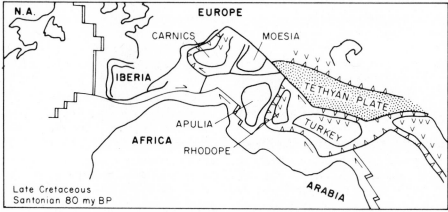

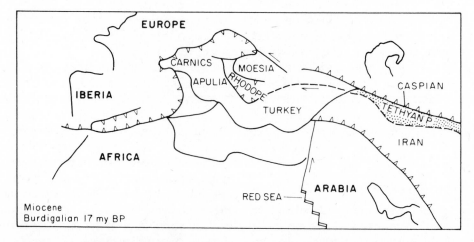

Fig. 18.8. Proposed plate boundary schemes in early Jurassic, late Cretaceous and late Tertiary times for the Alpine System (compiled from Dewey and coworkers, 1973; reproduced by permission of The Geological Society of America)

293

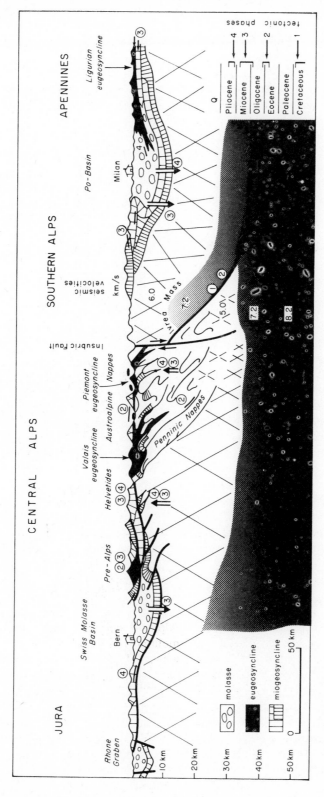

Fig. 18.9. Schematic cross-section through the Jura Mountains, Swiss Alps and northern Apennines. Location of section shown in Fig. 18.10. (After De Jong, 1973)

late Oligocene, the Palaeogene fold belt was uplifted, gravity sliding gave rise to the pre-Alpine, Penninic, and Ultrahelvetic nappes and molasse sedimentation began in the northern foredeep and the Po valley. A new deformation stage appeared in the late Oligocene and Miocene when the Helvetic folds were formed: uplift and molasse deposition continued. The main folding in the southern Alps was in the late Miocene, and in the Jura Mountains (causing a shortening of the cover by 15–20 km) in the early Pliocene. The end of molasse sedimentation and a levelling down of the Alps occurred in the Pliocene.

In reviewing the timing of orogenic events Trümpy (1973) points out that the main Palaeogene Alpine 'orogeny' was confined to a remarkably short time span between the late Eocene and the early Oligocene; within about 5–6 my there was a N–S crustal shortening of at least 300 km (giving a probable rate of shortening of about 5–6 cm/yr) and there was 'premetamorphic stripping of sedimentary cover, folding, thrusting, metamorphism, uplift and erosion'. This estimate of such severe crustal shortening brings out the question of the fate of the continental crust that once underlay many of the nappes because the plate tectonic model does not allow it to be dragged down into the mantle. The question may be answered by Helwig's (1976) idea that the excessive crustal shortening is compensated by the initial extreme thinning of the continental crust, as is known to exist today across the Red Sea region.

Alpine Metamorphism and Plate Structure

In the southern Alps and Austro-Alpine nappes (Fig. 18.9) Hercynian and Caledonian basement rocks were recrystallized under greenschist, low- and high-grade amphibolite facies conditions during pre-Alpine periods of metamorphism. These rocks are overlain unconformably by largely unmetamorphosed Mesozoic platform-type carbonate sediments.

Alpine regional metamorphism took place between the late Cretaceous and early Tertiary, during and after nappe formation in

rocks that are preserved in the Pennine Alps and Engadine and Tauern 'windows' (Fig. 18.10). According to Ernst (1971) there is a progressive increase in grade from an 'outer' laumontite-bearing zone (Niggli, 1970), through prehnite–pumpellyite rocks and lawsonite–jadeite blueschists and greenschists (N. D. Chatterjee, 1971), to an eclogite–albite amphibolite zone (Hunziker, 1970) containing rutile-bearing eclogitic blueschists and kyanite schists.

However the development of these zones was not synchronous. Ernst (1973) distinguishes three temporally-overlapping episodes:

1. An early Alpine, high-P/low-T event, syntectonic with nappe formation that gave rise to the eclogites and blueschists, etc.
2. A syn- to post-tectonic event that variably altered the products of episode 1 under prasinitic greenschist-type conditions.
3. A late Alpine, more localized, syn- to post-tectonic recrystallization under moderately high T and P, representing a more normal geothermal gradient.

The timing of the several metamorphisms varied across the Alpine belt. The early, more internal, high-pressure episode has a late Cretaceous age. In the Oligocene there was an early Alpine laumontization in the outer Helvetic zone at the same time as a late Alpine high-temperature event retrogressed blueschists in the Lepontine area. In general, the time of relatively high pressure metamorphism decreases externally where the lower grade, less deeply subducted sections are found (Ernst, 1975).

Ernst (1973) proposed that the contact between the southern Alps and Austro-Alpine nappes with their pre-Alpine metamorphics and the Pennine nappes etc. with their Alpine metamorphics is a major convergent plate junction—the Alpine Suture (Fig. 18.10). According to this interpretation the northern lithospheric slab with its ophiolites and Mesozoic shelf, slope and sea floor sediments contains the northern margin of a Mesozoic Tethyan plate, and it collided with a NNW-moving and overriding slab with significantly different geological make-up; only

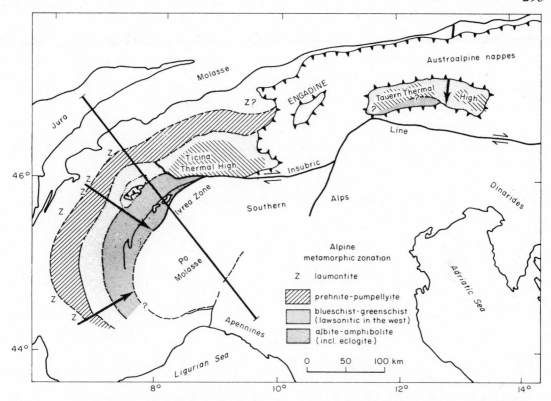

Fig. 18.10. Progressive metamorphic sequence across the Helvetic–Penninic Alps of Switzerland. Development of the zones was not synchronous—see text. Open arrows indicate inferred subduction direction and small arrows post-subduction strike slip motion (after Ernst, 1971). Section of Fig. 18.9 is shown. (Reproduced by permission of *Contr. Mineral. Petrol.*)

the overridden plate was subjected to Alpine high-pressure metamorphism.

The structural situation is complicated because there may have been 300 km of post-Oligocene dextral strike movement along the Insubric Line (Laubscher, 1971). However

Dewey and Bird (1970b) evisaged the Insubric Line to be the convergent plate junction. A totally different model was postulated by Oxburgh (1972) who suggested the Austro-Alpine nappes were 'flaked' off a *northward-descending* southern lithospheric plate.

Chapter 19

The Evolving Continents

In the previous chapters we have looked at the main features of Archaean, Proterozoic and Phanerozoic continental regions. From this survey two major points emerge.

1. Whether one considers Archaean greenstone belts, early Proterozoic dyke swarms, mid-Proterozoic anorthosites, rapakivi granites and alkaline complexes, late Proterozoic tillites or early–late Phanerozoic continental margins, the individual rock suites concerned and their associated structures on any one continent are essentially similar to those on another, although they may be diachronous. In fact the similarities are so striking that one need not hesitate to conclude that continental growth went through comparable stages in different places which means, in turn, that it is justifiable to talk in terms of a sequence of stages in continental evolution. What form these evolutionary stages took, *viz.* cyclic, unidirectional, repetitious, is widely debated.

2. Many rock groups are characteristic of particular periods of earth evolution, e.g. greenstone belts in the Archaean, massif anorthosites in the Proterozoic, granulites/charnockites and banded iron formations in the Precambrian. Conversely, red beds, tillites, massif anorthosites, dolomites, glaucophane-bearing rocks, eclogites, alkaline complexes, kimberlites and lead deposits are rare in the Archaean. Yet there are marked similarities between Archaean (greenstone belts) and Mesozoic–Cenozoic gold mineralization, volcanic rock units and chemistry and sedimentary facies. Having said this, the question arises as to the meaning of these occurrences in terms of continental evolution.

In this final chapter we shall bring together many of the features and relationships previously described in order to see what role they played in the evolution of the continents. We shall do this by considering the evolution of the sedimentary, magmatic, metamorphic and metallogenic record, and finally by reviewing some major factors that contributed to crustal evolution.

The Atmosphere and Hydrosphere

In earlier chapters we reviewed a great many sedimentary rocks but did not consider possible factors controlling their development, such as the evolution of the earth's atmosphere and oceans from the Archaean to the present.

A long-appreciated aspect of the terrestrial atmosphere is that the noble gases are depleted by factors of between 10^{-7} and 10^{-11} compared with their cosmic abundances (Fisher, 1976; Schidlowski, 1976). This led to the inference that the earliest atmospheric constituents were lost from the earth when it accreted and that the present atmosphere is of secondary degassing origin (H. D. Holland, 1963).

Degassing Models

There are two types of degassing model:

1. Catastrophic—the atmosphere was created instantaneously during a single major 'burp' event related to the process of core formation or impact melting (Chase and Perry, 1972; Fanale, 1972; J. C. G. Walker, 1976).

2. Continuous—the atmosphere evolved in a continuous degassing process throughout geological time (Fisher, 1976).

The Anoxygenic Secondary Atmosphere

It is widely held that most of the gases of the secondary atmosphere were derived from volcanic exhalations, largely during the Archaean (Rubey, 1955; Rutten, 1964; Berkner and Marshall, 1967). Here one has to make an assumption: if the early degassing products were similar to the exhalations of modern volcanoes, water vapour and CO_2 would have made up the bulk of the volatiles, followed by H_2S, CO, H_2, N_2, CH_4, NH_3, HF, HCl, Ar, etc. (White and Waring, 1963; Schidlowski, 1976). However there is negligible molecular oxygen in modern volcanic exhalations, and therefore it is commonly argued that the secondary atmosphere must have been anoxygenic in the Archaean, the free oxygen that we see today having evolved subsequently.

Geological evidence can be cited in favour of an Archaean anoxygenic (but not necessarily reducing) atmosphere:

1. There is a widespread weathered zone on the Archaean erosion surface beneath the basal unconformity of the 2300 my old Lower Huronian Supergroup in Canada (Frarey and Roscoe, 1970). The important feature here is that compared with modern soil profiles iron has been lost rather than accumulated and the ferric–ferrous ratio has been relatively decreased. These changes are taken to indicate that the ground water and atmosphere lacked free oxygen at the time. A similar conclusion was drawn by Rankama (1955) from a Svecofennian breccia/conglomerate in Finland, regarded as an *in situ* weathering product in which FeO predominated over Fe_2O_3. Rankama (1955) also suggested that the absence of atmospheric oxygen 2000 my ago would explain the occurrence of reduced native carbon as a finely disseminated pigment in surrounding phyllites.

2. In the Lower Huronian sediments there are 'drab beds' that lack reddish colouration (Frarey and Roscoe, 1970). Comparable modern clastic sandstones and siltstones with a low haematite content and ferric/ferrous iron oxide ratio are known to form in an anoxygenic environment below the water table.

3. Early Precambrian conglomerates contain well-preserved detrital pyrite and uraninite that are normally easily weathered by oxidizing processes, e.g. in the Witwatersrand and Transvaal Systems of South Africa, the Huronian Supergoup in Canada, the Bababudan pre-Dharwar greenstone belt in India and the Jacobina Series in Brazil.

4. The presence in the Archaean of banded iron formations that required a weathering transport of ferrous iron (Cloud, 1973). The basis of Cloud's argument is that free O_2 is a poison to organisms in the absence of oxygen-mediating enzymes and therefore the first ones to produce it were only able to survive by having an external oxygen acceptor. The Archaean iron formations may have acted as this acceptor (Cloud, 1965, 1968a). Ferrous iron was washed away from the continents as a result of an anoxygenic weathering cycle; the iron could only be transported in the ferrous state, ferrous salts being more soluble than ferric salts, and so the lack of oxygen in the sea water allowed extensive transport of the iron. However the ferrous iron required the addition of oxygen in order to be precipitated as ferric oxides or hydroxides, and it was the early organisms that provided this oxygen. The Archaean oceans were finally cleared of the large amount of dissolved ferrous iron when it was precipitated in the ferric state (in particular in the haematite–Fe_2O_3 facies) in the last and major period of banded iron formations about 2100 ± 100 my ago (Goldich, 1973).

Kimberley and Dimroth (1976) dispute the hypothesis that the Archaean atmosphere contained practically no free oxygen. They counteract the argument about the preservation of pyrite and uraninite in early conglomerates by suggesting that the minerals were precipitated from through-flowing groundwater and were concentrated diagenetically, citing the detrital uraninite in the upper Indus River as a modern analogue of uranium being readily removable by oxygenic groundwater. Regarding the common below-wave-base

Archaean iron formations, they propose that the primary sediment was largely aragonite with the iron source in the overlying non-calcareous mud. The iron was leached downwards because of organic acid production during organic decay in the mud and was exchanged for calcium in the carbonate. The main support for this model is the observation that the Archaean, like all younger, iron formations, are characteristically overlain by carbonaceous mudrocks such as shale, argillite and phyllite. Finally, the authors remark that the lack of Archaean red beds is hardly surprising given the relative lack of preserved shelf facies where subaerial oxidation or shallow-water chemical precipitation could take place.

The Evolution of the Oxygenic Atmosphere

Because degassing could not supply free molecular oxygen to the atmosphere–ocean system, the oxygen must have been derived from dissociation of oxides like CO_2 or H_2O, solar radiation providing the energy source for two possible photochemical reactions, inorganic (photodissociation) and organic (photosynthesis) (Berkner and Marshall, 1967; Schidlowski, 1971, 1976).

Inorganic Photodissociation The ultraviolet-rich short wavelengths between 1500 and 2100 Å of the solar spectrum provide the energy to dissociate water vapour in the upper atmosphere, the hydrogen escaping preferentially from the earth's gravity field leaving the atmosphere enriched in free oxygen (Urey, 1960; Berkner and Marshall, 1965). However, the oxygen-producing capacity of this process is considerably smaller (about 10^{-3}) than that of photosynthesis and it is therefore this second process that must have produced the bulk of the oxygen now present in the atmosphere (Schidlowski, 1971).

Organic Photosynthesis The low energy spectral range of visible light provided the energy for primitive organisms, particularly blue–green algae, to produce carbohydrates by photosynthesis from water and carbon dioxide, releasing oxygen as a byproduct:

$$CO_2 + H_2O \rightarrow CH_2O + O_2$$

The photosynthesis reaction requires a porphyrin body (chlorophyll) to be present as a catalyst—3000 my porphyrins are known from the Swaziland System in South Africa (Kvenvolden and Hodgson, 1969). Junge and coworkers (1975) calculate that photosynthesis began more than 3700 my ago, which agrees with the occurrence of BIF older than 3760 ± 70 my in the Isua area of West Greenland (Moorbath, O'Nions and Pankhurst, 1973; Allaart, 1976) if one accepts that BIF acted as the acceptors of free O_2 from the first photosynthesizing blue–green algae (Cloud, 1965, 1968a). The build-up of *appreciable* free oxygen in the atmosphere had to await the proliferation of blue–green algae in the early Proterozoic.

As photosynthetically-derived oxygen accumulated in the oceans, it began to escape into the atmosphere. The effect of ultraviolet radiation on this escaping O_2 was to convert some of it into atomic oxygen (O) and ozone (O_3), the reaction rates of which (with surface weathering materials) are many orders of magnitude greater than those of O_2. This means that even in the Archaean with its thin oxygenic atmosphere, surface oxidation rates may not have been negligible (Berkner and Marshall, 1967). The earliest record of red beds formed by oxidation processes is in the 2500 my old Dharwar greenstone belts of India (Srinivasan and Sreenivas, 1972).

It can be argued that by early Proterozoic times there was sufficient oxygen in the atmosphere and oceans, not only to give rise to the extensive red beds that became increasingly common throughout the Proterozoic, but also to allow oxidation of sulphur and hydrogen sulphide, a process that liberated sulphate ions in the water. The accumulation of the sulphate ions was responsible for the first anhydrite–gypsum beds in the mid-Proterozoic (Ronov, 1968), for example, in the Grenville Supergroup in Canada, the Belt–Purcell System in the American Cordillera, and the Bitter Springs Formation of the Amadeus Basin in Australia (Cloud, 1968a).

The Carbon Dioxide Balance

The input of CO_2 into the oceans and atmosphere, and its abundance relative to other constituents, have, no doubt, varied with geological time and this variation should be expressed in the composition and quantity of a variety of sediments, in particular the carbonates.

The appearance of biogenic oxygen in the early Archaean accelerated the oxidation of the juvenile CH_4 and CO, thus increasing the content of CO_2 in the atmosphere and its dissolution in the oceans, and an increase in the partial CO_2 pressure during the periods of major Archaean volcanism intensified weathering on land and made it possible for marine waters to carry larger amounts of carbonates in solution (Ronov, 1964, 1968). H. L. James (1966) pointed out that, if the partial pressure of CO_2 in the mid-Precambrian atmosphere was one hundred times greater than that in the present atmosphere (i.e. 0·03 atm compared with 0·0003 atm), then the equilibrium pH of surface waters would be changed from the present weakly alkaline (8·17) to weakly acid (6·1) (Rubey, 1955). If, as would be expected, the total volume of water on the earth's surface was less than at present (see Chase and Perry, 1972), the water would have been even more acidic. If such conditions did prevail, then surface waters would have had a much greater capacity for leaching iron and transporting it in solution than they have at present. Thus large quantities could have been moved in surface solution without accompanying detritus, such as Al_2O_3; Precambrian iron formations are characterized by an exceedingly low alumina content (Stanton, 1972).

J. C. G. Walker (1976) believes that terrestrial life evolved in the early Archaean when a primitive steam atmosphere had formed from the conversion of hydrocarbons largely to water and carbon dioxide according to the following type of reaction

$$CH_2 + 3Fe_3O_4 \rightarrow CO_2 + H_2O + 9FeO$$

Over a period of time the H_2O condensed to form sea water and the CO_2 dissolved in the water to form eventually carbonate sediments. According to this model one would expect to find sediments containing or consisting of carbonates in Archaean regions (Cameron and Baumann, 1972); the following examples can be cited:

Isua, West Greenland (Allaart, 1976)

Fiskenæsset, West Greenland (Windley, Herd and Bowden, 1973)

Limpopo Belt, southern Africa (Sohnge, Le Roex and Nel, 1948; R. G. Mason, 1973)

Tamil Nadu, S India (Naidu, 1963)

Scourian, NW Scotland (Coward and coworkers, 1969)

Labrador, Canada (Bridgwater and coworkers, 1975; Collerson, Jesseau and Bridgwater, 1976)

Lofoten, N Norway (Heier and Griffin, 1973)

Malagasy (Besairie, 1967)

Bahia, Brazil (Sighinolfi, 1974)

Greenstone Belts in general (Anhaeusser and coworkers, 1969; Schidlowski, Eichmann and Junge, 1975)

In order to explain the rarity of carbonate rocks compared with the abundance of silica with BIF Cloud (1973) proposed the following model. The early hydrosphere was probably saturated with monosilicic acid (H_4SiO_4), the polymerization of which gave rise to much SiO_2 in neutral to slightly acidic solutions. But very little carbonate ion is present in naturally acidic waters and so the potential components of carbonate sedimentation remained in solution as Mg^{2+}, Ca^{2+} and HCO_3^- ions. This model may further explain the adundant mid- to late Proterozoic carbonate sediments younger than the main BIF; they formed as a result of pH increase related to a reduction of CO_2 levels caused by the segregation of C and O_2.

Schidlowski, Eichmann and Junge (1975) concluded that a mass of organic carbon, close to 80% of that presently contained in the sedimentary shell of the earth, already existed 3000 my ago. Evidence of organic carbon in Archaean greenstone belts lies in the stromatolithic carbonates in the Bulawayan Group in Rhodesia (Bond, Wilson and Winnall, 1973) and at Steep Rock Lake, Ontario, Canada (Joliffe, 1966).

Isotope Chemistry of Sedimentary Carbonates as a Function of Time

Oxygen, carbon and strontium isotope variations of sedimentary carbonate rocks can be used to determine the corresponding isotopic compositions of sea water. These measurements are limited to largely unaltered or unmetamorphosed carbonate rocks, but such rocks are well preserved as far back as the early–mid-Proterozoic and the Archaean; the marbles in Archaean high-grade regions are unsuitable.

Fig. 19.1 shows the oxygen isotope composition of sedimentary carbonates as a function of time. There is a noticeable increase in ^{18}O through geological history, ascribed by Schidlowski, Eichmann and Junge (1975) to post-depositional exchange with oxygen from surface waters. Because there is a large excess of ^{18}O in crustal rocks (compared with that predicted by simple differentiation processes) and because there is a progressive increase in ^{18}O in cherts from the Archaean to the present,

Perry and Tan (1972) suggested that the ocean 3200 my ago was perhaps half the size of the present ocean. The increase in the ^{18}O in crustal rocks with time was due to exchange with increasing quantities of oxygen isotopes between ocean and surface sediments. However, from a study of oxygen isotope data Chase and Perry (1972) concluded that there has been negligible change in the volume of the oceans since the early Precambrian. They suggest that water can be recycled through the mantle, i.e. the amount of water outgassed remains about equal to the amount subducted. Using oxygen isotope variations in cherts and banded iron formations from the 3760 ± 70 my old supracrustal sequence at Isua, West Greenland, Oskvarek and Perry (1976) proposed that the early Archaean ocean had a temperature in the range of 89–146°C.

Notable carbon isotope variations in carbonates are reported by Nagy and coworkers (1974) and Schidlowski, Eichmann and Junge (1975) and the relevance of these ratios to early organic life is reviewed by Sylvester-Bradley (1975). Fig. 19.2 shows that with a few exceptions the $\delta\ ^{13}C$ values in marine carbonates have been remarkably constant

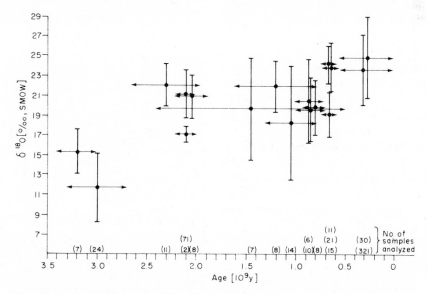

Fig. 19.1. Oxygen isotope composition of substantially unaltered sedimentary carbonates as a function of geological time. Circles indicate mean values, vertical bars the standard deviation and horizontal arrows the possible time range (after Schidlowski, Eichmann and Junge, 1975; reproduced by permission of Elsevier Scientific Publ.)

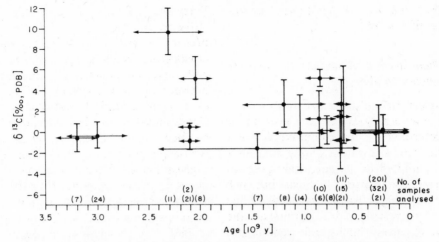

Fig. 19.2. Carbon isotope composition of substantially unaltered sedimentary carbonates as a function of geological time. Circles indicate mean values, vertical bars the standard deviation and horizontal arrows the possible time range (after Schidlowski, Eichmann and Junge, 1975; reproduced by permission of Elsevier Scientific Publ.)

throughout geological time. According to the model of Schidlowski, Eichmann and Junge (1975) this constancy implies that total sedimentary carbon was almost always partitioned between organic carbon and carbonate carbon in the ratio of 20:80, as at present.

The trend of the isotopic variations of Sr during earth history is shown in Fig. 19.3. The lowest measured strontium isotope ratios in the sedimentary carbonates are taken as the best approximation for $^{87}Sr/^{86}Sr$ of coeval well-mixed seawater (Veizer and Compston, 1974, 1976). The sharp increase in values in the period 2500–2000 my ago and the decrease in the Phanerozoic parallel the K_2O/Na_2O trend in sedimentary and igneous rocks established by Engel and coworkers (1974); these trends probably define major fractionation stages of the earth's crust (see later in this chapter under *Isotopes and Crustal Evolution*).

Atmospheric Evolution and the Development of Life Forms

There has been a close interdependence of atmospheric and biospheric activity throughout geological time. On the one hand, most of the free oxygen in the atmosphere resulted from biological activity through the photosynthesis reaction, and on the other, changes in atmospheric composition, in particular the progressive increase in the molecular oxygen content, triggered off major biological innovations which enabled life to advance and diversify (J. W. Schopf, 1974, 1975; Schidlowski, 1976).

The following milestones in biological evolution related to atmospheric conditions can be recorded, albeit tentatively (see Fig. 19.4). (For further aspects of Archaean and Proterozoic life forms see Chapters 2 and 7, respectively.)

1. In the absence of oxygen-mediating enzymes free O_2 is a poison to living cells. In the early Archaean free oxygen was produced by the photosynthesis reaction by the first organisms and the banded iron formations conveniently acted as the oxygen acceptor, which they themselves needed for their formation (Cloud, 1965, 1968a, 1973). By keeping the oxygen levels down to a safe minimum the BIF gave the early life forms time to adapt to their oxygeneous waste product. These first (blue–green algae) organisms were procaryotes which were relatively resistant to ultraviolet radiation as the atmosphere in the early Archaean lacked a radiation-protective ozone screen.

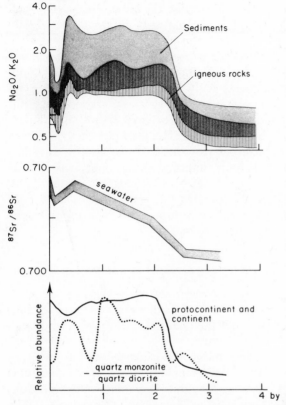

Fig. 19.3. Variations in K_2O/Na_2O ratios of sedimentary and igneous rocks, relative proportion of continental crust and protocrust, and quartz monzonite/quartz diorite ratio during geological history (after Engel and coworkers, 1974), and $^{87}Sr/^{86}Sr$ of seawater with time (after Veizer and Compston, 1976; figure from Veizer, 1976a; reproduced by permission of J. Wiley)

2. By 1500 ± 300 my ago the oxygen content of the atmosphere had increased to such an extent that a primitive form of oxidative metabolism was possible; the first oxygen-employing eucaryotes appeared at this time (Schopf, 1974). These were complex organisms with a nucleus capable of cell division enabling the genetic coding material (DNA) to be passed on to their descendants (Cloud, 1968a). But note that Knole and Barghoorn (1975) challenge this early date (Chapter 7).

Louis Pasteur discovered that many modern microbes undergo a fundamental change in their type of metabolism from fermentation to respiration when the oxygen content reaches 1% of the present atmospheric level (PAL)—known as the Pasteur level. Berkner and Marshall (1967) identified the Precambrian–Cambrian boundary with an oxygen level of 1% PAL, but this Pasteur level must have been reached by 1500 ± 300 my ago, the age of the oldest eucaryotes (Schidlowski, 1976).

The rise of the oxygen pressure of the ancient atmosphere to 1% PAL enabled life to pass the Pasteur level and change from fermentation, an anaerobic process, to respiration, a highly advanced form of aerobic metabolism (Schidlowski, 1971, 1976). With the appearance of respiration primitive eucaryotic organisms were able to evolve a nervous system to control the process and a circulatory system to distribute the oxygen.

By the time the O_2 content of the atmosphere had reached somewhere upward of 1% PAL towards the late Precambrian, an ozone layer had developed in the upper atmosphere

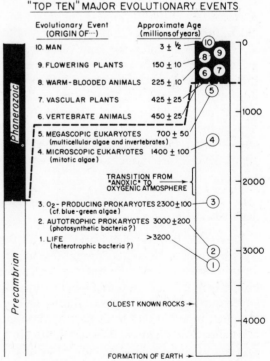

Fig. 19.4. The temporal distribution of important innovations that have occurred during the course of biological evolution (modified by Schopf after Schopf, 1974; reproduced by permission of D. Reidel Publishing Co.)

which effectively screened out the DNA-inactivating ultraviolet radiation in the 2400–2600 Å wavelength range, as a result of which the ocean waters were opened up to pelagic life (Cloud, 1968a).

3. By 1000–900 my ago primitive micro-organisms had evolved advanced techniques of sexual cell reproduction (J. W. Schopf, 1972) and about 700–600 my ago the first Metazoa appeared—the Ediacara fauna (for details see Chapter 7). These are complex multicellular organisms that require oxygen for their growth—3% PAL according to Cloud (1968a). The earliest fauna were soft bodied, i.e. jelly-fish, worms and sponges etc., and they probably contained collagen, the main structural protein in Metazoan tissues, which requires molecular oxygen for its synthesis (Towe, 1970), although collagen is particularly concerned with the formation of hard skeletons and shells, evidence for which appeared in the fossil record about 570 my ago at the Precambrian–Cambrian boundary.

4. According to the Berkner–Marshall (1967) model further increase in the oxygen content to 10% PAL enabled the land surface to be protected from DNA-damaging ultraviolet radiation and this is identified with the evolution of land plants, the first spores of which appeared in the mid-Silurian. The oxygen content in the atmosphere reached the present level by the Carboniferous by which time highly developed continental flora were flourishing.

Sedimentation

There are appreciable variations in the amounts, types and compositions of sedimentary rocks formed at different times in earth history (Fig. 19.11). A great many factors may have contributed to these variations, such as changes in the types of tectonic environment, atmospheric and hydrospheric evolution, erosion rates dependent on changes of acidity of soils due in turn to development of plant and other species, changes in the composition of igneous rocks forming the bedrocks that contributed detritus (dependent on the prevailing tectonic environment), and recycling processes accounting for the destruction, preservation and formation of different types of sediment. The aim here is to summarize current concepts on the predominant processes and trends that have given rise to these major changes with time. In this subject caution should be used as today we are looking at the remains, not the original extent, of sedimentary sequences. The main changes concerned have been ascribed to two contributory factors: recycling and evolutionary trends. The review of Veizer (1973) forms the basis of part of the following account. But first we must look at the volume and type of preserved sedimentary rocks.

Sedimentary Deposits in Relation to Age

An estimate of the average mass of sedimentary rocks that remains today for each period of the Phanerozoic is obtained by dividing the estimated mass by the duration of each Period; most authorities agree that there is a marked minimum for the Permian and distinct maxima for the Devonian and Cenozoic. According to Garrels and Mackenzie (1971) the mass of Precambrian sedimentary rock is only a little less than the total of the Phanerozoic but its distribution with age is poorly known. There are relatively few sediments of late Precambrian age (600–800 my ago) and this represents a genuine minimum as there is a high volume about 1000 my ago.

Fig. 19.5 shows the relative volume percent of various sedimentary rocks as a function of their age. The figure illustrates that the majority of Precambrian carbonates are dolomites, whilst evaporites and limestones are more typical of young periods, jaspilites are restricted to the Precambrian, and arkoses reached their maximum in the Proterozoic whereas mature sandstones increase in younger periods and there are more Archaean greywackes than later. During the Precambrian lutites (argillaceous rocks) were fairly constant in amount.

Recycling

The sedimentary recycling concept takes account of the fact that sediments have been

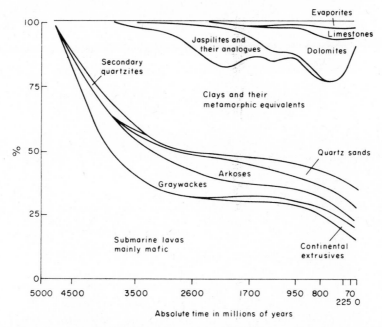

Fig. 19.5. Outline of the compositional evolution of sedimentary rocks (after Ronov, 1968; based on Ronov, 1964; reproduced by permission of Blackwell Scientific Publications Ltd.)

eroded and redeposited (i.e. recycled) throughout time. The point is brought home when one considers that at the present rate of erosion half the currently exposed sediments in the world will be destroyed in about 100 million years. Some rocks such as lutites and volcanogenics are more resistant to recycling than, for example, limestones which may have been recycled at least five times in the last 3000 my. Thus the recycling of sediments has led to differences in the proportion of rock types originally deposited compared with those seen today (Garrels, Mackenzie and Siever, 1972).

Sedimentological Trends

Besides changes due to recycling there are several prominent long-term changes in the composition and relative abundance of sedimentary rocks:

1. Banded iron formations of the oxide–carbonate–silicate–sulphide type occur in the oldest (>3760 my) Isua sediments, West Greenland, and in most Archaean greenstone belts. They reached their culminating stage about 2000–1800 my ago in early Proterozoic basins (Goldich, 1973), after which they failed to form—at least, in mid–late Proterozoic and Phanerozoic sedimentary piles there are no BIF of this type. On the other hand, iron formations of the chamosite–goethite–siderite type, unknown in the Precambrian, formed in the Phanerozoic.

2. The oldest continental red beds appeared in the Dharwar greenstone belts of India about 2500 my ago (Srinivasan and Sreenivas, 1972), but these occurrences are early rarities. Red beds did not form in abundance until 1800 my ago after which they became prominent in the late Proterozoic and Phanerozoic. They are regarded by Cloud (1968b) as complementary to the earlier BIF.

3. Carbonate rocks make up about 25% of the Phanerozoic sedimentary mass, only 5% of the Proterozoic and even less of the Archaean (Garrels, Mackenzie and Siever, 1972). The Mg/Ca ratio of carbonate sediments has broadly decreased with time, dolomites being characteristic of Archaean greenstone belts and Proterozoic basins, and limestones of the Phanerozoic (H. D. Holland,

1976). Ferruginous dolomites are common in the Archaean (e.g. magnetite marbles in gneissic areas of S India) and the iron content of these rocks generally decreased with time. Shallow-water pelitic–littoral biogenic limestones appeared in the Cambrian and about 150 my ago in the late Jurassic pelagic limestones made their appearance in deep-water oceanic environments created by post-Pangaea continental drift (Veizer, 1973).

4. There are other non-detrital sediments that changed with time. Manganese deposits associated with carbonate sediments are unknown in the Archaean; they appeared about 2300–2000 my ago in several continents. There are three records of sedimentary anhydrite–gypsum deposits in mid–late Proterozoic rocks, namely in the Grenville Supergroup in Canada, the Belt–Purcell System in the American Cordillera, and the Bitter Springs Formation (Amadeus Basin) in Australia (Cloud, 1968a; Veizer, 1973); such sedimentary deposits did not accumulate in abundance until the early Phanerozoic. Evaporites are virtually unknown in the Precambrian but occupy roughly 5% of the Phanerozoic mass (Garrels, Mackenzie and Siever, 1972). A 3 m thick coal–graphite bed occurs with carbonaceous shales in the 2000 my old Ketilidian sediments of South Greenland (Bondesen, 1970), but coal formation did not flourish until the late Devonian after the evolution of land plants. Table 19.1 illustrates the evolution of the non-detrital sedimentary rocks mentioned above.

5. Greywackes and turbidites are predominant in Archaean greenstone belts and arkoses are typical in early Proterozoic basins. Such immature detrital sediments gave way with time to more common, mature sandstones and ortho-quartzites. However this may not be a simple linear trend of evolution as the little-known sediments in Archaean high-grade regions belong to the orthoquartzite–carbonate (marble) association (Sutton, 1976).

Chemical Trends

There are several distinctive secular chemical variations in the composition of sedimentary rocks, in particular of shales and carbonates (Ronov and Migdisov, 1971; Veizer, 1973).

1. K_2O/Al_2O_3, Na_2O/Al_2O_3 and K_2O/Na_2O variations in detrital rocks, especially shales which are illustrated in Fig. 19.6. Veizer (1973) interprets the alkali variations as reflecting a continental crustal evolution from a more mafic one (plagioclase-dominant) in the Archaean to more felsic (with increase in potash feldspar) in younger periods, until the marked reversal in the Palaeozoic. Whilst this conclusion is consistent with the presence of extensive mafic volcanics in greenstone belts, it is invalid for the high-grade Archaean regions where plagioclase gneisses predominate.

2. Increasing total Fe/Al_2O_3, MnO/Al_2O, $MnO/total Fe$ and Fe^{2+}/Fe^{3+} with age of the rocks, especially carbonates. As the separation of Mn and Fe is caused, in particular, by the oxidation potential of Fe the trends may reflect the fact that the oxygen pressure was progressively increasing with time.

3. Increasing MgO/CaO and SiO_2/Al_2O_3 with age, especially for carbonates and pelites (Van Moort, 1973; H. D. Holland, 1976). According to Veizer (1973) the depositional environment of carbonate sediments has spread with time from semi-barred to littoral sequences in the Precambrian into more open milieux. The higher proportion of dolomites in the Precambrian may be related to the fact that the earlier sequences were more susceptible to dolomitization (Friedman and Sanders, 1967).

A more likely explanation, however, is that the decrease in the dolomite/calcite ratio reflects variations in the evolution of the ocean–atmosphere system, such as the input of CO_2 and Ca. As more CO_2 enters the oceans, so more Ca is deposited within $CaCO_3$; when more CO_2 is put into the system than Ca is released during weathering, as in the Archaean and Proterozoic, then Mg is also deposited as a carbonate phase in dolomites. It seems that the marked decrease in the dolomite/calcite ratio by the Phanerozoic was due to a decrease in the ratio of CO_2 supply to Mg and Ca demand, caused by a decrease in

Table 19.1. Evolution of non-detrital sedimentary rocks in geological history (after Veizer, 1973; reproduced by permission of *Contrib. Mineral. Petrol.*)

Pelagic (deep sea) environments	Open shelf (littoral–neritic) environments	Semi-barred environments	Terrestrial environments
Pelagic biogenic limestones (150 my–present)			
			Coal formations (350 my–present)
	Neritic–littoral biogenic limestones (600 my–present), phosphatic (?) in early stages	Salts (600 my–present)	
		Ca-sulphates (700 my–present)	
	Oolithic goethite–chamosite–siderite iron ores (700 my–present)		
	Manganese carbonates and oxides (700 my–present)		
	Inorganic and/or biochemical limestones (1000 my–present)		
	Ca-phosphates (1000 my–present)		
			Red beds (1800 my–present)
	Early diagenetic dolomites (2500–600 my)		
Banded iron ore formations and manganese silicates, carbonates and oxides (3400–1800 my)			

both the rate of juvenile degassing and of degassing of recycled CO_2 (Holland, 1976).

As a result of these variations dolomites reached their maximum thickness during the Proterozoic; examples are:

Dolomite Series, Transvaal System (S Africa)	2·6 km
Coronation Geosyncline (Canada)	1·2
MacArthur Basin (Australia)	4·2
Mount Isa Geosyncline (Australia)	3·0

Another factor in the formation of such thick early Proterozoic dolomites may be that

weathering of the common Archaean basic volcanic rocks gave rise to a mobility series of elements, $Mg > Ca > Na > K$, with the result that Mg and Ca were the predominant cations in the oceans (Krynine, 1960). In fact Ronov (1964) calculated that early Proterozoic sea-water contained between 8 and 9 times as much Ca and 2·5 times as much Mg as modern sea-water.

4. Engel and coworkers (1974) use the K_2O/Na_2O ratio of clastic sediments as an index of their increasing maturity, i.e. from the greywackes and subgreywackes in the Archaean to the argillaceous sandstones and shales characteristic of stable continental shelf

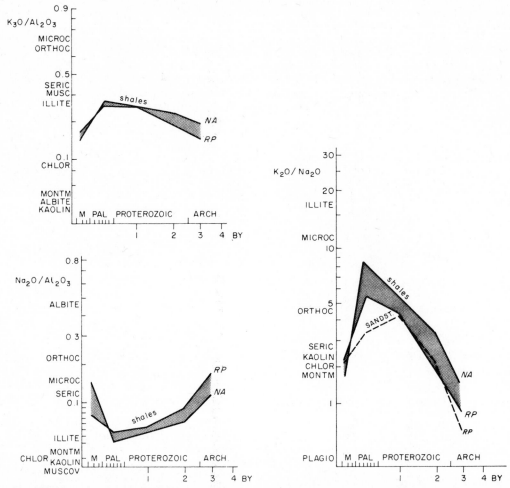

Fig. 19.6. Variations of K_2O/Al_2O_3, Na_2O/Al_2O_3 and K_2O/Na_2O ratios with age for shales and their partly metamorphosed equivalents. NA = North America, RP = Russian Platform (after Veizer, 1973, based on summary of Ronov and Migdisov, 1971). (Reproduced by permission of *Contr. Mineral. Petrol.*)

assemblages in the Proterozoic. Fig. 19.3 shows a plot of this ratio increasing as a function of decreasing age, demonstrating that the sediments were progressively derived from more and more fractionated igneous rocks. This ratio, they suggest, may be used as a guide to thickness, composition and stability of the source region.

Table 19.2 summarizes possible changes in chemical composition of sedimentary rocks by a variety of factors.

Magmatism

There are many who ascribe to the view that magmatic activity has been the major process contributing to the growth of the continents (Green, 1972a; Jakes, 1973; Engel and coworkers, 1974). Indeed if one considers just the Archaean period, which accounts for about one third of geological time, results from the last few years in Scotland, Greenland and Labrador suggest that sedimentary processes there played a very small role in the crustal development of these 'high-grade' regions (Bowes, Barooah and Khoury, 1971; Bridgwater and coworkers, 1973c; Collerson, Jesseau and Bridgwater, 1976), whilst it is well known that sediments make up a minor part of the predominantly volcanic greenstone belts of the world (Anhaeusser and coworkers, 1969; Glikson, 1970).

Table 19.2. Possible changes in the chemical composition of sedimentary rocks with increasing age caused by four factors (after Veizer, 1973; reproduced by permission of *Contr. Mineral. Petrol.*)

Process	Increasing with increasing age	Decreasing with increasing age
(a) Diagenetic and metamorphic effects	Al_2O_3, MgO, MgO/CaO, Fe^{2+}/Fe^{3+}, K_2O, K_2O/Na_2O, K_2O/Al_2O_3, Rb, Total Fe	CaO, Sr, Na_2O, Na_2O/Al_2O_3
(b) Evolution of sedimentary environments	MgO, MgO/CaO	CaO
(c) Changes in the chemical composition of atmosphere and hydrosphere	MnO, Total Fe, MnO/Total Fe, Fe^{2+}/Fe^{3+}, MgO, MgO/CaO, SiO_2	(CaO)
(d) Changes in chemical composition of the upper continental crust.	Na_2O/K_2O, MgO/CaO, Fe^{2+}/Fe^{3+}, Al_2O_3/SiO_2, K/Rb, Sr, Mn, Ti, Mg, Fe, Al, Ca, P	K, Na, Si, Rb

The problem that arises in a discussion of this kind is the paucity of synthesized petrochemical data; the report by Engel and coworkers (1974) is outstanding in this respect. It is only possible, therefore, to outline here some of the key points and petrogenetic trends that were important in crustal evolution.

Precambrian Activity

Most Precambrian cratons have suffered much erosion and therefore we would expect many high-level igneous rocks to have been removed. This means that we must be careful to compare rock complexes from the same approximate depth zones, amongst other variables, in order to study crustal fractionation as a function of time (Engel and coworkers, 1974).

Lambert (1971) and Glikson and Lambert (1973) demonstrated that K, Th, U and Rb decrease with depth and increasing grade from amphibolite to granulite facies in Australian gneisses and similar decreases are known in other cratons (Ramberg, 1951; Eade, Fahrig and Maxwell, 1966; Eade and Fahrig, 1971). The cause of this decrease is debatable. Sheraton (1970), Lambert (1971) and Tarney, Skinner and Sheraton (1972) believe that the granulite facies metamorphism was responsible for partial melting, depletion and migration of these elements into high level crustal

zones; however Holland and Lambert (1975) are of the opinion that the present chemical composition is inherited from the igneous parent and that there has been no 'depletion' in these elements. This opinion is corroborated by the discovery that many late Archaean tonalite gneisses (both in amphibolite and granulite grade) are derived by recrystallization of igneous tonalites (McGregor, 1973; Davies, 1976) that contained low values of the large ionic lithophile trace elements when they were intruded, therefore requiring no 'depletion' during the subsequent metamorphism.

A large proportion of Precambrian volcanic rocks have been metamorphosed and a similar cautionary note applies to the interpretation of their original compositions with respect to the 'incompatible elements' that are most mobile during metamorphism. Gunn (1976) in particular stresses this problem. With these reservations in mind let us consider some of the more important features of Precambrian magmatism.

As Engel and coworkers (1974) well document (Fig. 19.3), there is a tendency for the K_2O/Na_2O ratio of igneous rocks to increase progressively with time (until the Phanerozoic). The rocks in Archaean greenstone belts are notably low in potash, illustrated by the common occurrence of low-K tholeiites, komatiites and tonalitic plutons and the rarity of alkaline rocks, whilst the

310

high-grade Archaean regions are dominated by tonalitic to granodioritic gneisses and potash-poor granulites. In contrast Proterozoic belts tend to be characterized by reactivated potash granites and gneisses, granite plutons including rapakivi granites, appinites, and alkaline complexes. One explanation for this increase with time in the K_2O/Na_2O ratio of volcanic and plutonic rocks could be that many were derived by arc- and Cordilleran-type plate collisions in the Archaean, whereas in the Proterozoic they were due more to continental collision-type plate movements (Mitchell, 1976a).

Tholeiitic dolerite (diabase) dykes are prominent in Precambrian cratons. From a study of 32 swarms in Canada, Eade and Fahrig (1971) showed that dykes older than 1300 my tend to have lower K_2O, Na_2O and TiO_2 contents than younger dykes. The relatively low values of these oxides in Archaean dykes from Wyoming (Condie, Barsky and Mueller, 1969), Greenland (Rivalenti and Sighinolfi, 1971; Rivalenti, 1975) and Scotland (Tarney, 1973) are in agreement with these relations. In addition to these variations, Mueller and Rogers (1973) demonstrated that the MgO content decreases by a factor of 2 and the initial strontium isotope ratio increases from about 0·7025 to 0·706 in basic dykes ranging from 2700 to 700 my old in the Beartooth Mountains of Montana and Wyoming (Fig. 19.7). There are several possible causes of such long-term chemical changes: variations with time of crustal contamination or of the degree of partial melting of mantle material. Condie (1976b) favours the suggestion that the depth of melting increased progressively with decreasing age as expected from a falling geothermal gradient with time.

However such unidirectional evolutionary trends tend to hide the 'unique' character of a great many Precambrian rock suites. For example, ultramafic lavas are common in the Archaean but so far have not been found in the Proterozoic. Likewise, calc-alkaline volcanic rocks and calcic layered anorthosite–gabbro complexes typify many Archaean regions whereas alkaline complexes and more sodic

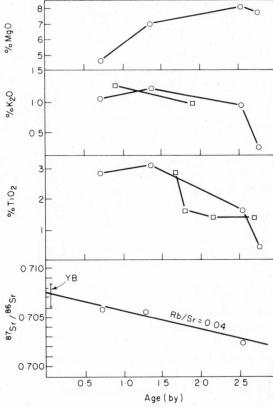

Fig. 19.7. MgO, K_2O, TiO_2 and initial strontium isotope ratios in Precambrian diabase dykes versus age. ○ = Beartooth Mountains, Montana; □ = Canadian Shield; YB = Yellowstone basalts (after Mueller and Rogers, 1973; reproduced by permission of The Geological Society of America)

anorthosites are characteristic of Proterozoic belts. Plateau basalts of continental tholeiitic type erupted onto the stabilized Proterozoic cratons but were apparently unable to form in the more mobile Archaean environments.

Many Precambrian rock groups may have formed as a result of collisional plate movements. For example, many 'granitic' rocks are not unlike the granitic complexes developed along Phanerozoic Cordilleran-type plate margins. This subject can be considered in two parts.

Firstly, the Archaean. The tonalites (since deformed into tonalitic gneisses) that make up the bulk of the Archaean high-grade regions are comparable to the early tonalites in the deep-seated parts of the batholiths in western

USA (Windley and Smith, 1976). In both environments they contain the remnants of gabbro–leucogabbro–calcic anorthosite composed predominantly of calcic plagioclase and hornblende and the complexes are bordered by a recrystallized shelf-type sedimentary assemblage of mica schists, quartzites and marbles together with amphibolitic metavolcanics in places; all these igneous and sedimentary rocks occur as rafts within the engulfing and intruding tonalites. The Archaean rocks that formed during the major period of growth or thickening of the continents about 3100–2800 my ago are little different from those that evolved at the leading edges of continental plates in the Phanerozoic.

An interesting comparison can be made between the 510 my old Caledonian migmatite complex in the Cashel district of Connemara in Eire (Leake, 1970) and the typical Archaean high-grade gneissic complexes. According to several plate tectonic models Connemara must have lain at a continental margin bordering an ocean in the Cambrian. The remarkable similarity of the rock associations and their chemistry, structure, age relations, etc. in the Caledonian and Archaean regions can perhaps best be brought home by saying that if an Archaean high-grade field geologist were to be dropped unwittingly into this part of Connemara, he would not realize that he was not on a typical Archaean complex. Early meta-sedimentary schists and gneisses were intruded firstly by gabbro–leucogabbro complexes (calcic plagioclase-hornblende) and then semiconcordantly by innumerable sheets of tonalite. The resulting conformable tectonic pile was deformed and metamorphosed to a high amphibolite grade, the tonalites being converted to gneisses.

Geochemical studies by Arth and Hanson (1974, 1975) established that tonalites (quartz diorites) and trondhjemites (potash–feldspar-free granodiorites) of the 3500 my old Morton Gneiss in Minnesota, several 2700–2750 my old plutons in Minnesota and the Mesozoic Craggy Peak pluton in California all have similar major element chemistry, Rb, Sr and rare earth contents. The Archaean plutons have low initial Sr isotope ratios of less than 0·7010 indicating derivation from mantle sources (Arth, 1976). All the rocks have low total REE contents, near chondritic, depleted heavy REE, and K/Rb and Sr/Ba ratios comparable to sub-alkaline tholeiites. The REE distributions are also similar to those of the Amîtsoq gneisses in Greenland and the Bonsall tonalite in California (B. Mason, 1975). The similarity in the chemical patterns suggests that they are associated with the primary origin of the magmas and that the rocks formed under similar physico-chemical conditions and had a similar mode of origin. T. H. Green and Ringwood (1968) suggested that such quartz dioritic or granodioritic rocks may be derived by transformation of basalt or gabbro along a subduction zone. Thus both the Archaean tonalitic gneisses and plutons could well have formed along some kind of primitive active continental margin.

Secondly, let us consider some mid-Proterozoic granitic complexes in relation to possible modern analogues. Hietanen (1975) compared the granitic rocks of southern Finland (Svecofennian) with those of the Sierra Nevada (Palaeozoic–Mesozoic). The earliest synkinematic plutonic rocks in both areas are sheets of trondhjemite intruded parallel to the bedding of meta-sedimentary gneisses. Early trondhjemites and tonalites contain remnants of hornblende gabbros (Hietanen, 1943). Arth (1976) points out that the gabbro–diorite-tonalite–trondhjemite suite of southern Finland could be explained by fractional crystallization of a mildly alkaline hornblende gabbro magma and this model would also fit the data from the Archaean tonalite suite in Minnesota. Later granitic phases in Finland are intrusive and enriched in potassium feldspar. There is an increase in the potassium content of the plutonic magmas from south to north and also with time and Hietanen (1975) suggests that these chemical variations in time and space are so similar to those from west to east in the Sierra Nevada that they likewise may have been generated by plate tectonic mechanisms (Fig. 6.5).

Another mid-Proterozoic granitic complex worth considering in the present context is the Julianehaab granite of South Greenland which was compared with the Andean batholiths of South America by Bridgwater, Escher and Watterson (1973a). Synplutonic basic dykes are an important feature of the Julianehaab granite (Watterson, 1968) and the main place in Phanerozoic fold belts where similar dykes are common is within the batholiths of British Columbia and Washington (Goodspeed, 1955; Roddick and Armstrong, 1959) and Peru (Cobbing and Pitcher, 1972). Prominent in the development of the latest phases of the Julianehaab granite was the formation of an appinitic suite including primary hornblende gabbros, diorites and microdiorites that were intruded and cooled only just before their enclosing granitic rocks (Walton, 1965). A common result of this nearly simultaneous intrusion of acid and basic magma was the formation of net-veined diorite complexes. Similar contemporaneous appinites and granites occur in:

1. The Peninsular Range batholith of southern California where, in the Bonsall tonalite, the hornblende gabbro complex described by Nishimori (1974) has net-veined relations with the engulfing tonalite and is itself cut by net-veined tonalite–diorite dykes (my observation); and
2. In the Caledonian belt of Scotland where an appinitic suite is intimately related to the Newer Granites (Mercy, 1965).

Walton's paper was published in 1965 before the modern plate tectonic concepts were applied to continental geology, but he recognized in essence the similarity of many of the key features of the Ketilidian mobile belt with those of the Caledonides—e.g. the appinite suite described above, the possible relation of the diorites to the high-level andesites since removed by erosion and, finally, that the Gardar alkaline intrusions and molasse-type sandstones represent a typical post-orogenic development of the mobile belt. When the major zones of the Ketilidian fold belt outlined by Bridgwater, Escher and Watterson (1973a) are compared with the expected sequence across a continent–continent collision mobile belt suggested by Dewey and Burke (1973), and when the appinitic suites and their relationships to surrounding granites are compared with those in the batholiths of western USA and the plutons of the Caledonides of Scotland, there is a strong body of evidence favouring the likelihood that the Ketilidian belt developed in response to some form of plate collision.

Phanerozoic Magmatism and Plate Tectonics

Compared with the Precambrian, a very different situation emerges when we begin to consider the connection between magmatism and tectonics in Phanerozoic belts. Mesozoic–Cenozoic fold belts can be related to some form of continental drift because much of the oceanic crust formed in that time is extant. There has been little difficulty in proposing and accepting plate tectonic models for early Palaeozoic belts like the Appalachians and Caledonides, in spite of the fact that their relevant oceanic magnetic tape recorder has been destroyed, because there is a marked regularity in the distribution of sedimentary and volcanic facies comparable with that of Mesozoic–Cenozoic belts and also because a few convincing remnants of obducted oceanic crust are preserved as ophiolite complexes. Because of a lack of regularity in supracrustal distribution and of ophiolites, relations in Hercynian fold belts are less readily comprehensible, as witnessed by the multitude of current tectonic models.

Bearing in mind the above comments it is not difficult to relate Phanerozoic magmatism with plate tectonic activity, either with certainty if it is young, or expectation if it is old. The environments concerned are intracontinental rifts and graben, oceanic (and marginal basin) accretion boundaries, island arcs and continental margins, and continent–continent junctions. Table 19.3 lists the average bulk compositions of basalts and andesites in modern arcs, rises and rifts. The sequence of magmatic events from early continental rifting to final continent–continent collision can be followed in Chapters 14 to 18.

Table 19.3. Average compositions of modern basalts and andesites (after Condie, 1976b; reproduced by premission of Pergamon Press)

	Low-K tholeiite		Continental rift tholeiite	Island tholeiite	High-Al tholeiite	Oceanic Alkali basalt	Cont. rift Alkali basalt	Arc andesite	Low-K andesite	High-K andesite	Shoshonite
	Rise	Arc									
SiO_2	49.8	51.1	50.3	49.4	51.7	47.4	47.8	57.3	59.5	60.8	52.9
TiO_2	1.5	0.83	2.2	2.5	1.0	2.9	2.2	0.58	0.70	0.77	0.85
Al_2O_3	16.0	16.1	14.3	13.9	16.9	18.0	15.3	17.4	17.2	16.8	17.2
Fe_2O_3*	10.0	11.8	13.5	12.4	11.6	10.6	12.4	8.1	6.8	5.7	8.4
MgO	7.5	5.1	5.9	8.4	6.5	4.8	7.0	3.5	3.4	2.2	3.6
CaO	11.2	10.8	9.7	10.3	11.0	8.7	9.0	8.7	7.0	5.6	6.4
Na_2O	2.75	1.96	2.50	2.13	3.10	3.99	2.85	2.63	3.68	4.10	3.50
K_2O	0.14	0.40	0.66	0.38	0.40	1.66	1.31	0.70	1.6	3.25	3.69
Cr	300	50	160	250	40	67	400	44	56	3	30
Ni	100	25	85	150	25	50	100	15	18	3	20
Co	32	20	38	30	50	25	60	20	24	13	~20
Rb	1	5	31	5	10	33	200	10	30	90	100
Cs	0.02	0.05	0.2	0.1	0.3	2	>3	~0.1	0.7	1.5	≥2
Sr	135	225	350	350	330	800	1500	215	385	620	850
Ba	11	50	170	100	115	500	700	100	270	400	850
Zr	85	60	200	125	100	330	800	90	110	170	150
La	3.9	3.3	33	7.2	10	17	54	3.0	12	13	15
Ce	12	6.7	98	26	19	50	95	7.0	24	23	32
Sm	3.9	2.2	8.2	4.6	4.0	5.5	9.7	2.6	2.9	4.5	3.2
Eu	1.4	0.76	2.3	1.6	1.3	1.9	3.0	1.0	1.0	1.4	0.95
Gd	5.8	4.0	8.1	5.0	4.0	6.0	8.2	4.0	3.3	4.9	4.2
Tb	1.2	0.40	1.1	0.82	0.80	0.81	2.3	1.0	0.68	1.1	0.50
Yb	4.0	1.9	4.4	1.7	2.7	1.5	1.7	2.7	1.9	3.2	1.7
U	0.10	0.15	0.4	0.18	0.2	0.75	0.5	0.4	0.7	2.2	1.3
Th	0.18	0.5	1.5	0.67	1.1	4.5	4.0	1.3	2.2	5.5	2.8
Th/U	1.8	3.3	3.8	3.7	5.9	6.0	8.0	3.2	3.1	2.5	2.2
K/Ba	105	66	32	32	12	28	16	58	49	68	36
K/Rb	1160	660	176	630	344	420	55	580	440	300	306
Rb/Sr	0.007	0.022	0.089	0.014	0.029	0.045	0.13	0.046	0.078	0.145	0.118
La/Yb	1.0	1.7	10	4.2	3.7	11	32	1.1	6.3	4.0	8.8

* Total Fe as Fe_2O_3.

Thermal Regimes

Types of Metamorphism and Metamorphic Belts

There are two points about thermal regimes in space and time that have important implications for continental evolution:

1. Various Phanerozoic fold belts have different metamorphic assemblages because they formed under different geothermal gradients. This refers to the metamorphic facies series, particularly with respect to the Alpine, Hercynian and Caledonian fold belts of Europe and the paired belts of island arcs (Miyashiro, 1961, 1972, 1973a,c; Zwart, 1967a,b, 1969).

2. Certain types of metamorphic rocks seem to be confined to specific periods of time. For example, blueschists are only found in Phanerozoic fold belts, even though they may have started to form in the late Proterozoic (Ernst, 1972), and granulite facies rocks only occur at the present erosion level in the Precambrian shields (Oliver, 1969; Heier, 1973). If Miyashiro (1973a) is correct in believing that there should be no great difference in the depth of erosion between Precambrian shields and many Phanerozoic fold belts, then the restriction of granulite facies rocks to the former suggests that metamorphism was more intense and geothermal gradients steeper in earlier periods of earth history.

Regional metamorphism, with which we are concerned, may be classified into three basic types representing different geothermal gradients (Miyashiro, 1972, 1973a).

Type	Characterized by	Average geothermal gradient
Low pressure	Andalusite	$>25°C\,km^{-1}$
Medium pressure	Kyanite (without glaucophane)	$\sim 20°C\,km^{-1}$
High pressure	Glaucophane and jadeite	$\sim 10°C\,km^{-1}$ or less

There are low-pressure metamorphic belts with stable andalusite of all ages from Archaean (Binns, Gunthorpe and Groves, 1976), through Proterozoic (Zwart, 1967a; Shimron and Zwart, 1970), to Tertiary (Miyashiro, 1973c), and likewise medium-pressure belts with kyanite from the Archaean

(Sreenivas and Srinivasan, 1974; Kalsbeek, 1976) to the Tertiary (Ernst, 1973). High-pressure belts formed largely in Phanerozoic time and their evolution and tectonic significance are now discussed in detail.

Phanerozoic Blueschists

The majority of glaucophane-bearing, high-pressure belts are of Mesozoic–Cenozoic age having formed in island arcs, Cordilleran and Alpine belts related to plate subduction, obduction and collision since the break up of Pangaea (Miyashiro, 1961; de Roever, 1965; Ernst, 1971, 1972; Coleman, 1972). There are some glaucophane rocks in Palaeozoic belts, especially in the early Japanese arcs (Ernst, 1972), the Urals (Hamilton, 1970) and the Caledonides of Scotland (Bloxam and Allen, 1959); those in the Caledonian Mona Complex of Wales are late Precambrian (Shackleton, 1969; D. S. Wood, 1974). The rare glaucophane rocks in the Archaean greenstone belts of the Dharwar Group in Mysore, India (Uadarajan, 1968) are regarded by Shackleton (1973a) as of local significance.

Ernst (1972) reviewed the occurrence and mineralogical evolution of blueschists with time. He pointed out that whilst the epidote-bearing greenschist–blueschist rocks occur throughout the Phanerozoic, lawsonite is rare in the Palaeozoic, and aragonite with jadeitic pyroxene and quartz are strictly confined to Mesozoic and Cenozoic terrains (Fig. 19.8). These index minerals indicate higher pressures in the formation of the younger blueschist rocks as indicated by experimental phase equilibrium studies.

The virtual absence of blueschists in the Precambrian cannot be ascribed simply to the fact that they have been removed by erosion from older terrains because there are very many well-preserved, otherwise comparable, greenschists in Archaean greenstone belts. The absence of blueschists in the Precambrian is consistent with the absence of ophiolite complexes, with which they are usually associated. The explanation of their appearance in the later part of earth history may be related to factors such as the decrease in the

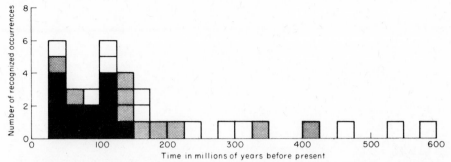

Fig. 19.8. Histogram showing incidence of blueschists of contrasting mineralogies with time. Open boxes represent epidote-bearing glaucophane schists, stippled pattern lawsonite (±epidote), and black boxes aragonite and/or jadeitic pyroxene and quartz (generally also lawsonite) (after Ernst, 1972; reproduced by permission of *American Journal of Science*)

geothermal gradient with time (see later this chapter).

Experimental phase equilibrium studies enable the P–T conditions to be worked out for high-pressure metamorphism. However, the precise conditions necessary for the formation of glaucophane, the characteristic mineral of blueschists, are not yet known; estimates of its appearance are 200°C at 5 kb and 350°C at 7 kb (Winkler, 1974). Other critical minerals are jadeite with quartz and aragonite, which lie on the high pressure side of the following reactions:

$$\text{jadeitic pyroxene} + \text{quartz} = \text{albite}$$

$$\text{aragonite} = \text{calcite}$$

Geothermal gradients estimated from these curves, combined with oxygen isotope geothermometry of glaucophane-bearing rocks (Taylor and Coleman, 1968), are about 10°C/km (Miyashiro, 1973c).

Paired Metamorphic Belts

The blueschists referred to above belong to a pair of parallel regional metamorphic belts of similar age but contrasting type (Miyashiro, 1973a,c). The high-pressure belt characterized by the presence of glaucophane in blueschists and ophiolite complexes lies near the trench on the oceanic side of a low-pressure belt characterized by andalusite together with granites, andesites and rhyolites. They are best developed in the circum-Pacific region where there are at least 14 pairs (Figs.

16.8, 19.9). The high-pressure metamorphism takes place at the site of the subduction zone where a relatively cold oceanic plate is rapidly thrust under an island arc or a continental margin giving rise to a very low geothermal gradient and low heat flow. The low-pressure belt represents the metamorphic complex further on land in an island arc or continental margin with a steep geothermal gradient and high heat flow, the heat transfer probably being caused by the rise of water, magmas and mantle materials from the descending oceanic slab (Miyashiro, 1973c). Hasebe, Fujii and Uyeda (1970), Toksöz, Minear and Julian (1971) and Oxburgh and Turcotte (1971) have made thermal models to explain the formation of such belts in relation to plate subduction.

According to Miyashiro (1972) paired metamorphic belts occur in three kinds of modern orogenic belts:

1. In continental margins such as the Cordilleran belt where the Pacific oceanic plates are being thrust beneath the American plates. There are, for example, Mesozoic pairs in the Franciscan and Sierra Nevada (Miyashiro, 1961; Hamilton, 1969; Ernst, 1971) and a late Palaeozoic pair in Chile (González-Bonorino, 1971).

2. Beneath ordinary island arcs where an oceanic plate is consumed under an arc. Examples are Mesozoic paired belts in New Zealand (Landis and Coombs, 1967) and possibly the late Palaeozoic and Mesozoic paired belts of southwest Japan.

3. Beneath reversed island arcs where the

316

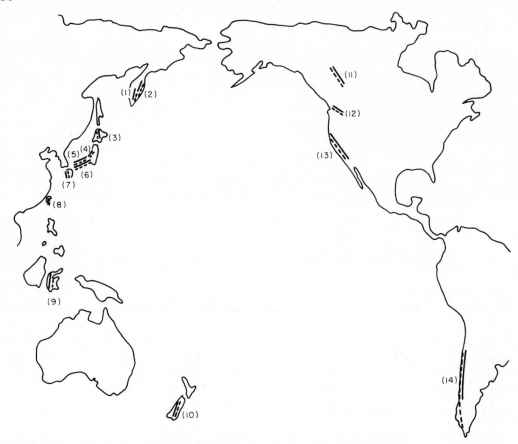

Fig. 19.9. Paired metamorphic belts in the circum-Pacific region. Thick broken lines represent high-pressure belts and thick full lines low-pressure belts (after Miyashiro, 1973c; reproduced by permission of Elsevier Scientific Publishing Co.)

plate of a marginal sea is thrust under the arc, such as beneath the New Hebrides Islands. Another example might be the paired belts of Hokkaido, Japan, but Okada (1974) presents evidence to show that these actually belong to two parallel pairs of ordinary belts.

Two factors may have controlled the common occurrence of high-pressure belts and paired belts in the circum-Pacific region (Miyashiro, 1973c).

1. The present rapid plate motion in the Pacific region compared with the slow motion in the Atlantic (Le Pichon, 1968) may have existed since the late Palaeozoic. A high velocity of underthrusting of a relatively cool lithospheric plate facilitates a lower geothermal gradient and the formation of a high-pressure blueschist metamorphic belt (Ernst, 1972).

2. The rate of consumption of an oceanic plate should be increased by the rate of accretion of new lithosphere at the mid-oceanic ridge. This means that blueschist metamorphic belts may form just before mid-oceanic ridges with a steep rate of accretion pass under an island arc or continental margin because of their high rate of plate convergence. The Sanbagawa and Franciscan high-pressure metamorphisms took place a few tens of millions of years before such ridge descent (Uyeda and Miyashiro, 1974).

Precambrian Granulites

Granulite facies rocks are found at the present surface level on every continent. Oliver (1969) made a compilation of such regions throughout the world and these are plotted in

Fig. 8.6, on the map of Piper (1974) that shows the palaeomagnetically constructed Proterozoic supercontinent. Using this reconstruction of the continents the granulite facies regions are more widely scattered than they are on the Permo-Triassic plot of Hurley and Rand (1969). Of course we would expect that the occurrence of granulites was once more widespread than these maps indicate since many must have been obliterated by later retrograde metamorphism and orogenic activity (there are some remnant inliers in young Phanerozoic fold belts such as the Coast Ranges of California) and some are, no doubt, hidden below younger cover deposits.

Exposed granulite rocks are of Precambrian age. There are no Phanerozoic whole-rock Rb/Sr or Pb/Pb ages of granulite facies rocks of significant extent, but present heat flow and crustal thickness suggest that conditions for granulite-grade metamorphism might prevail at depth up to this day (Heier, 1973).

There appear to have been two main periods of granulite formation, in the Archaean about 3000–2900 my and in the Proterozoic 1400–1000 my ago. Proterozoic granulites occur in Namaqualand in South Africa, the Grenville, the Fraser and Musgrave Ranges of Australia, Ceylon and NW Europe (NW Spain, Saxony, S Norway). Most granulites are of Archaean age: the Scourian of Scotland, W Greenland, Labrador coast, Superior Province of Canada, Lofoten in Norway, Aldan and Anabar Shields in USSR, India, Malagasy, in the Kasai, Sierra Leone–Ivory Coast, Gabon–Cameroons and Limpopo in Africa.

There is some disagreement about the geothermal gradients that may have operated in the early periods of earth history, and this applies to the P–T conditions controlling the formation of Precambrian granulites. Until recently cordierite was thought by petrographers to be a low-pressure mineral (e.g. Green and Ringwood, 1967), but this has since been shown to be wrong (Winkler, 1974), particularly from experimental work such as that of Schreyer and Yoder (1964), Hensen and Green (1972) and Currie (1974). In high-temperature rocks almandine garnet

and cordierite can coexist, their P–T range of stability being a function of the $FeO/MgO + FeO$ ratio of the rocks concerned. If the ratio is decreased, the stability field of garnet is reduced and that of cordierite extends towards higher pressures. Using the fact that this ratio is commonly lower than 0·8–0·9, Clifford (1974) demonstrates that many cordierite-bearing granulites in Africa formed under medium pressures.

Geothermal Gradients with Time

An interesting and speculative problem is whether certain types of gradients were typical or more common at different times of earth history. Saggerson (1973) speculated that the regional metamorphic P–T regimes have changed during geological time. Low pressure and a high thermal gradient were typical of the early Precambrian, intermediate pressure and an intermediate thermal gradient of the late Precambrian, and high pressure with a low gradient of the Palaeozoic–Mesozoic. Once the low pressure–high thermal gradient facies series was established, it reappeared at all subsequent times.

As pointed out in the previous section, high-pressure metamorphism formed under very low gradients was common during the Phanerozoic, being confined to convergent plate margins; data from the Precambrian, however, are less definite. Lambert (1976) notes that there are as yet virtually no detailed studies of the mineralogy of truly regionally metamorphosed Archaean rocks, and none which yields closely defined P–T conditions and therefore that we have a long way to go before we can assert that Archaean metamorphic gradients were truly of any particular character. A similar conclusion could equally apply to Proterozoic rocks. There is clearly a great opportunity for petrologists to undertake modern studies on the geochemistry and genesis of suitable Precambrian metamorphic rock suites.

The lack of Precambrian blueschists is explained if the steepness of the earth's geothermal gradient decreased with time (de Roever, 1965; Martin, 1969; Ernst, 1972).

One cannot, however, ignore the relevant tectonic environment when postulating such a secular variation in the geothermal gradient and so much depends on whether or not plate tectonics were operative during the Precambrian.

It is possible, however, to make some broad and interesting extrapolations. If, in the Archaean, the descending plates were thinner or the velocity of plate underthrusting faster, the inclination of the subducting plate would be shallower (Luyendyk, 1970), there would be an upward deflection of the isotherms (Oxburgh and Turcotte, 1970; Toksöz, Minear and Julian, 1971), and there would be less pressure for a given temperature, resulting in a steeper geothermal gradient which would prevent the formation of blueschists and facilitate that of low-pressure metamorphism (Ernst, 1972).

An increase in the temperature of the descending slab would have a similar effect. If oceanic plates were smaller in the early Precambrian, the time difference between plate accretion and subduction would be less, descending plates would be hotter and the geothermal gradient under the subduction zone therefore too high to allow the formation of high-pressure metamorphism (rapid plate underthrusting only facilitates a low gradient if the subducting plate is cold). The combination of these factors may well offer an explanation for the type of development of Archaean high-grade regions (widespread generation of tonalites and of high-temperature metamorphism). With thinner, smaller, faster moving and hotter lithospheric plates, the dip of the descending slabs would tend to be very shallow giving rise to an extraordinarily wide zone of igneous and metamorphic activity. A similar model was advanced by Uyeda and Miyashiro (1974) to account for the fact that the zone of late Mesozoic volcanic, plutonic and metamorphic rocks in eastern Asia is up to 3000 km wide, projected eastwards from the main subduction zones in Japan.

Returning to a more factual and agreed-upon point, the radiogenic heat production of the earth has decreased with time (Dickinson and Luth, 1971; Lambert, 1976). Fig. 19.10 shows heat production of K, U and Th with time and relative heat production rates for some important rock compositions. Heat generation by U and K (but not Th) and by the most important basic and ultrabasic rocks was two to three times higher in the Archaean than today. There are two ways of relating this high heat production to geothermal gradients during early crustal development.

1. If a high proportion of the heat released was dissipated through the continental crust, geothermal gradients would have been steepened giving rise to extensive granulite formation in the lower part of the crust in the early Precambrian (Heier, 1973).

2. If in the past most radiogenically-created thermal energy was released through active oceanic ridges as it is today, then ridge processes would have been two or three times faster than at present (Burke, Dewey and Kidd, 1976). These authors suggest that this could be achieved by faster ridge activity (for further discussion see Chapter 3) or by a greater ridge length which would be consistent with the widely held view that plates were smaller in the past. However two relevant points arise from this model. Firstly, oceanic thermal gradients were probably not much greater in the Archaean because

a) the temperatures and pressures that controlled the partial melting of pyrolite to form basalt were probably little different from those of today (Archaean and recent arc basalt types have similar trace element chemistry; Jahn, Shih and Murthy, 1974),

b) the greater amount of dissipated heat was released by faster spreading at larger ridges, and

c) the proportion of ultramafic lavas requiring local steep thermal gradients to initiate high temperature melting of pyrolite (D. H. Green, 1975) is low within Archaean volcanic piles.

Secondly, although the radioactive isotopes in the Archaean continental crust must also have been generating more heat than now, they do not seem to have given rise to any long-lasting persistent steep thermal gradients. Granulite formation was prominent in only certain areas during relatively brief periods, often about

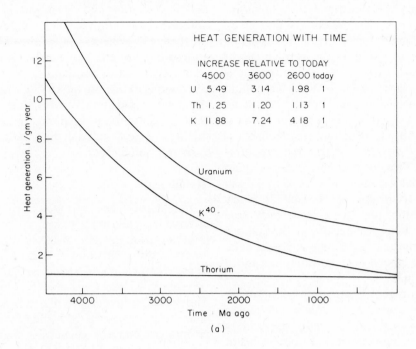

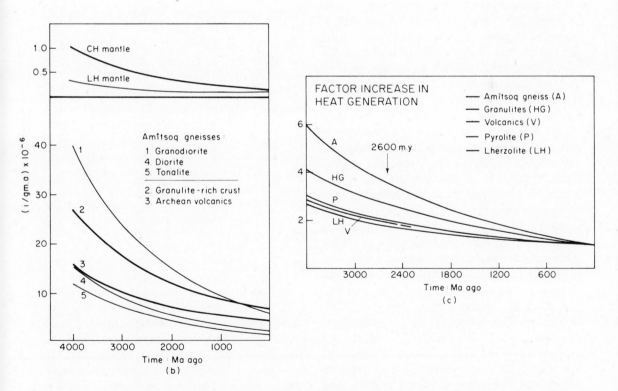

Fig. 19.10. Variation of heat generation with time (a) For K, U and Th. (b) For several rock types (CH = chondrodite, LH = lherzolite). (c) For specific crustal or mantle compositions, normalized to today = unity. (After Lambert, 1976; reproduced by permission of J. Wiley)

3000–2900 my ago. An argument in favour of less than steep early gradients is that steep gradients should give rise to narrow metamorphic zones, as in the Hercynides of Europe (Zwart, 1967a). In fact the reverse is the case in places such as West Greenland where Archaean rocks have the same monotonous amphibolite facies grade for tens, and even hundreds, of kilometres (Kalsbeek, 1976). Moreover, the fact that the predominantly mafic volcanic sequences in Archaean greenstone belts, conmonly exceeding 10 km in thickness (Glikson, 1971c, 1972), were heated so little that they have only low greenschist facies mineralogy throughout shows that thermal gradients in these belts could not have been high, certainly not more than 25°C/km.

From the foregoing discussion the following points emerge regarding geothermal gradients in earth history. There is only one widely accepted fact: high pressure metamorphism, which formed under very low gradients and gave rise to blueschists, only developed at plate-convergent boundaries and only during the last 800 my of geological time. There is a wide divergence of opinion as to the nature of thermal gradients in the Precambrian; in the Archaean, for example, they vary from 100°C/km to less than 25°C/km. It would appear that there is far too great a tendency to generalize about 'gradients in the Archaean' rather than to confine one's remarks to particular gradients in specific places; it can hardly be justifiable to pool together greenstone belts and granulites in one group formed under a single type of gradient. Low-pressure metamorphism definitely took place in greenstone belts in SW Australia because andalusite is stable and common (Binns, Gunthorpe and Groves, 1976), and yet medium-pressure Barrovian facies series with kyanite (350–600°C/4–6 kb) developed in the Dharwars greenstone belts of India (Sreenivas and Srinivasan, 1974). Pressures with gradients *as low as* 25–35°C/km operated in the formation of African granulites (Clifford, 1974), and 24°C/km was responsible for Scourian granulites (O'Hara, 1975), whilst in the Limpopo there were P–T conditions of 10 kb or more at 800°C (Chinner and Sweatman, 1968). In addition, medium-pressure metamorphism must have been necessary for the formation of widespread kyanite associated with eclogites in the Scourian of Scotland (J. V. Watson in discussion of Saggerson, 1973). From all this the picture that emerges is one of widely diverse geothermal gradients ranging from medium to steep set up in different places very early in earth history.

Metallogeny

Throughout the previous chapters an attempt has been made to indicate the kinds of mineralization prominent in various environments at different times in earth history. Just as the rocks have undergone secular trends or variations, so too have their enclosed ores. The modes of occurrence of ore deposits have changed in sympathy with the prevailing environments, some of which were characteristic of certain periods (Fig. 19.11). Not all types of mineralization evolved at the same time; in fact it was a very diachronous evolution. The aim here is to review these ore occurrences by putting them into the context of the long-term evolution of the continents.

The record of mineralization is related to changes in the earth's structure and behaviour throughout 4000 my of time (Bilibin, 1955; Watson, 1973b). The long-term variations in type and style of mineralization may be correlated with the evolution of the earth's crust from permobile in the Archaean, to a stabilized cratonic stage in the Proterozoic, to a final plate tectonic stage throughout the Phanerozoic.

The Archaean

During the Archaean period from 3800–2500 my ago there were two main tectonic environments, neither of which survived in an intact state. The few ores that exist in the high-grade regions occur within the gneisses in thin strips of metamorphosed volcanics, sediments and intruded layered complexes, i.e. iron formations at Isua (West Greenland) and, rarely, in the Messina Formation of the Limpopo belt (South Africa), Cu and Ni in meta-

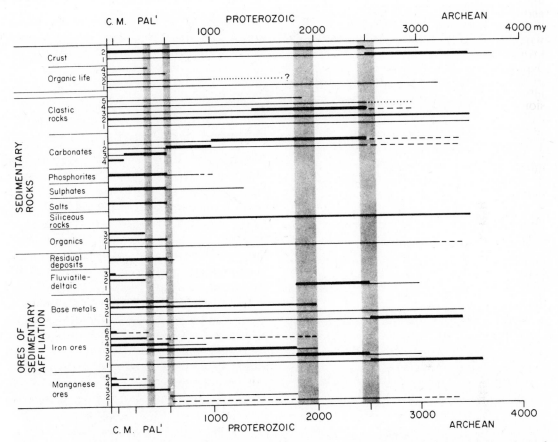

Fig. 19.11. Summary of variations in the composition of the crust, biosphere, sedimentary rocks and ores of sedimentary affiliation during geologic history (after Veizer, 1976b; reproduced by permission of Elsevier Scientific Publ.). Explanations:

Crust: 1. Archaean 'protocrust'. 2. Thick continental crust with 'granitic' upper layer.

Organic life: 1. Procaryota. 2. Eucaryota. 3. Metazoa. 4. Terrestrial life.

Clastic rocks: 1. Greywackes. 2. Shales and slates. 3. Arkoses. 4. Oligomictic conglomerates and orthoquartzites. 5. Red beds.

Carbonates: 1. Early diagenetic dolomites. 2. Inorganic and/or biochemical (?) limestones. 3. Littoral–neritic organogenic and organodetrital limestones and late diagenetic dolomites. 4. Deep sea limestones.

Organics: 1. Bituminous shales. 2. Oil and gas. 3. Coal.

Fluviatile–deltaic deposits: 1. Conglomerate–U–Au–pyrite type. 2. Sandstone–U–V–Cu type. 3. Placers and palaeoplacers.

Base metal deposits: 1. Volcanogenic and volcano–sedimentary type. 2. Sedimentary basinal type. 3. Sedimentary red bed type. 4. Carbonate–Pb–Zn type.

Iron ores: 1. Algoma type. 2. Superior type. 3. Clinton type. 4. Bilbao type. 5. Minette type. 6. River bed, bog iron, laterite and similar types.

Manganese ores: 1. 'Jaspilitic' type. 2. Volcano–sedimentary and orthoquartzite–siliceous shale–Mn carbonate types. 3. Carbonate association. 4. Orthoquartzite–glauconite–clay association. 5. Marsh and lake deposits

volcanic amphibolites at Pikwe (Botswana) in the Limpopo belt, and chromite in layered anorthositic complexes in West Greenland (Fiskenæsset), southern Africa (Limpopo belt–Messina Formation) and southern India (Sittampundi).

In contrast, the Archaean greenstone belts abound in ore deposits, a lot of which are economic. In many respects the belts have an 'oceanic' character and were individually relatively short lived. Summarizing the points brought out in Chapter 2 the main mineral deposits can be related to distinct rock groups in the stratigraphic pile and to the associated granites:

1. Chromite, nickel, asbestos, magnesite and talc in the lower ultramafic flows and intrusions.
2. Gold, silver, copper and zinc in the intermediate mafic-to-felsic volcanics.
3. Banded iron formations, manganese and barytes in the upper sediments.
4. Rare lithium, tantalum, beryllium, tin, molybdenum and bismuth in pegmatites associated with granite plutons.

The chromite, nickel, gold, silver, copper and zinc occur in ultramafic or mafic-to-felsic extrusive flows or sill-like intrusions. The copper–zinc and related gold–silver occur in deposits of the pyrite–sphalerite–chalcopyrite association which are characteristic of Archaean greenstone belts and which, although repeated, became scarcer in later geological time (Fig. 19.12) (Hutchinson, 1973). The ores bear certain resemblances to the Kuroko deposits of Japan (Horikoshi, 1969; Matsukuma and Horikoshi, 1970) and in this respect may represent a similar form of ore accumulation along more primitive plate-convergent boundaries about 3000 my ago (Sawkins, 1972).

The lack of Pb mineralization is noteworthy in these greenstone belt base metal deposits. One explanation for this that seems reasonable is that in this early period of earth history there was insufficient time for lead to generate from radioactive decay of uranium and thorium and so to accumulate in appreciable quantities in the crust (Sangster, 1972)—by Proterozoic time lead makes its first appearance in volcanogenic sulphide deposits.

The mineralization that we see today in Archaean terrains occurs, or can potentially occur, in rocks spanning the period from 3800–2500 my ago, which is a third of geological time. What is striking is the remarkable degree of similarity in rock and ore development over such a long period. In the succeeding Proterozoic stage of earth history the types and styles of mineralization

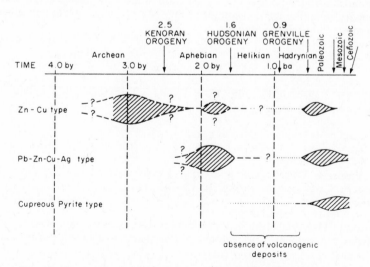

Fig. 19.12. Apparent time ranges and maxima of massive volcanogenic sulphide deposits (after Hutchinson, 1973; reproduced by permission of *Economic Geology*)

were substantially different from those of the Archaean.

The Proterozoic

There was a dramatic change in tectonic conditions between the Archaean and Proterozoic although the changeover was both transitional and diachronous. The Archaean crust cooled, was stabilized, uplifted, eroded and covered with new platform-type sediments over a period of several hundred million years, beginning as early as 3000 my in southern Africa and probably by 2600 my in the Atlantic region of the northern hemisphere. Unroofing of large areas of Archaean terrains made available many elements that were recycled at the earth's surface simultaneously with the evolution of stable cratonic environments more suitable for the deposition of extensive cover sequences. The formation of a thickened continental crust made it possible also for the first continental margin geosynclines to develop, e.g. the Coronation (Hoffman, 1973) and Mount Isa (Dunnet, 1976) geosynclines.

In these new tectonic environments there formed four principal types of sedimentary ore deposits (partly reviewed in Chapter 5):
1. Gold and uranium in conglomerates
2. Manganese and lead–zinc in carbonates
3. Banded iron formations
4. Clastic copper and minor cobalt in red beds which appeared between 2500 my ago (in the Nyanzian of Kenya) and 2000 my ago (in the Udokan Series of the USSR and in western Uganda)

The following are factors that may have influenced the accumulation of these deposits:
1. Extensive erosion of the Archaean basement made available several suitable elements. In particular, the Fe and the Au in the BIF were no doubt derived from the greenstone belts with their abundant basic volcanics; Viljoen, Saager and Viljoen (1970) showed that the gold placer province in southern Africa is centrally located in the Kaapvaal craton and thus near to the uplifted greenstone belts.
2. A low oxygen content of the prevailing atmosphere may have contributed to the deposition of:
a) Detrital uraninite which readily oxidizes on weathering.
b) Banded iron formations which used photosynthetically derived oxygen for the oxidation of ferrous iron in the seawater (Trudinger, 1971; Cloud, 1973). Jacobsen (1975) suggests the essentially non-oxidizing atmosphere during the Archaean and early Proterozoic did not allow the chemical breakdown of primary copper sulphides, thereby preventing subsequent sedimentary accumulations. The earliest breakdowns of the copper sulphides about 2000–1800 my ago are roughly contemporaneous with the first red beds and the peak of development of banded iron formations.
3. The CO_2 content in the oceans had increased to such an extent by the early Proterozoic that thick dolomite sequences were deposited. The accumulation of Mn deposits in these carbonates was facilitated by the increasing partial pressure of oxygen because the separation of Mn from Fe is achieved by a higher oxidation of iron (Veizer, 1976b).

The dolomites typically belong to a shallow marine intertidal lagoonal-to-littoral facies, often with algal reefs that formed in a carbonate shelf-bank setting at rifted continental margins (Hoffman, 1973; Dunnet, 1976). Such dolomites provide the host for Cu–Pb–Zn mineralization at McArthur River and Mount Isa in narrow basins, aulacogens or at triple junctions; these are the first examples of this type of rift-controlled mineralization that became common in the late Phanerozoic (Table 19.4).

There are also early Proterozoic Cu Pb Zn ores with a volcanogenic affiliation associated with intermediate-felsic calc-alkaline volcanics, such as the Errington and Vermilion Lake deposits of the Sudbury basin (W. C. Martin, 1957). These pyrite–galena–sphalerite–chalcopyrite deposits (Fig. 19.12) are of the same type and have the same mode of occurrence as those of California and of the Kuroko district of Japan, all of which formed in active island arcs or continental margins

Table 19.4. Sullivan-type Pb–Zn–Ag sulphide deposits in rifted environments (after Sawkins, 1976b; reproduced by permission of the Geological Association of Canada)

District	Host rock lithology	Composition	Age
Sullivan, British Columbia	Argillites	Ag, Pb, An	Proterozoic
Mt Isa, Queensland	Shales	Ag, Pb, Zn	Proterozoic
Rammelsberg, West Germany	Slates	Pb, Zn, Cu	Devonian
Meggen, West Germany	Slates	Pb, Zn	Devonian
Ducktown, Tennessee	Metamorphosed greywackes	Cu, Zn	Late Precambrian
Southern Appalachians	Argillites	Cu	Late Precambrian

(Hutchinson, 1973). Because these ores and felsic volcanics were derived from magmas that have been through a two-stage mantle fractionation process, the early Proterozoic examples may have formed in a similar plate tectonic environment.

Thus it seems possible that both types of early Proterozoic Cu–Pb–Zn deposits formed in the first active continental margins that are similar to those of the Mesozoic.

In the early to mid-Proterozoic major dyke swarms and layered complexes were intruded into either the Archaean basement itself, or the unconformable cover successions. Fractionation of the mantle-derived tholeiitic liquids gave rise to a distinctive group of metal concentrations, in particular nickel, chromium, platinum and copper.

During mid-Proterozoic time there was renewed tectonic activity giving rise to medium- to high-grade mobile belts. The principal mineral deposits associated with these belts are:
1. Uranium, lithium, beryllium and tin in late granites and pegmatites.
2. The well-known andesine–labradorite anorthosites contain important ilmenite–titaniferous magnetite deposits. There is a distinct Ti metallogenic association with these mid-Proterozoic anorthosites on all continents and in this respect they differ from the com-

monly chromite-bearing Archaean anorthosites.
3. Some of the oldest porphyry copper deposits may be found in the 1800 my old porphyry intrusions in the Echo Bay Group andesites of northern Canada (Badham, Robinson and Morton, 1972)—an early Proterozoic subduction zone?

In the late Proterozoic 1200–600 my ago one or several large continents began to break-up and much igneous activity, sedimentation and related mineralization were localized in major rifts, aulacogens and along continental margins. Examples of such mineralization are as follows:
1. COPPER IN VOLCANICS AND SEDIMENTS. Basalt flows were extensively extruded onto, or at the margins of, continents in this period with many located in rifts or failed arms. The most common ore in these volcanics is copper, as in the Coppermine River Group (1110 my) in NW Canada, the Keweenawan basalts (1100 my) in Michigan, and the slightly younger Bukoban lavas in East Africa (100–800 my).

Certain slightly younger clastic sediments contain enrichments in copper, but only some are economic. It is probable that the copper in these sediments was derived by recycling from that in the lavas (Watson, 1973b). The Katanga System in Zambia has well-known economic deposits of this type, and the sediments of the Belt Basin in northwestern USA contain anomalously high values of copper in the range of 100 ppm over thousands of square miles.

2. TIN IN GRANITES. Tin-bearing granites and pegmatites, often of alkaline to peralkaline character, reached prominence in the period 1100–550 my, particularly in Africa where they tend to be aligned in three belts

(Fig. 19.13) (Hunter, 1973). Within these belts the tin mineralization is located in the Kibaran (1100 my) and Damaran (550 my) orogenic belts, both of which belong to cratonic areas stable since 1500 my ago. Another belt passes through the Rhodônia district of western Brazil. The tin and related elements may have been derived from crustal sources if the granites were formed by remobilization of earlier basement granites

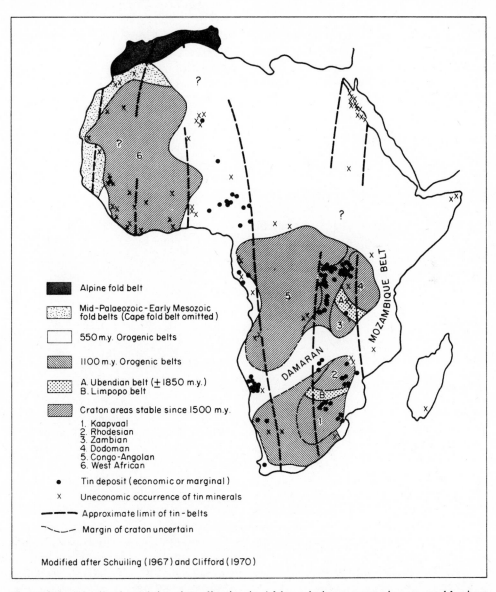

Alpine fold belt

Mid-Palaeozoic-Early Mesozoic fold belts (Cape fold belt omitted)

550 m.y. Orogenic belts

1100 m.y. Orogenic belts

A. Ubendian belt (± 1850 m.y.)
B. Limpopo belt

Craton areas stable since 1500 m.y.
1. Kaapvaal
2. Rhodesian
3. Zambian
4. Dodoman
5. Congo-Angolan
6. West African

• Tin deposit (economic or marginal)

x Uneconomic occurrence of tin minerals

— — Approximate limit of tin-belts

— · — Margin of craton uncertain

Modified after Schuiling (1967) and Clifford (1970)

Fig. 19.13. Distribution of tin mineralization in Africa relative to cratonic areas stable since ±1500 my ago and orogenic belts (after Hunter, 1973, reproduced by permission of *Minerals Sci. Engng.*)

and gneisses (Watson, 1973b), or from mantle sources if they formed in intraplate hot spots due to plume activity (Sillitoe, 1974). The fact that Sn deposits were persistently localized along several major lineaments from the early Proterozoic to the Palaeozoic may suggest that the Sn was derived from anomalous zones in the upper mantle (Clark and coworkers, 1976).

3. ALKALINE COMPLEXES. These are often associated, both temporally and spatially, with carbonatites formed in conjunction with rift activity and they contain deposits of U, Th, Nb, Be and Zr (Ilimaussaq, South Greenland), Ba and rare earths (Mountain Pass, California) and nepheline (Blus Mountain and Bancroft, Ontario).

Phanerozoic Plate Boundaries

Starting in the late Proterozoic a new type of tectonic pattern developed in the continents which eventually gave rise to Phanerozoic fold belts formed by continental drift. In the following discussion the formation of ores will be considered throughout Phanerozoic time in relation to the following principal tectonic environments; early domes, aulacogens and rifts in continents, accreting and consuming plate margins and transform faults (for summary see Fig. 19.14). For general reviews, see Mitchell and Garson (1976) and Evans (1976).

Domes, Aulacogens and Rifts We want to consider here the early structures and associated mineralization that developed in continental crust before separation took place and new oceanic crust was formed. These represent the incipient stages of continental fragmentation.

The first major structures to develop are topographic domes like one sees today in Africa (Gass, 1970a; Le Bas, 1971) and the earliest ores are tin deposits in anorogenic alkaline-to-peralkaline granites that lie in linear belts associated with the domes and whose formation stopped before the triple rift systems developed (Burke and Whiteman, 1973). Examples are the Younger Granites of Nigeria, and the granites of the Tibesti area of Chad and in Damaraland in South West Africa.

Although some rifts may open to give rise eventually to new oceans, some may not open and so remain as aulacogens which are failed rift arms that project from an oceanic margin into a continental plate (Burke and Dewey, 1973a). The main types of mineralization in such rift zones are:

1. Lead–Zinc. For details of this mineralization near the Red Sea and in the Benue Trough, the Oslo Graben and the Rio Grande Rift, see Chapter 14. These Pb–Zn deposits, often with Ag, are of the Sullivan type of Sawkins (1976b) which form during the attempts to fragment continental plates; they make their first appearance during the early Proterozoic (Mount Isa) and later examples are given in Table 19.4.

2. Niobium and REE concentrations in carbonatites in Tanzania, Uganda, Zambia, Malawi and Oka (Quebec).

Accreting Plate Margins With further extension new oceanic lithosphere is created along the axial zone of the rifts giving rise to oceanic ridges which we see today preserved on a minor scale in the Red Sea and in back-arc marginal basins. Continental drift gives rise to wide oceans. Examination of material from these spreading-centre ridges is possible from dredges and drill cores and from onland ophiolite complexes. Fig. 18.6 gives an idealized cross-section through the oceanic crust showing the stratigraphic location of the principal mineral deposits. Evidence of metal concentrations in the topmost sediments comes from currently active ridges like the East Pacific Rise and Red Sea as well as from correlated examples capping ophiolite sequences; the mineralization in the volcanic–plutonic part of the cross-section, however, is known only from 'fossil' ophiolites.

The oldest Phanerozoic ores formed at constructive plate boundaries are in the ophiolite complexes of western Newfoundland (Church and Stevens, 1971; Dewey and Bird, 1971; H. Williams, 1971) which were obducted in Ordovician time; asbestos, chromite

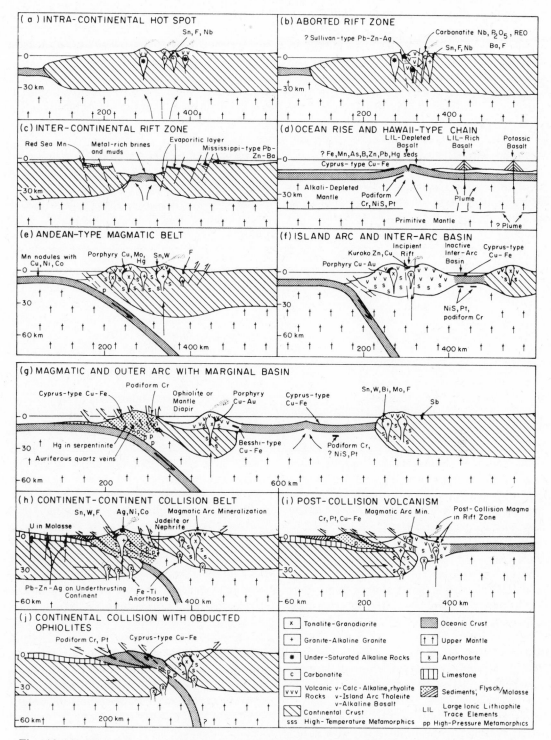

Fig. 19.14. Schematic cross-sections through plate boundary-related tectonic settings showing principal types of mineralization (after Mitchell and Garson, 1976; reproduced by permission of *Minerals, Science and Engineering*)

and nickel ores occur in the lower plutonics (Strong, 1974a,b).

Podiform chromite deposits are common in the lower serpentinized ultramafics of many ophiolite complexes in the eastern part of the Alpine fold belt (Chapter 18) (Engin and Hurst, 1970; Thayer, 1971; Peters and Kramers, 1974) and in arc environments in Cuba (Flint, Francisco and Guild, 1948) and the Philippines (Bryner, 1969).

Cupreous pyrite deposits (ochres) with minor zinc (but no lead) occur in or above the basaltic pillow lavas of several Phanerozoic ophiolite successions (Figs. 19.2, 19.15) (Hutchinson, 1973):

ner and coworkers, 1967). In the Red Sea there are sulphide-rich muds with sphalerite and lesser chalcopyrite, pyrite and marcasite overlain by layers enriched in Fe and Mn minerals; compaction of these metalliferous muds would give rise to deposits similar in several ways to the umbers in ophiolites (Sillitoe, 1972b).

Consuming Plate Margins When an oceanic lithospheric plate is carried down a subduction zone, new magmas are generated at successive pressure/temperature levels which arise through the overlying plate. These magmas give rise to plutonic and volcanic

Period	Locality	References
Early Palaeozoic	Whalesback, Tilt Cove, Notre Dame Bay, Betts Cove, Bay of Island complexes, etc. (Newfoundland)	Upadhyay and Strong (1973) Strong (1974a,b)
Late Palaeozoic	150+ deposits, Sanbagawa zones, etc. (Japan)	Tatsumi, Sekine and Kanehira (1970) Sillitoe (1972b)
	100+deposits (Urals, USSR)	Smirnov (1970)
Mesozoic	Troodos complex (Cyprus)	Hutchinson (1965) Constantinou and Govett (1972) Searle (1972)
	Kure (Turkey)	Suffel and Hutchinson (1973)
	Island Mtn (Franciscan, Calif.)	see Hutchinson (1973)
Cenozoic	Ergani Maden (Turkey)	Griffitts, Albers and Öner (1972)
	Phillipines	Bryner (1969)

The topmost sediments in ophiolites are associated with Mn–Fe rich sediments free of sulphides which lie unconformably in depressions on the basalt lava surface: these are the umbers of Cyprus (Constantinou and Govett, 1972; Robertson and Hudson, 1974; Robertson, 1975).

Layer 1 sediments on the flanks of the East Pacific Rise contain anomalously high concentrations of Mn and Fe, and lesser quantities of Cu, Zn, Pb, B, As, U and Hg (Boström and Peterson, 1966, 1969; D. Z. Piper, 1973; Boström and coworkers, 1974; Cronan, 1976); similar metal-enrichments occur elsewhere on active oceanic ridges (Boström and coworkers 1969; Boström and Fisher, 1971). High concentrations of Fe, Mn and a variety of non-ferrous metals also occur in hot brines in the Red Sea (Fig. 19.17) (Degens and Ross, 1969; Backer and Schoell, 1972; Hackett and Bischoff, 1973; Bignell, 1975) and beneath the Salton Sea, California (Skin-

rocks, as well as mineral deposits, which change in composition and type with increasing distance from the trench. The metals are emplaced as components of magmas that produced tholeiites nearest the ocean, calc-alkaline granitic batholiths and andesitic–rhyolitic lavas in the main arc, and alkaline rocks furthest from the ocean. The zonal distribution and types of mineral deposits across island arcs (e.g. Zonenshain and coworkers, 1974) and Cordilleran-type continental margins (e.g. Sillitoe, 1972a, 1976; Clark and coworkers, 1976) are substantially similar (Mitchell, 1976b). The general sequences across the American Cordilleran margin and the arcs of the western Pacific (omitting extensional marginal basins) are:

Oceanic side

Eastern Pacific

Fe (Au) Mo Au Pb Zn Sn Mo

Western Pacific

Au Cu Mo Au Pb Zn Sn W Sb Hg

329

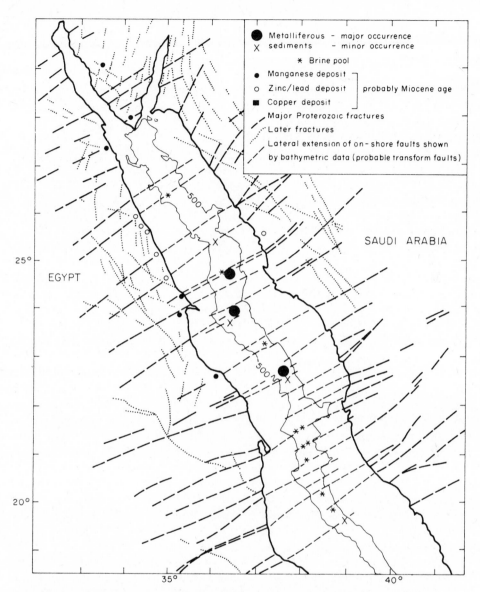

Fig. 19.15. Distribution of metalliferous sediments and brine pools in the Red Sea in relation to continental fractures, presumed transform faults and on-shore Miocene deposits (after Mitchell and Garson, 1976; reproduced by permission of *Minerals, Science and Engineering*)

There are various types of base metal sulphide ores at Phanerozoic active plate boundaries. According to Hutchinson (1973) the volcanogenic zinc–copper type (Fig. 19.12), is present in Palaeozoic belts (West Shasta district in California; Fleur de Lys Supergroup, W Newfoundland; Bathurst in Brisbane–New England eugeosyncline, NSW, Australia), but it did not survive into post-Palaeozoic times,

whereas the lead–zinc–copper–silver volcanogenic type reached a maximum in early Palaeozoic orogenic belts (New Brunswick; Newfoundland; NSW, Australia) and continued into younger belts (Triassic in East Shasta district of California; Jurassic in Foothill Copper Zinc belt of California; Tertiary in Kuroko district of Japan). Hutchinson (1973) suggests that the first type developed in early, stages of subduction along continental margins, and the second in later, more felsic, calc-alkaline volcanics.

Transform Faults Mitchell and Garson (1976) suggest that the following types of mineralization occur along transform faults, either in the oceanic crust where they offset ridges or along their extensions in continental fractures.

The location of the metalliferous brine pools in the Red Sea is controlled by transform faults (Garson and Krs, 1976). Also Zn, Pb, Cu and Mn deposits of mainly Miocene age occur near the Red Sea in continental fractures which extend onshore from presumed former transform faults (Fig. 19.15).

Serpentinized peridotites, gabbros and alkaline plutonic rocks have been derived from transform faults in the equatorial mid-Atlantic ridge (Bonatti, Honnorez and Ferrara, 1971). Some of the peridotites contain high Ni, Co, Ti and Cu values.

Alkaline complexes and carbonatites are aligned along marked lineaments in Angola, Brazil, SW Africa and Uruguay, the lineaments lying along transform faults centred on the Cretaceous poles of rotation for the South Atlantic (Marsh, 1973). Diamond-bearing kimberlite pipes are centred on these transform lineaments, especially along the Lucapa graben in Angola (Reis, 1972; Rodrigues, 1972).

Onland transcurrent faults that may be extensions of transform faults have localized porphyry copper deposits in the Philippines, porphyries with sulphide mineralization and Kuroko-type Cu–Pb–Zn deposits in the Red Sea region, and tin-bearing granites and lepidolite–tin pegmatites in Thailand.

Crustal Evolution

During the 3800 my dealt with in these pages the continents evolved to their present form; they passed through three distinct stages of growth which roughly correspond to the Archaean, Proterozoic and Phanerozoic. The tectonic patterns during these periods are a reflection of the conditions of aggregation, stabilization and fragmentation, respectively, that contributed to the growth of the continents. Here we shall look at some of the processes and factors that influenced this secular crustal evolution.

Crustal Thickening in the Archaean

Ideas about the thickness of the Archaean crust have changed a great deal in the last decade. Engel (1966) and Anhaeusser and coworkers (1969) envisaged a thin (15–25 km) crust but today estimates of as much as 70–80 km are made (Condie, 1973, 1976b; O'Hara, 1975). There could have been two types of *thin* Archaean crust. Firstly, an originally thin crust formed at an early stage of fractionation of the earth and, secondly, an attenuation developed along rifted continental margins. Today there is a zone of attenuation 60–80 km wide along the edge of rifted plate margins such as the Red Sea where the continental crust thins from about 30 km to zero (Kinsman, 1975b). If active ridge lengths were longer and plates smaller in the Archaean, as is argued elsewhere in this volume, then there would have been a more extensive tectonically thinned crust than there is today. The 'thickness' question is important because it concerns the problem of how and when the earth's crust developed in its early stages; we shall consider it from several standpoints.

Engel's idea of a thin crust was based on certain characteristics of the Barberton greenstone belt: the telescoped nature of the metamorphic aureoles at contacts with younger granites, the bimodal nature of the mineral facies, lack of blueschists, and predominance of chemically primitive tholeiites; he saw the most likely modern analogues as the emergent island arcs of the Philippines, New Caledonia and Kamchatka. The Barberton rocks were thought to have developed on a thin oceanic crust. Such conclusions, drawn from consideration of a greenstone belt and appropriate at the time, have today to be modified to take account of more recent data from the Archaean high-grade regions.

Stratigraphic Sections Greenstone belts are characterized by extremely thick successions of sedimentary and volcanic rocks—minimum estimates of greenstone belt thicknesses range from 8–17 km (Table 19.5). In most examples these are less than maximum estimates because the topmost parts of the successions have been removed by erosion and the lowest parts may be unexposed. Also, and most important, the exposure of many belts is so poor that strict structural and stratigraphic control may be less than adequate.

Estimates of stratigraphic thickness cannot be made by measuring sections in high-grade terrains because, where examined in detail in Greenland, Labrador and Scotland, it is clear that the conformability of the present rock units is caused by a combination of tectonic intercalation and *lit-par-lit* intrusion of granitic (now gneiss) sheets.

Chemical Parameters Geochemical data from igneous and metamorphic rocks can be used to estimate the thickness of the Archaean crust.

Using the Sugimura index estimated for island arcs, Naqvi, Rao and Narain (1974) calculated the rapid increase in crustal thickness of the Indian Shield between the Archaean and the Proterozoic from the composition of basalts and dolerites (Fig. 19.16).

There is a positive correlation between the potash contents of volcanic rocks (for specific silica values) at modern active continental margins and island arcs, and the depth to the subduction zone (Hatherton and Dickinson, 1969; Dickinson, 1970; Hatherton, 1974) and the crustal thickness (Condie, 1973, 1976b). Condie and Potts (1969) suggested that if the relationship with crustal thickness pertained in the Archaean, then the K_2O contents of the calc-alkaline volcanic rocks indicate that they

Table 19.5. Archaean crustal thicknesses and depths to subduction zone inferred from geochemical polarity indices* (after Condie, 1976b; reproduced by permission of Pergamon Press)

Archaean locations	T_M	T_{RB}	T_K	T	SB
Greenstone belts					
1. South Pass, Wyoming		20–30	25	30	105
2. Pickle Lake, Ontario†			12	≥15	45
3. Swayze, Ontario			19	20	80
4. Timmins–Kirkland Lake, Ontario			12	15	45
5. Noranda, Ontario	13	15–20	13	14	50
6. Birch–Uchi Lakes, Ontario	10		18	28	75
7. Kakagi–Eagle Lakes, Ontario†			17	20	70
8. Lake of the Woods, Ontario	8		20	28	85
9. Michipicoten, Ontario	11		25	35	105
10. Barberton (Schoongezicht Fm), South Africa	17	20–30	20	20	85
11. Yellowknife, Northwest Territories, Canada	13	20–30	25	31	105
12. Average greenstone belt			20	25	85
Granitic rocks					
1. Louis Lake batholith, Wyoming		>30	30	≥30	(170)
2. Laramie batholith, Wyoming		>30	35	≥35	(170)
3. Ancient tonalitic gneisses, South Africa		20–30	24	≥24	(105)

T_M = minimum measured thicknesses in greenstone belts.
T_{RB} = crustal thickness inferred from Rb–Sr index grid.
T_K = crustal thickness inferred from potash index.
T = probable total crustal thickness allowing for index errors.
SB = depth to subduction zone inferred from potash index.
* All data in kilometres.
† Sections dominantly composed of tholeiites.

332

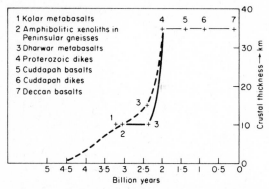

Fig. 19.16. Rapid thickening and evolution of the sialic crust of the Indian Shield calculated from the index ($SiO_2 - 47\,Na_2O + K_2O/Al_2O_3$) of Sugimura (1968) using basalts and dolerites (after Naqvi, Rao and Narain, 1974)

were erupted onto a crust that was about 10–25 km thick; because up to 17 km of these volcanics are now exposed, the total thickness of the Archaean crust may have approached that of the present day.

Archaean crustal thickness estimated by Condie (1973) from Rb/Sr distributions in volcanic rocks give values ranging from 15–35 km. An average crustal thickness of 25 km and a depth to a subduction zone of 85 km for Archaean greenstone belts were calculated, mainly from the calc-alkaline series (see Fig. 19.17). Table 19.5 compares Archaean crustal thicknesses derived by geochemical parameters with measured thicknesses of greenstone belts; the former are consistently the higher.

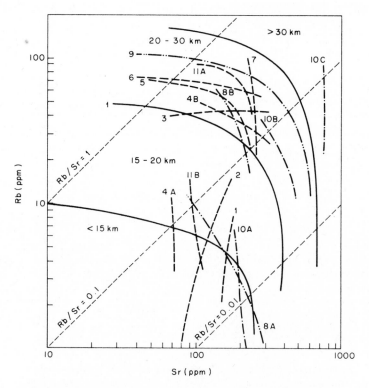

Fig. 19.17. Rb/Sr distributions in Archaean greenstone volcanic rocks and inferred crustal thickness. 1. Ely. 2. Chibougamau. 3. Rainy Lake. 4. Abajevis, A = Lower, B = Upper. 5. Lower Wawa, Ontario. 6. Clericy, Quebec. 7. Upper Wawa, Ontario. 8. Onver-wacht, A = Lower, B = Upper. 9. Schoonegezicht (Upper Fig Tree). 10. South Pass, Wyoming, A = low-K tholeiites, B = calc-alkaline volcanic rocks, C = high-K calc-alkaline volcanic rocks. 11. Yellowknife, A = second cycle, B = first cycle. (After Condie, 1976b; reproduced by permission of Pergamon Press)

Petrochemical data from metamorphic rocks allow additional constraints to be made about Archaean burial depths. Mostly these are granulite facies rocks that formed under deep-seated conditions. The significance of garnet/cordierite-bearing granulites formed under gradients of 33–34°C/km is considered in the section *Precambrian granulites* in this chapter. The P–T conditions under which layered igneous complexes at Fiskenæsset, Sittampundi, Rodil and Scourie were metamorphosed are reviewed in Chapter 1; estimates of crustal thickness are up to 45 km, perhaps even 75 km (O'Hara, 1975). Condie (1976b) points out that experimental data on Archaean granulites with high P–T assemblages suggest thicknesses of 30–40 km (Chinner and Sweatman, 1968) and because the thickness of crust today is about 35–40 km the original Archaean crust may have been in the order of 65–80 km.

Tectonic Regimes The traditional view of greenstone belts is that they formed by down-sagging of a supracrustal sequence and therefore their present form has been controlled largely by vertical tectonics (e.g. Anhaeusser and coworkers, 1969; Glikson, 1970). However, the discovery of large-scale early nappes in Rhodesia (Stowe, 1971; Coward, Lintern and Wright, 1976) and Botswana (Key, Litherland and Hepworth, 1976) provide an additional factor for consideration: stratigraphic duplication caused by horizontal tectonics. The question arises as to how many other greenstone belts contain nappe–thrust structures (Burke and coworkers, 1976).

From the work of McGregor (1973) a horizontal tectonic regime in the high-grade terrain of Greenland was proposed by Bridgwater, McGregor and Myers (1974a). They suggest that a combination of two processes led to an appreciable thickening of the sialic crust. Firstly, the stacking of thrust sheets and nappes which interleaved sialic Amîtsoq gneisses with Malene supracrustal rocks; and, secondly, the injection into the intercalated thrust pile of large numbers of concordant sheets of calc-alkaline dominantly tonalitic rocks of Nûk type. The relationship between this type of horizontal tectonic regime in a high-grade region with that in a low-grade greenstone belt region may be seen in a section from the Limpopo mobile belt northwards to the Rhodesian craton (Coward, Lintern and Wright, 1976). Like Bridgwater, McGregor and Myers (1974a), these authors suggest that the thrust nappe tectonics and granitic intrusions were a result of crustal collision, closure of an ocean basin and subduction of oceanic floor material. Accordingly, thickening of the Archaean crust was caused by some form of proto-plate movements.

In conclusion, the stratigraphic, petrochemical and tectonic evidence for late Archaean crustal thickening in greenstone belts and high-grade regions provides an adequate clue to the marked increase in lithospheric stability which took place by the early Proterozoic and which is discussed in the following two sections.

Cratonization During the Precambrian

There are several ways of demonstrating that the Archaean was a permobile era and that large stable cratons did not evolve until the beginning of the Proterozoic.

During the first third of geological time (roughly 1300 million years) in the Archaean there was such a high degree of tectonic activity that no rocks escaped deformation and metamorphism. But it would be wrong to think that permobile conditions operated for long or throughout the Archaean; the isochron age peaks are indicative of several short-lived, but widespread events.

So far it has not proved possible to define any well-structured continental–oceanic margins or platform-basement contacts in the Archaean. Admittedly there are a few unconformities between greenstone belts and higher grade gneissose/granitic rocks (Windley, 1973a; Bickle, Martin and Nisbet, 1975), but these are only local erosional remnants and are not at the borders of well-defined cratons. Nevertheless, geochemical differentiation had progressed sufficiently, even by 3700–3800 my ago, to produce the well-known bimodal granite–greenstone association

(Moorbath, 1975c). There was certainly a marked change in crustal conditions by the late Archaean in southern Africa (somewhat ahead of other continents—Windley, 1973a) because the formation of the greenstone belts in Rhodesia (2900 my for the Bulawayan—Hawkesworth, Moorbath and O'Nions, 1975) overlapped with the deposition of platform-type cover sediments of the Pongola (c. 2850 my—Hunter, 1974b) and Dominion Reef (c. 2800 my—Anhaeusser, 1973) sequences (Sutton, 1976).

There are only a few places in the world where pre-3200 my old rocks have been defined isotopically: Minnesota River Valley in the USA, Godthaab–Isua in West Greenland, northern Norway, Barberton–Swaziland–Rhodesia, and the Labrador coast of Canada. One of the main reasons for this is the fact that the peak of accretion activity took place in the late Archaean between 3100 and 2700 my.

Evidence and arguments supporting the idea that the most important period of crustal growth was in the permobile late Archaean and that by the early Proterozoic the continents had consequently achieved roughly or nearly their present size, thickness and degree of stability, are as follows:

1. There was extensive formation of granulites and charnockites and high amphibolite facies metamorphism in the period 3100–2800 my (Oliver, 1969; Fyfe, 1973; Heier, 1973), viz. Scotland, Greenland, Labrador, Aldan and Annabar Shields, Limpopo, W Africa, India, W Australia, etc., and the thickness of the crust in these areas must have been more than that of modern continental crust for it to be able to support such high temperature metamorphism in its lower part (Moorbath, 1975c).

2. The majority of greenstone belts formed between 3000 and 2700 my; only rarely did they continue to form in the early Proterozoic (Bell, Blenkinsop and Moore, 1975).

3. In some places such as W Greenland, and probably in others like Scotland, S India and the Limpopo, there was such an intensive intrusion of late Archaean tonalites that they make up the greater part of the present surface area. The low initial strontium isotope ratios of about 0·701 of the tonalites and greenstone belt volcanics suggest that the isochron ages represent the actual time of derivation and accretion of material from the mantle in the late Archaean (Moorbath, 1975a,b).

4. At least twice, maybe even three times, as much heat was generated by radioactive decay in the Archaean than today (Fig. 19.10). Burke, Dewey and Kidd (1976) suggested that if this heat were the sole source of formation of the continents through a two-stage fractionation process, then about two-thirds of the continental crust would have been produced by 2500 my ago; but with probable additional heat sources to promote mantle fractionation in the early Archaean (short-lived isotopes, core formation and meteorite impacts), it is likely that much, if not nearly all, of the present continental mass was produced during this permobile era. In his review of early Precambrian development Moorbath (1975c) put the figure at 50–60%.

5. By comparing the potash contents of modern and Archaean calc–alkaline volcanics, Condie and Potts (1969) and Condie (1973) inferred that the crust of North America approached or reached present-day thicknesses by about 2700 my ago, whilst Wise (1974) showed that there is a constant relationship between the thickness, and the volume of the continents.

6. The presence of thin strips of marble–orthoquartzite–K pelite in the Archaean (high-grade regions) suggests that shallow-water continental margin/platform conditions were in existence, although very poorly developed, in this period (Sutton, 1976). The Dharwars of India are the only greenstone belts to have a prominent marble–orthoquartzite association and for this reason Naqvi, Rao and Narain (1974) assign them to the early Proterozoic. Conditions obviously changed drastically between about 3000 and 2500 my because by the early Proterozoic kilometre-thick carbonate–quartzite sequences appear in, for example, the Witwatersrand–Transvaal series and the Coronation and Mount Isa geosynclines. From their world survey Engel and coworkers

(1974) calculated that the ratio of Archaean carbonate plus orthoquartzite to other Archaean sediments was about 1:100 whereas in the Proterozoic the ratio was 1:5. This can be regarded as an approximate estimate of the growth of stable shelf environments at this time.

7. There was an appreciable increase in amount of mature quartz sediments about 2500 ± 200 my ago, which was complementary to an abrupt decrease in relative volume of immature wackes, conglomerates and arkoses. The K_2O/Na_2O index of clastic sediments, roughly reflecting their degree of maturity, increased from less than 1·0 in Archaean greenstone belts to 1·5–5·0 in Proterozoic sequences (Engel and coworkers, 1974). Also the unstable tectonic environments of the Archaean would have been unsuitable hosts for deposition of mature sediments, whereas the broad platforms from the early Proterozoic onwards were ideal. These sedimentary relationships are reflected by the increasing areal extent of basins with time in southern Africa, considered by Anhaeusser (1973) to be directly proportional to the crustal stability (Fig. 5.1).

8. In the Archaean two different proto-plate tectonic regimes can be distinguished: back-arc marginal basins (greenstone belts) and continental margin areas (now in a high-grade state). The first recognizable sedimentary sequences comparable to those along modern continental margins occur in the Coronation Geosyncline which developed 2100–1750 my ago (Hoffman, Fraser and McGlynn, 1970; Hoffman, 1973). Its depositional and structural evolution so faithfully mimics that of modern Cordilleran-type geosynclines that it is difficult to escape the conclusion (bearing in mind that such structures are unknown in the Archaean) that plate tectonics at modern-type continental margins began in the early Proterozoic.

9. Aulacogens are sediment-filled failed rift arms that extend from an oceanic margin into a continent (Burke and Dewey, 1973a). They develop at triple junctions related to domal uplifts and are a key indication of the existence of stable continents and plate tectonics. No Archaean aulacogens are known; they first appear in the early Proterozoic in connection with the Coronation Geosyncline and are increasingly common in the late Proterozoic and Phanerozoic (Hoffman, Dewey and Burke, 1974).

10. Stable continental crust must have existed by 2800 my ago in southern Africa and by about 2500 my ago elsewhere for the enormously thick supracrustal piles to be deposited against or on them. Examples are:

Sequence	Age (my)	Thickness (km)
Pongola	c. 2850	10·7
Dominion Reef, Witwatersrand	2700–2800	11·9
Ventersdorp	>2300	5·0
Huronian	2300	c. 12·0
Labrador Trough	c. 2000	17·0
Coronation	2100	10·7
MacArthur	c. 1700	8·5
Mount Isa	c. 1700	22·2

11. The intrusion of extensive basic dyke swarms implies the existence of widespread areas of continental crust. The Ameralik dykes in West Greenland and the Saglek dykes in Labrador belong to the only notable swarms in the Archaean (McGregor, 1973; Collerson, Jesseau and Bridgwater, 1976); there are a few other thin amphibolite dykes elsewhere but they amount to very little. In contrast the Proterozoic abounds in major dyke swarms, particularly in the periods 2500–2000 my and 1300–600 my ago, reviewed in Chapters 4 and 7. The most recent age determinations given in these chapters demonstrate that there was a broad bimodal time distribution with some, but fewer, dykes intruded in the intervening period. The times of dyke intrusion in general coincide with 'lows' in the histogram of isochron ages of rocks from mobile belts and thus correlate with periods of crustal stability. It has been suggested that they also coincide with the hairpin bends of some polar wander curves which relate to large changes in the direction of movement of continental masses (Donaldson and coworkers, 1973).

The Proterozoic dykes broadly reflect the various occasions when different parts of the continents had become stable and began to fracture, perhaps during an early abortive attempt at break-up; it was a diachronous stabilization. The late Proterozoic dykes were related in some way to various initial stages of fragmentation of the continents (or of a supercontinent) independently indicated by the formation of many aulacogens at this time (e.g. eight in North America and three in Siberia: Burke and Dewey, 1973a).

12. Intrusive alkaline complexes, carbonatites and kimberlites are characteristic of stable cratons. They are all unknown in the Archaean but appear in the Proterozoic.

The above points, and there are no doubt more, all lead to the conclusion that there was a dramatic change in the stability of the continents in the early Precambrian. Table 19.6 summarizes the evidence from southern Africa. The probably thin and unstable early

Table 19.6. Changes with time of crustal development in southern Africa (modified after Anhaeusser, 1973; reproduced by permission of The Royal Society)

	Early Archaean (~4000–3400 Ma)	Late Archaean (~3000–1600 Ma) Proterozoic
Crust	Partly oceanic partly sialic with primitive island-arc-like centres nucleating to form protocontinental masses	Progressively thickened and stabilized protocontinental masses nucleated by addition of mantle-derived sialic constituents and welded together to form cratons and large continental blocks
Volcanic rock types	Sequential primitive peridotitic and basaltic komatiites, abyssal and island-arc-type tholeiitic basalts, and calc-alkaline to mildly alkalic volcanic and pyroclastic phases. Rapid, cylical alternations of magma types with stratigraphic height. Volcanism generally K_2O deficient	Extensive, thick, non-sequential, chemically diverse volcanism, including continental-type tholeiitic basalts, andesites, trachytes and rhyolites. Progressive enrichment in K_2O evident
Sedimentary rock types	Palagonitic oozes, thin, poorly developed laminated carbonates, chemical precipitates (banded cherts, thin banded iron formations—jaspilite) carbonaceous cherts and shales. Thick greywacke-shale successions, immature polymictic conglomerates, quartzites, sub-greywackes, sandstones and grits. Sediments derived from rapid erosion and deposited in unstable environments. Sediments both volcanogenic and quartzofeldspathic in origin	Extensive, thick units of conglomerates, orthoquartzites, arkoses, sandstones, shales, dolomite and limestone, chert, and banded and granular iron formations
Granites and igneous complexes	Early, extensive introduction of mantle derived, potash-deficient trondhjemitic, tonalitic, quartz dioritic and granodioritic magmas. Crustal stability increases progressively by granitic additions	Less extensive granitic influx manifest on craton surface, as thickening proceeds. Areas affected by anatexis and metasomatism. Local magmatic granite intrusion (granodiorites, adamellites, granites—reflect K_2O enrichment). Massive, basic differentiated bodies emplaced on stable cratons (e.g. Bushveld, Rooiwater, Usushwana). Extensive mafic dyke invasion

Archaean sialic crust was transformed by particularly rapid growth in the late Archaean into a thick crust by the early Proterozoic and therefore the boundary between these periods represents the most important tectonic turning point in earth history (Salop and Scheinmann, 1969; Ronov, 1972; Hoffman, 1973; Engel and coworkers, 1974).

Let us now consider possible reasons for this changeover in crustal conditions from the Archaean to the Proterozoic.

The Archaean–Proterozoic Boundary

Realizing that this boundary separates the permobile from the platform-geosynclinal stage of earth evolution, Salop and Scheinmann (1969) proposed that it was a tectonic reflection of a dramatic change in the energy regimes of the crust and upper mantle. The break-down of radioactive elements such as U^{235} and K^{40} was very high in the Archaean and decreased rapidly throughout the Proterozoic (Fig. 19.10). In other words, the heat flow responsible for the early main growth of the Archaean continents was very high, but by the time large Proterozoic shields had formed it had decreased substantially.

Following the connection between thermal and tectonic behaviour, Strong and Stevens (1974) suggested that the explanation for the contrasting geological regimes across the Archaean–Proterozoic transition lies in the intersection of phase boundaries with the geotherm (Fig. 19.18). As a result of cooling, geotherms migrate downward with time and because of degassing the peridotite solidus migrates upward, becoming progressively drier. During the Archaean, with a steep geothermal gradient and high water pressure there would be extensive melting which would contribute to rapid and widespread continental growth. At point A the geotherm and solidus would no longer intersect so that with further cooling and with a lower water pressure in the Proterozoic there would be no more melting, accounting for the restriction of magmatism at this time to smaller belts and lineaments, caused by local geothermal variations.

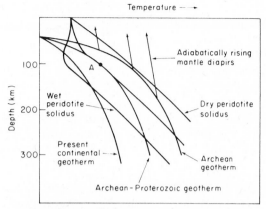

Fig. 19.18. Schematic relations between peridotite solidus at different water pressures and geotherms at different times in a cooling earth. Designed to show the nature of these effects, but not precise temperatures or water pressures (after Strong and Stevens, 1974; reproduced by permission of Macmillan Press)

With its high heat flow the Archaean can be expected to have been a period of rapid continental growth caused by some sort of protoplate tectonics (Dickinson and Luth, 1971; Burke, Dewey and Kidd, 1976). Substantial decrease in heat generation by radioactive decay during the Proterozoic limited growth by ridge and plate activity. The Archaean was dominated by the formation and aggregation of proto/mini-continents which had coalesced into one or a few large/supercontinents by the early Proterozoic, as illustrated in Fig. 3.10 (Engel and coworkers, 1974). The Archaean–Proterozoic boundary can thus be regarded as a threshold in the geothermal/tectonic evolution of the continents.

Isotopes and Crustal Evolution

It is possible to use the evolution of Sr and Pb isotopes as an indication of the fractionation rate of the continental crust. Two main models have been proposed to explain crustal growth rates throughout geological time.

Firstly, by relating a compilation of age determinations to the present areal distribution of continental basements Hurley and Rand (1969) worked out the extent of the continents underlain by rocks within successive intervals of crustal age. A recompilation

338

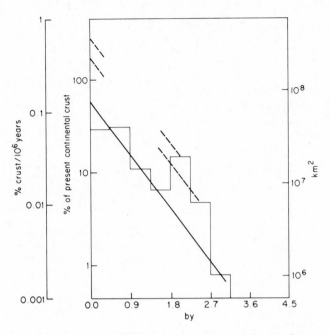

Fig. 19.19. Present-day areal distribution of the continental basement as a function of its geological 'age'. (Based on compilation of Hurley and Rand, 1969, after Veizer, 1976a; reproduced by permission of J. Wiley)

by Veizer (1976) is shown in Fig. 19.19. If areal increase with time is proportional to volume increase, it might be that the rate of continental crust generation has been increased exponentially from 3800 my ago until the present. The model makes the assumption that there has been no recycling of crust through the mantle, although this is doubtful on two accounts. By calculating Sr isotopic evolutionary curves for the continental crust and upper mantle Veizer (1976) suggested that the slope in Fig. 19.19 must have been caused by recycling and that two-thirds of the present-day areal extent of the upper continental crust was formed between 2700 and 1800 my ago. Also the conclusion that far more new material was added to the continental crust in the Phanerozoic than in the Archaean is at variance with the geological evidence—for example, in the Archaean it is likely that most of the volcanic rocks in greenstone belts and most of the granitic rocks in large areas such as the North Atlantic Craton were juvenile additions from the mantle (Moorbath, 1975a,b); at no other time in earth history was a greater proportion of new material added to the continents.

Secondly, Armstrong (1969) and Armstrong and Hein (1973) point out that Pb isotope data indicate that the continents have existed in nearly their present volume throughout the last few billion years. They propose a steady-state model according to which most of the continental crust was fractionated by 2500 my ago and since then crustal material has been recycled through the mantle by some form of subduction process, thus accounting for the Sr isotope evidence of continual addition of material to the continents. The calculations of Veizer (1976) affirm this model. He suggests that this early rapid growth was expressed by the most pronounced 'granite phase' of earth history in the late Archaean and concludes that about two thirds of the present areal extent of the upper continental crust (granitic layer) was formed by 1800 my ago (according to Jahn and Nyquist, 1976, by 2500–1700 my ago). This model is consistent with the geological data previously listed which indicate that stable platforms and rigid continental lithosphere with a thickness little different from that of today had evolved by the early Proterozoic. For more detailed recent discussion of the constraints on early crustal evolution imposed by Rb–Sr data the reader is referred to Jahn and Nyquist (1976).

Armstrong (1969) suggested that not only the volume of the continents but also that of the oceans has remained nearly constant for the last 2500 my. This idea was confirmed by the ingenious equilibrium freeboard model of the earth produced by Wise (1974) according to which the relative elevation of continents with respect to sea level has stayed the same throughout the Proterozoic and Phanerozoic. This means that the continents have had at least 90% of their present thickness for a minimum of 2500 my, a conclusion corroborating that of Condie (1973), who demonstrated that the seismically-determined thicknesses of cratonic areas for this age-range vary only between 38 and 40 km and are independent of age. The strontium isotope trend for seawater with geological time (Fig. 19.3) shows a marked increase in the period 2500–2000 my ago which probably defines the major stage of fractionation of the crust (Veizer, 1976).

A Summary

Here let us consider some of the main conclusions arrived at in this volume, in particular those relating to secular changes that may have contributed to the evolution of the continents.

1. During the first third of the geological record from 3800 to about 2500 my ago the earth evolved in a broadly similar manner giving rise to the granite–greenstone/tonalite (gneiss) association. In this permobile period an overall high radiogenic heat flow was expressed by a rapid growth and consequent thickening of continental crust. Addition of mantle-derived magmas to the continents gave rise to suites of volcanic and plutonic rocks whose chemical parameters are so remarkably similar to those of Mesozoic–Cenozoic times that it is unlikely that early continental growth did not take place by some form of subduction-controlled proto-plate collision mechanism (Fig. 3.10). The consistently low K_2O/Na_2O ratios of Archaean volcanic and plutonic rocks are suggestive of an arc–Cordilleran margin–back arc basin type of crustal growth. In other words, continental aggregation in the Archaean and dispersal in the Phanerozoic both depended on subduction mechanisms and thus both gave rise to similar magma products. Aggregation of many small proto-continental plates created a few large, more stable continental masses by the early Proterozoic. A great variety of geological, geochemical (major and trace element) and isotopic evidence plus a consideration of the constancy of freeboard of the continents with time all corroborate the idea that the continental lithosphere attained a degree of rigidity 2500 my ago that began to approximate that of today.

2. A major result of this newly attained stability was that active Cordilleran-type continental margins, as we understand them today, developed from about 2100 my ago (Coronation and Mount Isa geosynclines) as also did aulacogens, transcontinental basic dyke swarms, rift systems with alkaline complexes, and extensive thick intracontinental sedimentary basins.

3. The role of plate tectonics during the period 2500–1000 my ago is poorly understood and thus controversial. Some internal geosynclines like the Labrador Trough, typically with banded iron formations, have the character of narrow Red Sea-type rifts that underwent a small degree of subduction during closure. Some high-grade belts such as the Grenville and Ketilidian may have formed by collision of two continents with much basement reactivation. Some high-grade shear belts may be deep-seated equivalents of Proterozoic transform faults. Palaeomagnetic data indicate less than 15° of horizontal motion between cratons in Africa across Pan-African high-grade belts and some data, admittedly controversial, suggest the existence of a supercontinent in the mid–late Proterozoic. Whether or not one believes in plate collisional or intracontinental mobile belts in the Proterozoic it is clear that this period of

earth history was different in many respects from the Archaean.

4. The size of the oceans in the Archaean is under dispute but the atmosphere was largely anoxygenic allowing only the existence of primitive procaryotic organisms. By 2000 my ago the atmospheric oxygen content had increased considerably enabling banded iron formations to reach a peak of development and the first red beds to form. By the late Precambrian it had reached 1% of the present level allowing the formation of an ozone layer and the appearance of oxygen-employing eucaryotes and, eventually, of complex multicellular organisms.

The deposition of carbonate sediments increased drastically between the Archaean and the early Proterozoic and there was a marked decrease in their dolomite–calcite ratio between the Precambrian and the Phanerozoic. These and other chemical variations in sedimentary rocks reflect secular changes in the evolution of the ocean-atmospheric system.

5. Beginning about 1000 my ago a new stage of continental fragmentation, drift and collision began which led to the formation of the Appalachian–Caledonian–Hercynian–Uralian fold belts and the creation of a Pangaea landmass which itself broke up into continental segments that drifted throughout the Mesozoic–Cenozoic into their present positions. The different types of sedimentary, igneous and metamorphic rocks that developed along plate boundaries during the growth and closure of oceans provide a major key to the unravelling of the complex tectonic relationships found in modern and most ancient fold belts.

Finally, what about uniformitarianism? Similar physicochemical principles must have controlled the processes of sedimentation, magmatism and metamorphism throughout earth history. But these were influenced by two major factors: firstly, the evolving atmosphere and hydrosphere controlled the variation in composition of many types of sediment and, secondly, the difference between the Archaean and the Proterozoic tectonic regimes was responsible for a significant change in development of the rock record. Bearing these constraints in mind, the principles of uniformitarianism have applied throughout the evolution of the continents.

References

Aalto, K. R. (1971). Glacial marine sedimentation and stratigraphy of the Toby conglomerate (Upper Proterozoic), south-eastern British Columbia, North West Idaho and northeastern Washington, *Can. J. Earth Sci.*, **8**, 753–787.

Abbate, E., Bortolotti, V. and Passerini, P. (1970). Development of the Northern Appenines Geosyncline—olistostromes and olistoliths, *Sed. Geol.*, **4**, 521–557.

Ager, D. V. (1975). The geological evolution of Europe. *Proc. Geol. Assoc.*, **86**, 127–154.

Alderman, A. R. (1936). Eclogites from the vicinity of Glenelg, Inverness-shire, *Quart. Jl. geol. Soc. Lond.*, **92**, 488–533.

Allaart, J. H. (1967). Basic and intermediate igneous activity and its relationship to the evolution of the Julianehaab Granite, South Greenland. *Meddr. Grønland*, **175**, No. 1, 136 pp.

Allaart, J. H. (1976). The pre-3760 my old supracrustal rocks of the Isua area, central West Greenland, and the associated occurrence of quartz-banded ironstone. In B. F. Windley (Ed.), *The Early History of the Earth*, Wiley, London, 177–189.

Allaart, J. H., Bridgwater, D. and Henriksen, N. (1969). Pre-Quaternary geology of Southwest Greenland and its bearing on North Atlantic correlation problems. In *North Atlantic—Geology and Continental Drift, Mem. Amer. Ass. Petrol. Geol.*, **12**, 859–882.

Allard, G. O. (1970). The Dore Lake Complex of Chibougamau, Quebec, a metamorphosed Bushveld-type layered complex, Symp. on Bushveld Igneous Complex and other layered intrusions. *Geol. Soc. S. Afr. Sp. Publ.*, **1**.

Allsopp, H. L. (1964). Rubidium/strontium ages from the western Transvaal, *Nature*, **204**, 361–363.

Allsopp, H. L. (1965). Rb–Sr and K–Ar age measurements on the Great Dyke of southern Rhodesia, *J. geophys. Res.*, **70**, 977.

Allsopp, H. L., Ulrych, J. J. and Nicholaysen, L. O. (1968). Dating some significant events in the history of the Swaziland System by the Rb–Sr isochron method, *Can. J. Earth Sci.*, **5**, 605–619.

Alvarez, W. (1973). The application of plate tectonics to the Mediterranean region. In D. H. Tarling and S. K. Runcorn (Eds.), Vol. 2 of *Implications of Continental Drift to the Earth Sciences*, Academic Press, London, 893–908.

Alvarez, W., Cocozza, T. and Wezel, F. C. (1974). Fragmentation of the Alpine orogenic belt by microplate dispersal. *Nature*, **248**, 309–313.

Amstutz, G. C., Zimmermann, R. A. and Schot, E. H. (1971). The Devonian mineral belt of western Germany. In *Sedimentology of parts of central Europe*, Verlag Waldemar Kramer, Frankfurt-on-Main, 253–272.

Anderson, A. T. and Morin, M. (1969). Two types of massif anorthosites and their implications regarding the thermal history of the crust. In *Origin of Anorthosite and Related Rocks, Mem. New York State Mus. Sci. Serv.*, **18**, 57–70.

Anderson, R. N. (1975). Heat flow in the Mariana marginal basin, *J. Geophys. Res.*, **80**, 4043–4048.

Andrew-Jones, D. A. (1966). Geology and mineral resources of the northern Kambui schist belt and adjacent granulites. *Bull. geol. Surv. Sierra Leone*, **6**, 100 pp.

Andrews, J. R. (1973). Stratigraphic, structural and metamorphic features of the Archaean (pre-Ketilidian) rocks in the Frederikshaab district, south-west Greenland. In R. G. Park and J. Tarney (Eds.), *The Early Precambrian of Scotland and related rocks of Greenland*, Univ. of Keele, 179–188.

Andrews, J. R. and Emeleus, C. H. (1975). Structural aspects of

kimberlite dyke and sheet intrusions in south-west Greenland. In L. H. Ahrens, J. B. Dawson, A. R. Duncan and A. J. Erlank (Eds.), Vol. 9 of *Physics and Chemistry of the Earth*, Pergamon Press, Oxford, 43–50.

Anhaeusser, C. R. (1971a). The Barberton Mountain Land, South Africa—a guide to the understanding of the Archaean geology of western Australia. *Geol. Soc. Australia Sp. Publ.*, No. 3, 103–120.

Anhaeusser, C. R. (1971b). Cyclic volcanicity and sedimentation in the evolutionary development of Archaean greenstone belts of Shield areas. *Geol. Soc. Australia Sp. Publ.*, No. 3, 57–70.

Anhaeusser, C. R. (1973). The evolution of the early Precambrian crust of southern Africa. *Phil. Trans. R. Soc. Lond.*, **A273**, 359–388.

Anhaeusser, C. R. (1975). The geological evolution of the primitive earth: evidence from the Barberton Mountain Land. *Econ. Geol. Res. Unit*, Univ. Witwatersrand, Johannesburg. Inf. Circ. 98, 1–22.

Anhaeusser, C. R. (1976). The nature and distribution of Archaean gold mineralisation in southern Africa. *Minerals Sci. Engng.*, **8**, 46–84.

Anhaeusser, C. R., Roering, C., Viljoen, M. J. and Viljoen, R. P. (1968). The Barberton Mountain Land: a model of the elements and evolution of an Archaean fold belt. *Geol. Soc. S. Afr.*, Annexure to vol. 71, 225–254.

Anhaeusser, C. R., Mason, R., Viljoen, M. J. and Viljoen, R. P. (1969). A reappraisal of some aspects of Precambrian Shield geology. *Bull Geol. Soc. Amer.*, **80**, 2175–2200.

Anhaeusser, C. R., Fritze, K., Fyfe, W. S. and Gill, R. C. O. (1975). Gold in "primitive" Archaean volcanics, *Chem. Geol.*, **16**, 129–135.

Anwar, Y. M., El-Mahdy, O. R. and El-Dahhar, M. A. (1972). Geology and origin of the lead–zinc ores of the Um Gheig areas, Egypt. *Econ. Geol.*, **67**, p. 1002.

Argand, E. (1916). Sur l'arc des Alpes Occidentales. *Eclogae Geol. Helvetiae*, **14**, 145–191.

Argand, E. (1924). La tectonique de l'Asie. *13th Int. Geol. Congr. Brussels*, 171–372.

Armstrong, R. L. (1969). A model for the evolution of strontium and lead isotopes in a dynamic earth. *Rev. Geophys.*, **6**, 175–199.

Armstrong, R. L. and Hein, S. M. (1973). Computer simulation of Pb and Sr isotope evolution of the Earth's crust and upper mantle. *Geochim. Cosmochim. Acta*, **37**, 1–18.

Arriens, P. A. (1971). The Archaean geochronology of Australia, *Spec. Publ. geol. Soc. Aust.*, **3**, 11–24.

Arth, J. G. (1976). A model for the origin of the early Precambrian greenstone–granite complex of northeastern Minnesota. In B. F. Windley (Ed.), *The Early History of the Earth*, Wiley, London, 299–302.

Arth, J. G. and Hanson, G. N. (1972). Quartz diorites derived by partial melting of eclogite or amphibolite at mantle depths. *Contr. Mineral. Petrol.*, **37**, 161–174.

Arth, J. G. and Hanson, G. N. (1975). Geochemistry and origin of the early Precambrian crust of northeastern Minnesota. *Geochim. Cosmochim. Acta*, **39**, 325–362.

Asklund, B. (1931). Fennoskandias geologi (prekambrium). In W. Ramsay, *Geologiens grunder*, 3rd ed., pt II, Stockholm, 139–305.

Asmus, H. E. and Ponte, F. C. (1973). The Brazilian marginal

341

342

basins. In A. E. M. Nairn and F. G. Stehli (Eds.), *The South Atlantic*, Vol. 1, Plenum Press, New York, 87–134.

Aswathanarayana, U. (1968). Metamorphic chronology of the Precambrian provinces of South India, *Can. J. Earth Sci.*, **5**, 591–600.

Atwater, T. (1970). Implications of plate tectonics for the Cenozoic tectonic evolution of western North America. *Bull. geol. Soc. Amer.*, **81**, 3513–3536.

Audley-Charles. M. (1970). Triassic palaeogeography of the British Isles. *Quart. J. geol. Soc. Lond.*, **126**, 49–89.

Aumento, F. (1968). The Mid-Atlantic Ridge near 45°N. II: Basalts from the area of Confederation Peak. *Can. J. Earth Sci.*, **5**, 1–21.

Aumento, F. and Loubert, H. (1971). The Mid-Atlantic Ridge near 45°N. XVI: Serpentinized ultramafic intrusions. *Can. J. Earth Sci.*, **8**, 631–663.

Awramik, S. M. (1971). Precambrian columnar stromatolite diversity: reflection of Metazoan appearance. *Science*, **174**, 825–826.

Axelrod, D. I. (1972). Revolutions in the plant world. *Geol. Soc. Amer. Abstr. with Prog.*, **4**, p. 124.

Aymé, J. M. (1965). The Senegal salt basin. In D. C. Ion (Ed.), *Salt basins around Africa*, London Inst. Petroleum, 83–90.

Ayres, D. E. (1971). The hematite ores of Mount Tom Price and Mount Whaleback, Hamersley Iron Province. *Aust. Inst. Min. Metall.*, **No. 238**, 47–58.

Baadsgaard, H. (1973). U–Th–Pb dates on zircons from the early Precambrian Amîtsoq gneisses, Godthaab district, West Greenland. *Earth Plan. Sci. Lett.*, **19**, 22–28.

Baadsgaard, H., Lambert, R. St. J. and Krupicka, J. (1976). Mineral isotopic age relationships in the polymetamorphic Amîtsoq gneisses, Godthaab district, West Greenland. *Geochim. Cosmochim. Acta*, **40**, 513–528.

Backer, H. and Schoell, M. (1972). New deeps with brines and metalliferous sediments in the Red Sea. *Nature Phys. Sci.*, **240**, 153–158.

Badham, J. P. N. (1976a). General discussion on Proterozoic tectonics. *Phil. Trans. R. Soc. Lond.*, **A280**, 661–662.

Badham, J. P. N. (1976b). Orogenesis and metallogenesis with reference to the silver-nickel-cobalt arsenide ore association. In D. F. Strong (Ed.), *Metallogeny and Plate Tectonics, Geol. Ass. Can. Sp. Pap.* **14**, 559–572.

Badham, J. P. N., Robinson, B. W. and Morton, R. D. (1972). The geology and genesis of the Great Bear Lake silver deposits. *24th Int. geol. Congr. Montreal*, Sect. 4, 541–548.

Badham, J. P. N. and Halls, C. (1975). Microplate tectonics, oblique collisions and evolution of the Hercynian orogenic systems. *Geology*, **3**, 373–376.

Baer, A. J. (1976a). The Grenville Province in Helikian times: a possible model of evolution. *Phil. Trans. R. Soc. Lond.*, **A280**, 499–515.

Baer, A. J. (1976b). Preliminary report on the Borgia metaanorthosite (La Tuque, Quebec, Canada). *Can. J. Earth Sci.*, **13**, 84–91.

Baer, A. J., Emslie, R. F., Irving, E. and Tanner, J. C. (1974). Grenville geology and plate tectonics. *Geoscience, Canada*, **1**, 54–61.

Bahneman, K. P. (1971). In E. R. Morrison and J. F. Wilson, *Symp. on the granites, gneisses and related rocks: excursion guidebook. Geol. Soc. S. Afr., Rhodesian Branch*, 16–22.

Bailey, D. K. (1964). Crustal warping—a possible tectonic control of alkaline magmatism. *J. Geophys. Res.*, **69**, 1103–1111.

Bailey, D. K. (1970). Volatile flux, heat focussing and the generation of magma. *Geol. J. Spec.*, **Issue 2**, 177–186.

Bailey, D. K. (1974). Continental rifting and alkaline magmatism. In H. Sørensen (Ed.), *The Alkaline Rocks*, Wiley, London, 148–159.

Bailey, E. B. and McCallien, W. J. (1950). The Ankara mélange and the Anatolian thrust. *Nature*, **166**, 938–940.

Bailey, E. B. and McCallien, W. J. (1963). Liguria nappe, northern Apennines. *Roy. Soc. Edinb. Trans.*, **65**, 315–333.

Bailey, E. H., Irwin, W. P. and Jones, D. L. (1964). Franciscan and related rocks and their significance in the geology of western California. *Calif. Div. Mines and Geol. Bull.*, **183**, 89–112.

Bailey, E. H., Blake, M. C., Jr. and Jones, D. L. (1970). On-land Mesozoic oceanic crust in California Coast Ranges. *U.S. geol. Surv. Prof. Pap.* 700–C, 70 pp.

Baird, A. K., Baird, K. W. and Welday, E. E. (1974). Chemical trends across Cretaceous batholithic rocks of southern California. *Geology*, **2**, 493–495.

Bak, J., Sørensen, K., Grocott, J., Korstgaard, J. A., Nash, D. and Watterson, J. (1975). Tectonic implications of Precambrian shear belts in western Greenland. *Nature*, **254**, 566–509.

Baker, B. H., Williams, L. A. J., Miller, J. A. and Fitch, F. J. (1971). Sequence and geochronology of the Kenya Rift volcanics. *Tectonophysics*, **11**, 191–215.

Baker, B. H., Mohr, P. A. and Williams, L. A. J. (1972). Geology of the eastern Rift System of Africa. *Geol. Soc. Amer. Sp. Pap.* **136**, 67 pp.

Baker, P. E. (1968). Comparative volcanology and petrology of the Atlantic island arcs. *Bull. Volcan.*, **32**, 189–206.

Baker, P. E. (1972). Volcanism at destructive plate margins. *J. Earth Sci., Leeds*, **8**, 183–204.

Baker, P. E. (1973). Islands of the South Atlantic. In A. E. M. Nairn and F. G. Stehli (Eds.). *The Ocean Basins and Margins*, Vol. I *The South Atlantic*, Plenum Press, New York, 493–553.

Balasundaram, M. S. and Balasubrahmanyan, M. N. (1973). Geochronology of the Indian Precambrian. *Bull. geol. Soc. Malaysia*, **6**, 213–226.

Ballard, R. D. and Uchupi, E. (1975). Triassic rift structure in Gulf of Maine. *Bull. Amer. Assoc. Petrol. Geol.*, **59**, 1041–1072.

Banno, S. (1964). Petrologic studies on Sanbagawa crystalline schists in the Bessi–Ino district, central Sikoku, Japan. *Tokyo Univ. Fac. Sci. J. Sect. 2*, **15**, 203–219.

Baragar, W. R. A. (1968). Major element geochemistry of the Noranda volcanic belt, Ontario. *Can. J. Earth Sci.*, **4**, 773–790.

Baragar, W. R. A. and Goodwin, A. M. (1969). Andesites and Archaean volcanism of the Canadian Shield. Proc. Andesite Conf., A. R. McBirney (Ed.), State of Oregon, *Dept of Geol. and Mineral. 2nd Bull.*, **65**, 121–142.

Barazangi, M. and Dorman, J. (1969). World seismicity map of ESSA coast and geodetic survey epicenter data for 1961–7. *Bull. Seismol. Soc. Amer.*, **59**, 369–380.

Barber, C. (1974). The geochemistry of carbonatites and related rocks from two carbonatites, South Nyanza, Kenya, *Lithos*, **7**, 53–63.

Barberi, F., Bonatti, E., Marinelli, G. and Varet, J. (1974). Transverse tectonics during the split of a continent; data from the Afar rift. *Tectonophysics*, **23**, 17–29.

Barbosa, A. L. M. and Grossi Sad, J. H. (1973). Tectonic control of sedimentation and trace element distribution in iron ores of central Minas Gerais (Brazil). In *Genesis of Precambrian iron and manganese deposits*, Kiev Symp. 1970, Unesco, Paris, 125–131.

Bard, J. P., Capdevila, R., Matte, P. and Ribeiro, A. (1973). Geotectonic model for the Iberian Variscan orogen. *Nature Phys. Sci.* **241**, 50–52.

Barghoorn, E. S. and Tyler, S. A. (1965). Micro-organisms from the Gunflint chert. *Science*, **147**, 563–577.

Barghoorn, E. S. and Schopf, J. W. (1966). Micro-organisms three billion years old from the Precambrian of South Africa. *Science*, **152**, 758–763.

Barker, D. S. (1969). North American feldspathoidal rocks in space and time. *Bull. geol. Soc. Amer.*, **80**, 2369–2372.

Barker, F. and Peterman, Z. E. (1974). Bimodal tholeiitic-dacitic magmatism and the early Precambrian crust. *Precamb. Res.*, **1**, 1–12.

Barnard, P. D. W. (1973). Mesozoic floras. In N. F. Hughes (Ed.), *Organisms and Continents through Time, Pal. Ass. Sp. Pap. in Pal.* **12**, 175–187.

Barth, T. F. W. (1945). Studies on the igneous rock complex of the Oslo region. II. Systematic petrography of the plutonic rocks. *Skr. norske Videns. Akad. Mat.-naturv. kl. 1944*, No. 9.

Barth, T. F. W. (1955). Studies on the igneous rock complex of the Oslo region. XIV. Provenance of the Oslo magmas. *Skr. norske Vidensk. Akad. Mat.-naturv. kl. 1954*, No. 4.

Barth, T. F. W. and Reitan, P. H. (1963). The Precambrian of Norway. In K. Rankama (Ed.), *The Precambrian*, Vol. 1, Wiley Interscience, New York, 27–80.

343

Bartolini, C., Gehin, C. and Stanley, D. J. (1972). Morphology and recent sediments of the western Alboran basin in the Mediterranean Sea, *Marine Geol.*, **13**, 159–224.

Bass, M. N. (1970). North American feldspathoidal rocks in space and time: discussion. *Bull. geol. Soc. Amer.*, **81**, 3493–3500.

Bassov, V. A., Krimholz, G. Ya., Mesezhnicov, M. S., Saks, V. N., Shulgina, N. L. and Vakhrameen, V. A. (1972). The problem of continental drift during the Jurassic and Cretaceous in the light of palaeobiogeographical data. *24th Int. geol. Congr. Montreal, Sect. 7*, 257–264.

Bateman, P. C. and Dodge, F. C. W. (1970). Variations of major chemical constituents across the Central Sierra Nevada batholith. *Bull. geol. Soc. Amer.*, **81**, 409–420.

Baumgartner, T. R. and Van Andel, T. H. (1971). Diapirs of the continental margin of Angola, *Bull. geol. Soc. Amer.*, **82**, 793–802.

Bazin, D. and Hübner, H. (1969). Copper deposits in Iran. *Rep. geol. Surv. Iran*, No. 3, 232 pp.

Beach, A. (1976). The interrelations of fluid transport, deformation, geochemistry and heat flow in early Proterozoic shear zones in the Lewisian complex, *Phil. Trans. R. Soc. Lond.*, **A280**, 569–604.

Bearth, P. (1967). Die Ophiolithe der Zone von Zermatt–Saas Fee. *Beitr. geol. Karte Schweiz*, n.f. 132, 130 pp.

Bell, K., Blenkinsop, J. and Moore, J. M. (1975). Evidence for a Proterozoic greenstone belt from Snow Lake, Manitoba. *Nature*, **258**, 698–701.

Bell, R. T. (1970). The Hurwitz Group, a prototype for deposition on metastable cratons. In A. J. Baer (Ed.), *Symposium on Basins and Geosynclines of the Canadian Shield, Geol. Surv. Can. Pap. 70–40*, 159–168.

Bell, R. T. 1971. Boundary geology, Upper Nelson River area, Manitoba and northwestern Ontario. In A. C. Turnock (Ed.), *Geoscience Studies in Manitoba, Geol. Ass. Can. Spec. Pap. 9*, 11–39.

Bellido, B. E., de Montreuil, D. C. and Girard, P. D. (1969). Aspectos generales de la metalogenia del Peru, Lima, *Peru Serv. Geol. Min.*, 96 pp.

Belmonte, Y., Hirtz, P. and Wenger, R. (1965). The salt basins of the Gabon and the Congo (Brazzaville). In *Salt basins around Africa*, Inst. Pet., London, 55–74.

Bennett, E. M. (1965). Lead–zinc–silver and copper ore deposits of Mount Isa. In J. McAndrew (Ed.), *Geology of Australian Ore Deposits*, 8th Commonwealth Min. and Metall. Congr. Australia and New Zealand, 233–246.

Berger, W. H. (1972). Deep sea carbonates: dissolution facies and age-depth constancy. *Nature*, **236**, 392.

Berger, W. H. and Winterer, E. L. (1974). Plate stratigraphy and the fluctuating carbonate line. In K. J. Hsü and H. C. Jenkyns (Eds.), *Pelagic sediments: on land and under the sea, Int. Ass. Sed. Sp. Publ.*, **1**, 11–48.

Berkner, L. V. and Marshall, L. C. (1965). On the origin and rise of oxygen concentration in the earth's atmosphere. *J. atmosph. Sci.*, **22**, 225–261.

Berkner, L. V. and Marshall, L. C. (1967). The rise of oxygen in the Earth's atmosphere with notes on the Martian atmosphere. *Adv. Geophys.*, **12**, 309–331.

Bernoulli, D. and Peters, T. (1970). Traces of rhyolitic–trachytic volcanism in the Upper Jurassic of the southern Alps. *Eclogae Geol. Helvetiae*, **63**, 609–621.

Bernoulli, D., Laubscher, H. P., Trümpy, R. and Wenk, E. (1974). Central Alps and Jura mountains. In A. M. Spencer (Ed.), *Mesozoic–Cenozoic Orogenic Belts, Geol. Soc. Lond. Sp. Publ.*, **4**, 85–108.

Bernoulli, D. and Jenkyns, H. C. (1974). Alpine, Mediterranean and Central Atlantic Mesozoic facies in relation to the early evolution of the Tethys. In R. H. Dott Jr and R. H. Shaver (Eds.), *Modern and Ancient Geosynclinal Sedimentation, Soc. Econ. Pal. Mineral. Sp. Publ.*, **19**, 129–160.

Berrangé, J. P. (1970). The geology of two small layered hornblende peridotite (picrite) plutons in South Greenland. *Meddr. Grønland*, **192**, 1, 43 pp.

Berthelsen, A. and Bridgwater, D. (1960). On the field occurrence and petrography of some basic dykes of supposed Precambrian

age from the southern Sukkertoppen district, western Greenland. *Meddr. Grønland*, **123**, 1–42.

Besairie, H. (1967). The Precambrian of Madagascar. In K. Rankama (Ed.), *The Precambrian*, Vol. 3, Wiley Interscience, New York, 133–142.

Bezzi, A. and Picardo, G. B. (1971). Structural features of the Ligurian ophiolites: petrologic evidence for the 'oceanic' floor of the northern Apennines geosyncline; a contribution to the problem of the Alpine-type gabbro–peridotite association. *Mem. Soc. Geol. Ital.*, **10**, 53–63.

Bichan, R. (1969). Origin of chromite seams in the Hartley complex of the Great Dyke, Rhodesia. In H. D. B. Wilson (Ed.), *Magmatic Ore Deposits, Econ. Geol. Mono.*, **4**, 95–113.

Bichan, R. (1970). The evolution and structural setting of the Great Dyke, Rhodesia. In T. N. Clifford and I. G. Gass (Eds.), *African Magmatism and Tectonics*, Oliver and Boyd, 51–71.

Bickle, M. J., Martin, A. and Nisbet, E. G. (1975). Basaltic and peridotitic komatiites and stromatolites above a basal unconformity in the Belingwe greenstone belt, Rhodesia, *Earth Plan. Sci. Lett.*, **27**, 155–162.

Bignell, R. D. (1975). Timing, distribution and origin of submarine mineralization in the Red Sea. *Trans. Inst. Min. Metall.*, **84**, B1–6.

Bilibin, Yu. A. (1955). Metallogenic provinces and metallogenic epochs, Moscow, Gosgeoltekhizdat (in Russian; English translation by Dept. of Geology, City University of New York, 1968).

Binns, R. A., Gunthorpe, R. J. and Groves, D. I. (1976). Metamorphic patterns and development of greenstone belts in the eastern Yilgarn block, western Australia. In B. F. Windley (Ed.), *The Early History of the Earth*, Wiley, London, 303–313.

Bird, J. M. and Dewey, J. F. (1970). Lithosphere plate: continental margin tectonics and the evolution of the Appalachian orogen. *Bull., geol. Soc. Amer.*, **81**, 1031–1059.

Black, D. I. (1971). Polarity reversal and faunal extinction. In I. G. Gass, P. J. Smith and R. C. L. Wilson (Eds.), *Understanding the Earth*, Artemis Press. Sussex, 257–261.

Black, L. P., Moorbath, S., Pankhurst, R. J. and Windley, B. F. (1973). $^{207}Pb/^{206}Pb$ whole-rock age of the Archaean granulite facies event in West Greenland. *Nature Phys. Sci.*, **244**, 50–53.

Blake, M. C., Jr and Jones, D. L. (1974). Origin of Franciscan mélanges in northern California. In R. H. Dott Jr and R. H. Shaver (Eds.), *Ancient and Modern Geosynclinal Sedimentation, Soc. Econ. Pal. Mineral. Sp. Publ.*, **19**, 345–357.

Bliss, N. W. (1969). Thermal convection in the Archaean crust? *Nature*, **222**, 972–973.

Bliss, N. W. and Stidolph, P. A., (1969). A review of the Rhodesian basement complex. *Trans. geol. Soc. S. Afr. Sp. Publ.*, **2**, 305–333.

Bloxam, T. W. and Allen, J. B. (1959). Glaucophane schist, eclogite and associated rocks from Knockormal in the Girvan–Ballantrae complex, South Ayrshire. *Trans. R. Soc. Edinb.*, **64**, 1–27.

Boccaletti, M. and Guazzone, G. (1974). Remnant arcs and marginal basins in the Cainozoic development of the Mediterranean. *Nature*, **252**, 18–21.

Bogatikov, O. A. (1974). *Anorthosites of the USSR*, Acad. Nauk, Moscow, 122 pp (in Russian).

Bogdanov, N. A. (1973). Tectonic development of trenches in the western Pacific. In P. J. Coleman (Ed.), *The Western Pacific: island arcs, marginal seas, geochemistry*, Univ. W. Australia Press, Nedlands, 327–333.

Bonatti, E. (1975). Metallogenesis at oceanic spreading centers. *Ann. Rev. Earth Planet. Sci.*, **3**, 401–431.

Bonatti, E. and Joensuu, O. (1966). Deep sea iron deposits from the South Pacific. *Science*, **154**, 643–645.

Bonatti, E., Honnorez, J. and Ferrara, G. I. (1971). Ultramafic rocks. Peridotite–gabbro–basalt complex from the equatorial Mid-Atlantic Ridge. *Phil. Trans. R. Soc. Lond.*, **A268**, 385–402.

Bonatti, E., Fisher, D. E., Joensuu, O., Rydell, H. S. and Beyth, M. (1972). Iron–manganese–barium deposit from the northern Afar Rift (Ethiopia). *Econ. Geol.*, **67**, 717–730.

Bonatti, E., Zerbi, M., Kay, R. and Rydell, H. (1976). Metalliferous deposits from the Apennine ophiolites: Mesozoic equivalents of modern deposits from oceanic spreading centers. *Bull. geol. Soc. Amer.*, **87**, 83–94.

344

Bond, G., Wilson, J. F. and Winnall, N. J. (1973). Age of the Huntsman limestone (Bulawayan) stromatolites. *Nature*, **244**, 275–276.

Bondesen, E. (1970). The stratigraphy and deformation of the Precambrian rocks of the Graenseland area, South-West Greenland. *Meddr. Grønland*, **185**, 125 pp.

Bondesen, E., Pedersen, K. R. and Jørgensen, O. (1967). Precambrian organisms and the isotopic composition of organic remains in the Ketilidian of south-west Greenland. *Meddr. Grønland*, **164**, 4, 41 pp.

Borley, G. D. (1974). Oceanic islands. In H. Sørensen (Ed.), *The Alkaline Rocks*, Wiley, London, 311–330.

Bosellini, A. and Hsü, K. J. (1973). Mediterranean plate tectonics and Triassic palaeogeography. *Nature*, **244**, 144–146.

Bosellini, A. and Winterer, E. L. (1975). Pelagic limestone and radiolarite of the Tethyan Mesozoic: a genetic model. *Geology*, **3**, 279–282.

Boström, K. and Peterson, M. N. A. (1966). Precipitates from hydrothermal exhalations on the East Pacific Rise. *Econ. Geol.*, **61**, 1258–1265.

Boström, K. and Peterson, M. N. A. (1969). The origin of aluminium-poor ferromanganoan sediments in areas of high heat flow on the East Pacific Rise. *Marine Geol.*, **7**, 427–447.

Boström, K., Peterson, M. N. A., Joensuu, O. and Fisher, D. E. (1969). Aluminium-poor ferromanganoan sediments on active oceanic ridges. *J. geophys. Res.*, **74**, 3261–3270.

Boström, K. and Fisher, D. E. (1971). Volcanogenic uranium, vanadium and iron in Indian Ocean sediments. *Earth Plan. Sci. Lett.*, **11**, 95–98.

Boström, K., Joensuu, O., Valdes, S. and Riera, M. (1972). Geochemical history of S. Atlantic Ocean sediments since late Cretaceous. *Marine Geol.*, **12**, 85–121.

Boström, K., Joensuu, O., Kraemer, T., Rydell, H., Valdes, S., Gartner, S. and Taylor, G. (1974). New finds of exhalative deposits on the East Pacific Rise. *Geol. Fören. Stockholm Förh.*, **96**, 53–60.

Bott, M. H. P. (1971). *The Interior of the Earth*, E. Arnold, London, 316 pp.

Boucot, A. J., Berry, W. B. N. and Johnson, J. G. (1968). The crust of the earth from a Lower Palaeozoic point of view. In R. A. Phinney (Ed.), *The History of the Earth's Crust*, Princeton Univ. Press, New Jersey, 208–228.

Boulanger, J. (1959). Les anorthosites de Madagascar, *Annls. geol. Madagascar*, **26**, 71 pp.

Bowden, P. (1970). Origin of the younger granites of northern Nigeria. *Cont. Mineral. Petrol.*, **25**, 153–162.

Bowes, D. R. (1968). An orogenic interpretation of the Lewisian of Scotland. *23rd Int. geol. Congr.*, **4**, 225–236.

Bowes, D. R. (1971). Lewisian chronology. *Scott. J. Geol.*, **7**, 179–187.

Bowes, D. R. (1972). Geochemistry of Precambrian crystalline basement rocks, north-west highlands of Scotland. *24th Int. geol. Cong.*, Sect. 1, 97–103.

Bowes, D. R. (1976). Archaean crustal history in North-West Britain. In B. F. Windley (Ed.), *The Early History of the Earth*, Wiley, London, 469–479.

Bowes, D. R., Wright, A. E. and Park, R. G. (1964). Layered intrusive rocks in the Lewisian of the north-west highlands of Scotland. *Quart. J. geol. Soc. Lond.*, **120**, 153–192.

Bowes, D. R., Barooah, B. C. and Khoury, S. G. (1971). Original nature of Archaean rocks of northwest Scotland. *Spec. Publ. geol. Soc. Aust.*, **3**, 77–92.

Bowes, D. R. and Hopgood, A. M. (1973). Framework of Precambrian crystalline complex of northwestern Scotland. In R. T. Pidgeon, R. M. Macintyre, S. M. F. Sheppard and O. Van Breemen (Eds.), *Geochronology and isotope geology of Scotland: field guide and reference*, 3rd European Congr. of Geochronologists 1973, 1–14.

Bowes, D. R. and Hopgood, A. M. (1975). Framework of the Precambrian crystalline complex of the Outer Hebrides, Scotland. *Krystalinikum*, **11**, 7–23.

Brew, D. A. (1968). The role of volcanism in post-Carboniferous tectonics of southeastern Alaska and nearby regions, North America. *23rd Int. geol. Congr. Prague*, **2**, 107–121.

Briden, J. C. (1967). Recurrent continental drift of Gondwanaland. *Nature*, **215**, 1334–1339.

Briden, J. C. (1968). Palaeoclimatic evidence of a geocentric axial dipole field. In E. A. Phinney (Ed.), *History of the Earth's Crust*, Princeton Univ. Press, New Jersey, 178–194.

Briden, J. C. (1970a). Palaeomagnetic polar wander curve for Africa. In S. K. Runcorn (Ed.), *Palaeogeophysics*, Academic Press, London, 277–290.

Briden, J. C. (1970b) Palaeolatitude distribution of precipitated sediments. In S. K. Runcorn (Ed.), *Palaeogeophysics*, Academic Press, London, 437–444.

Briden, J. C. (1976). Application of palaeomagnetism to Proterozoic tectonics. *Phil. Trans. R. Soc. Lond.*, **A280**, 405–416.

Briden, J. C. and Irving, E. (1964). Palaeolatitude spectra of sedimentary palaeoclimatic indicators. In A. E. M. Nairn (Ed.), *Problems in Palaeoclimatology*, Interscience, New York, 199–224.

Briden, J. C., Henthorn, D. I. and Rex, D. C. (1971). Palaeomagnetic and radiometric evidence for the age of the Freetown igneous complex, Sierra Leone. *Earth Planet. Sci. Lett.*, **12**, 385–391.

Briden, J. C., Morris, W. A. and Piper, J. D. A. (1973). Regional and global implications, In *Palaeomagnetic studies in the British Caledonides, Geophys. J. R. astr. Soc.*, **34**, 107–134.

Bridgwater, D. and Watterson, J. S. (1967). Igneous intrusions and associated rocks of the mangerite–charnockite suite. *Nature*, **213**, 897.

Bridgwater, D. and Harry, W. T. (1968). Anorthosite xenoliths and plagioclase megacrysts in Precambrian intrusions of South Greenland. *Bull. Grønlands geol. Unders.*, **77**, 234 (also *Meddr. Grønland*, **185**, Nr. 2).

Bridgwater, D. and Coe, K. (1970). The role of stoping in the emplacement of the giant dykes of Isortoq, South Greenland. In G. Newall and N. Rast (Eds.), *Mechanism of Igneous Intrusion*, Seal House Press, Liverpool, 67–78.

Bridgwater, D. and Windley, B. F. (1973). Anorthosites, postorogenic granites, acid volcanic rocks, and crustal development in the North Atlantic Shield during the Mid-Proterozoic. In L. A. Lister (Ed.), *Symp. on Granites, Gneisses and Related Rocks, Geol. Soc. S. Afr. Sp. Publ.*, **3**, 307–318.

Bridgwater, D., Escher, A. and Watterson, J. (1973a). Tectonic displacements and thermal activity in two contrasting Proterozoic mobile belts from Greenland. *Phil. Trans. R. Soc. Lond.*, **A273**, 513–533.

Bridgwater, D., Escher, A. and Watterson, J. (1973b). Dyke swarms and the persistence of major geological boundaries in Greenland. In R. G. Park and J. Tarney (Eds.), *The Early Precambrian of Scotland and Related Rocks of Greenland*, Univ. of Keele, 137–142.

Bridgwater, D., Watson, J. and Windley, B. F. (1973c). The Archaean craton of the North Atlantic region. *Phil. Trans. R. Soc. Lond.*, **A273**, 493–512.

Bridgwater, D., McGregor, V. R. and Myers, J. S. (1974a). A horizontal tectonic regime in the Archaean of Greenland and its implications for early crustal thickening. *Precambrian Res.*, **1**, 179–198.

Bridgwater, D., Sutton, J. and Watterson, J. (1974b). Crustal downfolding associated with igneous activity. *Tectonophysics*, **21**, 57–77.

Bridgwater, D., Collerson, K. D., Hurst, R. W. and Jesseau, C. W. (1975). Field characters of the early Precambrian rocks from Saglek, coast of Labrador. *Geol. Surv. Can. Pap.*, **75–1**, pt A, 287–296.

Bridgwater, D. and Collerson, K. (1976). The major petrological and geochemical characters of the 3600 m.y. Uivak gneisses from Labrador. *Cont. Mineral. Petrol.*, **54**, 43–60.

Brinkman, R. (1969). *Geologic Evolution of Europe*. Ferdinand Enke Verlag, Stuttgart, 161 pp.

Bromley, A. V. (1975). Tin mineralisation of western Europe: is it related to crustal subduction? *Trans. Inst. Min. Metall. Sect. B*, **84**, 28–30.

Brook, M., Brewer, M. S. and Powell, D. (1976). Grenville age for rocks in the Moine of north-western Scotland, *Nature*, **260**, 515–517.

Brooks, C. and Hart, S. R. (1972). An extrusive basaltic komatiite from a Canadian metavolcanic belt. *Can. J. Earth Sci.*, **9**, 1250–1253.

Brooks, C. and Hart, S. R. (1974). On the significance of komatiite. *Geol.*, **2**, 107–110.

Brooks, J., Muir, M. D. and Shaw, G. (1973). Chemistry and morphology of Precambrian microorganisms. *Nature*, **244**, 215–217.

Brown, A. S. (1969). Mineralisation in British Columbia and the copper and molybdenum deposits. *Trans. Can. Inst. Mining Metall.*, **72**, 1–15.

Brown, D. A. (1968). Some problems of distribution of late Palaeozoic and Triassic terrestrial vertebrates. *Australian J. Sci.*, **30**, 439–445.

Brown, D. A., Campbell, K. S. W. and Crook, K. A. W. (1968). *The Geological Evolution of Australia and New Zealand*, Commonwealth and Int. Lib., Pergamon Press, 409 pp.

Brown, J. S. (1970). Mississippi Valley-type lead–zinc ores. *Mineral. Deposita*, **5**, 103–119.

Brummer, J. J. and Mann, E. L. (1961). Geology of the Seal Lake area, Labrador. *Bull. geol. Soc. Amer.*, **72**, 1361–1382.

Bryner, L. (1969). Ore deposits of the Philippines—an introduction to their geology. *Econ. Geol.*, **64**, 644–666.

Bullard, E., Everett, J. E. and Smith, A. G. (1965). The fit of the continents around the Atlantic. *Phil. Trans. R. Soc. Lond.*, **A258**, 41–51.

Bulman, O. M. B. (1971). Graptolite fauna distribution. In F. A. Middlemiss, P. F. Rawson and G. Newall (Eds.), *Faunal Provinces in Space and Time*, Seal House Press, Liverpool, 47–60.

Burchfiel, B. C. and Davis, G. A. (1972). Structural framework and evolution of the southern part of the Cordilleran orogen, western United States. *Amer. J. Sci.*, **272**, 97–118.

Burek, P. J. (1967). Korrelation revers magnetisierter Gesteinsfolgen im Oberen Buntsandstein SW Deutschlands. *Geol. Jahrb.*, **84**, 591–616.

Burger, A. J., Oosthuyzen, E. J. and Van Niekerk, C. B. (1967). New lead isotopic ages for minerals from granitic rocks, northern and central Transvaal. *Annls. geol. Surv. S. Afr.*, **6**, 85–89.

Burke, K. (1975). Atlantic evaporites formed by evaporation of water spilled from Pacific, Tethyan and southern oceans. *Geology*, **3**, 613–616.

Burke, K., Dessauvagie, T. F. J. and Whiteman, A. J. (1971). Opening of the Gulf of Guinea and geological history of the Benue Depression and Niger delta. *Nature Phys. Sci.*, **233**, 51–55.

Burke, K. and Whiteman, A. J. (1973). Uplift, rifting and the break-up of Africa. In D. H. Tarling and S. K. Runcorn (Eds.), *Implications of Continental Drift to the Earth Sciences*, Vol. 2, Academic Press, London, 735–756.

Burke, K. and Dewey, J. F. (1973a). Plume-generated triple junctions: key indicators in applying plate tectonics to old rocks. *J. Geol.*, **81**, 406–433.

Burke, K. and Dewey, J. F. (1973b). An outline of Precambrian plate development. In D. H. Tarling and S. K. Runcorn (Eds.), *Implications of Continental Drift to the Earth Sciences*, Vol. 2, Academic Press, London, 1035–1045.

Burke, K., Dewey, J. F. and Kidd, W. S. F. (1976). Dominance of horizontal movements, arc and microcontinental collisions during the later permobile Regime. In B. F. Windley (Ed.), *The Early History of the Earth*, Wiley Interscience, London, 113–129.

Burne, R. V. (1973). Palaeogeography of south-west England and Hercynian continental collision. *Nature Phys. Sci.*, **241**, 129–131.

Burrett, C. F. (1972). Plate tectonics and the Hercynian orogeny. *Nature*, **239**, 155–157.

Button, A. (1975). Geochemistry of the Malmani dolomite of the Transvaal Supergroup in the northeastern Transvaal. *Econ. Geol. Res. Unit*, Univ. Witwatersrand, Johannesburg, Inf. Circ. 97, 1–16.

Cahen, L. and Snelling, N. J. (1966). *The geochronology of equatorial Africa*, North-Holland Publ., Amsterdam, 195 pp.

Calvin, M. (1969). *Chemical Evolution*, Oxford, 278 pp.

Cameron, E. M. and Baumann, A. (1972). Carbonate sedimentation during the Archaean. *Chem. Geol.*, **10**, 17–30.

Campbell, I. H., McCall, G. J. H. and Tyrwhitt, D. S. (1970). The Jimberlana norite, western Australia—a smaller analogue of the Great Dyke of Rhodesia. *Geol. Mag.*, **107**, 1–12.

Cann, J. R. (1970a). Rb, Sr, Y, Zr and Nb in some ocean floor basaltic rocks. *Earth Plan. Sci. Lett.*, **10**, 7–11.

Cann, J. R. (1970b). New model for the structure of the ocean crust. *Nature*, **226**, 928–930.

Cann, J. R. (1974). A model for oceanic crustal structure developed. *Geophys. J. R. astr. Soc.*, **39**, 169–187.

Card, K. D., Church, W. R., Franklin, J. M., Frarey, M. J., Robertson, J. A., West, G. F. and Young, G. M. (1972). The southern Province. In *Variations in Tectonic Styles in Canada*, *Geol. Ass. Can. Sp. Pap. 2*, 335–380.

Casella, C. J. (1969). A review of the Precambrian geology of the eastern Beartooth Mountains, Montana and Wyoming. In L. H. Larsen (Ed.), *Igneous and Metamorphic Geology, Geol. Soc. Amer. Mem.*, **115**, 53–72.

Casey, R. (1971). Facies, faunas and tectonics in late Jurassic–early Cretaceous in Britain. In F. A. Middlemiss, P. F. Rawson and G. Newall (Eds.), *Faunal Provinces in Space and Time*, Seal House Press, Liverpool, 153–168.

Cawthorn, R. G. and Strong, D. F. (1974). The petrogenesis of komatiites and related rocks as evidence for a layered upper mantle. *Earth Plan. Sci. Lett.*, **23**, 369–375.

Chadwick, B. (1969). Patterns of fracture and dyke intrusion near Frederikshaab, South-West Greenland. *Tectonophys.* **8**, 247–264.

Chadwick, B., Coe. K., Gibbs, A. D., Sharpe, M. R. and Wells, P. R. A. (1974). Field evidence relating to the origin of 3000 M.yr. gneisses in southern West Greenland. *Nature*, **249**, 136.

Chadwick, B. and Coe, K. (1976). New evidence relating to Archaean events in southern West Greenland. In B. F. Windley (Ed.), *The Early History of the Earth*, Wiley Interscience, London, 203–212.

Challis, G. A. (1971). Alteration processes and ore minerals in New Zealand ultramafic rocks (abstract). *Proc. 12th Pacific Sci. Congr. Canberra, Australia*, **1**, 427.

Chaloner, W. G. and Creber, G. T. (1973). Growth maprings in fossil woods as evidence of past climates. In D. H. Tarling and S. K. Runcorn (Eds.), *Implications of Continental Drift to the Earth Sciences*. Vol. 1, Academic Press, London, 425–437.

Channell, J. E. T. and Tarling, D. H. (1975). Palaeomagnetism and rotation of Italy. *Earth Plan. Sci. Lett.*, **25**, 177–188.

Chappell, B. W. and White, A. J. R. (1970). Further data on an 'eclogite' from the Sittampundi complex, India. *Mineral. Mag.*, **37**, 555–560.

Charig, A. J. (1971). Faunal provinces on land: evidence based on the distribution of fossil tetrapods, with especial reference to the reptiles of the Permian and Mesozoic. In F. A. Middlemiss, P. F. Rawson and G. Newall (Eds.), *Faunal Provinces in Space and Time*, Seal House Press, Liverpool, 111–130.

Chase, C. G. and Perry E. C., Jr (1972). The oceans: growth and oxygen isotope evolution. *Science*, **177**, 992–994.

Chase, C. G. and Gilmer, T. H. (1973). Precambrian plate tectonics: the mid-continent gravity high. *Earth Plan. Sci. Lett.*, **21**, 70–78.

Chatterjee, N. D. (1971). Phase equilibria in the Alpine metamorphic rocks of the environs of the Dora–Maira Massif, western Italian Alps, Pts 1 and 2. *Neues Jahrb.*, **114**, 181–245.

Chatterjee, S. C. (1936). *The anorthosites of Bengal*, Univ. of Calcutta, 46 pp.

Cheney, E. S. and Stewart, R. J. (1975). Subducted greywacke in the Olympic Mountains, USA: implications for the origin of Archaean sodic gneisses. *Nature*, **258**, 60–61.

Chernov, V. M. (1973). The ferruginous–siliceous formations of the eastern part of the Baltic Shield. In *Genesis of Precambrian iron and manganese deposits*, Proc. Kiev. Symp. 1970, Unesco, Paris, 89–94.

Chinner, G. A. and Sweatman, T. R. (1968). A former association of enstatite and kyanite. *Mineral. Mag.*, **36**, 1052–1060.

Christensen, N. I. (1970). Composition and evolution of the oceanic crust, *Marine Geol.*, **8**, 139–154.

Chiristensen, N. I. and Salisbury, M. H. (1972). Sea floor spreading, progressive alteration of layer 2 basalts and associated changes in seismic velocities. *Earth Plan. Sci. Lett.*, **15**, 367–375.

Christiansen, R. L. and Lipman, P. W. (1972). Cenozoic volcanism and plate-tectonic evolution of the western United States. II. Late Cenozoic. *Phil. Trans. R. Soc. Lond.*, **A271**, 249–284.

Church, W. R. and Young, G. M. (1970). Discussion of the progress report of the Federal–Provincial Committee on Huronian stratigraphy. *Can. J. Earth Sci.*, **7**, 912–918.

Church, W. R. and Stevens, R. K. (1971). Early Palaeozoic ophiolite complexes of the Newfoundland Appalachians as mantle–oceanic crust sequences, *J. Geophys. Res.*, **76**, 1460–1466.

Church, W. R. and Gayer, R. A. (1973). The Ballantrae ophiolite. *Geol. Mag.*, **110**, 497–510.

Churkin, M., Jr (1974). Palaeozoic marginal ocean basin–volcanic arc systems in the Cordilleran fold belt. In R. H. Dott Jr and R. H. Shaver (Eds.), *Ancient and Modern Geosynclinal Sedimentation, Soc. Econ. Pal. Mineral. Sp. Publ.*, **19**, 174–192.

Cita, M. B. (1971). Palaeoenvironmental aspects of DSPD Legs I–IV. *Proc. 2nd Planktonic Conf.*, **1**, 251.

Clague, D. A. and Jarrard, R. D. (1973). Tertiary Pacific plate motion deduced from the Hawaiian–Emperor chain. *Bull. geol. Soc. Amer.*, **84**, 1135–1154.

Clar, E.(1973). Review of the structure of the Eastern Alps. In K. A. de Jong and R. Scholten (Eds.), *Gravity and Tectonics*, Wiley, New York, 253–270.

Clark, A. H. and 8 coworkers (1976). Longitudinal variations in the metallogenic evolution of the Central Andes. In D. F. Strong (Ed.), *Metallogeny and Plate Tectonics, Geol. Ass. Can. Sp. Pap.*, **14**, 23–58.

Clark, S. P. and Jäger, E. (1969). Denudation rate in the Alps from geochronologic and heat flow data. *Amer. J. Sci.*, **267**, 1143–1160.

Clarke, D. B. (1970). Tertiary basalts of Baffin Bay: possible primary magmas from the mantle. *Contr. Mineral. Petrol.*, **25**, 203–224.

Clifford, T. N. (1966). Tectono-metallogenic units and metallogenic provinces of Africa. *Earth Plan. Sci. Lett.*, **1**, 421–434.

Clifford, T. N. (1970). The structural framework of Africa. In T. N. Clifford and I. G. Gass (Eds.), *African magmatism and tectonics*, Oliver and Boyd, London, 1–26.

Clifford, T. N. (1971). Location of mineral deposits, In I. G. Gass, P. J. Smith and R. C. L. Wilson (Eds.), *Understanding the Earth*, Artemis Press, Sussex, 315–325.

Clifford, T. N. (1972). The evolution of the crust of Africa. Notes and *M. Serv. géol. Maroc No. 236*, 29–39 (Coll. intern. CNRS, Paris, No. 192).

Clifford, T. N. (1974). Review of African granulites and related rocks, *Geol. Soc. Amer. Sp. Pap.*, **156**, 49 pp.

Cloos, H. (1939). Hebung–Spaltung-Vulcanismus, *Geol. Rundsch.*, **30**, 405–527.

Cloud, P. E., Jr (1965). Significance of the Gunflint (Precambrian) microflora, *Science*, **148**, 27–35.

Cloud, P. E., Jr (1968a). Pre-Metazoan evolution and the origins of the Metazoa, In E. T. Drake (Ed.), *Evolution and Environment*, Yale Univ. Press, New York, 1–72.

Cloud, P. E., Jr (1968b). Atmospheric and hydrospheric evolution on the primitive earth, *Science*, **160**, 729–736.

Cloud, P. E., Jr (1973). Palaeological significance of the Banded Iron Formation, *Econ. Geol.*, **68**, 1135–1143.

Cloud, P. E., Jr, and Semikhatov, M. A. (1969). Proterozoic stromatolite zonation, *Amer. J. Sci.*, **267**, 1017–1061.

Cobbing, E. J. and Pitcher, W. S. (1972). The coastal batholith of central Peru, *J. geol. Soc. Lond.*, **128**, 421–460.

Cocks, L. R. M. and McKerrow, W. S. (1973). Brachiopod distributions and faunal provinces in the Silurian and Lower Devonian, In N. F. Hughes (Ed.), *Organisms and Continents through Time, Pal. Ass. Sp. Pap. in Pal.*, **12**, 291–304.

Cocks, L. R. M. and Toghill, P. (1973). The biostratigraphy of the Silurian rocks of the Girvan district, Scotland, *J. Geol. Soc. Lond.*, **129**, 209–243.

Cohen, L. H., Condie, K. C., Kuest, L. J., Mackenzie, G. S., Meister, F. H., Pushkar, P. and Steuber, A. M. (1963). Geology of the San Benito islands, Baja California, Mexico, *Bull. geol. Soc. Amer.*, **74**, 1355–1370.

Colbert, E. H. (1973). Continental drift and the distribution of fossil reptiles, In D. H. Tarling and S. K. Runcorn (Eds.), *Implications of Continental Drift to the Earth Sciences*, Vol. 1, Academic Press, London, 395–412.

Coleman, P. J. (1973). *The Western Pacific*; island arcs, marginal seas, geochemistry, P. J. Coleman (Ed.), Univ. W. Australia Press, 675 pp.

Coleman, P. J. (1975). On island arcs. *Earth Sci. Rev.*, **11**, 47–80.

Coleman, R. G. (1971). Plate tectonic emplacement of upper mantle peridotites along continental edges. *J. Geophys. Res.*, **76**, 1212–1222.

Coleman, R. G. (1972). Blueschist metamorphism and plate tectonics. *24th Int. geol. Congr. Montreal*, Sect. 2, 19–26.

Coleman, R. G. and Lanphere, M. A. (1971). Distribution and age of high-grade blueschists, associated eclogites and amphibolites from Oregon and California. *Bull. geol. Soc. Amer.*, **82**, 2397–2412.

Collerson, K. D., Jesseau, C. W. and Bridgwater, D. (1976). Crustal development of the Archaean gneiss complex, eastern Labrador. In B. F. Windley (Ed.), *The Early History of the Earth*, Wiley, London, 237–253.

Comninakis, P. E. and Papazachos, B. C. (1972). Seismicity of the eastern Mediterranean and some tectonic features of the mediterranean ridge. *Bull. geol. Soc. Amer.*, **83**, 1093–1102.

Compston, W. and Arriens, P. A. (1968). The Precambrian geochemistry of Australia. *Can. J. Earth Sci.*, **5**, 561–583.

Compston, W., McDougall, I. and Heier, K. S. (1968). Geochemical comparison of the Mesozoic basaltic rocks of Antarctica, South Africa, South America and Tasmania. *Geochim. Cosmochim. Acta*, **32**, 129–149.

Compston, W. and McElhinny, M. W. (1975). The Rb/Sr age of the Mashonaland dolerites of Rhodesia and its significance for palaeomagnetic correlation in southern Africa. *Precambrian Res.*, **2**, 305–315.

Condie, K. C. (1973). Archaean magmatism and crustal thickening. *Bull. geol. Soc. Amer.*, **84**, 2981–2992.

Condie, K. C. (1975). A mantle plume model for the origin of Archaean greenstone belts based on trace element distributions. *Nature*, **258**, 413–414.

Condie, K. C. (197a). The Wyoming Archean Province in the western United States. In B. F. Windley (Ed.), *The Early History of the Earth*, Wiley, London, 419–424.

Condie, K. C. (1976b). *Plate Tectonics and Crustal Evolution*, Pergamon Press, New York, 288 pp.

Condie, K. C., Barsky, C. K. and Mueller, P. A. (1969). Geochemistry of Precambrian diabase dykes from Wyoming. *Geochim. Cosmochim. Acta*, **33**, 1371–1388.

Condie, K. C. and Potts, M. J. (1969). Calc-alkaline volcanism and the thickness of the early Precambrian crust in North America. *Can. J. Earth Sci.*, **6**, 1179–1184.

Condie, K. C. and Harrison, N. M. (1976). Geochemistry of the Archaean Bulawayan Group, Midlands Greenstone belt, Rhodesia. *Precambrian Res.*, **3**, 253–271.

Condie, K. C. and Hunter, D. R. (1976). Trace element geochemistry of Archaean granitic rocks from the Barberton region, South Africa. *Earth Plan. Sci. Lett.*, **29**, 389–400.

Coney, P. J. (1971). Cordilleran tectonic transitions and motion of the North American plate. *Nature*, **233**, 462–465.

Coney, P. J. (1972). Cordilleran tectonics and North America plate motion. *Amer. J. Sci.*, **272**, 603–628.

Constantinou, G. and Govett, G. J. S. (1972). Genesis of sulphide deposits, ochre and umber of Cyprus. *Trans. Inst. Min. Metall.*, **81**, B34–46.

Cooke, D. L. and Moorhouse, W. W. (1969). Timiskaming volcanism in the Kirkland Lake area, Ontario, Canada. *Can. J. Earth Sci.*, **6**, 117–132.

Cooper, W. S. 1967. Coastal dunes of California. *Geol. Soc. Amer. Mem.*, **104**, 1–125.

Cordani, U. G., Melcher, G. C. and de Almeida, F. F. M. (1968). Outline of the Precambrian geochronology of South America. *Can. J. Earth Sci.*, **5**, 629–632.

Corliss, J. B. (1971). The origin of metal-bearing submarine hydrothermal solutions. *J. Geophys. Res.*, **76**, 8128–8138.

Cotton, R. E. (1965). H.Y.C. lead–zinc–silver deposit, MacArthur River. In J. McAndrew (Ed.), *Geology of Australian Ore Deposits*, 8th Commonwealth Mining and Metall. Congr. Australia and New Zealand, 197–200.

Coward, M. P. (1973). Heterogeneous deformation in the development of the Laxfordian complex of South Uist, Outer Hebrides. *J. geol. Soc. Lond.*, **129**, 139–160.

Coward, M. P., Francis, P. W., Graham, R. H., Myers, J. S. and Watson, J. (1969). Remnants of an early metasedimentary assemblage in the Lewisian of the Outer Hebrides. *Proc. Geol. Assoc.*, **80**, 387–408.

Coward, M. P., Graham, R. H., James, P. R. and Wakefield, J. (1973). A structural interpretation of the northern margin of the Limpopo orogenic belt, southern Africa. *Phil. Trans. R. Soc. Loc.*, **A273**, 487–491.

Coward, M. P. and James, P. R. (1974). The deformation patterns of two Archaean greenstone belts in Rhodesia and Botswana. *Precambrian Res.*, **1**, 235–258.

Coward, M. P., Lintern, B. C. and Wright, L. I. (1976). The pre-cleavage deformation of the sediments and gneisses of the northern part of the Limpopo belt. In B. F. Windley (Ed.), *The Early History of the Earth*, Wiley, London, 323–330.

Cowie, J. W. (1971). Lower Cambrian faunal provinces. In F. A. Middlemiss, P. F. Rawson and G. Newall (Eds.), *Faunal Provinces in Space and Time*, Seal House Press, Liverpool, 31–46.

Cowie, J. W. (1974). The Cambrian of Spitzbergen and Scotland. In C. H. Holland (Ed.), *Cambrian of the British Isles, Norden and Spitzbergen. Lower Palaeozoic rocks of the World*, Vol. 2, Wiley, London, 123–156.

Cox, C. B. (1973). Triassic tetrapods. In A. Hallam (Ed.), *Atlas of Palaeobiogeography*, Elsevier, Amsterdam, 213–223.

Cox, K. G. (1970). Tectonics and volcanism of the Karroo period and their bearing on the postulated fragmentation of Gondwanaland. In T. N. Clifford and I. G. Gass (Eds.), *African Magmatism and Tectonics*, Oliver and Boyd, Edinburgh, 211–236.

Cox, K. G. (1972). The Karroo volcanic cycle. *J. Geol. Soc. Lond.*, **128**, 311–336.

Cracroft, J. (1973). Vertebrate evolution and biogeography in the Old World tropics: implications of continental drift and palaeoclimatology. In D. H. Tarling and S. K. Runcorn (Eds.), *Implications of Continental Drift to the Earth Sciences*, Vol. 1, Academic Press, London, 373–393.

Crain, I. K. (1971). Possible direct causal relation between geomagnetic reversals and biological extinctions, *Bull. geol. Soc. Amer.*, **82**, 2603–2606.

Crawford, A. R. (1969). Reconnaissance Rb–Sr dating of the Precambrian rocks of southern Peninsula, India, *J. geol. Soc. India*, **10**, 117–166.

Crawford, A. R. and Daily, B. (1971). Probable non-synchroneity of late Precambrian glaciations. *Nature*, **230**, 111–112.

Creer, K. M. (1970a). A review of palaeomagnetism. *Earth Sci. Rev.*, **6**, 369–466.

Creer, K. M. (1970b). Palaeomagnetic survey of South American rocks, Pts. I–V. *Phil. Trans. R. Soc. Lond.*, **A267**, 457–558.

Creer, K. M. (1973). A discussion of the arrangement of palaeomagnetic poles on the map of Pangaea for epochs in the Phanerozoic. In D. H. Tarling and S. K. Runcorn (Eds.), *Implications of Continental Drift to the Earth Sciences*, Vol. 1, Academic Press, London, 47–76.

Creer, K. M., Irving, E. and Runcorn, S. K. (1954). The direction of the geomagnetic field in remote epochs in Great Britain. *J. Geomag. Geoelect.*, **6**, 163–168.

Cronan, D. S. (1976). Basal metalliferous sediments from the eastern Pacific. *Bull. geol. Soc. Amer.*, **87**, 928–934.

Cronan, D. S., and coworkers (1972). Iron-rich basalt sediments from the eastern Equatorial Pacific; Leg 16, Deep sea drilling project. *Science*, **175**, 61–63.

Crowell, J. C. and Frakes, L. A. (1970). Phanerozoic glaciation and the causes of Ice Ages. *Amer. J. Sci.*, **268**, 193–224.

Cullen, D. J. (1956). Pretoria Series formations near Kanye in the Bechuanaland Protectorate. *Geol. Mag.*, **95**, 456–464.

Currie, K. L. (1974). A note on the calibration of the garnet–cordierite geothermometer and geobarometer. *Contr. Mineral. Petrol.*, **44**, 35–44.

Dadet, P., Marchesseau, J., Millon, R. and Motti, E. (1970). Mineral occurrences related to stratigraphy and tectonics in

Tertiary sediments near Umm Lajj, eastern Red Sea area. *Phil. Trans. R. Soc. Lond.*, **A267**, 99–106.

Dagger, G. W. (1972). Genesis of the Mount Pleasant tungsten–molybdenum–bismuth deposit, New Brunswick, Canada, *Trans. Inst. Min. Metall. Sect. B*, **81**, 73–102.

Dallmeyer, R. D. (1975). The Palisades sill: a Jurassic intrusion? Evidence from $^{40}Ar/^{39}Ar$ incremental release ages, *Geology*, **3**, 243–245.

Dalrymple, G. B., Grommé, C. S. and White, R. W. (1975). Potassium–argon age and palaeomagnetism of diabase dikes in Liberia: initiation of central Atlantic rifting. *Bull. geol. Soc. Amer.*, **86**, 399–411.

Dalziel, I. W. D., Brown, J. M. and Warren, T. E. (1969). The structural and metamorphic history of the rocks adjacent to the Grenville Front near Sudbury, Ontario, and Mount Wright, Ontario. *Geol. Ass. Can. Sp. Pap.*, **5**, 207–228.

D'Argenio, B. (1970). Central and southern Italy Cretaceous bauxites. *Inst. Geol. Publ. Hungarici Ann.*, **54**, 3, 221–333.

Davidson, A. (1972). The Churchill Province. In R. A. Price and R. J. W. Douglas (Eds.), *Variations in Tectonic Styles in Canada*, *Geol. Ass. Can. Sp. Pap.*, **11**, 381–434.

Davies, F. B. (1974). A layered basic complex in the Lewisian, south of Loch Laxford, Sutherland. *J. geol. Soc. Lond.*, **130**, 279–284.

Davies, F. B. (1975a). Origin and ancient history of gneisses older than 2800 m.y. in the Lewisian complex. *Nature*, **258**, 589–591.

Davies, F. B. (1975b). Evolution of Proterozoic basement patterns in the Lewisian complex. *Nature*, **256**, 568–570.

Davies, F. B. (1976). Early Scourian structures in the Scourie–Laxford region and their bearing on the evolution of the Laxford Front. *J. geol. Soc. Lond.*, **132**, 543–554.

Davies, F B. and Windley, B. F. (1976). The significance of major Proterozoic high-grade linear belts in continental evolution. *Nature*, **263**, 383–385.

Davies, R. D., Allsopp, H. L., Erlank, A. J. and Manton, W. I. (1969). Sr-isotope studies of various layered mafic intrusions in southern Africa, *Geol. Soc. S. Afr. Sp. Publ.*, **1**, 576–593.

Davis, H. L. (1968). Papuan ultramafic belt. *23rd Int. geol. Congr. Prague*, **1**, 209–220.

Dawson, J. B. (1967). A review of the geology of kimberlite. In P. J. Wyllie (Ed.), *Ultramafic and Related Rocks*, Wiley, New York, 241–250.

Dawson, J. B. (1970). The structural setting of African kimberlite magmatism. In T. N. Clifford and I. G. Gass (Eds.), *African Magmatism and Tectonics*, Oliver and Boyd, Edinburgh, 321–336.

De, A. (1969). Anorthosites of the Eastern Ghats, India. In *Origin of Anorthosite and Related Rocks, New York State Mus. Sci. Serv. Mem.*, **18**, 425–434.

Deans, T. (1950). The Kupferschiefer and the associated lead–zinc mineralisation in the Permian of Silesia, Germany and England. *18th Int. geol. Congr. Pt 8*, 340–351.

Dearman, W. R. (1971). A general view of the structure of Cornubia. *Proc. Ussher Soc.*, **2**, 220.

Dearnley, R. (1963). The Lewisian complex of South Harris. *J. geol. Soc. Lond.*, **119**, 243–312.

Dearnley, R. (1966). Orogenic fold belts and a hypothesis of earth evolution. In L. H. Ahrens and coworkers (Eds.), *Physics and Chemistry of the Earth*, Vol. 7, Pergamon, 1–114.

Debelmas, J. and Lemoine, M. (1970). The Western Alps: palaeogeography and structure. *Earth Sci. Rev.*, **6**, 221–256.

Degens, E. T. and Ross, D. A. (Eds.) (1969). *Hot brines and recent heavy metal deposits in the Red Sea*, Springer-Verlag, New York, 600 pp.

De Jong, K. A. (1973). Mountain building in the Mediterranean region. In K. A. de Jong and R. Scholten (Eds.), *Gravity and Tectonics*, Wiley, New York, 125–139.

De Jong, K. A. and Scholten, R. (Eds.) (1973). *Gravity and Tectonics*, Wiley, New York, 502 pp.

De Roever, W. P. (1965). On the cause of the preferential distribution of certain metamorphic minerals in orogenic belts of different age. *Geol. Rund.*, **54**, 933–943.

Derry, D. R. (1961). Economic aspects of Archaean–Proterozoic boundaries. *Econ. Geol.*, **56**, 635–647.

Devaraju, T. C. and Sadashivaiah, S. (1969). Charnockites of the Satnur–Halaguru area, Mysore State. *Ind. Mineral.*, **10**, 67–88.

348

de Villiers, J. (1960). *The manganese deposits of the Union of South Africa*, Geol. Surv. S. Afr. Handbook 2, 280 pp.

de Waard, D. (1969). The anorthosite problem: the problem of the anorthosite–charnockite suite of rocks. In *Origin of Anorthosite and Related Rocks*, Mem. New York State Mus. Sci. Serv., **18**, 307–316.

de Waard, D. and Walton, M. S. (1967). Precambrian geology of the Adirondack Highlands, a reinterpretation. *Geol. Rundsch.*, **56**, 596–629.

de Waard, D. and Romey, W. D. (1969). Petrogenetic relationships in the anorthosite–charnockite series of Snowy Mountain Dome, south central Adirondacks. In *Origin of Anorthosite and Related Rocks*, Mem. New York State Mus. Sci. Serv., **18**, 307–316.

de Waard, D., Duchesne, J. C. and Michot, J. (1974). Anorthosites and their environments, *Géol. Domaines Cristallins, Liège*, 323–346.

Dewey, J. F. (1969). Evolution of the Appalachian/Caledonian orogen, *Nature*, **222**, 124–129.

Dewey, J. F. (1971). A model for the Lower Palaeozoic evolution of the southern margin of the early Caledonides of Scotland and Ireland, *Scott. J. Geol.*, **7**, 219–240.

Dewey, J. F. (1972). Plate tectonics, *Sci. Amer.*, **226**, 56–68.

Dewey, J. F. (1974). The geology of the southern termination of the Caledonides. In A. E. M. Nairn and F. G. Stehli (Eds.), *The Ocean Basins and Margins, The North Atlantic*, Vol. 2, Plenum Press, 205–231.

Dewey, J. F. and Kay, M. (1968). Appalachian and Caledonian evidence for drift in the North Atlantic. In R. A. Phinney (Ed.), *The History of the Earth's Crust*, Princeton Univ. Press, New Jersey, 161–167.

Dewey, J. F. and Bird, J. M. (1970a). Plate tectonics and geosynclines. *Tectonophysics*, **10**, 625–638.

Dewey, J. F. and Bird, J. M. (1970b). Mountain belts and the new global tectonics, *J. Geophys. Res.*, **75**, 2625–2647.

Dewey, J. F. and Horsfield, B. (1970). Plate tectonics, orogeny and continental growth, *Nature*, **225**, 521–525.

Dewey, J. F. and Pankhurst, R. J. (1970). Evolution of the Scottish Highlands and their radiometric pattern, *Trans. R. Soc. Edinb.*, **68**, 361–389.

Dewey, J. F. and Bird, J. M. (1971). Origin and emplacement of the ophiolite suite: Appalachian ophiolites in Newfoundland. *J. Geophys. Res.*, **76**, 3179–3206.

Dewey, J. F. and Burke, K. C. A. (1973). Tibetan, Variscan and Precambrian basement reactivation: products of continental collision, *J. Geol.*, **81**, 683–692.

Dewey, J. F., Pitman, W. C., III, Ryan, W. B. F. and Bonnin, J. (1973). Plate tectonics and the evolution of the Alpine System, *Bull. geol. Soc. Amer.*, **84**, 3137–3180.

Dewey, J. F. and Kidd, W. S. F. (1974). Continental collisions in the Appalachian–Caledonian orogenic belt: variations related to complete and incomplete suturing, *Geology*, **2**, 543–546.

Dickinson, B. B. and Watson, J. (1976). Variations in crustal level and geothermal gradient during the evolution of the Lewisian complex of northwest Scotland. *Precambrain Res.*, **3**, 363–734.

Dickinson, W. R. (1962). Petrogenetic significance of geosynclinal andesitic volcanism along the Pacific margin of North America. *Bull. geol. Soc. Amer.*, **73**, 1241–1256.

Dickinson, W. R. (1967). Tectonic development of Fiji. *Tectonophysics*, **4**, 543–553.

Dickinson, W. R. (1968). Circum-Pacific andesite types. *J. Geophys. Res.*, **73**, 2261–2269.

Dickinson, W. R. (1970). Relations of andesites, granites and derivative sandstones to arc–trench tectonics. *Rev. Geophys. Space Phys.*, **8**, 813–860.

Dickinson, W. R. (1971a). Plate tectonics in geologic history. *Science*, **174**, 107–113.

Dickinson, W. R. (1971b). Plate tectonic models of geosynclines. *Earth Plan. Sci. Lett.*, **10**, 164–174.

Dickinson, W. R. (1972). Evidence for plate tectonic regimes in the rock record. *Amer. J. Sci.*, **272**, 551–576.

Dickinson, W. R. (1973a). Widths of modern arc–trench gaps proportional to past duration of igneous activity in associated magmatic arcs. *J. Geophys. Res.*, **78**, 3376–3389.

Dickinson, W. R. (1973b). Reconstruction of past arc–trench systems from petrotectonic assemblages in the island arcs of the western Pacific. In P. J. Coleman (Ed.), *The Western Pacific; island arcs, marginal seas, geochemistry*, Univ. W. Australia Press, Nedlands, 569–601.

Dickinson, W. R. (1974a). Plate tectonics and sedimentation. In W. R. Dickinson (Ed.), *Tectonics and Sedimentation*, Soc. Econ. Pal. Mineral. Sp. Publ., **22**, 1–27.

Dickinson, W. R. (1974b). Sedimentation within and beside ancient and modern magmatic arcs. In R. H. Dott Jr and R. H. Shaver (Eds.), *Modern and Ancient Geosynclinal Sedimentation*, Soc. Econ. Pal. Mineral. Sp. Publ., **19**, 230–239.

Dickinson, W. R. and Luth., W. C. (1971). A model for plate tectonic evolution of mantle layers. *Science*, **174**, 400–404.

Dickinson, W. R. and Rich, E. I. (1972). Petrologic intervals and petrofacies in the Great Valley sequence, Sacramento Valley, California. *Bull. geol. Soc. Amer.*, **83**, 3007–3024.

Dietz, R. S. (1961). Continent and ocean basin evolution by spreading of the sea floor. *Nature*, **190**, 854–857.

Dietz, R. S. (1964). Sudbury structure as an astrobleme. *J. Geol.*, **72**, 412–434.

Dietz, R. S. and Holden, J. C. (1966). Miogeoclines in space and time, *J. Geol.*, **74**, 566–583.

Dietz, R. S. and Holden, J. C. (1970). Reconstruction of Pangaea: break-up and dispersion of continents, Permian to Recent, *J. Geophys. Res.*, **75**, 4939–4956.

Dietz, R. S., Holden, J. C. and Sproll, W. P. (1970). Geotectonic evolution and subsidence of the Bahama Platform. *Bull. geol. Soc. Amer.*, **81**, 1915–1928.

Dilley, F. C. (1971). Cretaceous foraminiferal biogeography. In F. A. Middlemiss, P. F. Rawson and G. Newall (Eds.), *Faunal Provinces in Space and Time*, Seal House Press, Liverpool, 169–190.

Dilley, F. C. (1973). Larger foraminifera and seas through time. In N. F. Hughes (Ed.), *Organisms and Continents through Time*, Pal Ass. Sp. Pap. in Pal., **12**, 155–168.

Dimroth, E. (1968). Sedimentary textures, diagenesis, and sedimentary environment of certain Precambrian ironstones. *N. Jb. Geol. Palaont. Abh.*, **130**, 247–274.

Dimroth, E. (1975). Paleo-environment of iron-rich sedimentary rocks. *Geol. Rundsch.*, **64**, 751–767.

Dimroth, E., Barager, W. R. A., Bergeron, R. and Jackson, G. D. (1970). The filling of the Circum-Ungava Geosyncline. In *Symp. on Basins and Geosyclines of the Canadian Shield*, Geol. Surv. Can. Pap., **70–40**, 45–142.

Dimroth, E., Boivin, P., Goulet, N. and Larouche, M. (1973). Tectonic and volcanological studies in the Rouyn–Noranda area. *Quebec Dept Nat. Res.*, Open File MS, 1–60.

Dingle, R. V. (1973). Mesozoic palaeogeography of the southern Cape, South Africa. *Palaeogeog. Palaeochim. Palaeoecol.*, **13**, 203–213.

Dingle, R. V. and Scrutton, R. A. (1974). Continental breakup and the development of post-Palaeozoic sedimentary basins around southern Africa. *Bull. geol. Soc. Amer.*, **85**, 1467–1474.

Dixon, C. J. and Pereira, J. (1974). Plate tectonics and mineralisation in the Tethyan region. *Mineral. Deposita*, **9**, 185–198.

Dmitriev, L., Barsukov, V. and Udintsen, G. (1971). Rift zones of the ocean and the problem of ore formation. In Proc. IMA–IAGOD meetings 1970, IAGOD vol., Tokyo Soc. Min. Geol. Japan, 65–69.

Doig, R. (1970). An alkaline rock province linking Europe and North America. *Can. J. Earth Sci.*, **7**, 22–28.

Donaldson, J. A., McGlynn, J. C., Irving, E. and Park, J. K. (1973). Drift of the Canadian Shield. In D. H. Tarling and S. K. Runcorn (Eds.), *Implications of Continental Drift to the Earth Sciences*, Academic Press, London, 3–18.

Donnelly, T. W. (1964). Evolution of eastern Greater Antillean island arc. *Bull. Amer. Ass. Petrol. Geol.*, **48**, 680–696.

Dorr, J. V. N. (1965). High-grade hematite ores of Brazil. *Econ. Geol.*, **60**, 2–46.

Dott, R. H. (1969). Circum-Pacific late Cenozoic structural rejuvenation: implications for sea floor spreading. *Science*, **166**, 874–876.

Doumani, G. A. and Long, W. E. (1962). The ancient life of the Antarctic, *Sci. Amer.*, **207**, 169–184.

Dewry, G. E., Ramsay, A. T. S. and Smith, A. G. (1974). Climatically controlled sediments, the geomagnetic field and trade wind belts in Phanerozoic time, *J. Geol.*, **82**, 531–533.

Dunham, K. C. (1953). Age relations of the epigenetic mineral deposits of Britain. *Trans. geol. Soc. Glasgow*, **21**, 395–429.

Dunham, K. C. (1966). Role of juvenile solutions, connate waters and evaporitic brines in the genesis of lead–zinc–fluorine–barium deposits. *Trans. Inst. Min. Metall. Sect. B*, **75**, 226–229.

Dunn, P. R., Thomson, B. P. and Rankama, K. (1971). Late Precambrian glaciation in Australia as a stratigraphic boundary. *Nature*, **231**, 498–502.

Dunnet, D. (1976). Some aspects of the Panantarctic cratonic margin in Australia. *Phil. Trans. R. Soc. Lond.*, **A280**, 641–654.

Du Toit, A. L. (1937). *Our Wandering Continents*, Oliver and Boyd, London.

Eade, K. E., Fahrig, W. F. and Maxwell, J. A. (1966). Composition of crystalline shield rocks and fractionating effects of regional metamorphism. *Nature*, **211**, 1245–1249.

Eade, K. E. and Fahrig, W. F. (1971). Geochemical evolutionary trends of continental plate—a preliminary study of the Canadian Shield. *Bull. Can. geol. Surv.*, **179**, 51 pp.

Eberlin, G. D. and Churkin, M., Jr (1970). Palaeozoic stratigraphy in the northwest coastal area of Prince of Wales Island, southeastern Alaska. *Bull. US geol. Surv.*, **1284**, 67 pp.

Eisbacher, G. H. (1974). Evolution of successor basins in the Canadian Cordillera. In R. H. Dott Jr and R. H. Shaver (Eds.), *Modern and Ancient Geosynclinal Sedimentation*, Soc. Econ. Pal. Mineral. Sp. Publ., **19**, 274–291.

Elders, W. A. (1968). Mantled feldspars from the granites of Wisconsin. *J. Geol.*, **76**, 37–49.

Emeleus, C. H. (1964). The Grønnedal–Ika alkaline complex, South Greenland. *Meddr. Grønland*, **172**, 3, 75 pp.

Emeleus, C. H. and Upton, B. G. J. (1976). The Gardar province in South Greenland. In A. Escher and W. S. Watt (Eds.), *Geology of Greenland*, Geol. Surv. Greenland, Copenhagen, 152–181.

Emiliani, C. (1966). Isotopic paleotemperatures, *Science*, **154**, 851–857.

Emslie, R. F. (1970). The geology of the Michikamau intrusion, Labrador. *Geol. Surv. Can. Pap.*, **68–57**, 88 pp.

Emslie, R. F. (1973). Some chemical characteristics of anorthosite suites and their significance. *Can. J. Earth Sci.*, **10**, 54–71.

Engel, A. E. J. (1966). The Barberton Mountain Land: clues to the differentiation of the earth. *Econ. Geol. Res. Unit, Univ. Witwatersrand, Johannesburg, Inf. Circ.*, **27**, 17 pp.

Engel, A. E. J., Engel, C. G. and Havens, R. G. (1965). Chemical characteristics of oceanic basalts and the upper mantle. *Bull. geol. Soc. Amer.*, **76**, 719–734.

Engel, A. E. J., Nagy, B., Nagy, L. A., Engel, C. G., Kremp, G. O. W. and Drew, C. M. (1968). Algal-like forms in Onverwacht Series, South Africa: oldest recognised lifelike forms on earth. *Science*, **161**, 1005–1008.

Engel, A. E. J. and Engel, C. G. (1970). Mafic and ultramafic rocks, in A. E. Maxwell (Ed.), *The Sea*, Wiley Interscience, 465–520.

Engel, A. E. J. and Kelm, D. L. (1972). Pre-Permian global tectonics: a tectonic test. *Bull. geol. Soc. Amer.*, **83**, 2325–2340.

Engel, A. E. J., Itson, S. P., Engel, C. G., Stickney, D. M. and Cray, E. J., Jr (1974). Crustal evolution and global tectonics: a petrogenic view, *Bull. geol. Soc. Amer.*, **85**, 843–858.

Engel, C. G. and Fisher, R. L. (1969). Lherzolite, anorthosite, gabbro and basalt dredged from the Mid-Indian ocean ridge. *Science*, **166**, 1136–1141.

Engel, C. G. and Fisher, R. L. (1975). Granitic to ultramafic rock complexes of the Indian Ocean ridge system, western Indian Ocean. *Bull. geol. Soc. Amer.*, **86**, 1553–1578.

Engin, T. and Hirst, D. M. (1970). The alpine chrome ores of the Andizlik–Zimparalik area, Fethiye, SW Turkey. *Trans. Inst. Min. Metall.*, **B79**, 16–27.

Ermanovics, I. F. and Davison, W. L. (1976). The Pikwitonei granulites in relation to the northwestern Superior Province of the Canadian Shield. In B. F. Windley (Ed.), *The Early History of the Earth*. Wiley, London, 331–347.

Ernst, W. G. (1965). Mineral parageneses in Franciscan metamorphic rocks, Panoche Pass, California, *Bull. geol. Soc. Amer.*, **76**, 879–914.

Ernst, W. G. (1970). Tectonic contact between the Franciscan mélange and the Great Valley Sequence—crustal expression of a late Mesozoic Benioff zone, *J. Geophys. Res.*, **75**, 868–901.

Ernst, W. G. (1971). Metamorphic zonations on presumably subducted lithospheric plates from Japan, California and the Alps, *Contr. Mineral. Petrol.*, **34**, 43–59.

Ernst, W. G. (1972). Occurrence and mineralogic evolution of blueschist belts with time. *Amer. J. Sci.*, **272**, 657–668.

Ernst, W. G. (1973). Interpretative synthesis of metamorphism in the Alps, *Bull. geol. Soc. Amer.*, **84**, 2053–2078.

Ernst, W. G. (1975). Introduction, In W. G. Ernst (Ed.), *Subduction Zone Metamorphism*, Dowden Hutchinson and Ross, Stroudsburg, Pa., 1–14.

Ernst, W. G. and Seki, Y. (1967). Petrologic comparison of the Franciscan and Sanbagawa metamorphic terranes, *Tectonophys.*, **4**, 463–478.

Ernst, W. G., Seki, Y., Onuki, H. and Gilbert, M. C. (1970). Comparative study of low-grade metamorphism in the California Coast Ranges and the Outer Metamorphic Belt of Japan, *Geol. Soc. Amer. Mem.*, **124**, 276 pp.

Escher, A., Escher, J. C. and Watterson, J. (1975). The reorientation of the Kangâmiut dike swarm, West Greenland, *Can. J. Earth Sci.*, **12**, 158–173.

Escher, A., Jack, S. and Watterson, J. (1976). Tectonics of the North Atlantic Proterozoic dyke swarm, *Phil. Trans. R. Soc. Lond.*, **A280**, 529–539.

Escher, J. C. and Myers, J. S. (1975). New evidence concerning the original relationship of early Precambrian volcanics and anorthosite in the Fiskenaesset region, southern West Greenland, *Rap. Grønlands geol. Unders.*, **75**, 72–76.

Eskola, P. (1963). The Precambrian of Finland, In K. Rankama (Ed.), *The Precambrian*, Vol. 1, Wiley Interscience, New York, 145–264.

Eugster, H. (1969). Inorganic cherts from the Magadi area, Kenya. *Contr. Mineral. Petrol.*, **22**, 1–31.

Evans, A. M. (1975). Mineralisation in geosynclines—the Alpine enigma, *Mineral. Deposita*, **10**, 254–260.

Evans, A. M. (1976). Mineralisation in Geosynclines. In K. H. Wolf (Ed.), *Handbook of Stratabound and Stratiform Ore Deposits*, Elsevier, Amsterdam, **4**, 1–29.

Evans, W. B., Wilson, A. A., Taylor, B. J. and Price, D. (1968). *Geology of the country around Macclesfield, Congleton and Middlewich*, Mem. geol. Surv. Great Britain, HMSO, 328 pp.

Evernden, J. F. and Kistler, R. W. (1970). Chronology of emplacement of Mesozoic batholithic complexes in California and western Nevada, *US geol. Surv. Prof. Pap.*, **623**, 42 pp.

Ewing, J. and Ewing, M. (1967). Sediment distribution on the mid-ocean ridges with respect to spreading of the sea-floor, *Science*, **156**, 1590–1592.

Ewing, J. and Ewing, M. (1970). Seismic reflection. In A. E. Maxwell (Ed.), *The Sea*, Vol. 4, pt. 1. Wiley Interscience, New York, 1–51.

Ewing, J., Windisch, C. and Ewing, M. (1970). Correlation of horizon A with JOIDES core hole results. *J. Geophys. Res.*, **75**, 5645–5653.

Fahrig, W. F. and Wanless, R. K. (1963). Age and significance of diabase dyke swarms of the Canadian Shield, *Nature*, **200**, 934–937.

Fahrig, W. F., Irving, E. and Jackson, G. D. (1971). Palaeomagnetism of the Franklin diabases, *Can. J. Earth Sci.*, **8**, 455–467.

Fahrig, W. F., Irving, E. and Jackson, G. D. (1973). Test of nature and extent of continental drift as provided by study of Proterozoic dike swarms of Canadian Shield. In M. G. Pitcher (Ed.), *Arctic Geology*, Amer. Ass. Petrol. Geol. Mem., **19**, p. 583.

Fairbridge, R. W. (1970). Ice age in the Saharan, *Geotimes*, **15**, 6, p. 18.

Fairbridge, R. W. (1971). Upper Ordovician glaciation in northwest Africa? Reply. *Bull. geol. Soc. Amer.*, **82**, 269–274.

Fanale, F. P. (1972). A case for catastrophic early degassing of the earth. *Chem. Geol.*, **8**, 79–105.

Farrar, E., Clark, A. H., Haynes, S. J., Quirt, G. S., Conn, H. and Zentilli, M. (1970). K–Ar evidence for the post-Palaeozoic migration of granitic intrusion foci in the Andes of northern Chile, *Earth Plan. Sci. Lett.*, **10**, 60–66.

Farrington, J. F. (1952). A preliminary description of the Nigerian lead–zinc field. *Econ. Geol.*, **47**, 583–608.

Feather, C. E. and Koen, G. M. (1975). The mineralogy of the Witwatersrand reefs, *Minerals Sci. Engng.*, **7**, 189–224.

Ferguson, J. (1964). Geology of the Ilimaussaq intrusion, South Greenland. *Meddr. Grønland*, **172**, 4, 181 pp.

Ferguson, J. and Pulvertaft, T. C. R. (1963). Contrasted styles of igneous layering in the Gardar Province of South Greenland. *Mineral. Soc. Amer. Sp. Pap.*, **1**, 11–21.

Fischer, A. G. (1964). Brackish oceans as a cause of the Permo-Triassic faunal crisis. In A. E. M. Nairn (Ed.), *Problems in Palaeoclimatology*, Interscience, New York, 705 pp.

Fisher, D. E. (1976). Rare gas clues to the origin of the terrestrial atmosphere. In B. F. Windley (Ed.), *The Early History of the Earth*, Wiley, London, 547–556.

Fitch, F. J. and Miller, J. A. (1964). The age of the paroxysmal-Variscan orogeny in England. *Quart. J. Geol. Soc. Lond.*, **120**, p. 159.

Fitton, J. G. (1971). The generation of magmas in island arcs. *Earth Plan. Sci. Lett.*, **11**, 63–67.

Fitton, J. G. and Hughes, D. J. (1970). Volcanism and plate tectonics in the British Ordovician. *Earth Plan. Sci. Lett.*, **8**, 223–228.

Flessa, K. W. and Imbrie, J. (1973). Evolutionary pulsations: evidence from Phanerozoic diversity patterns. In D. H. Tarling and S. K. Runcorn (Eds.), *Implication of Continental Drift to the Earth Sciences*, Vol. 1, Academic Press, London, 247–285.

Flint, D. E., Francisco, J. de A. and Guild, P. W. (1948). Geology and chromite deposits of the Camagüey district, Camagüey Province, Cuba. *Bull. US geol. Surv.*, **954B**, 36–62.

Flint, R. F., Sanders, J. E. and Rodgers, J. (1960). Diamictite; a substitute term for mixtite. *Bull. geol. Soc. Amer.*, **71**, 1809–1810.

Flores, G. (1970). Suggested origin of the Mocambique Channel. *Trans. geol. Soc. S. Afr.*, **73**, 1–16.

Floyd, P. A. (1972a). The tectonic environment of south-west England. *Proc. Geol. Assoc.*, **84**, 243–247.

Floyd, P. A. (1972b). Geochemistry, origin and tectonic environment of the basic and acidic rocks of Cornubia, England. *Proc. Geol. Assoc.*, **83**, 385–404.

Forbes, R. B., Hamilton, T., Tailleur, I. L., Miller, T. P. and Patton, W. W. (1971). Tectonic implications of blueschist facies metamorphism in Alaska. *Nature*, **234**, 106–108.

Ford, T. D. (1969). The stratiform ore-deposits of Derbyshire. In C. H. James (Ed.), *Sedimentary ores: Ancient and Modern, Dept. of Geol. Leicester Univ. Sp. Publ.*, **1**, 73–96.

Frakes, L. A. and Crowell, J. C. (1968). Late Palaeozoic glacial facies and the origin of the South Atlantic basin. *Nature*, **217**, 837–838.

Frakes, L. A. and Crowell, J. C. (1969). Late Palaeozoic glaciation. I: South America. *Bull. geol. Soc. Amer.*, **80**, 1007–1041.

Frakes, L. A. and Crowell, J. C. (1970). Late Palaeozoic glaciation. II: Africa exclusive of the Karroo basin. *Bull. geol. Soc. Amer.*, **81**, 2261–2285.

Frakes, L. A., Matthews, J. L. and Crowell, J. C. (1971). Late Palaeozoic glaciation. III: Antarctica. *Bull. geol. Soc. Amer.*, **82**, p. 1581.

Francis, P. W. and Rundle, C. C. (1976). Rates of production of the main magma types in the central Andes. *Bull. geol. Soc. Amer.*, **87**, 474–480.

Franks, S. and Nairn, A. E. M. (1973). The equatorial marginal basins of West Africa. In A. E. M. Nairn and F. G. Stehli (Eds.), *The Ocean Basins and Margins*, Vol. 1: The South Atlantic, Plenum Press, New York, 301–350.

Frarey, M. J. and Roscoe, S. M. (1970). The Huronian Supergroup north of Lake Huron. In A. J. Baer (Ed.), *Symp. on Basins and Geosynclines of the Canadian Shield, Geol. Surv. Can. Pap.*, **70–40**, 143–157.

Fraser, J. A., Donaldson, J. A., Fahrig, W. F. and Tremblay, L. P. (1970). Helikian basins and geosynclines of the northwestern Canadian Shield. In A. J. Baer (Ed.), *Symp. on Basins and Geosynclines of the Canadian Shield, Geol. Surv. Can. Pap.*, **70–40**, 213–238.

Fraser, J. A., Hoffman, P. F. and Irvine, T. N. (1972). The Bear Province. In *Variations in Tectonic Styles in Canada, Geol. Ass. Can. Sp. Pap.*, **11**, 453–504.

Friedman, G. M. and Sanders, J. E. (1967). Origin and occurrence of dolostones. In G. V. Chitingar, H. G. Bissell and R. W. Fairbridge (Eds.), *Carbonate Rocks*, Elsevier, Amsterdam, 267–348.

Friend, P. F. (1969). Tectonic features of Old Red sedimentation in North Atlantic borders. In M. Kay (Ed.), *North Atlantic—Geology and Continental Drift, Amer. Ass. Petrol. Geol. Mem.*, **12**, 703–710.

Frietsch, R. (1973). Precambrian iron ores of sedimentary origin in Sweden. In *Genesis of Precambrian iron and manganese deposits*, Proc. Kiev Symp. 1970, Unesco, Paris, 85–87.

Fripp, R. E. P. (1976). Gold metallogeny in the Archaean of Rhodesia. In B. F. Windley (Ed.), *The Early History of the Earth*, Wiley, London, 455–466.

Frisch, T. and Bridgwater, D. (1972). Minor intrusions of the rapakivi suite in southwest Greenland. *Geol. Soc. Amer. Abstr. with Prog.*, **4**, p. 513.

Fryer, B. J. (1972). Age determinations in the Circum-Ungava geosyncline and the evolution of Precambrian banded iron formations. *Can. J. Earth Sci.*, **9**, 652–663.

Fyfe, W. S. (1974). Archaean tectonics. *Nature*, **249**, p. 338.

Fyfe, W. S. and Leonardos, O. H., Jr (1973). Ancient metamorphic–migmatite belts of the Brazilian African coasts. *Nature*, **244**, 501–502.

Gabelman, J. W. (1968). Metallotectonic zoning in the North American Appalachian region, *23rd Int. geol. Congr. Prague*, **7**, 17–33.

Gabelman, J. W. (1976). Orogenic and taphrogenic mineralisation belts at continental margins. In D. F. Strong (Ed.), *Metallogeny and Plate Tectonics, Geol. Ass. Can. Sp. Pap.*, **14**, 273–300.

Gabelman, J. W. and Krusiewski, S. V. (1968). Regional metallotectonic zoning in Mexico. *Trans. Soc. Mining Eng.*, **241**, 113–128.

Gabelman, J. W. and Krusiewski, S. V. (1972). The metallotectonics of Europe. *24th Int. geol. Congr. Montreal, Sect. 4*, 88–97.

Gale, G. H. (1973). Palaeozoic basaltic komatiite and ocean floor basalts from northeastern Newfoundland. *Earth Plan. Sci. Lett.*, **18**, 22–28.

Garrels, R. M. and Mackenzie, F. T. (1971). *Evolution of Sedimentary Rocks*, Norton, New York, 397 pp.

Garrels, R. M., Mackenzie, F. T. and Siever, R. (1972). Sedimentary cycling in relation to the history of the continents and oceans. In E. C. Robertson (Ed.), *The Nature of the Solid Earth*, McGraw-Hill, New York, 93–121.

Garrison, R. E. and Fischer, A. G. (1969). Deep-water limestones and radiolarites of the Alpine Jurassic. In G. M. Friedman (Ed.), *Depositional Environments in Carbonate Rocks*, a Symposium, Tulsa, Oklahoma, *Soc. Econ. Pal. Mineral. Sp. Publ.*, **14**, 20–56.

Garson, M. S. and Livingstone, A. (1973). Is the South Harris Complex in North Scotland a Precambrian overthrust slice of oceanic crust and island arc? *Nature Phys. Sci.*, **243**, 74–76.

Garson, M. S. and Krs, M. (1976). Geophysical and geological evidence of the relationship of Red Sea transverse tectonics to ancient fractures. *Bull. geol. Soc. Amer.*, **87**, 169–181.

Gass, I. G. (1968). Is the Troodos massif of Cyprus a fragment of Mesozoic ocean floor? *Nature*, **220**, 39–42.

Gass, I. G. (1970a). Tectonic and magmatic evolution of the Afro-Arabian dome. In T. N. Clifford and I. G. Gass (Eds.), *African Magmatism and Tectonics*, Oliver and Boyd, Edinburgh, 285–300.

Gass, I. G. (1970b). The evolution of volcanism in the junction area of the Red Sea, Gulf of Aden and Ethiopian rifts. *Phil. Trans. R. Soc. Lond.*, **A267**, p. 369.

Gass, I. G. (1972a). The role of lithothermal systems in magmatic and tectonic processes. *J. Earth Sci. Leeds*, **8**, 261–274.

Gass, I. G. (1972b). *The role of magmatic processes in continental rifting and sea-floor spreading*, 4th Tomkeieff Memorial Lecture, Geol. Dept Univ. Newcastle upon Tyne, 1–16.

Gass, I. G. and Masson-Smith, D. (1963). The geology and gravity anomalies of the Troodos Massif, Cyprus. *Phil. Trans. R. Soc. Lond.*, **A255**, 417–467.

Gass, I. G., Mallick, D. I. J. and Cox, K. G. (1973). Volcanic islands of the Red Sea. *J. geol. Soc. Lond.*, **129**, 275–310.

Gass, I. G. and Smewing, J. D. (1973). Intrusion, extrusion and metamorphism at constructive margins: evidence from the Troodos Massif, Cyprus. *Nature*, **242**, 26–29.

Gastil, R. G. (1975). Plutonic zones in the Peninsular Ranges of southern California and northern Baja California. *Geology*, **3**, 361–363.

Gastil, R. G., Phillips, R. P. and Allison, E. C. (1975). Reconnaissance geology of the State of Baja California. *Geol. Soc. Amer. Mem.*, **140**, 170 pp.

Gates, T. M. and Hurley, P. M. (1973). Evaluation of Rb–Sr dating methods applied to the Matachewan, Abitibi, Mackenzie and Sudbury dike swarms in Canada. *Can. J. Earth Sci.*, **10**, 900–919.

Geijer, P. (1963). The Precambrian of Sweden. In K. Rankama (Ed.), *The Precambrian*, Vol. 1, Wiley-Interscience, New York, 81–144.

Geis, H. P. (1971). A short description of the iron–titanium provinces in Norway with special reference to those in production. *Mineral Sci. Engng.*, **3**, 13–24.

Geological Society, London, *Phanerozoic Time-scale*, 1964. *Quart. J. geol. Soc. Lond.*, **1205**, 260–262.

George, P. T. (1967). The Timmins district. Can. Inst. Min. Metall. Centennial field excursion guide NW Quebec–northern Ontario.

Ghisler, M. (1970). Pre-metamorphic folded chromite deposits of stratiform type in the early Precambrian of West Greenland. *Mineral. Deposita*, **5**, 223–236.

Ghisler, M. and Windley, B. F. (1967). The chromite deposits of the Fiskenaesset region, West Greenland. *Rapp. Grønlands geol. Unders.*, **12**, 1–39.

Gibb, R. A. (1975). Collision tectonics in the Canadian Shield? *Earth Plan. Sci. Lett.*, **27**, 378–382.

Gibb, R. A. and Walcott, R. I. (1971). A Precambrian suture in the Canadian Shield. *Earth Plan. Sci. Lett.*, **10**, 417–422.

Gidskehaug, A., Creer, K. M. and Mitchell, J. G. (1975). Palaeomagnetism and K–Ar ages of the south west African basalts and their bearing on the time of initial rifting of the South Atlantic ocean. *Geophys. J. R. astr. Soc.*, **42**, 1–20.

Giletti, B. J. and Day, H. W. (1968). Potassium–argon ages of igneous intrusive rocks of Peru. *Nature*, **220**, 570–572.

Gill, J. (1974). Magma. In *Yearbook of Technology*, McGraw-Hill, New York, 271–273.

Gill, R. C. O. and Bridgwater, D. (1976). The Ameralik dykes of West Greenland, the earliest known basaltic rocks intruding stable continental crust. *Earth Plan. Sci. Lett.*, **29**, 276–282.

Gill, W. D. (1965). The Mediterranean basin. In D. C. Ion (Ed.), *Salt Basins around Africa*, Inst. Pet., London, p. 101.

Gilluly, J. (1971). Plate tectonics and magmatic evolution. *Bull. geol. Soc. Amer.*, **82**, 2383–2396.

Gilluly, J. (1973). Steady plate motion and episodic orogeny and magmatism. *Bull. geol. Soc. Amer.*, **84**, 499–514.

Girdler, R. W., Fairhead, J. D., Searle, R. C. and Sowerbutts, W. T. C. (1969). Evolution of rifting in Africa. *Nature*, **224**, p. 1178.

Gittins, J., MacIntyre, R. M. and York, D. (1967). The ages of carbonatite complexes in eastern Canada. *Can. J. Earth Sci.*, **4**, 651–655.

Glaessner, M. F. (1971). Geographic distribution and time range of the Ediacara Precambrian fauna. *Bull. geol. Soc. Amer.*, **82**, 509–514.

Glikson, A. Y. (1970). Geosynclinal evolution and geochemical affinities of early Precambrian systems. *Tectonophys.*, **9**, 397–433.

Glikson, A. Y. (1971a). Archaean geosynclinal sedimentation near Kalgoorlie, Western Australia. *Geol. Soc. Australia Sp. Publ.*, 443–459.

Glikson, A. Y. (1971b). Structure and metamorphism of the Kalgoorlie system, south west of Kalgoorlie, Western Australia, *Geol. Soc. Aust. Sp. Publ.*, **3**, 121–133.

Glikson, A. Y. (1971c). Primitive Archaean element distribution patterns: chemical evidence and geotectonic significance. *Earth Plan. Sci. Lett.*, **12**, 309–320.

Glikson, A. Y. (1972). Early Precambrian evidence of a primitive ocean crust and island nuclei of sodic granite. *Bull. geol. Soc. Amer.*, **83**, 3323–3344.

Glikson, A. Y. (1976a). Stratigraphy and evolution of primary and secondary greenstones: significance of data from Shields of the southern hemisphere. In B. F. Windley (Ed.), *The Early History of the Earth*, Wiley, London, 257–277.

Glikson, A. Y. (1976b). Earliest Precambrian ultramafic–mafic volcanic rocks: ancient oceanic crust or relic terrestrial maria? *Geology*, **4**, 201–205.

Glikson, A. Y. (1976c). Archaean to Early Proterozoic Shield structures: relevance of plate tectonics. In D. F. Strong (Ed.), *Metallogeny and Plate Tectonics, Geol. Ass. Can. Sp. Pap.*, **14**, 487–516.

Glikson, A. Y. and Lambert, I. B. (1973). Relations in space and time between major Precambrian Shield units: an interpretation of Western Australian data. *Earth Plan. Sci. Lett.*, **20**, 395–403.

Glikson, A. Y. and Lambert, I. B. (1976). Vertical zonation and petrogenesis of the early Precambrian crust in western Australia. *Tectonophys.*, **30**, 55–89.

Goldich, S. S. (1973). Ages of Precambrian banded iron formations. *Econ. Geol.*, **68**, 1126–1134.

Goldich, S. S., Hedge, C. E. and Stern, T. W. (1970). Age of the Morton and Montivideo gneisses and related rocks, southwestern Minnesota. *Bull. geol. Soc. Amer.*, **81**, 3671–3696.

Goldich, S. S. and Hedge, C. E. (1974). 3800 Myr granitic gneisses in south western Minnesota. *Nature*, **252**, 467–468.

Goles, G. G. (1975). Basalts of unusual composition from the Chyulu hills, Kenya. *Lithos*, **8**, 47–58.

Goles, G. G. (1976). Some constraints on the origin of phonolites from the Gregory Rift, Kenya, and inferences concerning basaltic magmas in the Rift System. *Lithos*, **9**, 1–8.

González-Bonorino, F. (1971). Metamorphism of the crystalline basement of central Chile. *J. Petrol.*, **12**, 149–175.

González-Bonorino, F. and Aguirre, L. (1970). Metamorphic facies series of the crystalline basement of Chile. *Geol. Rund.*, **59**, 979–994.

Goodspeed, G. E. (1955). Relict dikes and relict pseudodikes, *Amer. J. Sci.*, **253**, 146–161.

Goodwin, A. M. (1962). Structure, stratigraphy and origin of iron formations, Michipicoten area, Algoma district, Ontario, Canada. *Bull. geol. Soc. Amer.*, **73**, 561–586.

Goodwin, A. M. (1968). Archaean protocontinental growth and early crustal history of the Canadian Shield. *23rd Int. geol. Congr.*, **1**, 69–89.

Goodwin, A. M. (1971). Metallogenic patterns and evolution of the Canadian Shield. *Geol. Soc. Aust. Sp. Publ.*, **3**, 157–174.

Goodwin, A. M. (1972). The Superior Province. In R. A. Price and R. J. W. Douglas (Eds.), *Variations in tectonic styles in Canada, Geol. Ass. Can. Spec. Pap.*, **11**, 527–564.

Goodwin, A. M. (1973a). Archaean iron-formations and tectonic basins of the Canadian Shield. *Econ. Geol.*, **68**, 915–933.

Goodwin, A. M. (1973b). Plate tectonics and evolution of Precambrian crust. In D. H. Tarling and S. K. Runcorn (Eds.), *Implications of Continental Drift to the Earth Sciences*, Vol. 2, Academic Press, London, 1047–1069.

Goodwin, A. M. and Ridler, R. H. (1970). The Abitibi orogenic belt. *Geol. Surv. Can. Pap.*, **70–40**, 1–24.

Goosens, P. J. (1969). Mineral index map, Republic of Ecuador, Quito, Ecuador, *Serv. Nac. Geol. Min.*

Gordon, P. S. L. (1973). The geology of the Selebi-Pikwe Ni–Cu deposits. In L. A. Lister (Ed.), *Symp. on Granites, Gneisses and Related Rocks, Geol. Soc. S. Afr. Sp. Publ.*, **3**, 167–188.

Gordon, W. A. (1975). Distribution by latitude of Phanerozoic evaporite deposits. *J. Geol.*, **83**, 671–684.

Gorokhov, I. M. (1964). Whole rock Rb–Sr ages of the Koresten granites, Dnieper migmatites, and metamorphosed mafic rocks of the Ukraine. *Geochem. for 1964*, 738–746.

Govett, G. J. S. (1966). Origin of banded iron formations. *Bull. geol. Soc. Amer.*, **77**, 1191–1212.

Gowda, S. S. (1970). Fossil blue–green algae and fungi from the Archaean complex of Mysore. *Abstr. Proc. 1st Int. Symp. Taxonomy and Biology of Blue Green Algae*, Madras, 40–41.

Grant, N. K. (1971). South Atlantic, Benue Trough, and Gulf of Guinea Cretaceous triple junction. *Bull. geol. Soc. Amer.*, **82**, 2295–2298.

Grant, N. K. (1973). Orogeny and reactivation to the west and southeast of the West African craton. In A. E. M. Nairn and F. G. Stehli (Eds.), *The Ocean Basins and Margins*, Vol. 1 *The South Atlantic*, Plenum Press, New York, 447–492.

Grant, J. A. and Goldich, S. S. (1972). Precambrian geology of the Minnesota River Valley between Morton and Montevideo, *Minn. geol. Surv. Guidebook Ser.*, **5**, 52 pp.

Green, D. C. and Baadsgaard, H. (1971). Temporal evolution and petrogenesis of an Archaean crustal segment at Yellowknife, N.W.T., Canada, *J. Petrol.*, **12**, 177–217.

Green, D. H. (1972a). Magmatic activity as the major process in the chemical evolution of the earth's crust and mantle. In A. R. Ritsema (Ed.), *The Upper Mantle*, Tectonophys., **13**, (1–4), 47–71.

Green, D. H. (1972b). Archaean greenstone belts may include terrestrial equivalents of lunar maria? *Earth Plan. Sci. Lett.*, **15**, 263–270.

Green, D. H. (1975). Genesis of Archaean peridotitic magmas and constraints on Archaean geothermal gradients and tectonics. *Geology*, **3**, 15–18.

Green, D. H. and Ringwood, A. E. (1967). The genesis of basaltic magmas. *Contr. Mineral. Petrol.*, **15**, 103–190.

Green, D. H., Nicholls, I. A., Viljoen, M. and Viljoen, R. (1975). Experimental demonstration of the existence of peridotitic liquids in earliest Archaean magmatism, *Geology*, **3**, 11–14.

Green, T. H. (1973). High pressure, hydrous crystallization of an island arc calc-alkaline andesite. In P. J. Goleman (Ed.), *The Western Pacific: Island Arcs, Marginal Seas, Geochemistry*, Univ. W. Australia Press, Nedlands, 497–502.

Green, T. H. and Ringwood, A. E. (1968). Genesis of the calc-alkaline igneous rock suite. *Contr. Mineral. Petrol.*, **18**, 105–162.

Greenbaum, D. (1972). Magmatic processes at ocean ridges: evidence from the Troodos Massif, Cyprus. *Nature Phys. Sci.*, **238**, 18–21.

Griffiths, D. H., King, R. F., Khan, M. A. and Blundell, D. J. (1971). Seismic refraction line in the Gregory rift. *Nature Phys. Sci.*, **229**, p. 69.

Griffitts, W. R., Albers, J. P. and Öner, O. (1972). Massive sulphide copper deposits of the Ergani–Maden area, southeastern Turkey. *Econ. Geol.*, **67**, 701–716.

Gross, G. A. (1965). Geology of iron deposits in Canada. I. General geology and evaluation of iron deposits. *Geol. Surv. Can. Econ. Geol. Rep.*, **22**, 181 pp.

Gross, G. A. (1967). Geology of iron deposits in Canada. II. Iron deposits in the Appalachian and Grenville regions of Canada. *Geol. Surv. Can. Econ. Geol. Rep.*, **22**, 111 pp.

Gross, G. A. (1968). Geology of iron deposits in Canada. III. Iron ranges of the Labrador Geosyncline. *Geol. Surv. Can. Econ. Geol. Rep.*, **22**.

Gross, S. O., 1968. Titaniferous ores of the Sanford Lake district, New York. Chapter 8 in Vol. 1 of J. D. Ridge (Ed.), *Ore Deposits of the United States 1933–1967* (Graton–Sales vol.), Amer. Inst. Min. Metall. Pet. Eng. N.Y., 140–154.

Grow, J. A. (1973). Crustal and upper mantle structure of the central Aleutian arc. *Bull. geol. Soc. Amer.*, **84**, 2169–2192.

Gulson, B. L. and Krogh, T. E. (1975). Evidence of multiple intrusion, possible resetting of U–Pb ages, and new crystallization of zircons in the post-tectonic intrusions (in rapakivi granites) and gneisses from South Greenland. *Geochim. Cosmochim. Acta*, **39**, 65–82.

Gundersen, J. N. and Schwartz, G. M. (1962). The geology of the metamorphosed Biwabik iron-formation, eastern Mesabi district, Minnesota. *Bull. Minn. geol. Surv.*, **43**.

Gunn, B. M. (1976). A comparison of modern and Archaean oceanic crust and island arc petrochemistry. In B. F. Windley (Ed.), *The Early History of the Earth*, Wiley, London, 389–403.

Hackett, J. P. and Bischoff, J. L. (1973). New data on the stratigraphy, extent and geologic history of the Red Sea geothermal brines. *Econ. Geol.*, **68**, 533–564.

Hagner, A. F. (1968). The titaniferous magnetite deposit at Iron Mountain, Wyoming. In J. D. Ridge (Ed.), *Ore Deposits of the United States 1933–1967*, Vol. 1 (Graton–Sales vol.), Chapter 13, *Amer. Inst. Min. Metall. Pet. Eng. N.Y.*, 665–680.

Haidutov, I. S. (1976). A greenstone belt–basement relationship in the Tanganyika Shield. *Geol. Mag.*, **113**, 53–60.

Hailwood, E. A. (1974). Palaeomagnetism of the Msissi norite (Morocco) and the Palaeozoic reconstruction of Gondwanaland. *Earth Plan. Sci. Lett.*, **23**, 376–386.

Hailwood, E. A. and Tarling, D. H. (1973). Palaeomagnetic evidence for a proto-Atlantic ocean. In D. H. Tarling and S. K. Runcorn (Eds.), *Implications of Continental Drift to the Earth Sciences*, Vol. 1, Academic Press, London, 37–46.

Hallam, A. (1967). The bearing of certain palaeozo-geographic data on continental drift. *Palaeogeog., Palaeochim., Palaeoecol.*, **3**, 201–241.

Hallam, A. (1969). Faunal realms and facies in the Jurassic. *Pal.*, **12**, 1–18.

Hallam, A. (1971). Mesozoic geology and the opening of the North Atlantic. *J. Geol.*, **79**, 129–157.

Hallam, A. (1973). Provinciality, diversity and extinction of Mesozoic marine invertebrates in relation to plate movements. In D. H. Tarling and S. K. Runcorn (Eds.), *Implications of Continental Drift to the Earth Sciences*, Vol. 1, Academic Press, London, 287–294.

Hallbauer, D. K. (1975). The plant origin of the Witwatersrand 'carbon', *Minerals Sci. Engng.*, **7**, 111–130.

Hallberg, J. A. (1971). Geochemistry of the Archaean basalt-dolerite association in the Coolgardie–Norseman area, western Australia. *Geol. Soc. Aust. Sp. Publ.*, **3**, p. 151 (Abstr.).

Hallberg, J. A. (1972). Geochemistry of Archaean volcanic belts in the Eastern Goldfields region of western Australia. *J. Petrol.*, **13**, 45–56.

Hallberg, J. A. and Williams, D. A. C. (1972). Archaean mafic and ultramafic rock associations in the Eastern Goldfields region, western Australia. *Earth Plan. Sci. Lett.*, **15**, 191–200.

Halls, H. C. (1975). Shock-induced remanent magnetisation in late Precambrian rocks from Lake Superior. *Nature*, **255**, 692–695.

Hamilton, W. (1969). Mesozoic California and the underflow of Pacific mantle. *Bull. geol. Soc. Amer.*, **80**, 2409–2430.

Hamilton, W. (1970). The Uralides and the motion of the Russian and Siberian platforms. *Bull. geol. Soc. Amer.*, **81**, 2553–2576.

Hamilton, W. and Krinsley, D. (1967). Upper Palaeozoic glacial deposits of South Africa and southern Australia. *Bull. geol. Soc. Amer.*, **78**, 783–799.

Hamilton, W. and Myers, W. B. (1966). Cenozoic tectonics of the western United States. *Rev. Geophys.*, **4**, 509–549.

Hamilton, W. and Myers, W. B. (1967). The nature of batholiths. *U.S. geol. Surv. Prof. Pap.*, **554C**, 30 pp.

Hanson, G. N. (1975). $^{40}Ar/^{39}Ar$ spectrum ages on Logan intrusions, a Lower Keweenawan flow, and mafic dykes in northeastern Minnesota–northwestern Ontario. *Can. J. Earth Sci.*, **12**, 821–835.

Harland, W. B. (1964a). Evidence of late Precambrian glaciation and its significance. In A. E. M. Nairn (Ed.), *Problems in Palaeoclimatology*, Interscience, New York, 119–149.

Harland, W. B. (1964b). Critical evidence for a great Infra-Cambrian glaciation, *Geol. Rundsch.*, **54**, 45–61.

Harland, W. B. (1967). Early history of the North Atlantic ocean and its margins, *Nature*, **216**, p. 464.

Harland, W. B. (1972). The Ordovician Ice Age. *Geol. Mag.*, **109**, 451–456.

Harland, W. B. (1974). The Precambrian–Cambrian boundary. In C. H. Holland (Ed.), *Lower Palaeozoic Rocks of the World, Vol. 2: Cambrian of the British Isles, Norden and Spitsbergen*, Wiley, London, 15–42.

Harland, W. B. and Rudwick, M. J. S. (1964). The great Infra-Cambrian ice age. *Sci. Amer.*, **211**, 28–36.

Harland, W. B., Herod, K. N. and Krinsley, D. H. (1966). The definition and identification of tills and tillites. *Earth Sci. Rev.*, **2**, 225–256.

Harland, W. B. and Herod, K. N. (1975). Glaciations through time. In A. E. Wright and F. Moseley (Eds.), *Ice Ages: Ancient and Modern*, Seal House Press, Liverpool, 189–216.

Härme, M. (1965). On the potassium migmatites of southern Finland. *Bull. Comm. geol. Finl.*, **219**, 43 pp.

Harpum, J. R. (1955). Recent investigations in pre-Karroo geology in Tanganyika, *C.R. Ass. Serv. geol. Afr.*, Nairobi Meeting (1954), 165.

Harris, A. L. and Pitcher, W. S. (1975). The Dalradian Supergroup. In A. L. Harris and coworkers (Eds.), *A correlation of the Precambrian Rocks in the British Isles*, Geol. Soc. Lond. Sp. Pap., **6**, 52–75.

Harris, P. G. (1969). Basalt type and African rift valley tectonism. *Tectonophys.*, **8**, 427–436.

Harris, P. G. (1970). Convection and magmatism with reference to the African continent. In T. N. Clifford and I. G. Gass (Eds.), *African Magmatism and Tectonics*, Oliver and Boyd, Edinburgh, 419–438.

Harrison, C. G. A. and Prospero, J. M. (1974). Reversals of the earth's magnetic field and climatic changes. *Nature*, **250**, 563–564.

Harrison, J. E. (1972). Precambrian Belt basin of northwestern United States: its geometry, sedimentation and copper occurrences. *Bull. geol. Soc. Amer.*, **83**, 1215–1240.

Hart, S. R., Glassley, W. E. and Karig, D. E. (1972). Basalts and sea-floor spreading behind the Mariana island arc. *Earth Plan. Sci. Lett.*, **15**, 12–18.

Hasebe, K., Fujii, N. and Uyeda, S. (1970). Thermal processes under island arcs. *Tectonophys.*, **10**, 335–355.

Hatcher, R. D., Jr (1972). Developmental model for the southern Appalachians. *Bull. geol. Soc. Amer.*, **83**, 2735–2760.

Hatcher, R. D., Jr (1974). North American Palaeozoic fold belts and deformational histories: a plate tectonic anomaly? *Amer. J. Sci.*, **274**, 135–147.

Hatfield, C. B. and Camp, M. J. (1970). Mass extinctions correlated with periodic galactic events. *Bull. geol. Soc. Amer.*, **81**, 911–914.

Hatherton, T. (1974). Active continental margins and island arcs. In C. A. Burk and C. L. Drake (Eds.), *The Geology of Continental Margins*, Springer-Verlag, Berlin, 93–103.

Hatherton, T. and Dickinson, W. R. (1969). The relationship between andesitic volcanism and seismicity in Indonesia, the Lesser Antilles, and other island arcs. *J. Geophys. Res.*, **74**, 5301–5310.

Haughton, S. H. (1969). Geological history of southern Africa. *Geol. Soc. S. Afr.*, 535 pp.

Hawkesworth, C. J., Moorbath, S. and O'Nions, R. K. (1975). Age relationships between greenstone belts and "granites" in the Rhodesian Archaean craton. *Earth Plan. Sci. Lett.*, **25**, 251–262.

Hawkins, J. W., Jr. (1974). Geology of the Lau Basin, a marginal sea behind the Tongo arc. In C. A. Burk and C. L. Drake (Eds.), *The Geology of Continental Margins*, Springer-Verlag, Berlin, 505–520.

Hayes, D. E. and Ewing, M. (1970). Pacific boundary structure. In A. E. Maxwell (Ed.), *New concepts of sea floor evolution*, Vol. 4, Pt. 2: *Regional observations*, Wiley, New York, 29–72.

Hays, J. D. (1971). Faunal extinctions and reversals of the earth's magnetic field. *Bull. geol. Soc. Amer.*, **82**, 2433–2447.

Hays, J. D. and Pitman, W. C., III (1973). Lithospheric plate motion, sea level changes and climatic and ecological consequences. *Nature*, **246**, p. 18.

Heezen, B. C. and Hollister, C. D. (1971). *The Face of the Deep*, Oxford Univ. Press, London, 659 pp.

Heier, K. S. (1973). Geochemistry of granulite facies rocks and problems of their origin. *Phil. Trans. R. Soc. Lond.*, **A273**, 429–442.

Heier, K. S. and Griffin, W. L. (1973). Geology and age relationships in Lofoten–Vesteralen. In R. G. Park and J. Tarney (Eds.), *The Early Precambrian of Scotland and related Rocks of Greenland* (abstract), Univ. of Keele, p. 189

Heirtzler, J. R. (1968). Sea floor spreading. *Sci Amer.*, **219**, 60–70.

Heirtzler, J. R., Le Pichon, X. and Baron, J. G. (1966). Magnetic anomalies over the Reykyanes Ridge. *Deep-Sea Res.*, **13**, 427–443.

Heirtzler, J. R. and Hayes, D. E. (1967). Magnetic boundaries in the north Atlantic ocean. *Science*, **157**, 185–187.

Helsley, C. E. (1969). Magnetic reversal stratigraphy of the Lower Triassic Moenkopi Formation of western Colorado. *Bull. geol. Soc. Amer.*, **80**, 2431–2450.

Helsley, C. E. and Steiner, M. B. (1968). Evidence for long periods

of normal polarity in the Cretaceous Period. *Geol. Soc. Amer. Sp. Pap.*, **121**, 133 pp.

Helsley, C. E. and Steiner, M. B. (1969). Evidence for long interval of normal polarity during the Cretaceous period. *Earth Plan. Sci. Lett.*, **5**, 325–332.

Helwig, J. (1976). Shortening of continental crust in orogenic belts and plate tectonics. *Nature*, **260**, 768–770.

Henderson, G. and Pulvertaft, T. C. R. (1967). The stratigraphy and structure of the Precambrian rocks of the Umanak area, West Greenland. *Meddr. dansk geol. Foren.*, **17**, 1–20.

Henderson, J. F. (1975). Sedimentology of the Archaean Yellowknife Supergroup of the Yellowknife District of Mackenzie. *Bull. geol. Surv. Can.*, **246**, 62 pp.

Henderson, J. F. and Brown, I. C. (1966). Geology and structure of the Yellowknife greenstone belt, district of Mackenzie. *Bull. Geol. Surv. Can.*, **141**, 1–87.

Henriksen, N. (1960). Structural analysis of a fault in south-west Greenland. *Bull. Grønlands geol. Unders.*, **26**, 41 pp (also *Meddr. Grønland*, **162**, Nr. 9).

Henriksen, N. (1969). Boundary relations between Precambrian fold belts in the Ivigtut area, southwest Greenland. *Geol. Ass. Can. Sp. Pap.*, **5**, 143–154.

Hensen, B. J. and Green, D. H. (1972). Experimental study of the stability of cordierite and garnet in pelitic compositions at high pressures and temperatures. *Contr. Mineral. Petrol.*, **35**, 331–354.

Hepworth, J. V. (1976). Discussion on late Proterozoic tectonics, *Phil. Trans. R. Soc. Lond.*, **A280**, 662–663.

Herak, M., Polsak, A., Gusic, I. and Babic, L. J. (1970). Dynamische und räumliche Sedimentations—begingungen der mesozoischen Karbonat-gesteine im Dinarischen Karstgebiet. *Geo. Bundesanstalt Wien Verh.*, 637–643.

Herz, N. (1966). Tholeiitic and alkaline volcanism in southern Brazil. In *Guidebook*, Int. Field Inst. Brazil. Chapter 5.

Herz, N. (1969). Anorthosite belts, continental drift and the anorthosite event, *Science*, **164**, 944–947.

Hess, H. H. (1960). Stillwater igneous complex, Montana. *Mem. geol. Soc. Amer.*, **80**, 230.

Hess, H. H. (1962). History of ocean basins. In A. E. J. Engel, H. L. James and B. F. Leonard (Eds.), *Petrologic Studies* (Buddington volume), *Geol. Soc. Amer.*, 599–620.

Hickman, M. H. (1974). 3500 m.y. old granite in southern Africa, *Nature*, **251**, 295–296.

Hietanen, A. (1943). Über das Grundgebirge des Kalantigebietes im südwestlichen Finnland. *Bull. Comm. geol. Finl.*, **130**, 1–105.

Hietanen, A. (1975). Generation of potassium-poor magmas in the northern Sierra Nevada and the Svecofennian of Finland. *J. Res. U.S. geol. Surv.*, **3**, 631–645.

Hilde, T. W. C. and Wageman, J. M. (1973). Structure and origin of the Japan Sea. In P. J. Coleman (Ed.), *The Western Pacific; island arcs, marginal seas, geochemistry*, Univ. W. Australia Press, Nedlands, 415–434.

Hjelmqvist, S. (1956). On the occurrence of ignimbrite in the Precambrian. *Sver. geol. Unders. Ser. C*, No. 542, 12 pp.

Hoering, T. C. (1967). The organic geochemistry of Precambrian rocks. In *Researches in Geochemistry*, Vol. 2, Wiley, New York, 87–111.

Hoffman, P. (1973). Evolution of an early Proterozoic continental margin: the Coronation Geosyncline and associated aulacogens of the northwestern Canadian Shield. *Phil. Trans. R. Soc. Lond.*, **A273**, 547–581.

Hoffman, P., Fraser, J. A. and McGlynn, J. C. (1970). The Coronation Geosyncline of Aphebian age, district of Mackenzie. In A. J. Baer (Ed.), *Symposium on Basins and Geosynclines of the Canadian Shield*, Geol. Surv. Can. Pap., **70–40**, 200–212.

Hoffman, P., Dewey, J. F. and Burke, K. (1974). Aulacogens and their genetic relation to geosynclines, with a Proterozoic example from Great Slave Lake, Canada. In R. H. Dott, Jr. and R. H. Shaver (Eds.), *Modern and Ancient Geosynclinal Sedimentation*, Soc. Econ. Pal. Mineral. Sp. Publ., **19**, 38–55.

Holland, C. H. (1971). Silurian faunal provinces? In F. A. Middlemiss, P. F. Rawson and G. Newall (Eds.), *Faunal Provinces in Space and Time*, Seal House Press, Liverpool, 61–76.

Holland, H. D. (1963). On the chemical evolution of the terrestrial and cytherean atmospheres. In P. J. Brancazio and A. G. W.

Cameron (Eds.), *The Origin and Evolution of Atmospheres and Oceans*, Wiley, New York, p. 86.

Holland, H. D. (1976). The evolution of sea water. In B. F. Windley (Ed.), *The Early History of the Earth*, Wiley, London, 559–567.

Holland, J. G. and Lambert, R. St J. (1975). The chemistry and origin of the Lewisian gneisses of the Scottish mainland: the Scourie and Inver assemblages and sub-crustal accretion. *Precambrian Res.*, **2**, 161–188.

Hollister, V. F. (1975). An appraisal of the nature and source of porphyry copper deposits. *Minerals Sci. Engng.*, **7**, 225–233.

Holmes, A. and Cahen, L. (1957). Geochronologie Africaine 1956, Acad. Roy. Sci. Coloniales (Brussels). *Cl. Sci. Nat. Med.*, Mem. in 8°, N.S., 5.

Hor, A. K., Hutt, D. K., Smith, J. V., Wakefield, J. and Windley, B. F. (1975). Petrochemistry and mineralogy of early Precambrian anorthositic rocks of the Limpopo belt, southern Africa. *Lithos*, **8**, 297–310.

Horikoshi, E. (1969). Volcanic activity related to the formation of the Kuroko-type deposits in the Kosaka district, Japan. *Mineral. Deposit.*, **4**, 321–345.

Horne, G. S. (1969). Early Ordovician chaotic deposits in the central volcanic belt of northeastern Newfoundland. *Bull. geol. Soc. Amer.*, **80**, 2451–2464.

Horowitz, A. (1970). The distribution of Pb, Ag, Sn, Te and Zn in sediments on active oceanic ridges. *Marine Geol.*, **9**, 241–259.

Hosking, K. F. G. (1966). Permo-Carboniferous and later primary mineralization of Cornwall and south west Devon. In *Present Views on some Aspects of the Geology of Cornwall and Devon*, (Comm. Vol. for 1964), *Roy geol. Soc. Cornwall*, 201–245.

Hospers, J. and van Andel, S. I. (1969). Palaeomagnetism and tectonics, a review. *Earth Sci. Rev.*, **5**, 5–44.

House, M. R. (1967). Fluctuations in the evolution of Palaeozoic invertebrates. In W. B. Harland and coworkers (Eds.), *The Fossil Record*, Geol. Soc. Lond., 41–54.

House, M. R. (1971a). Devonian faunal distributions. In F. A. Middlemiss, P. F. Rawson and G. Newall (Eds.), *Faunal Provinces in Space and Time*, Seal House Press, Liverpool, 77–94.

House, M. R. (1971b). Evolution and the fossil record. In I. G. Gass, P. J. Smith and R. C. L. Wilson (Eds.), *Understanding the Earth*, Artemis Press, Sussex, 263–286.

House, M. R. (1975). Faunas and time in the marine Devonian. *Proc. Yorks. geol. Soc.*, **40**, 459–490.

Houtz, R. E. and Phillips, K. A. (1963). Interim report on the economic geology of Fiji. *Geol. Surv. Fiji Econ. Rep.*, No. 1, 36 pp.

Hsü, K. J. (1967). Mesozoic geology of the Californian Coast Ranges: a new working hypothesis. In J. P. Schaer (Ed.), *Etages tectoniques*, Baconnière, Neuchâtel, Suisse, 216–296.

Hsü, K. J. (1968). Principles of mélange and their bearing on the Franciscan–Knoxville paradox. *Bull. geol. Soc. Amer.*, **79**, 1063–1074.

Hsü, K. J. (1970). The meaning of the word flysch—a short historical search. *Geol. Ass. Can. Sp. Pap.*, **7**, 1–11.

Hsü, K. J. (1971a). Franciscan mélanges as a model for eugeosynclinal sedimentation and underthrusting tectonics. *J. Geophys. Res.*, **76**, 1162–1170.

Hsü, K. J. (1971b). Origin of the Alps and western Mediterranean. *Nature*, **233**, 44–48.

Hsü, K. J. (1972). Alpine flysch in a Mediterranean setting. *24th Int. geol. Congr. Montreal*, Sect. 6, 67–74.

Hsü, K. J. (1974). Mélanges and their distinction from olistostromes, In R. H. Dott Jr and R. H. Shaver (Eds.) *Modern and Ancient Geosynclinal Sedimentation, Soc. Econ. Pal. Mineral. Sp. Publ.*, **19**, 321–333.

Hsü, K. J. and Ohrbom, R. (1969). Mélanges of San Francisco peninsula: geologic reinterpretation of type Franciscan. *Bull. Amer. Ass. Petrol. Geol.*, **53**, 1348–1367.

Hsü, K. J. and Schlanger, S. O. (1971). Ultrahelvetic flysch sedimentation and deformation related to plate tectonics. *Bull. geol. Soc. Amer.*, **82**, 1207–1218.

Hsü, K. J., Ryan, W. B. F. and Cita, M. B. (1973). Late Miocene desiccation of the Mediterranean. *Nature*, **242**, 240–244.

Hubbard, N. J. (1969). A chemical comparison of some oceanic ridge Hawaiian tholeiitic and Hawaiian alkali basalts. *Earth Plan. Sci. Lett.*, **5**, 346–352.

Hubregtse, J. J. M. W. (1976). Volcanism in the western Superior Province in Manitoba. In B. F. Windley (Ed.), *The Early History of the Earth*, Wiley, London, 279–287.

Hunter, D. R. (1970a). The ancient gneiss complex in Swaziland. *Trans. geol. Soc. S. Afr.*, **73**, 107–150.

Hunter, D. R. (1970b). The geology of the Usushwana complex in Swaziland. *Geol. Soc. S. Afr. Sp. Publ.*, **1**, 645–660.

Hunter, D. R. (1973). The localization of tin mineralisation with reference to southern Africa, *Minerals Sci. Engng.*, **5**, 53–77.

Hunter, D. R. (1974a). Crustal development in the Kaapvaal Craton. 1. The Archaean. *Precambrian Res.*, **1**, 259–294.

Hunter, D. R. (1974b). Crustal development in the Kaapvaal Craton. 2. The Proterozoic. *Precambrian Res.*, **1**, 295–326.

Hunziker, J. C. (1970). Polymetamorphism in the Monte Rosa, Western Alps. *Ecl. geol. Helv.*, **63**, 151–161.

Hurley, P. M. (1970). Distribution of age provinces in Laurasia. *Earth Plan. Sci. Lett.*, **8**, 189–196.

Hurley, P. M. (1972). Can the subduction process of mountain building be extended to Pan-African and similar orogenic belts? *Earth Plan. Sci. Lett.*, **15**, 305–314.

Hurley, P. M. (1973). On the origin of 450 ± 200 m.y. orogenic belts. In D. H. Tarling and S. K. Runcorn (Eds.), *Implications of Continental Drift to the Earth Sciences*, Vol. 2, Academic Press, London, 1083–1089.

Hurley, P. M. and Rand, J. R. (1969). Pre-drift continental nuclei. *Science*, **164**, 1229–1242.

Hurley, P. M., Pinson, J. W. H., Nagy, B. and Teska, T. M. (1972). Ancient age of the Middle Marker horizon, Onverwacht Group, Swaziland sequence, South Africa. *Earth Plan. Sci. Lett.*, **14**, 360–366.

Hurst, R. W., Bridgwater, D., Collerson, K. D. and Wetherill, G. W. (1975). 3600 m.y. Rb–Sr ages from the very early Archaean gneisses from Saglek Bay, Labrador. *Earth Plan. Sci. Lett.*, **27**, 393–403.

Hutchinson, R. W. (1965). Genesis of Canadian massive sulphides reconsidered by comparison to Cyprus deposits. *Bull. Can. Min. Metall.*, **58**, 972–986.

Hutchinson, R. W. (1973). Volcanic sulphide deposits and their metallogenic significance. *Econ. Geol.*, **68**, 1223–1246.

Hutchinson, R. W. and Engels, G. G. (1970). Tectonic significance of regional geology and evaporite lithofacies in northeastern Ethiopia. *Phil. Trans. R. Soc. Lond.*, **A267**, 313–329.

Hutchinson, R. W., Ridler, R. H. and Suffel, G. G. (1971). Metallogenic relationships in the Abitibi belt, Canada. A model for Archaean metallogeny. *Trans. Can. Inst. Min.*, **74**, 106–115.

Hutchinson, R. W. and Engels, G. G. (1972). Tectonic evolution in the southern Red Sea and its possible significance to older rifted continental margins. *Bull. geol. Soc. Amer.*, **83**, 2989–3002.

Illies, J. H. (1970). Graben tectonics as related to crust-mantle interaction. In J. H. Illies and S. Mueller (Eds.), *Graben Problems*, Schweizerbartsche Verlagsbuchhandlung, Stuttgart, 4–27.

Industria e Comercio de Mineiros S. A. (1966). Summary of the geology of the manganese deposits of the Serra do Navio. Chapter 2 in *Guidebook*, Int. Field Inst. Brazil.

Irvine, T. N. (1970). Crystallisation sequences in the Muskox intrusion and other layered intrusions. I. Olivine–pyroxene–plagioclase relations. In D. J. L. Visser and G. von Gruenewaldt (Eds.), *Symposium on The Bushveld Igneous Complex and other layered Intrusions*, Geol. Soc. S. Afr. Sp. Publ., **1**, 441–476.

Irvine, T. N. and Smith, C. H. (1967). The ultramafic rocks of the Muskox intrusion, Northwest Territories, Canada. In P. J. Wyllie (Ed.), *Ultramafic and related Rocks*, Wiley, New York, 38–49.

Irvine, T. N. and Smith, C. H. (1969). Primary oxide minerals in the layered series of the Muskox Intrusion. In H. D. B. Wilson (Ed.), *Magmatic Ore Deposits*, Econ. Geol. Monog., **4**, 76–94.

Irving, E. (1964). *Palaeomagnetism and its application to geological and geophysical problems*, Wiley, New York.

Irving, E. and Brown, D. A. (1966). Reply to Stehli's discussion of labyrinthodont abundance and diversity. *Amer. J. Sci.*, **264**, 488–496.

Irving, E. and Park, J. K. (1972). Hairpins and super-intervals. *Can. J. Earth Sci.*, **9**, 1318–1324.

Irving, E., Park, J. K. and Roy, J. L. (1972). Palaeomagnetism and the origin of the Grenville front. *Nature*, **236**, 344–346.

Irving, E., North, F. K. and Couillard, R. (1974a). Oil, climate and tectonics. *Can. J. Earth Sci.*, **11**, 1–17.

Irving, E., Emslie, R. F. and Ueno, H. (1974b). Upper Proterozoic palaeomagnetic poles from Laurentia and the history of the Grenville structural province. *J. Geophys. Res.*, **79**, 5491–5502.

Irving, E. and Lapointe, P. L. (1975). Palaeomagnetism of Precambrian rocks of Laurentia. *Geoscience, Canada*, **2**, 90–98.

Irving, E. and McGlynn, J. C. (1976). Proterozoic magnetostratigraphy and the tectonic evolution of Laurentia. *Phil. Trans. R. Soc. Lond.*, **A280**, 433–468.

Isachsen, Y. W. (1969). Origin of anorthosite and related rocks—a summarization. In *Origin of Anorthosite and related Rocks, Mem. New York State Mus. Sci. Serv.*, **18**, 435–445.

Isacks, B., Oliver, J. and Sykes, L. R. (1968). Seismology and the new global plate tectonics. *J. Geophys. Res.*, **73**, 5855–5899.

Isacks, B. and Molnar, P. (1971). Distribution of stresses in the descending lithosphere from a global survey of focal-mechanism solutions of mantle earthquakes. *Rev. Geophys. Space Phys.*, **9**, 103–174.

Jackson, E. D. (1969). Chemical variation in coexisting chromite and olivine in chromitite zones of the Stillwater Complex. In H. D. B. Wilson (Ed.), *Magmatic Ore Deposits, Econ. Geol. Monog.*, **4**, 41–70.

Jackson, S. A. and Beales, F. W. (1967). An aspect of sedimentary basin evolution: the concentration of Mississippi Valley-type ores during late stages of diagenesis. *Bull. Con. Petrol. Geol.*, **15**, 383–433.

Jacobs, J. A., Russell, R. D. and Wilson, J. T. (1974). *Physics and Geology*, McGraw-Hill, New York, 621 pp.

Jacobsen, J. B. E. (1975). Copper deposits in time and space. *Minerals Sci. Engng.*, **7**, 337–371.

Jacobsen, R. R. E., MacLeod, W. N. and Black, R. (1958). Ring-complexes in the Younger Granite Province of northern Nigeria. *Geol. Soc. Lond. Mem.*, **1**, 71 pp.

Jahn, B. M. and Shih, C. Y. (1974). On the age of the Onverwacht Group, Swaziland Sequence, South Africa. *Geochim. Cosmochim. Acta*, **38**, 873–885.

Jahn, B. M., Shih, C. Y. and Murthy, V. R. (1974). Trace element geochemistry of Archaean volcanic rocks. *Geochim. Cosmochim. Acta*, **38**, 611–627.

Jahn, B. M., Chen. P. Y. and Yen, T. P. (1976). Rb–Sr ages of granitic rocks in southeastern China and their tectonic significance. *Bull. geol. Soc. Amer.*, **86**, 763–776.

Jahn, B. M. and Nyquist, L. E. (1976). Crustal evolution in the early Earth–Moon system: constraints from Rb–Sr studies. In B. F. Windley (Ed.), *The Early History of the Earth*, Wiley, London, 55–76.

Jakes, P. (1973). Geochemistry of continental growth. In D. H. Tarling and S. K. Runcorn (Eds.), *Implications of Continental Drift to the Earth Sciences*, Academic Press, London, 999–1010.

Jakes, P. and White, A. J. R. (1969). Structure of the Melanesian arcs and correlation with distribution of magma types. *Tectonophys.*, **8**, 223–236.

Jakes, P. and Gill, J. (1970). Rare earth elements and the island arc tholeiitic series. *Earth Plan. Sci. Lett.*, **9**, 17–28.

Jakes, P. and White, A. J. R. (1972). Major and trace element abundances in volcanic rocks of orogenic areas. *Bull. geol. Soc. Amer.*, **83**, 29–40.

James, D. E. (1971). Plate tectonic model for the evolution of the central Andes. *Bull. geol. Soc. Amer.*, **82**, 3325–3346.

James, H. L. (1954). Sedimentary facies of iron formation. *Econ. Geol.*, **49**, 235–293.

James, H. L. (1966). Chemistry of the iron-rich sedimentary rocks. *U.S. geol. Surv. Prof. Pap.*, **440**, Chap. W.

James, H. L. (1969). Comparison between Red Sea deposits and older ironstone and iron formation. In E. T. Degens and D. A. Ross (Eds.), *Hot Brines and Recent Heavy Metal Deposits in the Red Sea*, Springer, New York, 525–532.

James, H. L. (1972). Subdivision of Precambrian: an interim scheme to be used by U.S. Geological Survey. *Bull. Amer. Ass. Petrol. Geol¹*, **56**, 1026–1030.

James, R. J. (1972). Evolution of the Cordilleran fold belt. *Bull. geol. Soc. Amer.*, **83**, 1989–2004.

Jamieson, B. G. and Clarke, D. B. (1970). Potassium and associated elements in tholeiitic basalts. *J. Petrol.*, **11**, 183–204.

Janardhanan, A. S. and Leake, B. E. (1975). The origin of the meta-anorthositic gabbros and garnetiferous granulites of the Sittampundi complex, Madras, India. *J. geol. Soc. India*, **16**, 391–408.

Jankovic, S. (1972). The origin of base-metal mineralisation on the mid-Atlantic ridge (based upon the pattern of Iceland). *24th Int. geol. Congr. Montreal Sect.*, **4**, 326–334.

Jansa, L. F. and Wade, J. A. (1975). Geology of the continental margin off Nova Scotia and Newfoundland. In *Offshore Geology of Eastern Canada*, Vol. 2, *Geol. Surv. Can. Pap.*, **74–30**, 51–105.

Jeans, P. J. F. (1973). Plate tectonic reconstruction of the southern Caledonides of Great Britain. *Nature*, **245**, p. 120.

Jenkyns, H. C. (1970a). Submarine volcanism and the Toarcian iron pisolites of western Sicily. *Ecl. Geol. Helv.*, **63**, 549–572.

Jenkyns, H. C. (1970b). Growth and disintegration of a carbonate platform. *Neues Jahrb. Geol. Pal. Monatsh.*, **6**, 325–344.

Jenkyns, H. C. (1970c). The Jurassic of western Sicily. In W. Alvarez and K. H. A. Gohrbandt (Eds.), *Geology and History of Sicily, Petrol. Expl. Soc. Libya*, 245–254.

Jenkyns, H. C. (1971). The genesis of condensed sequences in the Tethyan Jurassic. *Lethaia*, **4**, 327–352.

Jenkyns, H. C. (1974). Origin of red nodular limestones (Ammonitico Rosso, Knollenkalke) in the Mediterranean Jurassic: a diagenetic model. In K. J. Hsü and H. C. Jenkyns (Eds.), *Pelagic Sediments: on land and under the sea, Int. Ass. Sed. Sp. Publ.*, **1**, 249–272.

Johnson, A. E. (1972). Origin of Cyprus pyrite deposits. *24th Int. geol. Congr. Montreal, Sect. 4*, 291–298.

Johnson, G. A. L. (1973). Closing of the Carboniferous sea in western Europe. In D. H. Tarling and S. K. Runcorn (Eds.), *Implications of Continental Drift to the Earth Sciences*, Vol. 2, Academic Press, London, 843–850.

Johnstone, G. S. (1975). The Moine Succession. In A. L. Harris and coworkers (Eds.), *A Correlation of the Precambrian Rocks in the British Isles, Geol. Soc. Lond. Sp. Rep.*, **6**, 30–42.

Joliffe, A. W. (1966). Stratigraphy of the Steeprock group, Steeprock Lake, Ontario. *Precambrian Symp. geol. Ass. Can. Sp. Pap.*, **3**, 75–98.

Jones, D. L. and McElhinny, M. W. (1966). Palaeomagnetic correlation of basic intrusions in the Precambrian of Southern Africa. *J. Geophys. Res.*, **71**, 543–552.

Junge, C., Schidlowski, M., Eichmann, R. and Pietrek, H. (1975). Model calculations for the terrestrial carbon cycle: carbon isotope geochemistry and evolution of photosynthetic oxygen. *J. Geophys. Res.*, **80**, 4542–4552.

Kalliokoski, J. (1965). Geology of north-central Guayana Shield, Venezuela. *Bull. geol. Soc. Amer.*, **76**, 1027–1050.

Kalsbeek, F. (1970). The petrography and origin of gneisses, amphibolites and migmatites in the Qasigialik area, south west Greenland. *Meddr. Grønland*, **189**, 1–70 (also *Grønlands geol. Unders. Bull.*, **83**).

Kalsbeek, F. (1976). Metamorphism of Archaean rocks of West Greenland. In B. F. Windley (Ed.), *The Early History of the Earth*, Wiley, London, 225–235.

Kalsbeek, F. and Leake, B. E. (1970). The chemistry and origin of some basement amphibolites between Ivigtut and Frederikshaab, south west Greenland, *Meddr. Grønland*, **190**, 4, 1–36.

Kanasewich, E. R. (1968). Precambrian rift: genesis of stratabound ore deposits. *Science*, **161**, 1002–1005.

Kanasewich, E. R., Clowes, R. M. and McCloughan, C. M. (1968). A buried Precambrian rift in western Canada. *Tectonoyphys.*, **8**, 513–527.

Karig, D. E. (1970). Ridges and basins of the Tongo–Kermadec island arc system. *J. Geophys. Res.*, **75**, 239–254.

Karig, D. E. (1971). Origin and development of marginal basins in the western Pacific. *J. Geophys. Res.*, **76**, 2542–2561.

Karig, D. E. (1972). Remnant arcs. *Bull. geol. Soc. Amer.*, **83**, 1057–1068.

356

Karig, D. E. (1974). Evolution of arc systems in the western Pacific. *Ann. Rev. Earth Planet. Sci.*, **2**, 51–75.

Karig, D. E. and Glassley, W. E. (1970). Dacite and related sediment from the West Mariana ridge, Philippine Sea. *Bull. geol. Soc. Amer.*, **81**, 2143–2146.

Karig, D. E. and Moore, G. F. (1975). Tectonically controlled sedimentation in marginal basins. *Earth Plan. Sci. Lett.*, **26**, 233–238.

Kaseno, Y. (1972). On the origin of the Japan Sea basin. *24th Int. geol. Congr. Montreal, Sect. 8*, 37–42.

Katada, M. (1965). Petrography of Ryoke metamorphic rocks in northern Kiso district, central Japan. *J. Jap. Ass. Mineral Petrol. Econ. Geol.*, **53**, 77–90, 155–164, 187–204.

Kauffmann, E. G. (1972). Evolutionary rates and patterns of North American Cretaceous Mollusca. *24th Int. geol. Congr. Montreal, Sect. 7*, 174–189.

Kawano, Y., Yagi, K. and Aoki, K. (1961). Petrography and petrochemistry of the volcanics of Quaternary volcanoes of notheastern Japan. *Tohoku Univ. Sci. Rep. 3rd Ser.*, **7**, 1–46.

Kay, M. (1951). North American geosynclines. *Geol. Soc. Amer. Mem.*, **48**, 143 pp.

Kay, M. (1972). Dunnage mélange and Lower Palaeozoic deformation in northeastern Newfoundland. *24th Int. geol. Congr. Montreal, Sect. 3*, 122–133.

Kearey, P. (1976). A regional structural model of the Labrador Trough, northern Quebec from gravity studies, and its relevance to continent collision in the Precambrian. *Earth Plan. Sci. Lett.*, **28**, 371–378.

Keating, B., Helsley, C. E. and Pessagno, E. A., Jr (1975). Late Cretaceous reversal sequence. *Geology*, **3**, 73–76.

Kelly, V. C. (1968). Geology of the alkaline Precambrian rocks at Pajarito Mountain, Otero County, New Mexico. *Bull. geol. Soc. Amer.*, **79**, 1565–1572.

Kennedy, M. J. (1975). Repetitive orogeny in the northeastern Appalachians—new plate models based upon Newfoundland examples. *Tectonophys.*, **28**, 39–87.

Kennedy, W. Q. (1964). The structural differentiation of Africa in the Pan-African (±500 m.y.) tectonic episode. *Res. Inst. African Geol. Univ. Leeds 8th Ann. Rep.*, p. 48.

Kennedy, W. Q. (1965). The influence of basement structure on the evolution of the coastal (Mesozoic and Tertiary) basins of Africa. In D. C. Ion (Ed.), *Salt Basins around Africa, Lond. Inst. Petroleum*, 7–16.

Kent, P. E. (1973). The continental margin of Tanzania. In D. H. Tarling and S. K. Runcorn (Eds.), *Implications of Continental Drift to the Earth Sciences*, Academic Press, London, 949–952.

Key, R. M., Litherland, M. and Hepworth, J. V. (1976). The evolution of the Archaean crust of northeast Botswana. *Precambrian Res.*, **3**, 375–413.

Khan, M. A. (1975). The Afro-Arabian rift system. *Sci. Prog.*, **62**, 207–236.

Khan, M. A. and Mansfield, J. (1971). Gravity measurements in the Gregory rift. *Nature Phys. Sci.*, **229**, 72–75.

Kidd, R. B. (1976). The Messinian stage in Sicily. *Geotimes*, **Feb.**, 20–22.

Kimberley, M. M. and Dimroth, E. (1976). Basic similarity of Archaean to subsequent atmospheric and hydrospheric compositions as evidence in the distributions of sedimentary carbon, sulphur, uranium and iron. In B. F. Windley (Ed.), *The Early History of the Earth*, Wiley, London, 579–585.

Kimura, T. (1973). The old 'inner arc' and its deformation in Japan. In P. C. Coleman (Ed.), *The Western Pacific: island arcs, marginal seas, geochemistry*, Univ. W. Australia Press, Nedlands, 255–273.

King, B. C. (1970). Volcanicity and rift tectonics in East Africa. In T. N. Clifford and I. G. Gass (Eds.), *African Magmatism and Tectonics*, Oliver and Boyd, Edinburgh, 263–283.

King, B. C., Le Bas, M. J. and Sutherland, D. S. (1972). The history of the alkaline volcanoes and intrusive complexes of eastern Uganda and western Kenya. *J. Geol. Soc. Lond.*, **128**, 173–205.

King, P. B. (1966). The North American Cordillera. In *Tectonic History and Mineral Deposits of the Western Cordillera, Can. Inst. Min. Metall. Sp. Pap.*, **8**, 1–25.

King, P. B. (1970). The Precambrian of the United States of America: southeastern United States. In K. Rankama (Ed.), *The Precambrian*, Vol. 4, Wiley, New York, 1–71.

King, R. J. (1973). *The Mineralogy of Leicestershire*, Unpubl. PhD thesis, Univ. Leicester.

Kinsman, D. J. J. (1975a). Salt floors to geosynclines. *Nature*, **255**, 375–378.

Kinsman, D. J. J. (1975b). Rift valley basins and sedimentary history of trailing continental margins. In A. G. Fisher and S. Judson (Eds.), *Petroleum and Global Tectonics*, Princeton Univ. Press, 83–126.

Kirwan, J. L. (1966). The Sudbury irruptive and its history. *Can. Min. J.*, **87**, 7, 54–58.

Kistler, R. W. (1974). Phanerozoic batholiths in western North America. *Ann. Rev. Earth Plan. Sci.*, **2**, 403–418.

Klein, G. de V. (1969). Deposition of Triassic sedimentary rocks in separate basins, eastern North America. *Bull. geol. Soc. Amer.*, **80**, 1825–1832.

Klootwijk, C. T. and Van der Berg, J. (1975). The rotation of Italy: preliminary palaeomagnetic data from the Umbrian sequence, northern Apennines. *Earth Plan. Sci. Lett.*, **25**, 263–273.

Knole, A. H. and Barghoorn, E. S. (1975). Precambrian eukaryotic organisms: a reassessment of the evidence. *Science*, **190**, 52–54.

Kranck, E. H. (1969). Anorthosites and rapakivi, magmas from the lower crust. In *Origin of Anorthosite and related Rocks, Mem. New York State Mus. Sci. Serv.*, **18**, 93–97.

Kratz, K. O., Gerling, E. K. and Lobach-Zhuchenko, S. B. (1968). The isotope geology of the Precambrian of the Baltic Shield. *Can. J. Earth Sci.*, **5**, 657–660.

Krebs, W. and Wachendorf, H. (1973). Proterozoic–Palaeozoic geosynclinal and orogenic evolution of Central Europe. *Bull. geol. Soc. Amer.*, **84**, 2611–2630.

Krishnan, M. S. (1956). *Geology of India and Burma*, Higginbothams, Madras, 555 pp.

Krishnan, M. S. (1973). Occurrence and origin of the iron ores of India. In *Genesis of Precambrian Iron and Manganese Deposits*, Proc. Kiev. Symp. 1970 Unesco, Paris, 69–76.

Krogh, T. E. and Davis, G. L. (1972). Zircon U–Pb ages of Archaean metavolcanic rocks in the Canadian Shield. *Carnegie Inst. Washington Yearbook*, **70**, 241–242.

Kroner, A. (1976). Proterozoic crustal evolution in parts of southern Africa and evidence for extensive sialic crust since the end of the Archaean. *Phil. Trans. R. Soc. Lond.*, **A280**, 541–554.

Krynine, P. D. (1960). Primeval Ocean. *Bull. geol. Soc. Amer.*, **71**, p. 1911 (Abstr.).

Kulm, L. D. and Fowler, G. A. (1974). Oregon continental margin structure and stratigraphy: a test of the imbricate thrust model. In C. A. Burk and C. L. Drake (Eds.), *The Geology of Continental Margins*, Springer-Verlag, Berlin, 261–283.

Kumarapeli, P. S. and Saull, V. A. (1966). The St Lawrence Valley system: a North American equivalent of the East African rift valley system. *Can. J. Earth Sci.*, **3**, 639–658.

Kuno, H. (1960). High alumina basalt. *J. Petrol.*, **1**, 121–145.

Kuno, H. (1966). Lateral variation of basaltic magma across continental margins and island arcs. In W. H. Poole (Ed.), *Continental Margins and Island Arcs, Geol. Surv. Can. Pap.*, **66–15**, 317–336.

Kurtén, B. (1969). Continental drift and evolution. *Sci. Amer.*, **220**, 54–64.

Kushiro, I. (1968). Composition of magmas formed by partial zone melting of the Earth's upper mantle. *J. Geophys. Res.*, **73**, 619–634.

Kutina, J. (1972). Regularities in the distribution of hypogene mineralisation along rift structures. *24th Int. geol. Congr. Montreal, Sect. 4*, 65–73.

Kuznetsov, S. I., Ivanov, M. V. and Lyalikova, N. N. (1963). *Introduction to Geological Micro-biology*, McGraw-Hill, New York.

Kvenvolden, K. A. (1974). Natural evidence for chemical and early biological evolution. *Origins of Life*, **5**, 71–86.

Kvenvolden, K. A. and Hodgson, G. W. (1969). Evidence for prophyries in early Precambrian Swaziland System sediments. *Geochim. Cosmochim. Acta*, **33**, 1195–1202.

Ladd, W. L., Dickson, G. O. and Pitman, W. C., III (1973). The age of the south Atlantic. In A. E. M. Nairn and F. G. Stehli

(Eds.), *The Ocean Basins and Margins*, Vol. 1: *The South Atlantic*, Plenum Press, New York, 555–573.

Lambert, I. B. (1971). The composition and evolution of the deep continental crust. In J. E. Glover (Ed.), *Symp. on Archaean Rocks, Geol. Soc. Aust. Spec. Publ.* **3**, 419–428.

Lambert, I. B. and Wyllie, P. J. (1968). Stability of hornblende and a model for the low velocity zone. *Nature*, **219**, 1240–1241.

Lambert, I. B. and Saito, T. (1974). The Kuroko and associated ore deposits of Japan: a review of their features and metallogenesis. *Econ. Geol.* **69**, 1215–1256.

Lambert, R. St J. (1976). Archaean thermal regimes, crustal and upper mantle temperatures, and a progressive evolutionary model for the Earth. In B. F. Windley (Ed.), *The Early History of the Earth*, Wiley, London, 363–387.

Lambert, R. St J., Chamberlain, V. E. and Holland, J. G. (1976). The geochemistry of Archaean rocks. In B. F. Windley (Ed.), *The Early History of the Earth*, Wiley, London, 377–387.

Lambert, R. St J. and Holland, J. G. (1976). Amîtsoq gneiss geochemistry: preliminary observations. In B. F. Windley (Ed.), *The Early History of the Earth*, Wiley, London 191–201.

Landis, C. A. and Coombs (1967). Metamorphic belts and orogenesis in southern New Zealand. *Tectonophys.*, **4**, 501–518.

Landis, C. A. and Bishop, D. G. (1972). Plate tectonics and regional stratigraphic–metamorphic relations in the southern part of the New Zealand geosyncline. *Bull. geol. Soc. Amer.*, **83**, 2267–2284.

Larson, R. L. and Chase, C. G. (1972). Late Mesozoic evolution of the western Pacific ocean. *Bull. geol. Soc. Amer.*, **83**, 3627–3644.

Larson, R. L. and Pitman, W. C., III (1972). World-wide correlation of Mesozoic magnetic anomalies, and its implications. *Bull. geol. Soc. Amer.*, **83**, 3645–3662.

Larson, R. L. and Ladd, J. W. (1973). Evidence for the opening of the South Atlantic in the early Cretaceous. *Nature*, **246**, p. 209.

Laubscher, H. P. (1971). Das Alpen–Dinariden Problem und die Palinspastik der südlichen Tethys. *Geol. Rdsch.*, **60**, 813–833.

Laubscher, H. P. (1973). Jura Mountains. In K. A. de Jong and R. Scholten (Eds.), *Gravity and Tectonics*, Wiley, New York, 217–228.

Laurén, L. (1970). An interpretation of the negative gravity anomalies associated with the rapakivi granites and Jotnian sandstone in southern Sweden. *Geol. Foren. Stock. Forh.*, **92**, 1, 21–34.

Laurent, R. (1972). The Hercynides of South Europe: a model. *24th Int. geol. Congr. Montreal*, **3**, 363–370.

Laznicka, P. (1973). Development of nonferrous metal deposits in geological time. *Can. J. Earth Sci.*, **10**, 18–25.

Leake, B. E. (1969). Discrimination of ortho- and para-charnockitic rocks, anorthosites and amphibolites. *Ind. Mineral.*, **10**, 79–104.

Leake, B. E. (1970). The fragmentation of the Connemara basic and ultrabasic intrusions. In G. Newall and N. Rast (Eds.), *Mechanism of Igneous Intrusion*, Seal House Press, Liverpool, 103–122.

Le Bas, M. J. (1971). Peralkaline vulcanism, crustal swelling and rifting. *Nature Phys. Sci.*, **230**, p. 85.

Leblanc, M. (1976). Proterozoic oceanic crust at Bon Azzer. *Nature*, **261**, 34–35.

Leelanandam, C. (1967). Occurrence of anorthosites from the charnockitic area of Kondapalli, Andhra Pradesh. *Bull. geol. Soc. India*, **4**, 5–7.

Lefèvre, C. (1973). Les caractères magmatiques du volcanisme plio-quaternaire des Andes dans le Sud du Pérou. *Contr. Mineral. Petrol.*, **41**, 259–272.

Legg, C. (1972). The tin belt of the southern Province. *Geol. Surv. Zambia Econ. Rep.*, **29**, 58 pp.

Leggo, P. J., Aftalion, M. and Pidgeon, R. T. (1971). Discordant zircon U/Pb ages from the Uganda basement. *Nature Phys. Sci.*, **231**, 81–84.

Lehmann, E. (1952). Beitrag zur Beurteilung der paläozoischen Eruptivgesteine West deutschlands. *Zeit. Deut. Geol. Ges.*, **104**, p. 219.

Le Pichon, X. (1968). Sea-floor spreading and continental drift. *J. Geophys. Res.*, **73**, 3661–3697.

Lepp, H. and Goldich, S. S. (1964). Origin of Precambrian iron formations. *Econ. Geol.*, **59**, 1025–1060.

Lipman, P. W., Prostka, H. J. and Christiansen, R. L. (1971). Evolving subduction zones in the western United States, as interpreted from igneous rocks. *Science*, **174**, 821–825.

Lipman, P. W., Prostka, H. J. and Christiansen, R. L. (1972). Cenozoic volcanism and plate tectonic evolution of the western United States. I. Early and Middle Cenozoic. *Phil. Trans. R. Soc. Lond.*, **A271**, 217–248.

Lister, G. F. (1966). The composition and origin of selected iron-titanium deposits. *Econ. Geol.*, **61**, 275–310.

Litherland, M. (1973). Uniformitarian approach to Archaean 'schist relics'. *Nature Phys. Sci.*, **242**, 395–398.

Livingstone, D. E. and Damon, P. E. (1968). The age of stratified Precambrian rock sequences in central Arizona and northern Sonora. *Can. J. Earth Sci.*, **5**, 763–772.

Lloyd, F. E. and Bailey, D. K. (1975). Light element metasomatism of the continental mantle: the evidence and the consequences. In L. H. Ahrens and coworkers (Eds.), *Physics and Chemistry of the Earth*, Vol. 9, Pergamon, Oxford, 389–416.

Long, R. E., Sundaralingam, K. and Maguire, P. K. H. (1973). Crustal structure of the East African rift zone. *Tectonophys.*, **20**, 269–281.

Lowell, J. D. and Guilbert, J. (1970). Lateral and vertical alteration mineralisation zoning in porphyry ore deposits. *Econ. Geol.*, **65**, 373–408.

Lowrie, W. and Alvarez, W. (1975). Palaeomagnetic evidence for rotation of the Italian Peninsula. *J. Geophys. Res.*, **80**, 1579–1592.

Ludwig, W. J., Nafe, J. E. and Drake, C. L. (1970). Seismic refraction. In A. E. Maxwell (Ed.), *The Sea*, Vol. 4, pt. 1, Wiley, 53–84.

Lundqvist, T. (1968). Precambrian geology of the Los-Harme region, southern Sweden. *Sveriges geol. Unders.*, **23**, 235 pp.

Luyendyk, B. P. (1970). Dips of downgoing lithospheric plates beneath island arcs. *Bull. geol. Soc. Amer.*, **81**, 3411–3416.

Luyendyk, B. P., Forsyth, D. and Phillips, J. D. (1972). Experimental approach to the palaeocirculation of the oceanic surface waters. *Bull. geol. Soc. Amer.*, **83**, 2649–2664.

McAlester, A. L. (1970). Animal extinctions, oxygen consumption and atmospheric history. *J. Pal.*, **44**, 405–409.

McBirney, A. R., Aoki, K. I. and Bass, M. N. (1967). Eclogites and jadeite from the Motague fault zone, Guatemala. *Amer. Mineral.*, **52**, 908–918.

McBirney, A. R. and Gass, I. G. (1967). Relations of oceanic volcanic rocks to mid-oceanic rises and heat flow. *Earth Plan. Sci. Lett.*, **2**, 265–276.

McBirney, A. R. and Williams, H. (1969). Geology and petrology of the Galapagos Islands. *Geol. Soc. Amer. Mem.*, **118**, 197 pp.

McCall, G. J. H. (1971). Some ultrabasic and basic igneous rock occurrences in the Archaean of western Australia. *Geol. Soc. Australia Sp. Publ.*, **3**, 429–442.

McCall, G. J. H. and Leishman, J. (1971). Clues to the origin of Archaean eugeosynclinal peridotites and the nature of serpentinization. *Geol. Soc. Australia Sp. Publ.*, **3**, 281–299.

McCall, G. J. H. and Peers, R. (1971). Geology of the Binneringie dyke, western Australia. *Geol. Rund.*, **60**, 1174–1263.

McClay, K. R. and Campbell, I. H. (1976). The structure and shape of the Jimberlana intrusion, western Australia, as indicated by an investigation of the Bronzite complex. *Geol. Mag.*, **113**, 129–139.

McConnell, R. B. (1972). Geological development of the Rift System of eastern Africa. *Bull. geol. Soc. Amer.*, **83**, 2549–2572.

McElhinny, M. W. (1970). Formation of the Indian Ocean. *Nature*, **228**, 977–979.

McElhinny, M. W. (1971). Geomagnetic reversals during the Phanerozoic. *Science*, **172**, 157–159.

McElhinny, M. W. (1973). *Palaeomagnetism and plate tectonics*, Cambridge Univ. Press, 358 pp.

McElhinny, M. W. and Luck, G. R. (1970). Palaeomagnetism and Gondwanaland. *Science*, **168**, 830–832.

McElhinny, M. W. and Briden, J. C. (1971). Continental drift during the Palaeozoic. *Earth Plan. Sci. Lett.*, **10**, 407–416.

McElhinny, M. W. and Burek, P. J. (1971). Mesozoic palaeomagnetic stratigraphy, *Nature*, **232**, 98–102.

McElhinny, M. W., Giddings, J. W. and Embleton, B. J. J. (1974). Palaeomagnetic results and late Precambrian glaciations. *Nature*, **248**, 557–561.

McElhinny, M. W. and Embleton, B. J. J. (1976). Precambrian and early Palaeozoic palaeomagnetism in Australia. *Phil. Trans. R. Soc. Lond.*, **A280**, 417–432.

McGlynn, J. C. and Henderson, J. B. (1970). Archaean volcanism and sedimentation in the Slave structural province. *Geol. Surv. Can. Pap.*, **70–40**, 31–44.

McGlynn, J. C. and Henderson, J. B. (1972). The Slave Province. *Geol. Ass. Can. Sp. Pap.*, **11**, 505–526.

McGlynn, J. C. and Irving, E. (1975). Paleomagnetism of early Aphebian diabase dykes from the Slave structural province, Canada. *Tectonophys.*, **26**, 23–38.

McGlynn, J. C., Irving, E., Bell, K. and Pullaiah, G. (1975). Palaeomagnetic poles and a Proterozoic supercontinent. *Nature*, **255**, 318–319.

MacGregor, A. M. (1951). Some milestones in the Precambrian of southern Africa. *Proc. geol. Soc. S. Afr.*, **54**, 27–71.

McGregor, V. R. (1973). The early Precambrian gneisses of the Godthaab district, West Greenland. *Phil. Trans. R. Soc. Lond.*, **A273**, 343–358.

MacIntyre, R. (1971). Apparent periodicity of carbonatite emplacement in Canada. *Nature*, **230**, 79–81.

McIver, J. R. and Lenthall, D. H. (1973). Mafic and ultramafic extrusions of the Onverwacht Group in terms of the system XO–YO–R$_2$O$_3$–ZO$_2$. *Econ. Geol. Res. Unit. Univ. Witwatersrand, Johannesburg, Inf. Circ.*, **80**, 8 pp.

McKee, E. H. and Noble, D. C. (1974). Rb–Sr age of the Cardenas lavas, Grand Canyon, Arizona. US geol. Surv. Guidebook *Geology of N. Arizona*, 87–96.

McKenzie, D. P. (1970). Plate tectonics of the Mediterranean region. *Nature* **226**, 239–243.

McKenzie, D. P. and Morgan, W. J. (1969). Evolution of triple junctions. *Nature*, **224**, 125–133.

McKenzie, D. P. and Sclater, J. G. (1971). The evolution of the Indian Ocean since the late Cretaceous. *Geophys. J. R. Astron. Soc.*, **24**, 437–528.

McKenzie, K. G. and Hussainy, S. U. (1968). Relevance of a freshwater Cytherid (Crustacea, Osteracoda) to the continental drift hypothesis. *Nature*, **220**, 806–808.

McKerrow, W. S. and Ziegler, A. M. (1972a). Palaeozoic oceans. *Nature Phys. Sci.*, **240**, 92–94.

McKerrow, W. S. and Ziegler, A. M. (1972b). Silurian paleogeographic development of the Proto-Atlantic ocean. *24th Int. geol. Congr. Montreal, Sect. 6*, 4–10.

McLeod, W. N. (1966). The geology and iron deposits of the Hamersley Range area, western Australia. *Bull. geol. Surv. W. Australia*, **117**.

Malpas, J. (1974). A restored section of oceanic crust and mantle in western Newfoundland. *Geol. Soc. Amer. Abstr. with Prog.*, **5**, 2, 191–192.

Maltman, A. J. (1975). Ultramafic rocks in Anglesey—their non-tectonic emplacement. *J. geol. Soc. Lond.*, **131**, 593–606.

Marchetti, M. P. (1957). The occurrence of slide and flowage materials (olistostromes) in the Tertiary series of Italy. *20th Int. geol. Congr. Mexico City, Sect. 5*, **1**, 209–225.

Margulis, L. (1974). The classification and evolution of prokaryotes and eukaryotes. In R. C. King (Ed.), *Handbook of Genetics*, Vol. 1, Plenum Press, New York, 1–41.

Markhinin, E. K. (1968). Volcanism as an agent of formation of the earth's crust. In L. Knopoff, Drake, C. L. and Hart, P. J. (Eds.), *The Crust and Upper Mantle of the Pacific Area, Amer. Geophys. Un. Mon.*, **12**, 412–423.

Marlow, M. S. and Scholl, D. W. (1973). Comparative geologic histories of the Aleutian and Lesser Antilles island arcs. *Geol. Soc. Amer. Abstr. with Prog.*, **5**, 77.

Marsh, J. S. (1973). Relationships between transform directions and alkaline rock lineaments in Africa and South America. *Earth Plan. Sci. Lett.*, **18**, 317–323.

Martignole, J. and Schrijver, K. (1970a). Tectonic setting and evolution of the Morin Anorthosite, Grenville Province, Quebec. *Bull. geol. Soc. Finl.* **42**, 165–209.

Martignole, J. and Schrijver, K. (1970b). The level of anorthosites and its tectonic pattern. *Tectonophys.*, **10**, 403–409.

Martin, H. (1969). Problems of age relations and structure in some metamorphic belts of southern Africa. *Geol. Ass. Can. Sp. Pap.*, **5**, 17–26.

Martin, W. C. (1957). Errington and Vermillion Lake mines, In *Structural Geology of Canadian Ore Deposits* (Congress Volume), Vol. 2, Can. Inst. Mining Metall., 363–376.

Mason, B. (1975). Mineralogy and geochemistry of two Amîtsoq gneisses from the Godthaab region, West Greenland. *Rapp. Grønlands geol. Unders.*, **71**.

Mason, R. G. (1958). A magnetic survey off the west coast of the United States between latitudes 32° and 36°N, longitudes 121° and 128°W. *Geophys. J.*, **1**, 320–329.

Mason, R. G. (1973). The Limpopo mobile belt, southern Africa. *Phil. Trans. R. Soc. Lond.*, **A273**, 463–486.

Mason, R. G. and Raff, A. D. (1961). Magnetic survey off the west coast of North America, 32°N latitude to 42°N latitude. *Bull. geol. Soc. Amer.*, **72**, 1259–1266.

Matsuda, T., Nakamura, K. and Sugimura, A. (1967). Late Cenozoic orogeny in Japan. *Tectonophys.*, **4**, 349–366.

Matsuda, T. and Uyeda, S. (1971). On the Pacific-type orogeny and its model—extension of the paired belts concept and possible origin of marginal seas. *Tectonophys.*, **11**, 5–27.

Matsukuma, T. and Horikoshi, E. (1970). Kuroko deposits in Japan, a review. In T. Tatsumi (Ed.), *Volcanism and Ore Genesis*, Univ. Tokyo Press, 153–179.

Mattinson, J. M. (1975). Early Palaeozoic ophiolite complexes of Newfoundland: isotopic ages of zircons. *Geology*, **3**, 181–183.

Maxwell, C. H. (1972). Geology and ore deposits of the Alegria district, Minas Gerais, Brazil. *U.S. geol. Surv. Prof. Pap.*, **341–J**, 72 pp.

Maxwell, J. C. (1959). Turbidites, tectonics and gravity transport, northern Apennine Mountains, Italy. *Bull. Amer. Ass. Petrol. Geol.*, **43**, 2701–2719.

May, P. R. (1971). Pattern of Triassic–Jurassic diabase dikes around the North Atlantic in the context of predrift position of the continents. *Bull. geol. Soc. Amer.*, **82**, 1285–1292.

Melson, W. G. and Thompson, G. (1970). Layered basic complexes in oceanic crust, Romanche Fracture, Equatorial Atlantic Ocean. *Science*, **168**, 817–820.

Menard, H. W. and Dietz, R. S. (1952). Mendocino submarine escarpment. *J. Geol.*, **60**, 266–278.

Mendelsohn, F. (Ed.) (1961). *The Geology of the Northern Rhodesian Copper Belt*, Roan Antelope Copper Mines Ltd and MacDonald Press, London.

Menzies, M. (1976). Rifting of a Tethyan continent—rare earth evidence of an accreting plate margin. *Earth Plan. Sci. Lett.*, **28**, 427–438.

Mercy, E. L. P. (1965). Caledonian igneous activity. In G. Y. Craig (Ed.), *The Geology of Scotland*, Oliver and Boyd, Edinburgh, 230–269.

Meyerhoff, A. A. (1970). Continental drift: implications of palaeomagnetic studies, meteorology, physical oceanography and climatology. *J. Geol.*, **78**, 1–51.

Meyerhoff, A. A. and Teichert, C. (1971). Continental drift. III. Late Palaeozoic glacial centers and Devonian–Eocene coal distribution. *J. Geol.*, **79**, 285–321.

Michot, J. (1972). Anorthosites et recherche pluridisciplinaire. *Ann. Soc. geol. Belg.*, **95**, 5–43.

Middlemost, E. A. K. (1970). Anorthosites; a graduated series. *Earth Sci. Rev.*, **6**, 257–265.

Middlemost, E. A. K. (1973). Evolution of volcanic islands. *Lithos*, **6**, 123–132.

Milligan, G. C. (1960). Geology of the Lynn Lake district. *Manitoba Dept Mines Nat. Res. Pub.*, 57–1, 317 pp.

Milner, A. R. and Panchen, A. L. (1973). Geographic variation in the tetrapod faunas of the Upper Carboniferous and Lower Permian. In D. H. Tarling and S. K. Runcorn (Eds.), *Implications of Continental Drift to the Earth Sciences*, Vol. 1, Academic Press, London, 353–368.

Minear, J. H. and Toksöz, M. N. (1970). Thermal regime of a downgoing slab and new global tectonics. *J. Geophys. Res.*, **75**, 1397–1419.

Misch, P. H. (1966). Tectonic evolution of the northern Cascades of Washington State. In *Tectonic History and Mineral Deposits of the Western Cordillera, Can. Inst. Mining Metall. Sp. Vol.*, **8**, 101–148.

Mitchell, A. H. G. (1970). Facies of an early Miocene volcanic arc, Malekula Island, New Hebrides, *Sediment*, **14**, 201–244.

Mitchell, A. H. G. (1973). Metallogenic belts and angle of dip of Benioff zones. *Nature*, **245**, 49–52.

Mitchell, A. H. G. (1974). South west England granites: magmatism and tin mineralization in a post-collision tectonic setting. *Trans. Inst. Min. Metall. Sect. B*, **83**, 95–97.

Mitchell, A. H. G. (1976a). Evolution and global tectonics. Petrogenic view: discussion. *Bull. geol. Soc. Amer.*, **86**, p. 1487.

Mitchell, A. H. G. (1976b). Tectonic settings for emplacement of subduction-related magmas and associated mineral deposits. In D. F. Strong (Ed.), *Metallogeny and Plate Tectonics, Geol. Ass. Can. Sp. Pap.*, **14**, 3–22.

Mitchell, A. H. G. and Reading, H. G. (1969). Continental margins, geosynclines and ocean floor spreading. *J. Geol.*, **77**, 629–646.

Mitchell, A. H. G. and Bell, J. D. (1970). Volcanic episodes in island arcs. *Geol. Soc. Lond. Proc. No.* 1662, 9–12.

Mitchell, A. H. G. and Reading, H. G. (1971). Evolution of island arcs. *J. Geol.*, **79**, 253–284.

Mitchell, A. H. G. and Garson, M. S. (1972). Relationship of porphyry copper and circum-Pacific tin deposits to palaeo-Benioff zones. *Trans. Inst. Min. Metall.*, **81**, B10–25.

Mitchell, A. H. G. and Bell, J. D. (1973). Island-arc evolution and related mineral deposits. *J. Geol.*, **81**, 381–405.

Mitchell, A. H. G. and McKerrow, W. S. (1975). Analogous evolution of the Burma Orogen and the Scottish Caledonides. *Bull. geol. Soc. Amer.*, **86**, 305–315.

Mitchell, A. H. G. and Garson, M. S. (1976). Mineralization at plate boundaries. *Minerals Sci. Engng.*, **8**, 129–169.

Miyashiro, A. (1958). Regional metamorphism of the Gosaisyo-Takanuki district in the central Abukuma Plateau. *Tokyo Univ. Fac. Sci. J., Sect.* 2, **11**, 219–272.

Miyashiro, A. (1961). Evolution of metamorphic belts. *J. Petrol.*, **2**, 277–311.

Miyashiro, A. (1967). Orogeny, regional metamorphism and magmatism in the Japanese islands. *Meddr. dansk geol. Foren.*, **17**, 390–446.

Miyashiro, A. (1972). Metamorphism and related magmatism in plate tectonics. *Amer. J. Sci.*, **272**, 629–656.

Miyashiro, A. (1973a). *Metamorphism and Metamorphic Belts*, Allen and Unwin, London, 492 pp.

Miyashiro, A. (1973b). The Troodos ophiolite complex was probably formed in an island arc. *Earth Plan. Sci. Lett.*, **19**, 218–224.

Miyashiro, A. (1973c). Paired and unpaired metamorphic belts. *Tectonophys.*, **17**, 241–254.

Miyashiro, A., Shido, F. and Ewing, M. (1970). Petrologic models for the Mid-Atlantic ridge. *Deep Sea Res.*, **17**, 109–123.

Monger, J. W. H. and Hutchinson, W. W. (1970). Metamorphic map of the Canadian Cordillera. *Can. geol. Surv. Pap.*, **70–33**, 61 pp.

Monger, J. W. H., Souther, J. G. and Gabrielse, H. (1972). Evolution of the Canadian Cordillera: a plate tectonic model. *Amer. J. Sci.*, **272**, 577–602.

Moorbath, S. (1969). Evidence for the age of deposition of the Torridonian sediments of north-west Scotland. *Scott. J. Geol.*, **5**, 154–170.

Moorbath, S. (1975a). Evolution of Precambrian crust from strontium isotopic evidence. *Nature*, **254**, 395–398.

Moorbath, S. (1975b). Geological interpretation of whole-rock isochron dates from high grade gneiss terrains. *Nature*, **255**, 391.

Moorbath, S. (1975c). The geological significance of early Precambrian rocks. *Proc. Geol. Ass.*, **86**, 259–279.

Moorbath, S. (1976). Age and isotope constraints for the evolution of Archaean crust. In B. F. Windley (Ed.), *The Early History of the Earth*, Wiley, London, 351–360.

Moorbath, S., Welke, H. and Gale, N. H. (1969). The significance of lead isotope studies in ancient, high-grade metamorphic basement complexes, as exemplified by the Lewisian rocks of north-west Scotland. *Earth Plan. Sci. Lett.*, **6**, 245–256.

Moorbath, S. and Park, R. G. (1971). The Lewisian chronology of

the southern region of the Scottish mainland. *Scott. J. Geol.*, **8**, 51–74.

Moorbath, S., O'Nions, R. K., Pankhurst, R. J., Gale, N. H. and McGregor, V. R. (1972). Further rubidium–strontium age determinations on the very early Precambrian rocks of the Godthaab district, West Greenland. *Nature Phys. Sci.*, **240**, 78–82.

Moorbath, S., O'Nions, R. K. and Pankhurst, R. J. (1973). Early Archaean age for the Isua iron formation, West Greenland. *Nature*, **245**, 138–139.

Moorbath, S., Powell, J. L. and Taylor, P. N. (1975a). Isotopic evidence for the age and origin of the 'grey gneiss' complex of the southern Outer Hebrides, Scotland. *J. geol. Soc. Lond.*, **131**, 213–222.

Moorbath, S., O'Nions, R. K. and Pankhurst, R. J. (1975b). The evolution of early Precambrian crustal rocks at Isua, West Greenland—geochemical and isotopic evidence. *Earth Plan. Sci. Lett.*, **27**, 229–239.

Moorbath, S. and Pankhurst, R. J. (1976). Further rubidium–strontium age and isotope evidence for the nature of the late Archaean plutonic event in West Greenland. *Nature*, **262**, 124–125.

Moore, J. G. (1959). The quartz diorite boundary line in the western United States. *J. Geol.*, **67**, 198–210.

Moore, J. G. and Calk, L. (1971). Sulphide spherules in vesicles of dredged pillow basalt. *Amer. Min.*, **56**, 476–488.

Moore, J. W. and Lanphere, M. A. (1971). The age of porphyry-type copper mineralization in the Bingham mining district, Utah—a refined estimate. *Econ. Geol.*, **66**, 331–334.

Moores, E. M. (1969). Petrology and structure of the Vourinos ophiolite complex of northern Greece. *Geol. Soc. Amer. Sp. Publ.*, **118**, 74 pp.

Moores, E. M. (1973). Geotectonic significance of ultramafic rocks. *Earth Sci. Rev.*, **9**, 241–258.

Moores, E. M. and Vine, F. J. (1971). The Troodos Massif, Cyprus, and other ophiolites as oceanic crust: evaluations and implications. *Phil. Trans. R. Soc. Lond.*, **A268**, 443–466.

Moores, E. M. and MacGregor, I. D. (1973). Types of alpine ultramafic rocks and their implications for fossil plate interactions. *Geol. Soc. Amer. Mem.*, **132**, 209–223.

Moores, E. M. and Jackson, E. D. (1974). Ophiolites and oceanic crust. *Nature*, **250**, 136.

Morgan, W. J. (1968). Rises, trenches, great faults, and crustal blocks. *J. Geophys. Res.*, **73**, 1959–1982.

Morgan, W. J. (1971). Convection plumes in the lower mantle. *Nature*, **230**, 42.

Morley, L. W. and Larochelle, A. (1964). Palaeomagnetism as a means of dating geological events. In F. F. Osborne (Ed.), *Geochronology in Canada, Roy. Soc. Can. Sp. Publ.* 8, Univ. Toronto Press, 39–51.

Morse, S. A. (1969). The Kiglapait layered intrusion, Labrador. *Geol. Soc. Amer. Mem.*, **112**, 146 pp.

Moseley, F. (1969). The Upper Cretaceous ophiolite complex of Masirah island, Oman. *Geol.. J.*, **6**, 293–306.

Moshikin, V. N. and Dagelaiskaja, I. N. (1972). The Precambrian anorthosites of the USSR. *24th Int. geol. Congr. Montreal, Sect.* 2, 329–333.

Mueller, P. A. and Rogers, J. J. W. (1973). Secular chemical variation in a series of Precambrian mafic rocks, Beartooth Mountains, Montana and Wyoming. *Bull. geol. Soc. Amer.*, **84**, 3645–3652.

Mueller, S. (1970). Geophysical aspects of graben formation in continental rift systems. In J. H. Illies and S. Mueller (Eds.), *Graben Problems*, Schweizerbartsche Verlag, Stuttgart, 27–37.

Muir, M. D. and Grant, P. R. (1976). Micropalaeontological evidence from the Onverwacht Group, South Africa. In B. F. Windley (Ed.), *The Early History of the Earth*, Wiley, London, 595–604.

Mutti, E. (1974). Examples of ancient deep-sea fan deposits from circum-Mediterranean geosynclines. In R. H. Dott Jr and R. H. Shaver (Eds.), *Modern and Ancient Geosynclinal Sedimentation*, *Soc. Econ. Pal. Mineral. Sp. Publ.*, **19**, 92–105.

Myers, J. S. (1971a). Zones of abundant Scourie dyke fragments and their significance in the Lewisian complex of western Harris, Outer Hebrides. *Proc. Geol. Ass.*, **82**, 365–378.

360

Myers, J. S. (1971b). The late Laxfordian granite–migmatite complex of western Harris, Outer Hebrides. *Scott. J. Geol.*, **7**, 234–284.

Myers, J. S. (1976). Granitoid sheets, thrusting and Archaean crustal thickening in West Greenland. *Geology*, **4**, 265–268.

Nagy, B. and Nagy, L. A. (1969). Early Precambrian Onverwacht microstructures: possibly the oldest fossils on earth? *Nature*, **223**, 1226–1229.

Nagy, B., Kunen, S. M., Zumberg, J. E., Long, A., Moore, C. B., Lewis, C. F., Anhaeusser, C. R. and Pretorius, D. A. (1974). Carbon content and carbonate ^{13}C abundance in the Early Precambrian Swaziland sediments of South Africa. *Precambrian Res.*, **1**, 43–48.

Nagy, B., Zumberg, J. E. and Nagy, L. A. (1975). Abiotic graphitic microstructures in micaceous metaquartzite about 3760 million years old from southwestern Greenland: implications for early Precambrian microfossils. *Proc. Nat. Acad. Sci. USA*, **72**, 1206–1209.

Naidu, P. R. J. (1963). A layered complex in Sittampundi, Madras State, India. *Min. Soc. Amer. Sp. Pap.*, **1**, 116–123.

Nairn, A. E. M. and Stehli, F. G. (1973). A model for the South Atlantic. In A. E. M. Nairn and F. G. Stehli (Eds.), *The Ocean Basins and Margins*, Vol. 1: The South Atlantic, Plenum Press, New York, 1–24.

Nakazawa, K. and Runnegar, B. (1973). The Permian–Triassic boundary: a crisis for bivalves? *Can. Soc. Petrol. Geol. Mem.*, **2**, 608–621.

Naldrett, A. J. (1970). Ultramafic and related mafic rocks of the Abitibi orogen. *Geol. Surv. Can. Pap.*, **70–40**, 24–29.

Naldrett, A. J. and Gasparrini, E. L. (1971). Archaean nickel sulphide deposits in Canada: their classification, geological setting and genesis with some suggestions as to exploration. *Geol. Soc. Aust. Sp. Publ.*, **3**, 201–226.

Naldrett, A. J. and Mason, G. D. (1968). Contrasting Archaean ultramafic igneous bodies in Dundonald and Clergue townships, Ontario. *Can. J. Earth Sci.*, **5**, 111–143.

Nalivin, D. V. (1960). *The Geology of the USSR* [Trans. S. I. Tomkeieff], Pergamon Press, New York.

Naqvi, S. M. (1967). The banded pyritiferous cherts from Ingaldhal and adjoining areas of Chitaldrug schist belt, Mysore. *Bull. Nat. Geophys. Res. Inst. Hyderabad*, **5**, 173–181.

Naqvi, S. M. (1973). Geological structure and aeromagnetic and gravity anomalies in the central part of the Chitaldrug schist belt, Mysore, India. *Bull. geol. Soc. Amer.*, **84**, 1721–1732.

Naqvi, S. M. (1976). Physico-chemical conditions during the Archaean as indicated by Dharwar geochemistry. In B. F. Windley (Ed.), *The Early History of the Earth*, Wiley, London, 289–298.

Naqvi, S. M. and Hussain, S. M. (1972). Petrochemistry of early Precambrian metasediments from the central part of the Chitaldrug schist belt, Mysore, India. *Chem. Geol.*, **10**, 109–135.

Naqvi, S. M. and Hussain, S. M. (1973a). Geochemistry of Dharwar metavolcanics and composition of the primeval crust of the peninsular India. *Geochim. Cosmochim. Acta*, **37**, 159–164.

Naqvi, S. M. and Hussain, S. M. (1973b). Relation between trace and major element composition of Chitaldrug metabasalts, Mysore, India, and the Archaean mantle. *Chem. Geol.*, **11**, 17–30.

Naqvi, S. M., Rao, V. D. and Narain, H. (1974). The protocontinental growth of the Indian shield and the antiquity of the rift valleys. *Precambrian Res.*, **1**, 345–398.

Nautiyal, S. P. (1965). Precambrian of Mysore Plateau. *53rd Indian Sci. Congr. Calcutta, Sect. Geol. Geograph.*, 1–14.

Neill, W. M. (1973). Possible continental rifting in Brazil and Angola related to the opening of the South Atlantic. *Nature Phys. Sci.*, **245**, 104–107.

Neprochnov, Y. P. (1968). Structure of the earth's crust of epicontinental seas: Caspian, Black and Mediterranean. *Can. J. Earth Sci.*, **5**, 1037–1043.

Nesbitt, R. W. (1971). Skeletal crystal forms in the ultramafic rocks of the Yilgarn block, western Australia; evidence for an Archaean ultramafic liquid. *Geol. Soc. Aust. Sp. Publ.*, **3**, 331–350.

Newell, N. D. (1971). An outline history of tropical organic reefs. *Amer. Mus. Novitates No. 2465*, 1–37.

Nicholas, A. (1972). Was the Hercynian orogenic belt of Europe of the Andean type? *Nature*, **236**, 221–223.

Nicolaysen. L. E., de Villiers, J. W. L., Burger, A. J. and Strelow, F. W. E. (1958). New measurements relating to the absolute age of the Transvaal System and of the Bushveld igneous complex. *Trans. geol. Soc. S. Afr.*, **61**, 137–163.

Nielsen, D. R. and Stoiber, R. E. (1973). Relationship of potassium content in andesitic lavas and depth to the seismic zone. *J. Geophys. Res.*, **78**, 6887–6892.

Nielsen, T. F. D. (1975). Possible mechanism of continental breakup in the North Atlantic. *Nature*, **253**, 182–184.

Niggli, E. (1970). Alpine Metamorphose und alpine Gebirgsbildung. *Fortschr. Mineral.*, **47**, 16–26.

Nisbet, E. and Pearce, J. A. (1973). TiO_2 and a possible guide to past oceanic spreading rates. *Nature*, **246**, 468–470.

Nishimori, R. K. (1974). Cumulate anorthositic gabbros and peridotites and their relation to the origin of the calc-alkaline trend of the Peninsular Ranges batholith. *Geol. Soc. Amer. Abstr. with Prog.*, **6**, 129–230.

Nixon, P. H. (1973). Kimberlitic volcanoes in East Africa. *Overseas Geol. Miner. Res.*, **41**, 119–138.

Noble, J. A. (1970). Metal provinces of the western United States. *Bull. geol. Soc. Amer.*, **81**, 1607–1624.

Noble, D. C. (1972). Some observations on the Cenozoic volcano-tectonic evolution of the Great Basin, western United States. *Earth Plan. Sci. Lett.*, **17**, 142–150.

Noble, D. C., McKee, E. H., Farrar, E. and Petersen, U. (1974). Episodic Cenozoic volcanism and tectonism in the Andes of Peru, *Earth Plan. Sci. Lett.*, **21**, 213–220.

Noltimier, H. C. (1974). The geophysics of the North Atlantic Basin. In A. E. M. Nairn and F. G. Stehli (Eds.), *The Ocean Basins and Margins*, Vol. 2, Plenum Press, New York, 539–588.

Nwachukwu, S. O. (1972). The tectonic evolution of the southern portion of the Benue Trough, Nigeria. *Geol. Mag.*, **109**, 411–419.

Oberhauser, R. (1968). Beiträge zur Kenntnis der Tektonik und der Paläogeographie während der Oberkreide und dem Paläogen im Ostalpenraum. *Jahrb. Geol. Bundes. Wien*, **111**, 115–145.

Obradovich, J. D. and Peterman, Z. E. (1968). Geochronology of the Belt Series, Montana. *Can. J. Earth Sci.*, **5**, 737–747.

O'Hara, M. J. (1961). Petrology of the Scourie dyke, Sutherland. *Min. Mag.*, **32**, 848–865.

O'Hara, M. J. (1968). The bearing of phase equilibria studies in synthetic and natural systems on the origin and evolution of basic and ultrabasic rocks. *Earth Sci. Rev.*, **4**, 69.

O'Hara, M. J. (1975). Great thickness and high geothermal gradient of Archaean crust: the Lewisian of Scotland. *Abstr. Int. Conf. on Geothermometry and Geobarometer*, Penn. State Univ., USA, Oct. 1975.

Okado, H. (1974). Migration of ancient arc-trench systems. In R. H. Dott Jr and R. H. Shaver (Eds.), *Ancient and Modern Geosynclinal Sedimentation*, Soc. Econ. Pal. Mineral. Sp. Publ., **19**, 311–320.

Olade, M. A. (1975). Evolution of Nigeria's Benue Trough (aulacogen): a tectonic model. *Geol. Mag.*, **112**, 575–583.

Oliver, J. and Isacks, B. (1967). Deep earthquake zones, anomalous structures in the upper mantle and the lithosphere. *J. Geophys. Res.*, **72**, 4259–4275.

Oliver, R. L. (1969). Some observations on the distribution and nature of granulite facies terrains. *Geol. Soc. Australia Sp. Publ.*, **2**, 259–268.

Olson, J. L. and Pray, L. C. (1954). The Mountain Pass rare earth deposits. *Calif. Div. Mines Bull.*, **170**, Ch. 8, 23–29.

O'Nions, R. K. and Pankhurst, R. J. (1974). Rare earth element distribution in Archaean gneisses and anorthosites, Godthaab area, West Greenland. *Earth Plan. Sci. Lett.*, **22**, 328–338.

Oosthuyzen, E. J. and Burger, A. J. (1964). Radiometric dating of intrusives associated with the Waterberg System. *Ann. geol. Surv. S. Afr.*, **3**, 87–106.

Opdyke, N. D. (1962). Palaeoclimatology and continental drift. In

S. K. Runcorn (Ed.), *Continental Drift*, Academic Press, New York, 41–65.

Oskvarek, J. D. and Perry, E. C. Jr (1976). Temperature limits on the early Archaean ocean from oxygen isotope variations in the Isua supracrustal sequence, West Greenland. *Nature*, **259**, 192–194.

Oxburgh, E. R. (1968). An outline of the geology of the central eastern Alps. *Proc. geol. Ass.*, **79**, 1–46.

Oxburgh, E. R. (1972). Flake tectonics and continental collision. *Nature*, **239**, 202–204.

Oxburgh, E. R. (1974). Eastern Alps. In A. M. Spencer (Ed.), *Mesozoic–Cenozoic Orogenic Belts*, Geol. Soc. Lond. Sp. Publ., **4**, 109–126.

Oxburgh, E. R. and Turcotte, D. L. (1970). Thermal structure of island arcs. *Bull. geol. Soc. Amer.*, **81**, 1665–1688.

Oxburgh, E. R. and Turcotte, D. L. (1971). Origin of paired metamorphic belts and crustal dilation in island arc regions. *J. Geophys. Res.*, **76**, 1315–1327.

Packham, G. H. and Falvey, D. A. (1971). An hypothesis for the formation of marginal seas in the western Pacific. *Tectonophys.*, **11**, 79–109.

Padget, P., (1973). Evolutionary aspects of Precambrian of northern Sweden. In M. G. Pitcher (Ed.), *Arctic Geology*, Amer. Ass. Petrol. Geol. Mem., **19**, 431–439.

Page, B. M. (1972a). Oceanic crust and mantle fragment in subduction complex near San Luis Obispo, California. *Bull. geol. Soc. Amer.*, **83**, 957–972.

Page, B. M. (1972b). Cannibalism in Franciscan mélanges of California. *Geol. Soc. Amer. Abstr. with Prog.*, **4**, 620.

Page, L. R. and coworkers (1953). Pegmatite investigations 1942–1945, Black Hills, South Dakota. *US geol. Surv. Prof. Pap.*, **247**, 228 pp.

Palmer, H. C. and Carmichael, K. M. (1973). Palaeomagnetism of some Grenville Province rocks. *Can. J. Earth Sci.*, **10**, 1175–1190.

Pankhurst, R. J. (1974). Rb–Sr whole rock chronology of Caledonian events in NE Scotland. *Bull. geol. Soc. Amer.*, **85**, 345–350.

Pankhurst, R. J., Moorbath, S. and McGregor, V. R. (1973). Late event in the geological evolution of the Godthaab district, West Greenland. *Nature Phys. Sci.*, **243**, 24–26.

Pannella, G. (1972). Precambrian stromatolites as palaeontological clocks. *24th Int. geol. Congr. Montreal, Sect. 1*, 50–57.

Park, R. G. (1970). Observations on Lewisian chronology. *Scott. J. Geol.*, **6**, 379–399.

Park, R. G. (1973). The Laxfordian belts of the Scottish mainland. In R. G. Park and J. Tarney (Eds.), *The Early Precambrian of Scotland and Related Rocks of Greenland*, Univ. Keele, 65–76.

Park, R. G. and Cresswell, D. (1972). Basic dykes in the early Precambrian (Lewisian) of NW Scotland: their structural relations, conditions of emplacement and orogenic significance. *24th Int. geol. Congr. Montreal, Sect. 1*, 238–145.

Parsons, G. E. (1961). *Niobium-bearing complexes east of Lake Superior*, Ont. Dept. Mines, Geol. Rep. No. 3.

Pasteels, P. and Michot, J. (1975). Geochronologic investigation of the metamorphic terrain of southwestern Norway. *Norsk geol. Tids.*, **55**, 111–134.

Pautot, G., Auzende, J. M. and Le Pichon, X. (1970). Continuous deep sea salt layer along North Atlantic margins related to early phase of rifting. *Nature*, **227**, 351–354.

Pautot, G., Renard, V., Daniel, J. and Du Pont, J. (1973). Morphology, limits, origin and age of salt layer along the South Atlantic African margin. *Bull. Amer. Ass. Petrol. Geol.*, **57**, 1658–1671.

Pavlovsky, Ye. V. (1971). Early stages in development of the Earth's crust. *Int. Geol. Rev.*, **13**, 318–331.

Peach, B. M. (1907). *The geological structure of the north-west Highlands of Scotland*, Mem. geol. Surv. U.K.

Pearce, J. A. and Cann, J. R. (1971). Ophiolite origin investigated by discriminant analysis using Ti, Zr and Y. *Earth Plan. Sci. Lett.*, **12**, 339–349.

Pearce, J. A. and Cann, J. R. (1973). Tectonic setting of basic volcanic rocks determined using trace element analyses. *Earth Plan. Sci. Lett.*, **19**, 290–300.

Pedersen, K. R. and Lam, J. (1970). Precambrian organic compounds from the Ketilidian of south-west Greenland, Pts 1–3. *Meddr. Grønland*, **185**, Nr. 5–7, 57 pp.

Perry, E. C. and Tan, F. C. (1972). Significance of oxygen and carbon isotope variations in early Precambrian cherts and carbonate rocks of southern Africa. *Bull. geol. Soc. Amer.*, **83**, 647–664.

Peterman, Z. E., Hedge, C. E. and Tourtelot, H. A. (1970). Isotopic composition of Sr in seawater throughout Phanerozoic time. *Geochim. Cosmochim. Acta*, **34**, 105–120.

Peterman, Z. E. and Hedge, C. E. (1971). Related strontium isotopic and chemical variations in oceanic basalts. *Bull. geol. Soc. Amer.*, **82**, 493–500.

Peters, T. and Kramers, J. D. (1974). Chromite deposits in the ophiolite complex of northern Oman. *Mineral. Deposita*, **9**, 253–259.

Petrascheck, W. E. (1973). Some aspects of the relations between continental drift and metallogenic provinces. In D. H. Tarling and S. K. Runcorn (Eds.), *Implications of Continental Drift to the Earth Sciences*, Vol. 1, Academic Press, London, 567–572.

Petrascheck, W. E. (1976). Mineral zoning and plate tectonics in the Alpine-Mediterranean area. In D. F. Strong (Ed.), *Metallogeny and Plate Tectonics*, Geol. Ass. Can. Sp. Pap., **14**, 351–358.

Pettijohn, F. J. (1970). The Canadian Shield; a status report, 1970. In A. J. Baer (Ed.), *Symp. on Basins and Geosynclines of the Canadian Shield*, Geol. Surv. Can. Pap., **70–40**, 239–265.

Pfeiffer, H. (1971). Die Variszische Hauptbeiregung (Sogenannte sudetische Phase) im Umkreis der ausseren Kristallinzone der variszischen Bogens. *Geologie*, **20**, 9, 945–958.

Pflug, H. D. (1966). Structured organic remains from the Fig Tree Series of the Barberton Mountain Land. *Econ. Geol. Res. Unit Univ. Witwatersrand, Johannesburg, Inf. Circ.*, **28**, 1–14.

Phemister, T. C. (1945). The Coast Range batholith near Vancouver, British Columbia. *Quart. J. geol. Soc. Lond.*, **101**, 37–88.

Phillips, J. D. and Forsyth, O. (1972). Plate tectonics, palaeomagnetism and the opening of the Atlantic. *Bull. geol. Soc. Amer.*, **83**, 1579–1600.

Phillips, W. E. A., Stillman, C. J. and Murphy, T. A Caledonian plate tectonic model. *J. geol. Soc. Lond.*, **132**, 579–609.

Phinney, W. C. (1970). Anorthosite occurrences in Keweenawan rocks of northeastern Minnesota. In Y. W. Isachsen (Ed.), *Origin of Anorthosite and Related Rocks, Mem. New York State Mus. Sci. Serv.*, **18**, 135–147.

Picha, F. (1974). Ancient submarine canyons of the Carpathian miogeosyncline. In R. H. Dott Jr and R. H. Shaver (Eds.), *Modern and Ancient Geosynclinal Sedimentation*, Soc. Econ. Pal. Mineral. Sp. Publ., **19**, 126–127.

Pichamuthu, C. S. (1971). Precambrian geochronology of Peninsular India. *J. geol. Soc. India*, **12**, 262–273.

Pichamuthu, C. S. (1974). The Dharwar craton. *J. geol. Soc. India*, **15**, 339–346.

Pidgeon, R. T. (1973). Discordant U–Pb isotopic system in zircons from a granite northwest of Frederikshaab Isblink, Fiskenaesset region. *Rapp. Grønlands geol. Unders.*, **51**, 28–30.

Piper, D. Z. (1973). Origin of metalliferous sediments from the East Pacific Rise. *Earth Plan. Sci. Lett.*, **19**, 75–82.

Piper, J. D. A. (1972). A palaeomagnetic study of the Bukoban System, Tanzania, *Geophys. J. Roy. astr. Soc.*, **28**, 111–127.

Piper, J. D. A. (1973a). Latitudinal extent of late Precambrian glaciations. *Nature*, **244**, 342–344.

Piper, J. D. A. (1973b). Geological interpretation of palaeomagnetic results from the African Precambrian. In D. H. Tarling and S. K. Runcorn (Eds.), *Implications of continental drift to the Earth Sciences*, Vol. 1, Academic Press, London, 19–32.

Piper, J. D. A. (1974). Proterozoic crustal distribution, mobile belts and apparent polar movement. *Nature*, **251**, 381–384.

Piper, J. D. A. (1975). Proterozoic supercontinent: time duration and the Grenville problem. *Nature*, **256**, 519–520.

Piper, J. D. A. (1976a). Definition of pre-2000 m.y. apparent polar movements. *Earth Plan. Sci. Lett.*, **28**, 470–478.

Piper, J. D. A. (1976b). Palaeomagnetic evidence for a Proterozoic supercontinent. *Phil Trans. R. Soc. Lond.*, **A280**, 469–490.

Piper, J. D. A., Briden, J. C. and Lomax, K. (1973). Precambrian Africa and South America as a single continent. *Nature*, **245**, 244–248.

Pitman, W. C., III, Talwani, M. and Heirtzler, J. R. (1971). Age of the North Atlantic from magnetic anomalies. *Earth Plan. Sci. Lett.*, **11**, 195–200.

Pitman, W. C., III and Talwani, M. (1972). Sea floor spreading in the North Atlantic. *Bull. geol. Soc. Amer.*, **83**, 619–646.

Pitman, W. C., III and Hayes, J. D. (1973). Upper Cretaceous spreading rates and the great transgression. *Eos*, **54**, 240.

Platt, R. G. and Myers, J. S. (in press). Corona structures from the Fiskenaesset complex, West Greenland. *Progr. Experimental Petr.*, **3**.

Playford, P. E. and Cockbain, A. E. (1969). Algal stromatolites: deep water forms in the Devonian of western Australia. *Science*, **165**, 1008–1010.

Plumstead, E. P. (1973). The late Palaeozoic Glossopteris flora. In A. Hallam (Ed.), *Atlas of Palaeobiogeography*, Elsevier, Amsterdam, 187–206.

Polkanov, A. A. (1956). Geologiya Khoglandiya–Jotniya baltüskogo shchita (stratigrafiya, tektonika, kinematika, magmatizm). *Trudy lab. Geol. Dokem. Byl.*, **6**.

Poole, F. G. (1964). Palaeowinds in the western United States. In A. E. M. Nairn (Ed.), *Problems in Palaeoclimatology*, Wiley, London, 394–405.

Pouba, Z., (1970). Relation between iron and copper–lead–zinc mineralization in submarine volcanic ore deposits in the Jeseniky Mountains, Czechoslovakia. *Proc. Int. Mineral Ass. Symp. on Genesis of Ore Deposits*, Tokyo, p. 87 (Abstr.).

Pretorius, D. A. (1975). The depositional environment of the Witwatersrand goldfields: a chronological review of speculations and observations. *Minerals Sci. Engng.*, **7**, 18–47.

Priem, H. N. A., Boelrijk, N. A. I. M., Hebeda, E. H., Verdurmen, E. A. T. L., Verschure, R. H. and Bon, E. H. (1971). Granitic complexes and associated tin mineralization of 'Grenville' age in Rondônia, western Brazil. *Bull. geol. Soc. Amer.*, **82**, 1095–1102.

Priem, H. N. A., Boelrijk, N. A. I. M., Hebeda, E. H., Verdurmen, E. A. T. L. and Verschure, R. H. (1973). Age of the Precambrian Roraima Formation in northeastern South America: evidence from isotopic dating of Roraima pyroclastic volcanic rocks in Surinam. *Bull. geol. Soc. Amer.*, **84**,, 1677–1684.

Puffett, W. P. (1969). The Reany Creek Formation, Marquette County, Michigan. *Bull. US geol. Surv.*, **1274–F**, 24 pp.

Pullaiah, G. and Irving, E. (1975). Paleomagnetism of the contact aureole and late dikes of the Otto Stock, Ontario and its application to early Proterozoic apparent polar wandering. *Can. J. Earth Sci.*, **12**, 1609–1618.

Pulvertaft, T. C. R. (1973). Recumbent folding and flat-lying structure in the Precambrian of northern West Greenland. *Phil. Trans. R. Soc. Lond.*, **A273**, 535–546.

Pyke, D. R., Naldrett, A. J. and Eckstrand, O. R. (1973). Archaean ultramafic flows in Munro Township, Ontario. *Bull. geol. Soc. Amer.*, **84**, 955–978.

Quirt, S., Clark, A. H., Farrar, E. and Sillitoe, R. H. (1971). Potassium–argon ages of porphyry copper deposits in northern and central Chile. *Geol. Soc. Amer. Abstr. Washington Mtg*, 676–677.

Raaben, M. E. (1969). Columnar stromatolites and late Precambrian stratigraphy. *Amer. J. Sci.*, **267**, 1–18.

Rabinowitz, P. D. and Ryan, W. B. F. (1970). Gravity anomalies and crustal shortening in the eastern Mediterranean. *Tectonophys.*, **10**, 585–608.

Radhakrishna, B. P. (1967). Reconsideration of some problems in the Archaean complex of Mysore. *J. geol. Soc. India*, **8**, 102–109.

Radhakrishna, B. P. (1974). Peninsular gneiss complex of the Dharwar craton—a suggested model for its evolution. *J. geol. Soc. India*, **15**, 439–456.

Radhakrishna, B. P. (1976). Mineralisation episodes in the Dharwar craton of peninsular India. *J. geol. Soc. India*, **17**, 79–88.

Rajasekaran, K. C. (1972). Omphacite from eclogite, near Talapalaiyam, Bhavani Taluq, Coimbatore district, Tamil Nadu State. *J. Ind. Acad. Geosci.*, **15**, 1–11.

Ramadurai, S., Sankaran, M., Selvan, T. A. and Windley, B. F. (1975). The stratigraphy and structure of the Sittampundi complex, Tamil Nadu, India. *J. geol. Soc. India*, **16**, 409–414.

Ramberg, H. (1951). Remarks on the average chemical composition of granulite facies and amphibolite to epidote amphibolite facies gneisses in West Greenland. *Dansk geol. Foren. Meddr.*, **12**, 27–34.

Ramsay, J. G. (1963a). Structural investigations in the Barberton Mountain Land, eastern Transvaal. *Trans. geol. Soc. S. Afr.*, **66**, 353–398.

Ramsay, J. G. (1963b). Stratigraphy, structure and metamorphism in the western Alps. *Proc. Geol. Ass.*, **74**, 357–391.

Ramsbottom, W. H. C. (1970). Carboniferous faunas and palaeogeography of the South-West of England region. *Proc. Ussher Soc.*, **2**, 144–157.

Rankama, K. (1955). Geologic evidence of chemical composition of the Precambrian atmosphere. In A. Poldervaart (Ed.), *Crust of the Earth*, Geol. Soc. Amer. Sp. Pap., **62**, 651–664.

Rankin, D. W., Espenshade, G. H. and Shaw, K. W. (1973). Stratigraphy and structure of the metamorphic belt in northwestern Virginia—a study from the Blue Ridge across the Brevard Far H zone to the Sauratown Mountains anticlinorium. *Amer. J. Sci.*, **273A**, 1–40.

Raup, D. M. (1972). Taxonomic diversity during the Phanerozoic. *Science*, **177**, 1065–1071.

Rast, N. (1969). Orogenic belts and their parts. In P. E. Kent, G. E. Satterthwaite and A. M. Spencer (Eds.), *Time and Place in Orogeny*, Geol. Soc. Lond. Sp. Publ., **3**, 197–214.

Read, H. H. (1957). *The Granite Controversy*, T. Murby, London.

Read, H. H. and Watson, J. (1975). *Introduction to Geology*, Vol. 2: *Earth History*, MacMillan, London.

Reading, H. G. (1972). Global tectonics and the genesis of flysch successions. *24th Int. geol. Congr. Montreal, Sect. 6*, 59–66.

Reading, H. G. and Walker, R. G. (1966). Sedimentation of Eocambrian tillites and associated sediments in Finnmark, Northern Norway. *Palaeogeog. Palaeoclim. Palaeoecol.*, **2**, 177–212.

Reid, G. C., Isaksen, I. S. A., Holzer, T. E. and Crutzen, P. J. (1976). Influence of ancient solar-proton events on the evolution of life. *Nature*, **259**, 177–179.

Reinhardt, B. M. (1969). On the genesis and emplacement of ophiolites in the Oman Mountains geosyncline. *Schweiz. Mineral. Petrol. Mitt.*, **49**, 1–30.

Reis, B. (1972). Preliminary note on the distribution and tectonic control of kimberlites in Angola. *24th Int. geol. Congr. Montreal, Sect. 4*, 276–281.

Reyment, R. A. (1969). Ammonite biostratigraphy, continental drift and oscillatory transgressions. *Nature*, **224**, 137–140.

Reyment, R. A. and Tait, E. A. (1972). Faunal evidence for the origin of the South Atlantic. *24th Int. geol. Congr. Montreal, Sect. 7*, 316–323.

Rezanov, I. A. and Chamo, S. S. (1969). Reasons for absence of a 'granitic' layer in basins of the South Caspian and Black Sea types. *Can. J. Earth Sci.*, **6**, 671–678.

Rhodes, F. H. T. (1967). Permo-Triassic extinction. In W. B. Harland and coworkers (Eds.), *The Fossil Record*, Geol. Soc. London, 57–76.

Richter, E. M. (1973). Dynamical models for sea floor spreading. *Rev. Geophys. Space Phys.*, **11**, 223–287.

Ridge, J. D. (1972). Annotated bibliographies of mineral deposits in the western hemisphere. *Mem. geol. Soc. Amer.*, **131**, 681 pp.

Riding, R. (1974). Model of the Hercynian fold belt. *Earth Plan. Sci. Lett.*, **24**, 125–135.

Ridler, R. H. (1970). Relationship of mineralisation to volcanic stratigraphy in the Kirkland–Larder Lakes area, Ontario. *Proc. Geol. Ass. Can.*, **21**, 33–42.

Rikitake, T., Miyamura, S., Tsubokawa, I., Murauchi, S., Uyeda, S., Kuno, H. and Gorai, M. (1968). Geophysical and geological data in and around the Japan arc. *Can. J. Earth Sci.*, **5**, 1101–1118.

Ringwood, A. E. (1974). The petrological evolution of island arc systems. *J. geol. Soc. Lond.*, **130**, 183–204.

Rivalenti, G. (1975). Chemistry and differentiation of mafic dikes in an area near Fiskenaesset, West Greenland. *Can. J. Earth Sci.*, **12**, 721–730.

Rivalenti, G. (1976). Geochemistry of meta-volcanic amphibolites from south west Greenland. In B. F. Windley (Ed.), *The Early History of the Earth*, Wiley, London, 213–223.

Rivalenti, G. and Rossi, A. (1972). The geology and petrology of the Precambrian rocks to the northeast of the fjord Qagssit, Frederikshaab district, south west Greenland. *Meddr. Grønland*, **192**, 5, 1–98.

Rivalenti, G. and Sighinolfi, G. P. (1971). Geochemistry and differentiation phenomena in basic dykes of the Frederikshaab district, southwest Greenland. *Atti. Soc. Tosc. Sci. Nat. Mem. Ser. A*, **77**, 358–380.

Robert, M. (1940). La glaciation du Kundelungu au Katanga (Congo Belge). *Congr. geol. intern., C.R. 17, Moscow 1937*, **6**, 99.

Roberts, D. G. (1975). Evaporite deposition in the Aptian South Atlantic Ocean. *Marine Geol.*, **18**, M65–71.

Robertson, A. H. F. (1975). Cyprus umbers: basalt–sediment relationships on a Mesozoic ocean ridge. *J. geol. Soc. Lond.*, **131**, S11–32.

Robertson, A. H. F. and Hudson, J. D. (1973). Cyprus umbers: chemical precipitates on a Tethyan ocean ridge. *Earth Plan. Sci. Lett.*, **18**, 93–101.

Robertson, A. H. F. and Hudson, J. D. (1974). Pelagic sediments in the Cretaceous and Tertiary history of the Troodos Massif, Cyprus. *Int. Ass. Sediment. Sp Publ.*, **1**, 403–436.

Robertson, A. H. F. and Fleet, A. J. (1976). The origins of rare earths in metalliferous sediments of the Troodos Massif, Cyprus. *Earth Plan. Sci. Lett.*, **28**, 385–394.

Robertson, I. D. M. (1973). The geology of the country around Mount Towla, Gwanda district. *Bull. geol. Surv. Rhodesia*, **68**.

Robertson, I. D. M. (1974). Explanation of the geological map of the country south of Chibi. *Rep. Geol. Surv. Rhodesia*, **41**.

Robertson, I. D. M. and van Bremmen, O. (1970). The southern satellite dykes of the Great Dyke, Rhodesia. *Geol. Soc. S. Afr. Sp. Publ.*, **1**, 621–644.

Robinson, P. L. (1973). Palaeoclimatology and continental drift. In D. H. Tarling and S. K. Runcorn (Eds.), *Implications of Continental Drift to the Earth Sciences*, Vol. 1, Academic Press, London, 451–476.

Roddick, J. A. and Armstrong, J. E. (1959). Relict dykes in the Coast Mountains, near Vancouver, British Columbia. *J. Geol.*, **67**, 603–613.

Rodgers, J. (1970). *The Tectonics of the Appalachians*, Wiley, New York.

Rodgers, J. (1971). The Taconic orogeny. *Bull. geol. Soc. Amer.*, **82**, 1141–1178.

Rodrigues, B. (1972). Major tectonic alignments of alkaline complexes in Angola. In T. F. J. Dessauvagie and A. J. Whiteman (Eds.), *African Geology*, Univ. of Ibadan, 149–153.

Roedder, E. (1967). Environment of deposition of stratiform (Mississippi Valley type) ore deposits from studies of fluid inclusions. In J. S. Brown (Ed.), *Genesis of Stratiform Lead–Zinc–Barite–Fluorite Deposits, Econ. Geol. Monog.*, **3**, 371–378.

Roering, C. (1968). Non-orogenic granites and the age of the Precambrian Pongola Sequence. *Econ. Geol. Res. Unit, Univ. Witwatersrand, Johannesburg, Inf. Circ.* **46**, 20 pp.

Romer, A. S. (1966). *Vertebrate Palaeonotology* (3rd edit.), Univ. Chicago Press.

Romey, W. D. (1968). An evaluation of some 'differences' between anorthosite in massifs and in layered complexes. *Lithos*, **1**, 230–241.

Rona, P. A. (1970). Comparison of continental margins of eastern North America at Cape Hatteras and northwestern Africa at Cape Blanc. *Bull. Amer. Ass. Petrol. Geol.*, **54**, 129–157.

Ronov, A. B. (1964). Common tendencies in the chemical evolution of the earth's crust, ocean and atmosphere. *Geochem. Int.*, **1**, 713–737.

Ronov, A. B. (1968). Probable changes in the composition of sea water during the course of geological time. *Sedimentology*, **10**, 25–43.

Ronov, A. B. (1972). Evolution of rock composition and geochemical processes in the sedimentary shell of the earth. *Sedimentology*, **19**, 157–172.

Ronov, A. B. and Migdisov, A. A. (1971). Geochemical history of the crystalline basement and the sedimentary cover of the Russian and North American platforms. *Sedimentology*, **16**, 137–185.

Roper, H. (1956). The manganese deposits at Otjosondu, SW Africa, *20th Int. geol. Congr. on Manganese*, **2**, 113–122.

Roscoe, S. M. (1968). Huronian rocks and uraniferous conglomerates in the Canadian Shield. *Geol Surv. Can. Pap.*, **68–40**, 205 pp.

Rose, E. R. (1969). St Urbain area and Lac Allard–Romaine River area. In *Geology of Titanium and Titaniferous Deposits of Canada, Geol. Surv. Can. Econ. Geol. Rep. No. 25*, 77–85, 96–102.

Ross, C. P. (1970). The Precambrian of the United States of America: northwestern United States, the Belt Series. In K. Rankama (Ed.), *The Precambrian*, Wiley, New York, 145–252.

Rousell, D. H. (1965). Geology of the Cross Lake area. *Manitoba Dept Mines nat. Res. Publ.*, 62–64.

Roy, S. (1966). *Syngenetic manganese formations of India*, Jadavpur Univ., Calcutta, 219 pp.

Roy, S. (1968). Mineralogy of the different genetic types of manganese deposits. *Econ. Geol.*, **6**, 760–786.

Rubey, W. W. (1955). Development of the hydrosphere and atmosphere, with special reference to probable composition of the early atmosphere. *Geol. Soc. Amer. Sp. Pap.*, **62**, 631–650.

Ruckmick, J. C. (1963). Iron ores of Cerro Bolivar, Venezuela. *Econ. Geol.*, **58**, 218–236.

Rudnik, V. A., Sobotovich, E. B. and Terent'yev, V. M., 1969. Archaean age of the oldest rocks of the Aldan complex. *Dokl. Acad. Sci. USSR*, **188**, 93–96.

Runcorn, S. K. (1956). Palaeomagnetic comparisons between Europe and North America. *Proc. geol. Ass. Can.*, **8**, 77–85.

Runcorn, S. K. (1961). Climatic change through geological time in the light of the palaeomagnetic evidence for polar wandering and continental drift. *Quart. J. Roy. Meterol. Soc.* **87**, 373, p. 282.

Runcorn, S. K. (1962). Palaeomagnetic evidence for continental drift and its geophysical cause. In S. K. Runcorn (Ed.), *Continental Drift*, Academic Press, New York, 1–39.

Runcorn, S. K. (1964). Palaeowind directions and palaeomagnetic latitudes. In A. E. M. Nairn (Ed.), *Problems in Palaeoclimatology*, Wiley, London, 409–419.

Runcorn, S. K. (1965). A symposium on Continental Drift. I. Palaeomagnetic comparisons between Europe and North America. *Phil. Trans. R. Soc. Lond.*, **A258**, 1–11.

Rutland, R. W. R. (1973a). On the interpretation of Cordilleran orogenic belts. *Amer. J. Sci.*, **273**, 811–849.

Rutland, R. W. R. (1973b). Tectonic evolution of the continental crust of Australia. In D. H. Tarling and S. K. Runcorn (Eds.), *Implications of Continental Drift to the Earth Sciences*, Vol. 2, Academic Press, London, 1011–1033.

Rutten, M. G. (1964). Geologic data on atmospheric history. *Palaeogeog. Palaeoclimatol. Palaeoecol.*, **2**, 47–57.

Ryan, W. B. F., and coworkers, (1970). Deep-sea drilling project, leg 13. *Geotimes*, Dec. 15, 12–15.

Ryan, W. B. F., Stanley, D. J., Hersey, J. B., Fahlquist, D. A. and Allan, T. D. (1971). The tectonics and geology of the Mediterranean Sea. In A. E. Maxwell (Ed.), *The Sea*, Vol. 4, Wiley, New York, 387–492.

Sadler, P. M. (1974). Trilobites from the Gorran quartzites, Ordovician of South Cornwall. *Palaeontology*, **17**, 71–93.

Saggerson, E. P. (1973). Metamorphic facies series in Africa: a contrast. In L. A. Lister (Ed.), *Symp. on Granites, Gneisses and Related Rocks, Geol. Soc. S. Afr. Sp. Publ.*, **3**, 227–234.

Saito, T. (1968). Ore deposits and mechanism of its formation of Uchinotai western ore body, Kosaka mine, Akita Prefecture, Japan. *Mining Geol. (Tokyo)*, **18**, 241–156 (in Japanese).

Saito, T. (1974). Distribution and geological setting of the Kuroko deposits. *Mining Geol. (Tokyo) Sp. Iss.*, **6**, 1–9.

Saito, T., Burkle, L. H. and Ewing, M. (1966). Lithography and palaeontology of the reflective layer Horizon A. *Science*, **154**, 1173–1176.

Saito, T. and Funnell, B. M. (1970). Pre-Quarternary sediments and microfossils in the oceans. In A. E. Maxwell (Ed.), *The Sea*, Vol. 4, Wiley, New York, 183–204.

Sakai, H., Osaki, S. and Tsukagishi, M. (1970). Suplhur and oxygen geochemistry of sulphate in the black ore deposits of Japan. *Geochem., J.*, **4**, 27–39.

Salop, L. I. (1964). Precambrian geochronology and some features of the early stage of the geological history of the earth. *Int. geol. Congr. India, Rep. 22nd Session*, pt 10, 131–149.

Salop, L. I. (1972a). A unified stratigraphic scale of the Precambrian. *24th Int. geol. Congr. Montreal, Sect. 1*, 253–259.

Salop, L. I. (1972b). Two types of Precambrian structures: gneiss folded ovals and gneiss domes. *Int. Geol. Rev.*, **14**, 1209–1228.

Salop, L. I. and Scheinmann, Y. M. (1969). Tectonic history and structure of platforms and shields. *Tectonophys.*, **7**, 565–597.

Sangster, D. F. (1972). Precambrian volcanogenic massive sulphide deposits in Canada: a review. *Geol. Surv. Can. Pap.*, **72–22**, 44 pp.

Saravanan, S. (1969). Origin of iron ores of Kanjamalai, Salem district, Madras state. *Indian Mineral.*, **10**, 236–244.

Sarkar, S. N. (1972). Present status of Precambrian geochronology of Peninsular India. *24th Int. geol. Congr. Montreal, Sect. 1*, 260–702.

Satyanarayana, K., Rao, D. K., Naqvi, S. M. and Hussain, S. M. (1973). On the provenance of Dharwar schists, Shimoga, Mysore. *Bull. Geophys. Res.*, **11**, 36–41.

Savolahti, A. (1962). The rapakivi problem and the rules of idiomorphism in minerals. *Bull. Comm. geol. Finl.*, **204**, 34–111.

Sawkins, F. J. (1972). Sulphide ore deposits in relation to plate tectonics. *J. Geol.*, **80**, 377–397.

Sawkins, F. J. (1976a) Widespread continental rifting; some considerations of timing and mechanism. *Geology*, **4**, 427–430.

Sawkins, F. J. (1976b). Massive sulphide deposits in relation to geotectonics. In D. F. Strong (Ed.), *Metallogeny and Plate Tectonics, Geol. Ass. Can. Sp. Pap.*, **14**, 221–242.

Scandone, P. (1975). Triassic seaways and the Jurassic Tethys ocean in the central Mediterranean area. *Nature*, **256**, 117–119.

Scarpelli, W. (1973). The Serra do Navio manganese deposit (Brazil). In *The Genesis of Precambrian Iron and Manganese Deposits*, Proc. Kiev Symp. 1970, Unesco, Paris, 217–228.

Schau, M. (1970). Stratigraphy and structure of the type area of the Upper Triassic Nicola Group in south–central British Columbia. *Geol. Ass. Can. Sp. Pap.*, **6**, 123–135.

Schenk, P. E. (1965). Depositional environment of the Gowganda Formation (Precambrian) at the south end of Lake Timagami, Ontario. *J. Sed. Petrol.*, **35**, 309–318.

Schenk, P. E. (1969). Carbonate–sulphate–redbed facies and cyclic sedimentation of the Windsorian stage (middle Carboniferous), Maritime Provinces. *Can. J. Earth Sci.*, **6**, 1037–1066.

Schermerhorn, L. J. G. (1966). Terminology of mixed coarse–fine sediments. *J. Sed. Petrol.*, **36**, 831–835.

Schermerhorn, L. J. G. (1974). Late Precambrian mixtites: glacial and/or nonglacial? *Amer. J. Sci.*, **274**, 673–824.

Schermerhorn, L. J. G. (1975). Tectonic framework of late Precambrian supposed glacials. In A. E. Wright and F. Moseley (Eds.), *Ice Ages: Ancient and Modern*, Seal House Press, Liverpool, 2410274.

Schermerhorn, L. J. G. and Stainton, W. I. (1963). Tilloids in the West Congo geosyncline. *Quart. J. geol. Soc. Lond.*, **119**, 201–241.

Schidlowski, M. (1971). Probleme der atmosphärischen Evolution im Präkambrium. *Geol. Rund.*, **60**, 1373–1384.

Schidlowski, M. (1976). Archaean atmosphere and evolution of the terrestrial oxygen budget. In B. F. Windley (Ed.), *The Early History of the Earth*, Wiley, London, 525–535.

Schidlowski, M., Eichmann, R. and Junge, C. E. (1975). Precambrian sedimentary carbonates: carbon and oxygen isotope geochemistry and implications for the terrestrial oxygen budget. *Precambrain Res.*, **2**, 1–69.

Schilling, J. G. (1971). Sea floor evolution: rare earth evidence. *Phil. Trans. R. Soc. Lond.*, **A268**, 663–706.

Schneider, E. D. (1972). Sedimentary evolution of rifted continental margins. *Geol. Soc. Amer. Mem.*, **132**, 109–118.

Scholl, D. W. and Marlow, M. S. (1974). Sedimentary sequence in modern Pacific trenches and the deformed circum-Pacific eugeosyncline. In R. H. Dott Jr and R. H. Shave (Eds.), *Modern and Ancient Geosynclinal Sedimentation, Soc. Econ. Pal. Mineral. Sp. Publ.*, **19**, 193–211.

Scholz, C. H., Barazangi, M. and Sbar, M. L. (1971). Late Cenozoic evolution of the Great Basin, western United States, as an ensialic interarc basin. *Bull. geol. Soc. Amer.*, **82**, 2979–2990.

Schopf, J. W. (1972). Evolutionary significance of the Bitter Springs (late Precambrian) microflora. *24th Int. geol. Congr. Montreal, Sect. 1*, 68–77.

Schopf, J. W. (1974). The development and diversification of Precambrian life. *Origins of Life*, **5**, 119–135.

Schopf, J. W. (1975). Precambrian paleobiology: problems and perspectives. *Ann. Rev. Earth Plan. Sci.*, **3**, 212–249.

Schopf, J. W. (1976). Evidence of Archaean life: a brief appraisal. In B. F. Windley (Ed.), *The Early History of the Earth*, Wiley, London, 589–593.

Schopf, J. W., Oehler, D. Z., Horodyski, R. J. and Kvenvolden, K. A. (1971). Biogenecity and significance of the oldest known stromatolites. *J. Pal.*, **45**, 477–485.

Schopf, J. M. (1973). Coal, climate and global tectonics. In D. H. Tarling and S. K. Runcorn (Eds.), *Implications of Continental Drift to the Earth Sciences*, Vol. 1, Academic Press, London, 609–622.

Schreyer, W. and Yoder, H. S. (1964). The system Mg–cordierite–H_2O and related rocks. *Neues. Jahrb. Mineral. Abh.*, **101**, 271–342.

Schuiling, R. D. (1967). Tin belts on the continents around the Atlantic ocean. *Econ. Geol.*, **62**, 540–550.

Scotese, C. R. (1974). The evolution and biogeography of Lower Palaeozoic crinoids in relation to the tectonic history of the proto-Atlantic. *Geol. Soc. Amer. Abstr. with Prog.*, **6**, 6, 543–544.

Scott, W. M., Modzeleski, V. E. and Nagy, B. (1970). Pyrolusis of early Precambrian Onverwacht organic matter. *Nature*, **225**, 1129–1130.

Scrutton, R. A. (1973). The age relationship of igneous activity and continental break-up. *Geol. Mag.*, **110**, 227–234.

Sdzuy, K. (1972). Das Kambrium der acadobaltischen Fauenprovinz. *Zbl. Geol. Paläeont.*, Teil II Jahrg. H.1/2.

Searle, D. L. (1972). Mode of occurrence of the cupiferous pyrite deposits of Cyprus. *Trans. Inst. Min. Metall.*, **81**, B189–197.

Sederholm, J. J. (1923). On migmatites and associated Precambrian rocks of south western Finland, Part I. *Bull. Comm. géol. Finl.*, **58**, 154 pp.

Seely, D. R., Vail, P. R. and Walton, G. G. (1974). Trench slope model. In L. A. Burk and C. L. Drake (Eds.), *The Geology of Continental Margins*, Springer-Verlag, Berlin, 249–260.

Seguin, M. (1969). Configuration et nature du mode tectonique en bordure de la partie centrale de la fosse du Labrador. *Can. J. Earth Sci.*, **6**, 1365–1379.

Sellwood, B. W. and Hallam, A. (1974). Bathonian volcanicity and North Sea rifting. *Nature*, **252**, 27–28.

Sellwood, B. W. and Jenkyns, H. C. (1975). Basins and swells and the formation of an epeiric sea (Pliensbachian–Bajocian of Great Britain). *J. geol. Soc. Lond.*, **131**, 373–388.

Semenenko, N. P. (1967). Absolute geochronology and histroy of formation of folded Precambrian zones in the east European platform. In A. P. Vinogradov (Ed.), *Chemistry of the Earth's Crust*, Israel Progr. Sci. Trans., Jerusalem.

Semenenko, N. P., Rodionov, S. P., Usenko, I. S., Lichak, I. L. and Tsarovsky, I. D. (1960). Stratigraphy of the Precambrian of the Ukrainian Shield, *21st Int. geol. Congr. Norden*, Pt 9, 108–115.

Semenenko, N. P., Scherbak, A. P., Vinogradov, A. P., Tougarinov, A. I., Eliseeva, G. D., Lotlovskay, F. I. and Demidenko, S. G. (1968). Geochronology of the Ukrainian Precambrian. *Can. J. Earth Sci.*, **5**, 661–671.

Semenenko, N. P., Shcherbak, N. P. and Bartnitsky, E. N. (1972). Geochronology, stratigraphy and tectonic structure of the Ukrainian Shield. *24th Int. geol. Congr. Montreal*, Sect. 1, 363–370.

Seyfert, C. K. and Sirkin, L. A. (1973). *Earth History and Plate Tectonics*, Harper and Row, New York, 504 pp.

Shackleton, R. M. (1969). The Precambrian of North Wales. In A. Wood (Ed.), *The Precambrian and Lower Palaeozoic rocks of Wales*, Wales Univ. Press, Cardiff, 1–22.

Shackleton, R. M. (1973a). Problems of the evolution of the continental crust. *Phil. Trans. R. Soc. Lond.*, **A273**, 317–320.

Shackleton, R. M. (1973b). Correlation of structures across Precambrian orogenic belts in Africa. In D. H. Tarling and S. K. Runcorn (Eds.), *Implications of Continental Drift to the Earth Sciences*, Vol. 2, Academic Press, London, 1091–1095.

Shackleton, R. M. (1976a). Shallow and deep level exposures of Archaean crust in India and Africa. In B. F. Windley (Ed.), *The Early History of the Earth*, Wiley, London, 317–321.

Shackleton, R. M. (1976b). Pan-African structures. *Phil. Trans. R. Soc. Lond.*, **A280**, 491–497.

Shagam, R. (1960). Geology of central Aragua, Venezuela. *Bull. geol. Soc. Amer.*, **71**, 249–302.

Sharp, R. P. (1964). Wind-driven sand in Coachella Valley, California. *Bull. geol. Soc. Amer.*, **75**, 785–803.

Shatski, N. S. (1955). The origin of the Pachelma trench. Comparative tectonics of ancient platforms. *Bull. Moscow Soc. Naturalists, Geol. Sect., Pap. No. 5*, **30**, 5–26.

Shatski, N. S. (1961). *Vergleichende Tecktonik alter Tafeln*, Akademie Verlag, Berlin, 220 pp.

Sheldon, R. P. (1964). Paleolatitudinal and paleogeographic distribution of phosphorite. *US geol. Surv. Prof. Pap.*, **501c**, 106–113.

Sheraton, J. W. (1970). The origin of the Lewisian gneisses of northwest Scotland, with particular reference to the Drumbeg area, Sutherland. *Earth Plan. Sci. Lett.*, **8**, 301–310.

Sheridan, R. E. and Drake, C. L. (1968). Seaward extension of the Canadian Appalachians. *Can. J. Earth Sci.*, **5**, 337–373.

Shiraki, K. (1966). Some aspects of the geochemistry of chromium. *J. Earth Sci. Nagoya Univ.*, **14**, 10–55.

Sidorenko, A. V. (1969). A unified geohistoric approach to the study of the Precambrian and post-Precambrian. *Dokl. Acad. Sci. USSR*, **186**, 36–37.

Sidorenko, A. V., Rozen, O. M., Gimmel'farb, G. B. and Tenyakov, V. A. (1969). Abundance of carbonate sediments in the Precambrian. *Dokl. Acad. Sci. USSR*, **189**, 127–129.

Sighinolfi, G. P. (1974). Geochemistry of early Precambrian carbonate rocks from the Brazilian Shield: implications for Archean carbonate sedimentation. *Contr. Mineral. Petrol.*, **46**, 189–214.

Sighinolfi, G. P. and Gorgoni, C. (1975). Genesis of massif-type anorthosites—the role of high-grade metamorphism. *Contr. Mineral. Petrol.*, **51**, 119–126.

Sillitoe, R. H. (1972a). Relation of metal provinces in western America to subduction of oceanic lithosphere. *Bull. geol. Soc. Amer.*, **83**, 813–818.

Sillitoe, R. H. (1972b). Formation of certain massive sulphide deposits at sites of sea floor spreading. *Trans. Inst. Min. Metall.*, **81**, B141–148.

Sillitoe, R. H. (1972c). A plate tectonic model for the origin of porphyry copper deposits. *Econ. Geol.*, **67**, 184–197.

Sillitoe, R. H. (1974). Tin mineralisation above mantle hot spots. *Nature*, **248**, 497–499.

Sillitoe, R. H. (1976). Andean mineralisation: a model for the metallogeny of convergent plate margins. In D. F. Strong (Ed.), *Metallogeny and Plate Tectonics, Geol. Ass. Can. Sp. Pap.*, **14**, 59–100.

Simmons, G. (1964). Gravity survey and geological interpretation, northern New York. *Bull. geol. Soc. Amer.*, **75**, 81–98.

Simonen, A. and Vorma, A. (1969). Amphibole and biotite from rapakivi. *Bull. Comm. géol. Finl.* **238**, 28 pp.

Simpson, E. S. W. and Otto, J. D. T. (1960). On the Precambrian anorthosite mass of southern Angola. *21st Int. geol. Congr. Norden, Pt 13*, 216–227.

Simpson, J. F. (1966). Evolutionary pulsations and geomagnetic polarity. *Bull. geol. Soc. Amer.*, **77**, 197–204.

Singewald, Q. D. (1950). Mineral resources of Columbia. *Bull. US geol. Surv.*, **946B**, 204 pp.

Skevington, D. (1973). Ordovician graptolites. In A. Hallam (Ed.), *Atlas of Palaeobiogeography*, Elsevier, Amsterdam, 28–35.

Skinner, B. J., White, D. E., Rose, H. J. and Mays, R. E. (1967). Sulphides associated with the Salton Sea geothermal brines. *Econ. Geol.*, **62**, 316–330.

Skinner, W. R. (1969). Geologic evolution of the Beartooth Mountains, Montana and Wyoming, Pt 8: Ultramafic rocks in the Highline Trail Lakes area, Wyoming. In L. H. Larsen (Ed.), *Geol. Soc. Amer. Mem.*, **115**, 19–52.

Smewing, J. D., Simonian, K. O. and Gass, I. G. (1975). Metabasalts from the Troodos Massif, Cyprus: genetic implications deduced from petrography and trace element geochemistry. *Contr. Mineral. Petrol.*, **51**, 49–64.

Shimron, A. E. and Zwart, H. J. (1970). The occurrence of low pressure metamorphism in the Precambrian of the Middle East and North East Africa. *Geol. en Mijnb.*, **49**, 369–374.

Smirnov, V. I. (1970). Pyritic deposits. Pts 1 and 2. *Int. geol. Rev.*, **12**, 881–908, 1039–1058.

Smith, A. G. (1971). Alpine deformation and the oceanic areas of the Tethys, Mediterranean and Atlantic. *Bull. geol. Soc. Amer.*, **82**, 2039–2070.

Smith, A. G. and Hallam, A. (1970). The fit of the southern continents. *Nature*, **225**, 139–144.

Smith, A. G., Briden, J. C. and Drewry, G. E. (1973). Phanerozoic world maps. *Sp. Pap. Palaeont.*, **12**, 1–42.

Smith, A. J. (1963). Evidence for a Talchir (Lower Gondwana) glaciation: striated pavement and boulder bed at Irai, central India. *J. Sed. Petrol.*, **33**, 739–750.

Smith, C. H. (1962). Notes on the Muskox intrusion, Coppermine River area, district of Mackenzie. *Geol. Surv. Can. Pap.*, **61–25**, 16 pp.

Smith, M. R. and Jensen, M. L. (1974). A theory of intracontinental tectonic adjustment. *Geol. Soc. Amer. Abstr. with Prog.*, **6**, 3, p. 255.

Smith, T. E. and Longstaffe, F. J. (1974). Archaean rocks of schoshonitic affinities at Bijou Point, Northwestern Ontario. *Can. J. Earth Sci.*, **11**, 1407–1413.

Smithson, S. B. (1965). The nature of the 'granitic' layer of the crust in the southern Norwegian Precambrian. *Norsk geol. Tids.*, **45**, 113–133.

Smithson, S. B., Murphy, D. J. and Houston, R. S. (1971). Development of an augen gneiss terrain. *Contr. Mineral. Petrol.*, **33**, 184–190.

Söhnge, P. G., Le Roex, H. D. and Nel, H. J. (1948). The geology of the country around Messina. *Geol. Surv. S. Afr. Explan.*, Sheet 46 (Messina).

Sokolov, B. S. (1972). The Vendian stage in earth history. *24th Int. geol. Congr. Montreal, Sect. 1*, 78–84.

Sonnenfeld, P. (1974). The Upper Miocene evaporite basins in the Mediterranean region—a study in palaeo-oceanography. *Geol. Rund.*, **63**, 1133–1172.

Sonnenfeld, P. (1975). The significance of Upper Miocene (Messinian) evaporites in the Mediterranean Sea. *J. Geol.*, **83**, 287–311.

Sørensen, H. (1966). On the magmatic evolution of the alkaline igneous province of South Greenland. *Rapp. Grønlands geol. Unders.*, **7**, 19 pp.

Souther, J. G. (1970). Volcanism and its relationship to recent crustal movements in the Canadian Cordillera. *Can. J. Earth Sci.*, **7**, 553–567.

Souther, J. G. (1972). Initial deposits in the Cordilleran geosyncline: evidence of a late Precambrian (>850 m.y.) continental separation. *Bull. geol. Soc. Amer.*, **83**, 1345–1360.

Sowerbutts, W. T. C. (1969). Crustal structure of the East African plateau and rift valleys from gravity measurements. *Nature*, **223**, 143–146.

Spall, H. (1972). Palaeomagnetism and Precambrian continental drift. *24th Int. geol. Congr. Montreal, Sect. 3*, 172–179.

Spencer, A. M. (1971). Late Precambrian glaciation in Scotland. *Geol. Soc. Lond. Mem.*, **6**, 98 pp.

Spencer, A. M. (Ed.) (1974). *Mesozoic–Cenozoic Orogenic Belts: Data for orogenic studies, Geol. Soc. Lond. Sp. Publ.*, **4**, 809 pp.

Spencer, A. M. (1975). Late Precambrian glaciations in the North Atlantic region. In A. E. Wright and F. Moseley (Eds.), *Ice Ages: Ancient and Modern*, Seal House Press, Liverpool, 217–240.

Spooner, C. M. and Fairbairn, H. W. (1970). Strontium

366

87/strontium 86 initial ratios in pyroxene granulite terranes. *J. Geophys. Res.*, **75**, 6706–6713.

Spooner, E. T. C. and Fyfe, W. S. (1973). Sub-sea floor metamorphism, heat and mass transfer. *Contr. Mineral. Petrol.*, **42**, 287–304.

Sreenivas, B. L. and Srinivasan, R. (1974). Geochemistry of granite–greenstone terrain in South India. *J. Geol. Soc. India*, **15**, 390–406.

Srinivasan, R. and Sreenivas, B. L. (1968). Sedimentation and tectonics in the Dharwars (Archaean) of Mysore. *Ind. Mineralogist*, **9**, 47–58.

Srinivasan, R. and Sreenivas, B. L. (1971). Flood basalts from Dharwars of Mysore, India. *Bull. Vulcanol.*, **35**, 824–840.

Srinivasan, R. and Sreenivas, B. L. (1972). Dharwar stratigraphy. *J. geol. Soc. India*, **13**, 75–85.

Stanley, D. J. (1974). Modern flysch sedimentation in a Mediterranean island arc setting. In R. H. Dott, Jr and R. H. Shaver (Eds.), *Modern and Ancient Geosynclinal Sedimentation*, Soc. Econ. Pal. Mineral. Sp. Publ., **19**, 240–259.

Stanley, D. J., Gehin, C. E. and Bartolini, C. (1970). Flysch-type sedimentation in the Alboran Sea, western Mediterranean. *Nature*, **228**, 979.

Stanley, K. O., Jordan, W. M. and Dott, R. H., Jr (1971). New hypothesis of early Jurassic palaeogeography and sediment dispersal for western United States. *Bull. Amer. Ass. Petrol. Geol.*, **55**, 10–19.

Stanton, R. L. (1955). Lower Palaeozoic mineralization near Bathurst, NSW. *Econ. Geol.*, **50**, 681–674.

Stanton, R. L. (1972). *Ore Petrology*, McGraw-Hill, New York, 713 pp.

Stauder, W. S. J. (1968). Tensional character of earthquake foci beneath the Aleutian Trench with relation to sea-floor spreading. *J. Geophys. Res.*, **73**, 7693–7701.

Stehli, F. G. (1973). Review of palaeoclimate and continental drift. *Earth Sci. Rev.*, **9**, 1–18.

Steiner, J. and Grillmair, E. (1973). Possible galactic causes for periodic and episodic glaciations. *Bull. geol. Soc. Amer.*, **84**, 1003–1018.

Steinmann, A. (1926). Die ophiolitischen Zones in den Mediterranean Kettengebirgen. *Proc. 14th Int. geol. Congr.*, Sect. 2, 638–667.

Stewart, A. D. (1975). 'Torridonian' rocks of western Scotland. In A. L. Harris (Ed.), *A Correlation of Precambrian Rocks in the British Isles*, Geol. Soc. Lond. Sp. Rep., **6**, 43–51.

Stewart, A. D. and Irving, E. (1974). Palaeomagnetism of sedimentary rocks from NW Scotland and its comparison with the polar wandering path relative to Laurentia. *Geophys. J.*, **37**, 51–73.

Stewart, J. H. (1972). Initial deposits in the Cordilleran Geosyncline: evidence of a late Precambrian (<850 m.y.) continental separation. *Bull. geol. Soc. Amer.*, **83**, 1345–1360.

Stewart, J. H. (1976). Late Precambrian evolution of North America: plate tectonics implication. *Geology*, **4**, 11–15.

Stewart, J. W. (1970). Precambrian alkaline–ultramafic/carbonatite volcanism at Qagssiarssuk, South Greenland. *Bull. Grønlands geol. Unders.*, **84**, 70 pp.

Stockwell, C. H. (1933). Great Slave Lake–Coppermine River area, northwest Territories. *Geol. Surv. Can. Sum. Rep. 1932*, pt C, 37–63.

Stockwell, C. H., McGlynn, J. C., Emslie, R. F., Sanford, B. V., Norris, A. W., Donaldson, J. A., Fahrig, W. F. and Currie, K. L. (1970). Geology of the Canadian Shield. In *Geology and Economic Minerals of Canada*, Geol. Surv. Can. Econ. Geol. Rep., **1**, 43–150.

Stowe, C. W. (1971). Summary of the tectonic development of the Rhodesian Archaean craton. *Geol. Soc. Australia Sp. Publ.*, **3**, 377–383.

Stowe, C. W. (1973). The older tonalite gneiss complex in the Selukwe area, Rhodesia. *Geol. Soc. S. Afr. Sp. Publ.*, **3**, 85–96.

Stowe, C. W. (1974). Alpine-type structures in the Rhodesian basement complex at Selukwe. *J. geol. Soc. Lond.*, **130**, 411–426.

Strong, D. F. (1974a). Plate tectonic setting of Newfoundland mineral deposits. *Geoscience Can.*, **1**, 20–30.

Strong, D. F. (1974b). Plate tectonic setting of Appalachian-Caledonian mineral deposits as indicated by Newfoundland examples. *Trans. Ass. Inst. Min. Eng.*, **256**, 121–128.

Strong, D. F., Dickson, W. L., O'Driscoll, C. F., Kean, B. F. and Stevens, R. K. (1974). Geochemical evidence for an east-dipping Appalachian subduction zone in Newfoundland. *Nature*, **248**, 37–39.

Strong, D. F. and Stevens, R. K. (1974). Possible thermal explanation of contrasting Archaean and Proterozoic geological regimes. *Nature*, **249**, 545–546.

Stueber, A. M., Heimlich, R. A. and Ikramuddin, M. (1976). Rb–Sr ages of Precambrian mafic dykes, Bighorn Mountains, Wyoming. *Bull. geol. Soc. Amer.*, **87**, 909–914.

Subramaniam, A. P. (1956). Mineralogy and petrology of the Sittampundi complex, Salem district, Madras State, India. *Bull. geol. Soc. Amer.*, **67**, 317–390.

Subramanian, A. P. (1959). Charnockites of the type area near Madras—a reinterpretation. *Amer. J. Sci.*, **257**, 321–353.

Suffel, F. G. and Hutchinson, R. W. (1973). Massive cupriferous pyrite deposits at Küre, Turkey. *Econ. Geol.*, **68**, 140–141.

Sugimura, A. (1960). Zonal arrangement of some geophysical and petrological features in Japan and its environs. *J. Fac. Sci. Univ. Tokyo*, Sect. 2, **12**, 133–153.

Sugimura, A. (1968). Composition of primary magmas and seismicity of earth's mantle in island arcs. *Geol. Soc. Can. Pap.*, **66–15**, 337–346.

Sugimura, A. and Uyeda, S. (1973). *Island Arcs; Japan and its environs*, Elsevier, Amsterdam, 247 pp.

Sugisaki, R. (1976). Chemical characteristics of volcanic rocks: relation to plate movements. *Lithos*, **9**, 17–30.

Sun, S. S. and Hanson, G. N. (1975). Evolution of the mantle: geochemical evidence from alkali basalt. *Geology*, **3**, 297–302.

Suppe, J. and Armstrong, R. L. (1972). Potassium–argon dating of Franciscan metamorphic rocks. *Amer. J. Sci.*, **272**, 217–233.

Sutton, J. (1963). Long term cycles in the evolution of the continents. *Nature*, **198**, 731–735.

Sutton, J. (1967). The extension of the geological record into the Precambrian. *Proc. Geol. Ass.*, **78**, 493–534.

Sutton, J. (1976). Tectonic relationships in the Archaean. In B. F. Windley (Ed.), *The Early History of the Earth*, Wiley, London, 99–104.

Sutton, J. and Watson, J. V. (1951). The pre-Torridonian metamorphic history of the Loch Torridon and Scourie areas of the north-west Highlands and its bearing on the chronological classification of the Lewisian. *J. geol. Soc. Lond.*, **106**, 241–296.

Sutton, J. and Watson, J. V. (1969). Scourian–Laxfordian relationships in the Lewisian of north-west Scotland. *Geol. Ass. Can. Sp. Pap.*, **5**, 119–128.

Sutton, J. and Watson, J. V. (1974). Tectonic evolution of continents in early Proterozoic times. *Nature*, **247**, 433–435.

Swanson, D. A. (1969). Lawsonite blueschist from north–central Oregon. *US geol. Surv. Prof. Pap.*, **650B**, 8–11.

Swett, K. and Smit, D. K. (1972). Cambro-Ordovician shelf sedimentation of western Newfoundland, northwest Scotland and central east Greenland, *24th Int. geol. Congr. Montreal*, Sect. 6, 33–41.

Swinden, H. S. and Strong, D. F. (1976). A comparison of plate tectonic models of metallogenesis in the Appalachians, the North American Cordillera and the East Australian Palaeozoic. In D. F. Strong (Ed.), *Metallogeny and Plate Tectonics*, Geol. Ass. Can. Sp. Pap., **14**, 441–470.

Sykes, L. R., Oliver, J. and Isacks, B. (1970). Earthquakes and tectonics. In A. E. Maxwell (Ed.), *The Sea*, Vol. 4, Wiley, New York, 353–420.

Sylvester-Bradley, P. C. (1971a). Environmental parameters for the origin of life. *Proc. Geol. Ass.*, **82**, 87–136.

Sylvester-Bradley, P. C. (1971b). Dynamic factors in animal palaeogeography. In F. A. Middlemiss, P. F. Rawson and G. Newall (Eds.), *Faunal Provinces in Space and Time*, Seal House Press, Liverpool, 1–18.

Sylvester-Bradley, P. C. (1974). Fossil evidence for the rotation of Spain during the evolution of the Mediterranean sea. *Ann. geol. Surv. Egypt*, **4**, 251–262.

Sylvester-Bradley, P. C. (1975). The search for protolife. *Phil. Trans. R. Soc. Lond.*, **B189**, 213–233.

Symons, D. T. A. (1967). Palaeomagnetism of Precambrian rocks near Cobalt, Ontario. *Can. J. Earth Sci.*, **4**, 1161–1170.

Talbot, C. J. (1973). A plate tectonic model for the Archaean crust. *Phil. Trans. R. Soc. Lond.*, **A273**, 413–428.

Taliaferro, N. L. (1943). Franciscan–Knoxville problem. *Bull. Amer. Ass. Petrol. Geol.*, **27**, 109–119.

Talwani, M., Le Pichon, X. and Ewing, M. (1965). Crustal structure of the mid-oceanic ridges. 2. Computed model from gravity and seismic refraction data. *J. Geophys. Res.*, **70**, p. 341.

Tappan, H. (1968). Primary production, isotopes, extinctions and the atmosphere. *Palaeogeog. Palaeoclimat. Palaeocol.*, **4**, 187–210.

Tappan, H. and Loeblich, A. R., Jr (1971). Geobiologic implications of phytoplankton evolution and time-space distribution. *Geol. Soc. Amer. Sp. Pap.*, **127**, 247–340.

Tarling, D. H. (1971a). Gondwanaland, palaeomagnetism and continental drift. *Nature*, **229**, 17–21.

Tarling, D. H. (1971b). *Principles and Applications of Palaeomagnetism*, Chapman and Hall, London, 164 pp.

Tarling, D. H. and Tarling, M. P. (1971). *Continental Drift*, Bell, London, 112 pp.

Tarney, J. (1963). Assynt dykes and their metamorphism. *Nature*, **199**, 672–674.

Tarney, J. (1973). The Scourie dyke suite and the nature of the Inverian event in Assynt. In R. G. Park and J. Tarney (Eds.), *The Early Precambrian of Scotland and Related Rocks of Greenland*, Univ. Keele, 105–118.

Tarney, J. (1976). Geochemistry of Archaean high-grade gneisses, with implications as to the origin and evolution of the Precambrian crust, In B. F. Windley (Ed.), *The Early History of the Earth*, Wiley, London, 405–417.

Tarney, J., Skinner, A. C. and Sheraton, J. W. (1972). A geochemical comparison of major Archaean gneiss units from northwest Scotland and East Greenland, *24th Int. geol. Congr. Montreal, Sect. 1*, 162–174.

Tarney, J., Dalziel, I. W. D. and De Wit, M. J. (1976), Marginal basin 'Rocas Verdes' complex from S. Chile: a model for Archaean greenstone belt formation. In B. F. Windley (Ed.), *The Early History of the Earth*, Wiley, London, 131–146.

Tatsumi, T., Sekine, Y. and Kanehira, K. (1970). Mineral deposits of volcanic affinity in Japan: metallogeny. In T. Tatsumi (Ed.), *Volcanism and Ore Genesis*, Univ. Tokyo Press, 3–47.

Taylor, H. P., Jr and Coleman, R. G. (1968). O^{18}/O^{16} ratios of coexisting minerals in glaucophane-bearing metamorphic rocks. *Bull. geol. Soc. Amer.*, **79**, 1727–1756.

Taylor, P. N. (1975). An early Precambrian age for migmatitic gneisses from Vikan i Bø, Vesterålen, North Norway. *Earth Plan. Sci. Lett.*, **27**, 35–42.

Taylor, R. B. (1964). Geology of the Duluth gabbro complex near Duluth, Minnesota. *Bull. geol. Surv. Minn.*, **44**, 63 pp.

Termier, H. and Termier, G. (1952). *Histoire géologique de la Biosphere*, Masson, Paris, 194 pp.

Thayer, T. P. (1969). Peridotite–gabbro complexes as keys to petrology of mid-ocean ridges. *Bull. geol. Soc. Amer.*, **80**, 1515–1522.

Thayer, T. P. (1971). Authigenic, polygenic and allogenic ultramafic and gabbroic rocks as hosts for magmatic ore deposits. *Geol. Soc. Australia Sp. Publ.*, **3**, 239–251.

Thomas, M. D. and Tanner, J. G. (1975). Cryptic suture in the eastern Grenville Province. *Nature*, **256**, 392–394.

Thompson, R. (1974). Palaeomagnetism. *Sci. Prog.*, **61**, 349–373.

Thorpe, R. S. (1972). Ocean floor basalt affinity of Precambrian glaucophane schist from Anglesey. *Nature Phys. Sci.*, **240**, 164–166.

Tilling, R. I. (1973). The Boulder batholith, Montana: a product of two contemporaneous but chemically distinct magma series. *Bull. geol. Soc. Amer.*, **84**,

Toksöz, M. N., Minear, J. W. and Julian, B. R. (1971). Temperature field and geophysical effects of a downgoing slab. *J. Geophys. Res.*, **76**, 1113–1138.

Towe, K. M. (1970). Oxygen-collagen priority and the early Metazoan fossil record, *Proc. N.Y. Acad. Sci.*, **65**, 781–788.

Travis, G. A., Woodall, R. and Bartram, G. D. (1971). The geology of the Kalgoorlie goldfield. *Geol. Soc. Australia Sp. Publ.*, **3**, 175–190.

Tremblay, L. P. (1967). Geology of the Beaverlodge mining area, Saskatchewan. *Mem. geol. Surv. Can.*, **367**, 468 pp.

Trendall, A. F. (1968). Three great basins of Precambrian banded iron formation: a systematic comparison. *Bull. geol. Soc. Amer.*, **79**, 1527–1544.

Trendall, A. F. (1972). Revolution in earth history. *J. geol. Soc. Australia*, **19**, 287–311.

Trendall, A. F. (1973a). Iron formations of the Hamersley Group of western Australia: type examples of varved Precambrian evaporites. In *Genesis of Precambrian iron and Manganese Deposits*, Proc. Kiev. Symp. 1970, Unesco, Paris, 257–270.

Trendall, A. F. (1973b). Time-distribution and type-distribution of Precambrian iron formations in Australia. In *Genesis of Precambrian Iron and Manganese Deposits*, Proc. Kiev. Symp. 1970, Unesco, Paris, 49–57.

Trendall, A. F. and Blockley, J. G. (1970). The iron formation of the Precambrian Hamersley Group, western Australia. *Bull. geol. Surv. W. Australia*, **119**, 366 pp.

Trudinger, P. A. (1971). Microprobes, metals and minerals. *Minerals Sci. Engng.*, **3**, 13–25.

Trümpy, R. (1960). Paleotectonic evolution of the central and western Alps. *Bull. geol. Soc. Amer.*, **71**, 843–908.

Trümpy, R. (1973). The timing of orogenic events in the central Alps. In K. A. de Jong and R. Scholten (Eds.), *Gravity and Tectonics*, Wiley, New York, 229–252.

Tugarinov, A. I., Bergaman, I. A. and Gavrilova, L. K. (1973). The facial nature of the Krivoyrog iron formation. In *Genesis of Precambrian Iron and Manganese Deposits*, Proc. Kiev. Symp. 1970, Unesco, Paris, 35–39.

Turek, A. and Compston, W. (1971). Rubidium–strontium geochronology in the Kalgoorlie region. *Geol. Soc. Australia Sp. Publ.*, **3**, 72 (Abstr.).

Turneaure, F. S. (1971). The Bolivian tin–silver province. *Econ. Geol.*, **66**, 212–225.

Turner, S. (1970). Timing of the Appalachian/Caledonian orogen contraction. *Nature*, **227**, 90.

Uadarajan, S. (1968). Occurrence of glaucophane schist in the Nuggihalli schist belt. Mysore State, and its significance. *Proc. 55th Ind. Sci. Congr.*, pt III, 216.

Uffen, R. J. (1963). Influence of the earth's core on the origin and evolution of life. *Nature*, **198**, 143.

Upadhyay, H. D. and Strong, D. F. (1973). Geological setting of the Betts Cove copper deposits, Newfoundland: an example of ophiolite sulphide mineralization. *Econ. Geol.*, **68**, 161–167.

Upton, B. G. J. (1974). The alkaline province of southwest Greenland. In H. Sørensen (Ed.), *The Alkaline Rocks*, Wiley, London and New York, 221–237.

Upton, B. G. J., Aspen, P., Graham, A. and Chapman, N. A. (1976). Pre-Palaeozoic basement of the Scottish Midland Valley. *Nature*, **260**, 517–518.

Urey, H. C. (1960). Primitive planetary atmospheres and the origin of life. In M. Florkin (Ed.), *Aspects of the Origin of Life*, Pergamon, Oxford, 8–14.

Uyeda, S. and Vacquier, V. (1968). Geothermal and geomagnetic data in and around the island arc of Japan. In L. Knopoff (Ed.), *The Crust and Upper Mantle of the Pacific Area, Washington D.C. Amer. Geophys. Un., Geophys. Monog.*, **12**, 349–366.

Uyeda, S. and Miyashiro, A. (1974). Plate tectonics and the Japanese islands: a synthesis. *Bull. geol. Soc. Amer.*, **85**, 1159–1170.

Vacquier, V. (1959). Measurement of horizontal displacements along faults in the ocean floor. *Nature*, **183**, 452–453.

Vacquier, V., Raff, A. D. and Warren, R. E. (1961). Magnetic survey off the west coast of North America, 40°N latitude to 50°N latitude. *Bull. geol. Soc.*, **72**, 1267–1270.

Vail, J. R. (1970). Tectonic control of dykes and related irruptive rocks in eastern Africa. In T. N. Clifford and I. G. Gass (Eds.), *African Magmatism and Tectonics*, Oliver and Boyd, Edinburgh, 337–354.

Vail, J. R. and Dodson, M. H. (1969). Geochronology of Rhodesia. *Trans. geol. Soc. S. Afr.*, **72**, 79–110.

Valencio, D. A. (1974). The South American palaeomagnetic data and the main episodes of the fragmentation of Gondwana. *Phys. Earth Plan. Inter.*, **9**, 221.

Valentine, J. W. (1973a). Plate and provinciality, a theoretical history of environmental discontinuties. *Palaeont. Sp. Pap.*, **12**, 79–92.

Valentine, J. W. (1973b). *Evolutionary paleoecology of the marine Biosphere*, Prentice-Hall, New Jersey, 511 pp.

Valentine, J. W. and Moores, E. M. (1970). Plate tectonic regulation of faunal diversity and sea level: a model. *Nature*, **228**, 657–659.

Valentine, J. W. and Moores, E. M. (1972). Global tectonics and the fossil record. *J. Geol.*, **80**, 167–184.

Van Andel, T. H. and Bowin, C. O. (1968). Mid-Atlantic Ridge between 22° and 23°N latitude and the tectonics of mid-ocean rises. *J. Geophys. Res.*, **73**, 1279–1298.

Van Breemen, O. (1970). Geochronology of the Limpopo orogenic belt, southern Africa, *J. Earth Sci.*, **8**, 57–62.

Van Breemen, O., Aftalion, M. and Pidgeon, R. T. (1971). The age of the granitic injection complex of Harris, Outer Hebrides. *Scott. J. Geol.*, **7**, 139–152.

Van Breemen, O. and Upton, B. G. J. (1972). Age of some Gardar intrusive complexes, South Greenland. *Bull. geol. Soc. Amer.*, **83**, 3381–3390.

Van Breemen, O., Aftalion, M. and Allaart, J. H. (1974). Isotopic and geochronologic studies on granites from the Ketilidian mobile belt of South Greenland. *Bull. geol. Soc. Amer.*, **85**, 403–412.

Vance, J. A. (1968). Metamorphic aragonite in the prehnite-pumpellyite facies, northwest Washington. *Amer. J. Sci.*, **266**, 299–315.

Van Der Voo, R. (1969). Palaeomagnetic evidence for the rotation of the Iberian peninsula. *Tectonophys.*, **7**, 5–56.

Van Houten, F. B. (1968). Iron oxides in red beds. *Bull. geol. Soc. Amer.*, **79**, 399–416.

Van Houten, F. B. (1974). Northern Alpine molasse and similar Cenozoic sequences of southern Europe. In R. H. Dott Jr and R. H. Shaver (Eds.), *Modern and Ancient Geosynclinal Sedimentation*, Soc. Econ. Pal. Mineral. Sp. Publ., **19**, 260–273.

Van Moort, J. C. (1973). The magnesium and calcium content of sediments, especially pelites, as a function of age and degree of metamorphism. *Chem. Geol.*, **12**, 1–37.

Van Niekerk, C. B. and Burger, A. J. (1964). The age of the Ventersdorp System. *Ann. geol. Surv. S. Afr.*, **3**, 75–86.

Van Niekerk, C. B. and Burger, A. J. (1969). A note on the minimum age of the acid lava of the Onverwacht Series of the Swaziland System. *Trans. geol. Soc. S. Afr.*, **72**, 9–21.

Vartiainen, H. and Wooley, A. R. (1974). The age of the Sokli carbonatite, Finland and some relationships of the North Atlantic alkaline igneous province. *Bull. geol. Soc. Finl.*, **46**, 81–91.

Veevers, J. J., Jones, J. G and Talent, J. A. (1971). Indo-Australian stratigraphy and the configuration and dispersal of Gondwanaland, *Nature*, **229**, 383–388.

Veizer, J. (1973). Sedimentation in geologic history: recycling vs. evolution or recycling with evolution. *Contr. Mineral. Petrol.*, **38**, 261–278.

Veizer, J. (1976a). $^{87}Sr/^{86}Sr$ evolution of seawater during geologic history and its significance as an index of crustal evolution. In B. F. Windley (Ed.), *The Early History of the Earth*, Wiley, London, 569–578.

Veizer, J. (1976b). Evolution of ores of sedimentary affiliation through geologic history: relations to the general tendencies in evolution of the crust, hydrosphere, atmosphere and biosphere. In K. H. Wolf (Ed.), *Handbook of strata-bound and stratiform ore deposits*, Elsevier, v. 3, 1–41.

Viezer, J. and Compston, W. (1974). $^{87}Sr/^{86}Sr$ composition of seawater during the Phanerozoic. *Geochem. Cosmochim. Acta*, **38**, 1461–1484.

Veizer, J. and Compston, W. (1976). $^{87}Sr/^{86}Sr$ in Precambrian carbonates as an index of crustal evolution. *Geochim. Cosmochim. Acta*, **40**, 905–914.

Venkatasubramaniam, V. S., Iyer, S. S. and Pal, S. (1971). Studies on Rb–Sr geochronology of the Precambrian formation of Mysore State, India. *Amer. J. Sci.*, **270**, 43–53.

Venkatasubramaniam, V. S. (1974). Geochronology of the Dharwar craton—a review. *J. geol. Soc. India*, **15**, 463–468.

Vidal, G. (1972). Algal stromatolites from the late Precambrian of Sweden. *Lethaia*, **5**, 353–368.

Viljoen, M. J. and Viljoen, R. P. (1967). A reassessment of the Onverwacht Series in the Komati river valley, *Econ. Geol. Res. Unit Witwatersrand Univ., Johannesburg, Inf. Circ.*, **36**, 1–35.

Viljoen, M. J. and Viljoen, R. P. (1969). A collection of 9 papers on many aspects of the Barberton granite–greenstone belt, South Africa. *Geol. Soc. S. Afr. Sp. Publ.*, **2**, 295 pp.

Viljoen, R. P., Saager, R. and Viljoen, M. J. (1970). Some thoughts on the origin and processes responsible for the concentrations of gold in the early Precambrian of southern Africa. *Mineral. Deposita*, **5**, 164–180.

Vincent, P. M. (1970). The evolution of the Tibesti volcanic province. In T. N. Clifford and I. G. Gass (Eds.), *African Magmatism and Tectonics*, Oliver and Boyd, Edinburgh, 301–320.

Vine, F. J. and Matthews, P. M. (1963). Magnetic anomalies over ocean ridges. *Nature*, **199**, 947–949.

Vine, F. J. and Hess, H. H. (1970). Sea floor spreading. In A. E. Maxwell (Ed.), *The Sea*, Vol. 4, Wiley, New York, 587–622.

Vine, F. J. and Moores, E. M. (1972). A model for the gross structure, petrology, and magnetic properties of oceanic crust. *Geol. Soc. Amer., Mem.* **132**, 195–205.

Vine, F. J., Poster, C. K. and Gass, I. G. (1973). Aeromagnetic survey of the Troodos igneous massif, Cyprus. *Nature Phys. Sci.*, **244**, 34–38.

Viswanathan, S. (1974a). Basaltic komatiite occurrences in the Kolar gold field of India. *Geol. Mag.*, **111**, 353–354.

Viswanathan, S. (1974b). Contemporary trends in geochemical studies of early Precambrian greenstone–granite complexes. *J. Geol. Soc. India*, **15**, 347–379.

Viswanathan, S. (1975). Rocks of unusual chemistry in the charnockite terrains of India, and their geological significance. *Geol. Mag.*, **112**, 63–69.

Vogt, P. R. and Ostenso, N. A. (1967). Steady-state crustal spreading. *Nature*, **215**, 810–817.

Vogt. P. R. and Higgs, R. H. (1969). An aeromagnetic survey of the eastern Mediterranean and its interpretation. *Earth Plan. Sci. Lett.*, **5**, 439–448.

Vokes, F. M. (1973). Metallogeny possibly related to continental break-up in south west Scandinavia. In D. H. Tarling and S. K. Runcorn (Eds.), *Implications of Continental Drift to the Earth Sciences*, Vol. 1, Academic Press, London, 573–579.

Volborth, A. (1962). Rapakivi-type granites in the Precambrian complex of Gold Butte, Clark County, Nevada. *Bull. geol. Soc. Amer.*, **73**, 813–832.

Von Eckermann, H. (1936). The Loos–Hamra region. *Geol. Foren. Stockh. Forh.*, **405**, 58, 129–343.

Von Eckermann, H. (1968). New contributions to the interpretation of the genesis of the Norra Kärr alkaline body in southern Sweden. *Lithos*, **1**, 76–88.

Von Huene, R. (1974). Modern trench sediments. In L. A. Burk and C. L. Drake (Eds.), *The Geology of Continental Margins*, Springer-Verlag, Berlin, 207–211.

Vorma, A. (1975). On two roof pendants in the Wiborg rapakivi massif, southeastern Finland. *Bull. geol. Surv. Finl.*, **272**, 1–86.

Wager, L. R. and Brown, G. M. (1968). *Layered Igneous Rocks*, Oliver and Boyd, Edinburgh, 588 pp.

Wager, L. R. and Deer, W. A. (1938). A dyke swarm and crustal flexure in East Greenland. *Geol. Mag.*, **75**, 39–45.

Walcott, R. I. (1972). Gravity, flexure, and the growth of sedimentary basins at a continental edge. *Bull. geol. Soc. Amer.*, **83**, 1845–1848.

Walker, J. C. G. (1976). Implications for atmospheric evolution of the inhomogeneous accretion model of the origin of the earth. In B. F. Windley (Ed.), *The Early History of the Earth*, Wiley, London, 535–546.

Walker, J. C. G., Turekian, K. K. and Hunten, D. M. (1970). An estimate of the present-day deep-mantle degassing rate from data on the atmosphere of Venus. *J. Geophys. Res.*, **75**, 3558–3561.

Walker, T. R. (1967). Formation of red beds in ancient and modern deserts. *Bull. geol. Soc. Amer.*, **78**, 353–368.

Walker, T. R. (1974). Formation of red beds in moist tropical climates: a hypothesis. *Bull. geol. Soc. Amer.*, **85**, 633–638.

Walter, M. R. (1970). Stromatolites used to determine the time of nearest approach of Earth and Moon. *Science*, **170**, 1331–1332.

Walter, M. R. (1972a). Stromatolites and the biostratigraphy of the Australian Precambrian and Cambrian. *Pal. Ass. Sp. Pap.*, **11**, 190 pp.

Walter, M. R. (1972b). A hot-spring analog for the depositional environment of Precambrian iron formations of the Lake Superior region. *Econ. Geol.*, **67**, 965–980.

Walter, M. R. and Preiss, W. V. (1972). Distribution of stromatolites in the Precambrian and Cambrian of Australia. *24th Int. geol. Congr. Montreal, Sect. 1*, 85–93.

Walton, B. J. (1965). Sanerutian appinitic rocks and Gardar dykes and diatremes, north of Narssarssuaq, South Greenland. *Meddr. Grønland*, **179**, 1–66.

Walton, E. K. (1956). Two Ordovician conglomerates in South Ayrshire. *Trans. geol. Soc. Glasgow*, **22**, 133–156.

Wardlaw, N. C. and Nicholls, G. O. (1972). Cretaceous evaporites of Brazil and West Africa and their bearing on the theory of continental separation. *24th Int. geol. Congr. Montreal, Sect. 6*, 43–55.

Waterhouse, J. B. and Bonham-Carter, G. F. (1975). Global distribution and character of Permian biomes based on brachiopod assemblages. *Can. J. Earth Sci.*, **12**, 1085–1146.

Watson, J. V. (1969). The Precambrian gneiss complex of Ness, Lewis, in relation to the effects of Laxfordian regeneration. *Scott. J. Geol.*, **5**, 269–285.

Watson, J. V. (1973a). Effects of reworking on high-grade gneiss complexes. *Phil. Trans. R. Soc. Lond.*, **A273**, 443–456.

Watson, J. V. (1973b). Influence of crustal evolution on ore deposition. *Trans. Inst. Min. Metall. Sect. B*, **82**, B107–114.

Watson, J. V. (1975). The Lewisian complex. In *A Correlation of Precambrian Rocks in the British Isles, Geol. Soc. Lond. Sp. Rep.*, **6**, 15–29.

Watson, J. V. (1976). Mineralisation in Archaean provinces. In B. F. Windley (Ed.), *The Early History of the Earth*, Wiley, London, 443–453.

Watt, W. S. (1966). Chemical analyses from the Gardar igneous province, South Greenland. *Rep. Grønlands geol. Unders.*, **6**, 92 pp.

Watt, W. S. (1969). The coast-parallel dike swarm of SW Greenland in relation to the opening of the Labrador Sea. *Can. J. Earth Sci.*, **6**, 1320–1321.

Watters, B. R. (1976). Possible late Precambrian subduction zone in south west Africa. *Nature*, **259**, 471–473.

Watterson, J. (1965). Plutonic development of the Ilordleq area, South Greenland. I. Chronology, and the occurrence and recognition of metamorphosed basic dykes. *Meddr. Grønland*, **172**, 7, 147 pp.

Watterson, J. (1968). Plutonic development of the Ilordleq area, South Greenland. II. Late kinematic basic dykes. *Meddr. Grønland*, **185**, 3, 104 pp.

Wegener, A. (1929). *Die Enstehung der Kontinente und Ozeane* (4th edit.), Vieweg und Sohn, Braunschweig.

Weimer, R. J. (1970). Rates of deltaic and intrabasin deformation, Upper Cretaceous of Rocky Mountain region. In J. P. Morgan (Ed.), *Deltaic Sedimentation, Modern and Ancient, Soc. Econ. Pal. Mineral. Sp. Publ.*, **15**, 270–292.

Welin, E. (1970). Den svecofenniska orogena zonen i norra Sverige; en preliminär diskussion. *Geol. Fören. Stockh. Förh.*, **92**, 433–451.

Welin, E., Christiansson, K. and Nillson, Ö. (1971). Rb–Sr radiometric ages of extrusive and intrusive rocks in northern Sweden, I. *Sverige geol. Unders. Ser. C*, No. 666, Arsb. 65, 1–38.

Wensink, H. (1973). The Indo–Pakistan subcontinent and the Gondwana reconstructions based on palaeomagnetic results. In D. H. Tarling and S. K. Runcorn (Eds.), *Implications of Continental Drift to the Earth Sciences*, Academic Press, London, 103–116.

Westbrook, G. K., Bott, M. H. P. and Peacock, J. H. (1973). Lesser Antilles subduction zone in the vicinity of Barbados. *Nature Phys. Sci.*, **244**, 118–120.

Wheatley, C. J. V. (1971). Aspects of metallogenesis within the southern Caledonides of Great Britain and Ireland. *Trans. Inst. Min. Metall.*, **80**, B211–223.

Wheeler, E. P., Jr (1960). Anorthosite–adamellite complex of Nain, Labrador. *Bull. geol. Soc. Amer.*, **71**, 1755–1762.

Wheeler, E. P., Jr (1965). Fayalitic olivine in northern Newfoundland–Labrador, *Can. Mineral.*, **8**, 339–346.

Wheeler, H. E. (1940). Permian volcanism in western North America. *Proc. 6th Pacific Sci. Congr.*, Berkeley, California, Univ. Calif. Press, **1**, 369–376.

Wheeler, J. O. and Gabrielse, H. (1972). The Cordilleran structural province. In R. A. Price and R. J. W. Douglas (Eds), *Variations in Tectonic Styles in Canada, Geol. Ass. Can. Sp. Pap.*, **11**, 1–82.

Wheeler, J. O., Charlesworth, H. A. K., Monger, J. W. H., Muller, J. E., Price, R. A., Ressor, J. E., Roddick, J. A. and Simony, P. S. (1974). Western Canada. In A. M. Spencer (Ed.), *Mesozoic–Cenozoic Orogenic Belts, Geol. Soc. Lond. Sp. Publ.*, **4**, 591–624.

Whitaker, J. H. McD. (1974). Ancient submarine canyons and fan valleys. In R. H. Dott Jr and R. H. Shaver (Eds.), *Modern and Ancient Geosynclinal Sedimentation, Soc. Econ. Pal. Mineral. Sp. Publ.*, **19**, 106–125.

White, A. J. R., Jakes, P. and Christie, D. M. (1971). Composition of greenstones and the hypothesis of sea-floor spreading in the Archaean. *Geol. Soc. Aust. Sp. Publ.*, **3**, 47–56.

White, D. E. and Waring, G. A. (1963). Volcanic emanations. *U.S. geol. Surv. Prof. Pap.*, **440-K**.

White, W. S. (1966). Geologic evidence for crustal structure in the western Lake Superior basin. *Amer. geophys. Un. Monog.*, **10**, 28–41.

Whiteman, A. J., Rees, G., Naylor, D. and Pegrum, R. M. (1975a). North Sea troughs and plate tectonics. *Norges geol. Unders.*, **316**, 137–161.

Whiteman, A., Naylor, D., Pegrum, R. and Rees, G. (1975b). North Sea troughs and plate tectonics. *Tectonophys.*, **26**, 39–54.

Whiteside, H. C. M. (1970). Volcanic rocks of the Witwatersrand Triad. In T. N. Clifford and I. G. Gass (Eds.), *African Magmatism and Tectonics*, Oliver and Boyd, Edinburgh, 73–87.

Whittington, H. B. and Hughes, C. P. (1972). Ordovician geography and faunal provinces deduced from tribolite distribution. *Phil. Trans. R. Soc. Lond.*, **B263**, 235–278.

Whittington, H. B. and Hughes, C. P. (1973). Ordovician trilobite distributions and geography. *Palaeont. Sp. Pap.*, **12**, 235–240.

Wiebols, J. H. (1955). A suggested glacial origin for the Witwatersrand conglomerates. *Trans. geol. Soc. S. Afr.*, **58**, 367–382.

Wilkinson, J. M. and Cann, J. R. (1974). Trace elements and tectonic relationships of basaltic rocks in the Ballantrae igneous complex, Ayrshire. *Geol. Mag.*, **11**, 35–41.

Wilkinson, J. F. G., Duggan, M. B., Herbert, H. K. and Kalocsai, G. I. Z. (1975). The Salk Lick Creek layered intrusion, east Kimberley region, western Australia. *Contr. Mineral. Petrol.*, **50**, 1–23.

Willemse, J. (1969). The geology of the Bushveld igneous complex, the largest repository of magmatic ore deposits in the world. *Econ. Geol. Mono.*, **4**, 1–22.

Williams, A. (1969). Ordovician of the British Isles. In M. Kay (Ed.), *North Atlantic—Geology and Continental Drift, Amer. Ass. Petrol. Geol. Mem.*, **12**, 236–264.

Williams, A. (1973). Distribution of brachiopod assemblages in relation to Ordovician palaeogeography. *Palaeont. Sp. Pap.*, **12**, 241–269.

Williams, A. (1976). Plate tectonics and biofacies evolution as factors in Ordovician correlation. In *The Ordovician System*, Proc. Pal. Ass. Symp. Birmingham Sept. 1974, Univ. Wales Press and Nat. Mus. Wales, 29–66.

Williams, D. (1969). Ore deposits of volcanic affiliation. In C. H. James (Ed.), *Sedimentary Ores, Ancient and Modern, Geology Dept. Univ. Leicester Sp. Publ.*, **1**, 197–206.

Williams, D. A. C. and Hallberg, J. A. (1973). Archaean layered intrusions of the eastern Goldfields region, western Australia. *Contr. Mineral. Petrol.*, **38**, 45–70.

Williams, G. E. (1975). Late Precambrian glacial climate and the earth's obliquity. *Geol. Mag.*, **112**, 441–465.

Williams, H. (1971). Mafic ultramafic complexes in western

Newfoundland Appalachians and the evidence for their transportation: a review and interim report. *Proc. Geol. Ass. Can.*, **24**, 9–25.

Williams, H. (1975). Structural succession, nomenclature, and interpretation of transported rocks in western Newfoundland. *Can J. Earth Sci.*, **12**, 1874–1894.

Williams, H. and Smyth, W. R. (1973). Metamorphic aureoles beneath ophiolite suites and Alpine peridotites: tectonic implications with West Newfoundland examples. *Amer. J. Sci.*, **273**, 594–621.

Williams, H., Kennedy, M. J. and Neale, E. R. W. (1974). The northeastward termination of the Appalachian orogen. In A. E. M. Nairn and F. G. Stehli (Eds.), *The Ocean Basins and Margins*, Vol. 2: *The North Atlantic*, Plenum Press, New York, 79–123.

Williams, H. and Stevens, R. K. (1974). The ancient continental margin of eastern North America. In C. A. Burk and C. L. Drake (Eds.), *The Geology of Continental Margins*, Springer-Verlag, New York, 781–796.

Williams, H. R. and Williams, R. A. (1976). The Kasila Group, Sierra Leone, an interpretation of new data. *Precambrian Res.*, **3**, 505–508.

Wilson, A. F. (1958). Advances in the knowledge of the structure and petrology of the Precambrian rocks of SW Australia. *J. Roy. Soc. W. Australia*, **41**, 57–83.

Wilson, H. D. B. and Brisbin, W. C. (1961). Regional structure of the Thompson–Moak Lake nickel belt. *Trans. Can. Inst. Min. Metall.*, **64**, 470–477.

Wilson, H. D. B., Andrews, P., Moxham, R. L. and Ra'mlal, K. (1965). Archaean volcanism in the Canadian Shield. *Can. J. Earth Sci.*, **2**, 161–175.

Wilson, H. D. B., Morrice, M. G. and Ziehlke, D. V. (1974). Archaean continents. *Geoscience, Can.*, **1**, 12–20.

Wilson, J. F. (1973a). The Rhodesian Archaean craton—an essay in cratonic evolution. *Phil. Trans. R. Soc. Lond.*, **A273**, 389–411.

Wilson, J. F. (1973b). The granites and gneisses of the Mashaba area, Rhodesia. *Geol. Soc. S. Afr. Sp. Publ.*, **3**, 79–84.

Wilson, J. T. (1963). Evidence from islands on the spreading of ocean floors. *Nature*, **197**, 536–538.

Wilson, J. T. (1965). A new class of faults and their bearing on continental drift. *Nature*, **207**, 343–347.

Wilson, J. T. (1966). Did the Atlantic close and then re-open? *Nature*, **211**, 676.

Wilson, J. T. (1973). Continental drift, transcurrent and transform faulting. In A. E. Maxwell (Ed.), *The Sea*, Vol. 4, Wiley, New York, 623–644.

Wilson, N. W. (1965). Geology and mineral resources of part of the Gola forest, southeastern Sierra Leone. *Bull. geol. Surv. Sierra Leone*, **4**, 102 pp.

Winchester, J. A. (1974). The zonal pattern of regional metamorphism in the Scottich Caledonides. *J. Geol. Soc.*, **130**, 509–524.

Winchester, J. A. and Floyd, P. A. (1976). Geochemical magma type discrimination: application to altered and metamorphosed basic igneous rocks. *Earth Plan. Sci. Lett.*, **28**, 459–469.

Windley, B. F. (1965). The composite net-veined diorite intrusives of the Julianehaab district, South Greenland. *Bull. Grønlands geol. Unders.*, **58**, 60 pp.

Windley, B. F. (1969). Anorthosites of southern West Greenland. *Mem Ass. Amer. Petrol. Geol.*, **12**, 899–915.

Windley, B. F. (1973a). Crustal development in the Precambrian. *Phil. Trans. R. Soc. Lond.*, **A273**, 321–341.

Windley, B. F. (1973b). Archaean anorthosites: a review with the Fiskenaesset complex, West Greenland, as a model for interpretation. *Geol. Soc. S. Afr. Sp. Publ.*, **3**, 312–332.

Windley, B. F. (in press). Layered igneous complexes in Archaean high-grade regions. Symp. Vol., *The Precambrian*, Moscow, 1975.

Windley, B. F. and Bridgwater, D. (1971). The evolution of Archaean low- and high-grade terrains. *Geol. Soc. Australia Sp. Publ.*, **3**, 33–46.

Windley, B. F., Herd, R. K. and Bowden, A. A. (1973). The Fiskenaesset complex, West Greenland, Pt. 1: A preliminary study of stratigraphy, petrology and whole rock chemistry from

Qeqertarssuatsiaq. *Grønlands geol. Unders. Bull.*, **106**, 1–80 (aslo *Meddr. Grønland*, **196**, 2).

Windley, B. F. and Smith, J. V. (1974). The Fiskenaesset complex, West Greenland, Pt. 2: General mineral chemistry from Qeqertarssuatsiaq. *Grønlands geol. Unders. Bull.*, **108**, 1–54 (also *Meddr. Grønland*, **196**, 4).

Windley, B. F. and Smith, J. V. (1976). Archaean high-grade complexes and modern continental margins. *Nature*, **260**, 671–675.

Wise, D. U. (1974). Continental margins, freeboard and volumes of continents and ocean through time. In C. A. Burk and C. L. Drake (Eds.), *The Geology of Continental Margins*, Springer-Verlag, Berlin, 45–58.

Wolfhard, M. R. and Ney, C. S. (1976). Metallogeny and plate tectonics in the Canadian Cordillera. In D. F. Strong (Ed.), *Metallogeny and Plate Tectonics, Geol. Ass. Can. Sp. Pap.*, **14**, 359–390.

Wood, B. J. (1975). The influence of pressure, temperature and bulk composition on the appearance of garnet in orthogneisses—an example from South Harris, Scotland. *Earth Plan. Sci. Lett.*, **26**, 299–311.

Wood, D. S. (1966). *The Rhodesian basement*, 10th Ann. Rep. Inst. African Geol. Univ. Leeds, 18–19.

Wood, D. S. (1970). *The tectonic evolution of Africa*, Geol. Soc. Amer. 1970 Mtg., Abstr., p. 725.

Wood, D. S. (1973). Patterns and magnitudes of natural strain in rocks. *Phil. Trans. R. Soc. Lond.*, **A274**, 373–382.

Wood, D. S. (1974). Ophiolites, mélanges, blueschists, and ignimbrites: early Caledonian subduction in Wales? In R. H. Dott Jr and R. H. Shaver (Eds.), *Modern and Ancient Geosynclinal Sedimentation, Soc. Econ. Pal. Mineral. Sp. Publ.*, **19**, 334–344.

Woodall, R. and Travis, G. A. (1969). *The Kambalda nickel deposits, western Australia*, 9th Commonwealth Min. Metall. Congr., Min. Pet. Geol. Sect., 1–17.

Woodside, J. and Bowin, C. (1970). Gravity anomalies and inferred crustal structure in the eastern Mediterranean Sea. *Bull. geol. Soc. Amer.*, **81**, 1107–1122.

Worst, B. G. (1960). The Great Dyke of southern Rhodesia. *Bull. geol. Surv. S. Rhod.*, **47**.

Wright, A. E. (1969). Precambrian rocks of England, Wales and South East Ireland. In M. Kay (Ed.), *North Atlantic Geology and Continental Drift, Amer. Ass. Petrol. Geol. Mem.*, **13**, 93–109.

Wright, J. B. (1970). Distribution of volcanic rocks about mid-ocean ridges and the Kenya rift valley. *Geol. Mag.*, **107**, 125–131.

Wright, J. B. and McCurry, P. (1973). Magmas, mineralization and sea-floor spreading. *Geol. Rund.*, **62**, 116–125.

Wright, L. A., Troxel, B. W., Williams, E. G., Roberts, M. T. and Diehl, P. E. (1974). Precambrian sedimentary environments of the Death Valley region, east California. In *Guidebook: Death Valley Region, California and Nevada*, Shoshone, California, Death Valley Publ. Co., 27–35.

Wyllie, P. J. (1971). *The Dynamic Earth*, Wiley, New York, 416 pp.

Wyllie, P. J. (1976). *The Way the Earth Works*, Wiley, New York, 296 pp.

Wynne-Edwards, H. R. (1972a). Grey gneiss complexes and the evolution of the continental crust. *24th Int. geol. Congr., Sect. 1*, p. 175.

Wynne-Edwards, H. R. (1972b). The Grenville Province. In R. A. Price and R. J. W. Douglas (Eds.), *Variations in Tectonic Styles in Canada, Geol. Ass. Can. Sp. Pap.*, **11**, 263–334.

Wynne-Edwards, H. R. (1976). Proterozoic ensialic orogenesis: the millipede model of ductile plate tectonics. *Amer. J. Sci.*, **276**, 927–953.

Wynne-Edwards, H. R. and Hasan, Z. (1970). Intersecting orogenic belts across the North Atlantic. *Amer. J. Sci.*, **268**, 289–308.

Yagi, K., Kawano, Y. and Aoki, K. (1963). Types of Quaternary volcanic activity in northeastern Japan. *Bull. Volcanol.*, **26**, 223–235.

Yardley, B. W. D. and Black, J. D. (1976). Sapphirine in the Sittampundi complex, India: a discussion. *Min. Mag.*, **40**, 523–524.

Young, G. M. (1970). An extensive early Proterozoic glaciation in North America? *Palaeogeog. Palaeoclimat. Palaeoecol.*, **7**, 85–101.

Zambrano, J. J. and Urien, C. M. (1970). Geological outline of the basins in southern Argentina and their continuation off the Atlantic shore. *J. Geophys. Res.*, **75**, 1363–1396.

Zen, E-an (1972). The Taconide Zone and the Taconic orogeny in the western part of the northern Appalachian orogen. *Geol. Soc. Amer. Sp. Pap.*, **135**, 72 pp.

Ziegler, P. A. (1975). Geologic evolution of North Sea and its tectonic framework. *Bull. Amer. Ass. Petrol. Geol.*, **59**, 1073–1097.

Zijderveld, J. D. A., De Jong, J. A. and Van Der Voo, R. (1970a).

Rotation of Sardinia: palaeomagnetic evidence from Permian rocks. *Nature*, **226**, 933–934.

Zijderveld, J. D. A., Hazeu, G. J. A., Nardin, M. and Van Der Voo, R. (1970b). Shear in the Tethys and the palaeomagnetism in the southern Alps, including new results. *Tectonophys.*, **10**, 639–661.

Zonenshain, L. P., Kuzmin, M. I., Kovalenko, V. I. and Saltykovsky, A. J. (1974). Mesozoic structural–magmatic pattern and metallogeny of the western part of the Pacific belt. *Earth Plan. Sci. Lett.*, **22**, 96–109.

Zwart, H. J. (1976a). The duality of orogenic belts. *Geol. Mijnb.*, **46**, 283–309.

Zwart, H. J. (1967b). Orogenesis and metamorphic facies series in Europe. *Meddr. dansk geol. Foren.*, **17**, 504-516.

Zwart, H. J. (1969). Metamorphic facies series in the European orogenic belts and their bearing on the causes of orogeny. *Geol. Ass. Can. Sp. Pap.*, **5**, 7–16.

Index

374